THE UNITED STATES ENERGY ATLAS

To Helen M. Cuff, artist, entrepreneur, and my wife. Her understanding and support made my graduate studies possible and were essential during projects like this. Thanks, Helen.

D.J.C.

To my wife, Edna Carter Young, and my mother, Drama Lillian Young, for their love, encouragement, and support.

W.J.Y.

THE UNITED STATES ENERGY ATLAS

Second Edition

David J. Cuff/William J. Young

Macmillan Publishing Company
A Division of Macmillan, Inc.
New York

Collier Macmillan Publishers
London

Macmillan Publishing Company
A Division of Macmillan, Inc.
866 Third Avenue, New York, N. Y. 10022

Collier Macmillan Canada, Inc.

Library of Congress Catalog Card Number: 85–4867

Printed in the United States of America

printing number
1 2 3 4 5 6 7 8 9 10

Library of Congress Cataloging in Publication Data

Cuff, David J.
   The United States energy atlas.

   Bibliography: p.
   Includes index.
   1. Power resources—United States. I. Young,
William J. (William Jack), 1935–   . II. Title.
TJ163.25.U6C83 1985   333.79′0973   85–4867
ISBN 0-02-691240-6

Editorial and Production Staff
Charles E. Smith, Publisher
Elyse I. Dubin, Project Editor
Mort I. Rosenberg, Production Manager

Contents

Preface to the Second Edition

In the five years since the first edition of this book there have been several important changes that make a second edition necessary.

First, the government agencies responsible for regular or occasional inventories of resources have revealed their latest estimates, thereby making it necessary to revise the amounts and distribution of coal, crude oil and natural gas, nuclear fuels, geothermal heat, and hydropower potential according to the most recent data.

Second, there have been significant developments that affect both nonrenewable and renewable energy prospects. One is the reduced demand for crude oil that has characterized the early 1980s. This, combined with federal policy, has caused not only a reduction of oil imports, but a lessening of U.S. dependence on OPEC nations. At the same time, a dramatic decline in the nuclear power industry in the United States has greatly reduced demand for fuel and altered our estimates of the life of uranium resources. For coal, the realities of recent production rates require us to revise our projections of how coal resources will last in those states where significant amounts are being mined.

Developments in renewable energy include the completion of futuristic solar thermal and photovoltaic power plants that we can now illustrate with maps and photographs. Most significant are the sprouting forests of windmills in various parts of the country, mostly in California. We map and illustrate this phenomenon as a sequel to the sections that analyze and map wind power potential in different regions of the country. As well, the use of fuel wood has grown sharply, and has been accompanied by data that allows us to document this important aspect of the nation's energy supply. New data also makes possible an analysis of alcohol fuel production that could not be made at the time of the first edition.

In this edition, the chapter on solar energy begins with measurement of radiation, and an explanation of how earth–sun relationships and the design of solar collectors make it necessary to refer, not to conventional maps of solar radiation, but to maps showing energy received by a tilted surface. These maps, just recently available, are included to make the chapter a more practical reference for the solar home builder. A new chapter on Ocean Energy appears in this edition, bringing together ocean thermal energy, tidal potential, ocean currents, and a brief discussion of wave energy and salinity gradients. This chapter includes new maps of ocean thermal potential, tidal prospects, and ocean current possibilities.

Finally, the authors have changed their view of how best to interpret and present certain information. In this edition tables are combined with maps, and maps with bar graphs, to make presentations that are more compact, comprehensive, and complete. Source documents are referred to on every illustration and listed at the end of each chapter. To provide additional depth in certain technical or policy subjects, we provide a number of explanatory essays dealing with topics not elaborated in the first edition. These include the Strategic Petroleum Reserve and the importance of tax policy to wind power investments.

Acknowledgments

A book like this draws much of its data from government documents. Fortunately for us, the documents section in Temple's Paley Library continues to be staffed by helpful and competent professionals. In particular, we thank Karen Aubitz, Mark Jacobs, and Barbara Wright for their efficiency and patience.

Published documents may, however, be dated by the time they are catalogued and shelved. And some valuable publications are not acquired—even by a federal depository library. Furthermore, a great deal of vital information is not published. Authors who need to track down the latest resource data and the status of research projects and energy developments are obliged, therefore, to telephone and write the experts in government agencies and trade associations, as well as utility company and project managers in the field. In doing this we met many who responded eagerly to our questions and made our work possible.

Craig Cranston of the Department of Energy provided us an early press release on oil and gas data that was extremely helpful, and the staff of the *Oil and Gas Journal* in Tulsa guided us once again to their invaluable tabulations of worldwide oil and gas resources and production.

For information on nuclear power plant status we are grateful to Ellen Nunnelee of the Atomic Industrial Forum, Brian Norris of the Nuclear Regulatory Commission, and Michael Totten of Critical Mass.

Latest events in geothermal research and development were clarified by Jim Reynolds, Ralph Burr, and Jim Bresee of the Department of Energy, and many sources of information were turned up for us by Russ Woessner and Andrew Goldman of the Department of Energy's Conservation and Renewable Energy Inquiry and Referral Service.

Steve Rubin of the Solar Energy Research Institute gave invaluable guidance to solar and wind information, and his colleague at SERI, Eugene Maxwell, was very helpful with solar radiation data. We thank Kirby Holte of Southern California Edison Company for a flood of information and photographs on solar installations, and Steve Ridenour of Temple's department of Mechanical Engineering for his advice on solar hot water heating. Our geographic analysis of active solar heating still leans heavily on work by John Duffie and William Beckman of the Solar Energy Laboratory at the University of Wisconsin, whom we thank for their generosity a few years ago.

For their early guidance on wind data we are still indebted to Lowell Kravitz of General Electric and Jack Reed of Sandia Laboratories. Keeping abreast of the explosive developments in wind energy was made possible by Linda and Thomas Gray of the American Wind Energy Association and Kathleen Gray and Dorothy Ward of the California Energy Commission.

Our original treatment of ocean thermal energy benefited from the advice of Robert Cohen and Lloyd Lewis of the Department of Energy. In dealing with latest developments we were aided by William Richards of the Department of Energy, Rick Kutler of Meridian Corporation, and Art Griffin of TRW Incorporated.

For information on biomass we are grateful to Jim Bones, Stan Bean, and Dwight Hair of the U.S. Forest Service. Our search was steered also by Jack Ranney at Oak Ridge National Laboratory, Russ O'Connell and Patricia Pickering at the Department of Energy, and Fred Potter of Information Resources Incorporated. Special thanks are due to Sinyan Shen, of Argonne National Laboratories, who was most generous with his time and the results of his project on biomass potential.

In conceiving the first and second editions of this book, we have enjoyed the enthusiastic support of Charles E. Smith, Vice President and Publisher at Macmillan Publishing Company. While working on this second edition, we have had the pleasure of dealing with Elyse Dubin, Editorial Supervisor at Macmillan. She expertly and graciously orchestrated the preparation and assembly of all the materials in this complex work. We are very glad she was there.

THE UNITED STATES ENERGY ATLAS

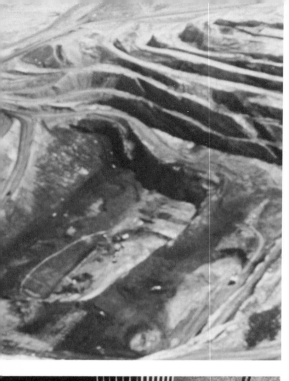

Introduction

- Organization
- Methods of Presentation
- Nonrenewable Resources
- Renewable Resources

More than a decade after the interruption of world oil supplies that brought energy to the forefront of national attention, the energy outlook has changed dramatically. Instead of the widespread shortages of the seventies, there is in the eighties a worldwide oil glut. This dramatic change in world oil supplies is the result of a sharp reduction in oil use, increasingly more efficient consumption patterns, and a push for increased use of renewable energy resources. According to Lester Brown, President of the Worldwatch Institute, the amount of oil required to produce $1,000 worth of goods and services decreased from 2.15 barrels in 1979 to 1.74 barrels in 1983, a 19 percent decline. At the same time, world consumption of oil fell from the all-time high of 23.8 billion barrels in 1979 to 20.5 billion barrels in 1983, a 14 percent decrease (Brown, 1984).

In the United States, oil consumption has dropped from 6.3 billion barrels annually in 1973 to 5.5 billion barrels in 1983, a 12 percent decrease. More significant, in commercial and heating oil use there has been a 46 percent decrease over the same period (Department of Energy, April 1984). In addition gasoline consumption in the United States has decreased 15 percent since 1978 because of more fuel-efficient automobiles.

At the same time, the use of renewable fuels is on the rise in the United States. In some parts of the country, wood is steadily displacing fuel oil as a source of residential heat. Wind generators and geothermal wells are becoming significant sources of energy in California. Moreover, approximately 300,000 households in the United States use solar energy as their main water-heating fuel (Department of Energy, April 1984). Figures 1 and 2 provide a historical perspective of energy use in the United States by source.

Although domestic and world dependency on oil is projected to continue to fall for the remainder of this century, overall energy consumption in the United States is expected to increase by about one-fourth by the year 2000. (Department of Energy, Oct. 1983) Therefore, other sources of energy will have to be developed to replace the use of conventional oil and natural gas. It is imperative that energy supplies *within the United States* be clearly understood: first, to demonstrate the inevitability of the domestic petroleum shortage; and second, to delineate alternatives to crude oil

3

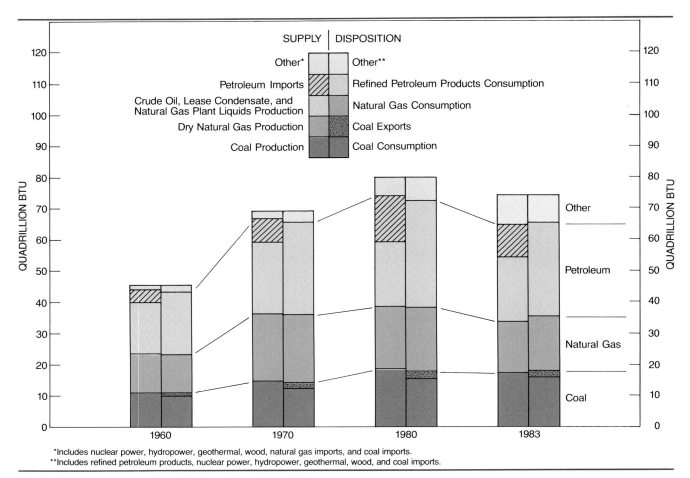

Fig. 1 U.S. energy supply and disposition, 1960–1983.
Source: Department of Energy. *Annual Energy Review,* Washington: Government Printing Office, April 1984.

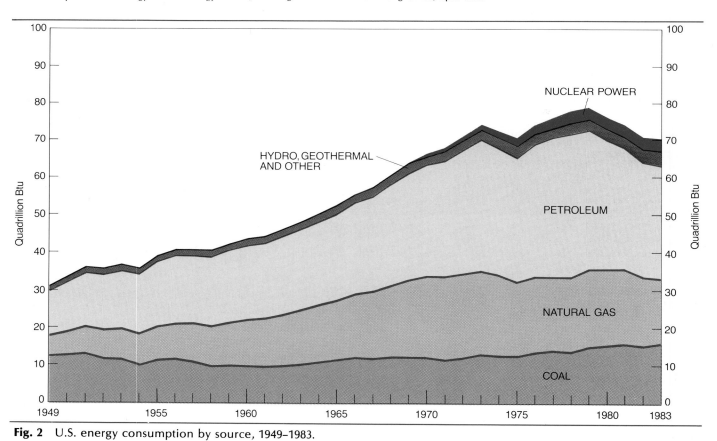

Fig. 2 U.S. energy consumption by source, 1949–1983.
Source: Department of Energy. *Annual Energy Review,* Washington: Government Printing Office, April 1984.

and natural gas. An understanding of the broad spectrum of possible energy sources is not easily attained because resource information for mineral fuels and for renewable energy sources resides in a variety of government documents, trade journals, scientific journals, monographs, and reports to the government by research contractors. Furthermore, each document has its own form of presentation, and employs units that are unique to the resource being studied. It is rare to find the amounts of different resources all expressed in common units which make comparisons easy. While a great deal has been written about energy policies and new technologies, very little has been written to promote an understanding of the magnitudes and locations of energy resources within the country.

This atlas strives to present a complete review of both renewable and nonrenewable energy resources. It will serve as a reference for those who need detailed information on a specific resource and for those who need an overview of the various possibilities. Specifically, the atlas provides the following.

- A detailed analysis of the amounts of mineral fuels and their locations within the country
- Clarification of the uncertainties that cloud the amounts of mineral fuels
- Highlights of the production and transportation of fuels
- An assessment of the renewable energy sources, with emphasis on their geographic aspects

- A regional and national overview of the energy attainable from both renewable and nonrenewable sources, with resource amounts expressed in energy units to facilitate comparisons.
- A profile of each of the nation's ten economic regions that captures their energy resource character for planning purposes.
- An introduction to selected energy futures, and an estimate of the impact of future demands upon the existing resources.

Technical and policy matters that bear on the use of fuels or the transition to renewable energy sources are not included, except to the extent that they illuminate some resource amounts or the possibility of their extraction. The opposite emphasis is seen in a number of works that study energy policy and exclude any consideration of resource amounts (see, for instance, Lovins and Lovins, 1984). This atlas will be complementary to such studies, and also to those recent publications that do compare energy resources but do not provide full information on resource amounts or their locations in the country (see Schurr, 1979, and National Academy of Sciences, 1980).

Figure 3 shows that a great deal of the energy that flows into and through the United States economy is not used, but is rejected. Thus, *primary energy* is roughly twice the nation's *end-use energy*. Of paramount importance are strategies that will re-use energy that is rejected from power plants and engines of all sorts, and strategies that will diminish demand

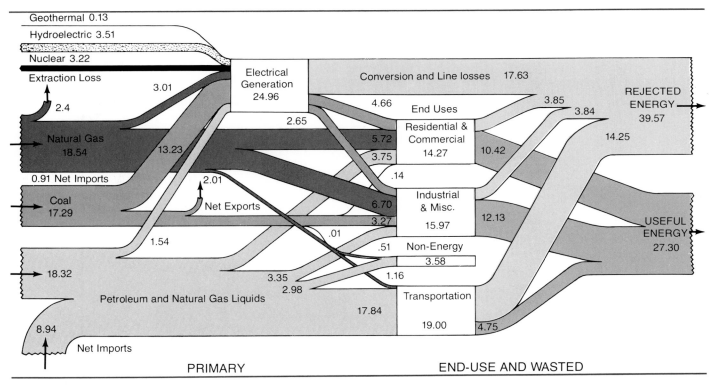

Fig. 3 Primary energy inputs and end-use sectors in the United States, 1983.
Source: Derived from Department of Energy. *Annual Energy Review,* Washington: Government Printing Office, 1983, and Schurr, Sam H. *Energy in America's Future: The Choices Before Us,* Washington: Resources for the Future, 1979.

for the end-use energy. A barrel of oil that is gained through conservation is just as real as one gained through production in a frontier area or by some synthetic fuel process—and it is less costly both in dollar and energy expenditure. Recognizing the urgency of conservation, this atlas nevertheless is concerned largely with the left side of Figure 1: the sources of primary energy. It also deals not only with the traditional fuels that now account for roughly 95 percent of the nation's primary energy, but also with renewable and semirenewable sources that may soon displace much of the mineral fuel and provide energy that is more healthful and more sustainable.

ORGANIZATION

The atlas has three parts and an Appendix. The first deals with nonrenewable (mineral) sources of energy. The second deals with renewable sources. The third part is a summary and overview, and is followed by the Appendix.

One. Nonrenewable Sources

One chapter is devoted to each of the following nonrenewable resources: coal, crude oil and natural gas, shale oil and tar sands, nuclear fuels, and geothermal heat. Geothermal heat is grouped with nonrenewable sources because the accessible occurrences of usable high temperatures can be exhausted by development.

Two. Renewable Sources

Solar radiation plays a major role as a primary energy source. It is discussed first, and is followed by chapters on wind and hydropower, energy flows that depend upon solar radiation. Another group of solar-dependent energy sources appears in the following chapter, *Energy from the Ocean,* which includes ocean thermal gradients, tidal power, currents, wave energy, and brief mention of salinity gradients, an energy potential that does not depend on solar radiation. Finally, a chapter is devoted to energy in biomass.

Three. Conclusion

The overview and comparison of renewable and nonrenewable energy sources maps energy amounts thought to be recoverable, and shows how the energy amounts from different sources coincide in certain states and regions. Energy profiles for economic regions are presented here, as are a few projections of energy needs for the future.

Appendix

The appendix provides a glossary of terms, a geologic time scale, and a table of factors for converting energy units and other measurements. *References,* which list the source documents used by the authors are located at the end of each chapter; suggested readings are located at the end of the book.

METHODS OF PRESENTATION

Resource amounts and their locations are expressed graphically in a variety of maps and graphs. The most prevalent types are previewed in Figure 4.

Segmented Circles for Relative Amounts

The traditional scale circle with segments (pie chart style) is an invaluable tool for showing amounts of an energy source in different locations as well as the relative size of the components that make up the total at any one location. Circle areas are made proportional to the amounts represented. A guide to the amounts is provided in a legend of nested circles as shown in Figure 4. The key to circle segments is the divided bar that accompanies each map of this type and shows how the components apply as a national average.

Discrete Symbols for Absolute Amounts

It is difficult to judge actual amounts from circles. For this reason, important quantities are mapped by a collection of symbols, each of which represents a certain tonnage or volume. Tons, whether coal or nuclear fuel, are shown by blocks; volumes of oil are shown by barrels; and volumes of gas are shown by cylinders. A quick impression can be gained from the size of the blockpile or barrel pile and a fairly accurate estimate can be obtained by counting symbols in the accumulation (Figure 4). Important amounts are tabulated for the reader who needs precise data.

Bar Graphs for Relative and Absolute Amounts

In a bar graph, amounts can be ranked and compared compactly. Often the quantities shown this way in the atlas have been mapped in an earlier illustration and are presented again in bar graph form to emphasize comparisons. With the exception of one graph in the coal chapter, all bar lengths are scaled in *linear* fashion (the length of 200 units is twice that for 100 units) so that relative magnitudes are clear.

PART ONE: NONRENEWABLE RESOURCES

The energy sources surveyed here, with the exception of geothermal heat, are mineral fuels: coal, crude oil and natural gas, and the fissionable metals, uranium and

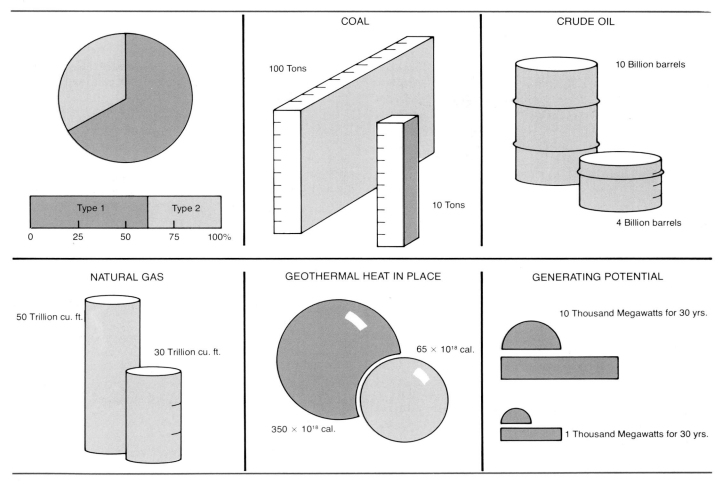

COAL

100 Tons

10 Tons

CRUDE OIL

10 Billion barrels

4 Billion barrels

Type 1   Type 2

0   25   50   75   100%

NATURAL GAS

50 Trillion cu. ft.

30 Trillion cu. ft.

GEOTHERMAL HEAT IN PLACE

65 × 10^{18} cal.

350 × 10^{18} cal.

GENERATING POTENTIAL

10 Thousand Megawatts for 30 yrs.

1 Thousand Megawatts for 30 yrs.

Fig. 4 Symbolization used in the atlas: segmented circles for relative amounts; five symbols used for absolute amounts of resources.

thorium. The term *nonrenewable* is applied broadly to all mineral substances; but it applies in a most absolute way to the mineral fuels. A metal such as copper is re-usable, and may be recycled in order to reduce the need to mine the virgin ore. The mineral fuels, however, are irrevocably lost as their chemical or physical energy is converted to heat.

The mineral fuels have two important traits in common. First, their occurrence is controlled by the types of rock that occur in different areas, making an understanding of major rock types very useful. Second, the mineral fuels occur in limited amounts which are not all of the same degree of certainty or availability. This matter must be clarified by careful use of resource terminology.

Rock Types and Occurrence of Mineral Fuels

Accumulations of crude oil, natural gas, and coal, like concentrations of metallic minerals in ores, do not occur randomly in the earth's crust but show a strong tendency to associate with rocks of certain kinds. Some metallic ores, especially those of abundant metals such as iron and aluminum, occur in such a variety of rocks that generalization is difficult. For fossil fuels, however, the relationship is simple.

IGNEOUS ROCKS Igneous rocks form from the cooling of molten rock, which is called magma if it cools within the earth, and lava if it cools at the surface. For example, deeply buried magma of a certain chemical composition will become a mass of granite, and will have the large crystals and coarse texture permitted by slow cooling at depth. Molten rock reaching the surface as lava in a volanic flow or eruption cools quickly to form a finer-texture rock, such as *basalt* which now covers much of the states of Washington and Oregon. A great number of rock textures and chemical compositions exist in rocks classified as igneous, but all originate in high temperature magma or lava.

SEDIMENTARY ROCKS Unlike igneous rocks, sedimentary rocks form entirely at or near the earth's surface. Practically all are due to materials settling and collecting quietly in oceans, lakes, or river beds. A sandstone, for example, is composed of sand grains compressed and cemented together. Shale is composed of compacted clay particles. Limestones are made of shell fragments in some cases, or of chemically-precipitated lime in other cases. The materials deposited become rock as they are subjected to pressure of overlying materials, which are usually younger sediments.

7

METAMORPHIC ROCKS As the name implies, metamorphic rocks are the result of some kind of alteration of an existing rock. The sedimentary rock sandstone, for instance, may be altered to a dense hard siliceous rock called quartzite. Similarly, a shale can be altered to slate; and heat and pressure can turn a limestone into marble. Although the most familiar examples of metamorphics were originally sedimentary, igneous and metamorphic rocks are also altered. Frequently the process of metamorphism occurs at great depth and under high temperatures and pressures that accompany the compressive forces of mountain building. As a result, the earlier rock may be wholly or partially melted and allowed to crystalize.

CRYSTALLINE ROCKS VERSUS SEDIMENTARY, AND THE OCCURRENCE OF FOSSIL FUELS For simplicity, igneous and metamorphic rocks may be considered together and called, for lack of a better term, *crystalline rocks.* The crystalline texture in igneous rocks is due to formation of crystals directly from magma or lava: in metamorphics it is due to crystals forming from re-melted materials or from percolating fluids rich in minerals.[1] All igneous and most metamorphic rocks have a history of high temperatures, pressures, and often violent deformation. Since hydrocarbon fuels are obviously susceptible to destruction at high temperatures, they would not form in igneous rocks and would not survive in most metamorphics.

Sedimentary rocks are the logical hosts for fossils fuels because they have a history that is relatively quiet, without excessive temperature or pressure. Equally important, however, is the environment in which both fossil fuels and sedimentary rocks are born. Coal is formed from plant materials that have fallen into swampy areas near an ocean shore: as sea level changes, silts and clay are deposited and eventually enclose the coal in a sedimentary sandwich. Natural gas and crude oil are thought to form from the alteration of marine plant and animal organisms whose organic remains collect on the ocean floor, along with the sediments that become rocks. Rocks of sedimentary type, then, are formed in the same environments as coal and petroleum, and have a history consistent with the preservation of the delicate hydrocarbons.

The locations of sedimentary versus crystalline rocks can be understood with the aid of a cross-section (Fig. 5). This figure shows crystalline rocks to be the continent's basement, exposed at the surface—to the exclusion of sedimentary—in two kinds of areas. One is broad low "shield" areas, such as the Canadian Shield north of the Great Lakes; and the other is where basement rocks poke through sedimentary rock in mountain ranges such as the Southern Rockies in Colorado, the Sierra Nevada's, and Ozark Dome in Missouri, and the Black Hills of South Dakota.

Areas not identified as crystalline (Fig. 5) may be assumed to have some, and possibly very thick, sedimentary rocks; those areas, therefore, are where fossil fuels may be expected. In the following pages, the amounts of coal or petroleum mapped will be influenced by the presence or absence of sedimentary rocks in the regions or states for which resource data is gathered.

ROCK TYPES AND OCCURRENCE OF NUCLEAR FUELS Although uranium and thorium are not abundant metals, some of their ores occur in sedimentary rocks, and quite different ores occur in igneous and metamorphic rocks. There is no simple relationship between areas of occurrence and the patterns of Figure 5. The various types of deposits and their locations will be discussed in a later chapter.

Resource Terminology, and The Life of Resources

HOW MUCH DO WE HAVE LEFT? Despite frequent references to mineral resources being *finite* amounts, the answer to the question of how much remains is quite *indefinite* because of two uncertainties. The first is due to incomplete exploration. There is, and always will be, *some* possibility of undiscovered uranium oxide deposits, coal beds, or petroleum accumulations. The amounts remaining can never be predicted accurately. The second is due to changing economic factors and technological advance that both can alter the definition of what constitutes a usable resource. Both these variables must be kept in mind in any review of mineral resources, whether fuels, metallics, or other types.

The following scheme, used by the U.S. Geological Survey and initiated by its recent director, Vincent E. McKelvey (McKelvey and Wang, 1969), is used in this book to organize resource amounts for coal, crude oil and natural gas, nuclear fuels, and geothermal heat.

The rectangular diagram in Figure 6 comprehends *all* of a particular resource that exists within a given area, such as a country or continent. Some of that resource quantity is better known than other parts of it. To reflect that, *identified* amounts are assigned to the left side while *undiscovered* amounts occupy the right side of the diagram. Some of the resources, whether *identified* or *undiscovered,* are more easily recovered from the earth because they are rich concentrations, near the surface, easily treated after mining, or have some combination of these characteristics. Such resources are located near the top of the diagram, while the resources less easily recovered are placed toward the bottom. The upward direction, then, is one of increasing economic feasibility, and leftward is the direction of increasing geologic certainty.

Because the diagram takes into consideration both geological knowledge and economic factors, it is an indispensable aid to classifying resources and assigning them unambiguous terms. *Reserves,* shown by the shaded area, are those amounts that are well-defined through geologic exploration and development, and at the same time are *recoverable* at current prices and with current technology. Other resources of the same substance may be disqualified as reserves for one of two major reasons: They may be attractive economically but as yet undiscovered (position X); or they

Fig. 5 U.S. areas that are dominantly crystalline rocks, therefore unfavorable to fossil fuels. Included is a cross-sectional view showing how crystalline rocks are overlain in some areas by sedimentary rocks which are favorable to fossil fuels.

Source: Derived from plate 74 of the *National Atlas of the United States of America*, Washington: U.S. Geological Survey, 1970, and *Loebeck's Physiographic Diagram of the United States*, 1957.

may be well-defined but not easily recovered because of low grade or difficult access (position Y). Some undiscovered resources are not now attractive economically (position Z) and are thus the most remote and the least likely to be used.

Although the definition of reserves is firm, the actual magnitude of reserves changes with time. Obviously,

	IDENTIFIED	UNDISCOVERED
ECONOMIC	RESERVES	X
SUBECONOMIC	Y	Z

← Increasing Economic Feasibility (vertical axis, upward)

← Increasing Geologic Assurance →

Fig. 6 Scheme for organizing mineral resources of differing degrees of certainty and economic feasibility.

production (consumption) will diminish reserves, but there are also two other general processes that tend to augment reserves. First is the *discovery* of resources which allow the reserve category to expand to the right. Equally important is the influence that economic and technological changes can have upon the lower limit of the reserves category. For example, when copper ores were more abundant in 1920 the average ores mined and considered economic contained 1.5 percent copper. In the 1970s, with usable ores more scarce, and with bigger machinery available for handling large quantities of rock, the ores mined averaged around 0.6 percent copper. Rock that was *not* reserve in 1920 *became* reserve as the boundary between economic and noneconomic shifted downward in the diagram. Rising prices for mineral fuels can have a similar effect, as will be shown in connection with crude oil.

HOW LONG WILL IT LAST? Raising the question of the *life* of a resource is disturbing, because of the implicit assumption that we plan to use all of the resource in question. Nevertheless, the exercise is useful because comparing resource amounts with rates of use is one of the best ways to bring meaning to large and unfamiliar numbers such as billions of tons of coal or trillions of cubic feet of natural gas.

There is, of course, no way of predicting resource life with confidence. As the foregoing section pointed out, there is uncertainty about resource amounts remaining to be used. In addition, there is uncertainty about future rates of use. One device used in this atlas is the *static reserve index* which is simply current reserves divided by current annual rate of use (i.e., rate of production or mining of the raw resource). A more thorough approach is to project both resource amounts and use rates into the future. Resource amounts may be assumed to include all resources that may be useful in the future, including those amounts now undiscovered and those identified but noneconomic (Fig. 6). The use rate may be assumed to remain constant, or to grow at a certain rate: in either case, the accumulated production through a period of years can be estimated. This accumulated production can be compared with total resources remaining to show the expected status at the end of the period chosen.

If rates of use increase continually at a fixed percentage rate, the impact upon resource amounts is staggering. For instance, if production increases steadily at a rate of 6 percent per year, then the cumulative amount used will *double in 12 years.* At such a rate, as one geologist put it, the world's petroleum will be half gone by the year 2003, and all gone by the year 2015 (Skinner, 1979). Or, to put in another way, if the world's petroleum resource is actually twice what we thought it to be, the reprieve is only 12 years!

Apparently, ever-increasing annual rates of resource use tend to make less significant the question of how much of a resource remains, because vast amounts can be consumed in a very few years. Happily, the increasing rates of energy use can be avoided in the future, and can be replaced by annual rates that level off and fall. Some projections of energy futures, both expansionist and more conservative, are referred to in Part Three of this atlas.

PART TWO: RENEWABLE RESOURCES

Renewable sources of energy are either the primary energy flows that reach the earth's surface, or the various forms of energy that depend upon a primary flow and can be utilized year after year if properly managed.

Figure 7 represents schematically the three fundamental sources from which energy flows continuously to the earth's surface: geothermal heat, gravitational forces, and solar radiation.

Geothermal heat originates in the hot semi-molten mantle rock at the base of the crust and in radioactive decay of elements within the crust. The supply of heat, especially from the deep source, is practically unlimited; but the portions that are in high-temperature occurrences accessible for exploitation are not, and for this reason geothermal heat is treated here as a nonrenewable resource. Work by the U.S. Geological Survey has defined and measured the resource so thoroughly that a substantial chapter can be devoted to interpretation of geothermal heat in the country.

Gravitational forces that lead to rise and fall of tides are due to influence of the moon and the sun, and are completely divorced from solar radiation. Usable mostly for electrical generation, the resource does not appear to be large, either on world or national scale.

Solar Radiation

Radiation can be employed for its thermal and its photovoltaic effects. In addition, a number of familiar physical and biological processes may be exploited—all of which are dependent upon solar radiation.

Fig. 7 Renewable energy flows from three sources: geothermal heat, gravitational forces, and solar radiation.

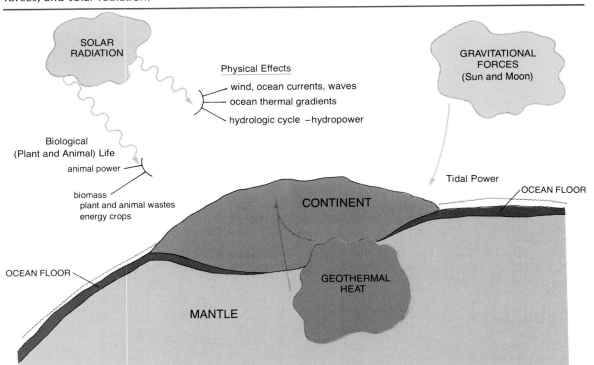

PHYSICAL EFFECTS Unequal heating of the earth's surface leads to winds, which can be harnessed for power and which, themselves, contribute to ocean currents and wave pulsations. Heating by the sun also causes surface waters of tropical and subtropical oceans to be warmer than deep waters, making a thermal gradient or contrast that can be used for power generation through ocean thermal energy conversion (OTEC). Solar radiation provides the energy by which water is evaporated from the oceans and dropped on highlands as water that can be used for hydroelectric power generation. Separate chapters are devoted to solar radiation, wind, hydroelectric power, and ocean energy.

BIOLOGICAL EFFECTS Because photosynthesis is essential to plant life, it is clear that all natural and domestic food chains are based on solar radiation. The energy afforded by harnessing draft animals such as horses is plant energy converted. The energy in coal and petroleum represents vestiges of the very large amounts of plant life that has existed on the planet throughout its history. Further opportunities exist today in the various organic materials, both plant and animals, that can be produced continuously through photosynthesis. These materials, broadly considered *biomass*, include wastes and crops grown specifically for their energy content. They are dealt with in a separate chapter.

The Question of Renewable Energy Amounts

For mineral fuels, and even for some dimensions of geothermal energy, finite amounts of energy in place and energy recoverable can be estimated—though with some nagging uncertainties about definition of "usable resource." For renewable sources the definition of total potential is less clear, although it can be approached. The total potential of hydroelectric is the easiest to define, at least conceptually, because river systems embrace only so much hydrostatic head, and only so many installations can be tolerated if certain conditions are specified. Total wind power can be defined if certain areas are delineated as suitable, and if those areas are then theoretically "saturated" with wind machines of a specified type and generating capacity. Nonetheless, new technologies, especially those that would capture light winds or allow more densely packed installations, can make earlier estimates obsolete.

The total potential of ocean thermal gradients can be approached in a similar fashion, by defining what would constitute saturation of suitable ocean areas. Most elusive is the potential of biomass, because in *energy crop* endeavors, as opposed to utilizing wastes, the energy use will compete with other uses of the land or of the crop. Future prices of various commodities, not only competing fuels, can therefore influence how much biomass energy can be produced in a given region and in the nation as a whole.

It is important to recognize that *energy delivered* through the use of radiation, wind, or falling water, leads to savings in mineral fuels that are larger than may appear at first glance. Ten thousand British Thermal Units (BTUs) of energy delivered through a solar heating system does not simply avert the burning of equivalent amounts of fuel oil or natural gas. If the furnace is 80 percent efficient, then the fuel needed to supply the 10,000 BTUs is 12,500 BTUs. In the case of electrical power generation the savings are more dramatic, because only 35 to 40 percent of fuel energy is converted to electrical in a conventional power plant. *Fuel savings* is the term used to denote the amounts of fuel energy (or actual amounts of fuel) that would be displaced by the use of some renewable energy supply.

NOTE

1. In fact, some sedimentary rocks, especially the carbonates, have crystalline texture, too; thus the term is not ideal.

References

Brown, Lester, *et al.* *The State of the World, 1984. A Worldwatch Institute report on progress toward a sustainable society.* New York: W.W. Norton, 1984.

Lovins, Amory, and L. Hunter Lovins. *Energy Unbound: Your Invitation to Energy Abundance.* San Francisco: Sierra Club Books, 1984.

McKelvey, Vincent, and F. H. Wang. *World Subsea Mineral Resources,* U.S. Geological Survey *Miscellaneous Investigation I–632.* Washington, DC: Government Printing Office, 1969.

National Academy of Sciences. *Energy in Transition, 1985–2010.* New York: W.H. Freeman, 1980.

Schurr, Sam H. *Energy in America's Future: The Choices Before Us.* Washington, DC: Resources for the Future, 1979.

U.S. Department of Energy. *Annual Energy Review.* Washington, DC: Government Printing Office, April 1984.

U.S. Department of Energy. *Energy Projections to the Year 2010.* Washington, DC: Government Printing Office, October 1983.

Skinner, Brian J. *Earth Resources* (2nd edition), Englewood Cliffs, N.J.: Prentice-Hall Inc., 1979.

U.S. Geological Survey. *National Atlas of the United States of America,* Washington, D.C., 1970.

Lobeck, Armin K. *Physiographic Diagram of the United States,* Maplewood, New Jersey: Hammond Inc., 1957.

Part One

NONRENEWABLE RESOURCES

Chapters 1 through 5 deal with nonrenewable energy sources: coal, crude oil and natural gas, oil shales and tar sands, nuclear fuels, and geothermal heat.

All of these must be found through exploration—and then, in some way, extracted from the earth. Their resource amounts are therefore subject to the uncertainties of limited geologic knowledge and changing economic factors, which are recognized by the *resource-reserve diagram* (see Introduction). That organizing device is used throughout this part of the book to maintain a uniform view of the various nonrenewable energy sources, whether they are mineral fuels or geothermal heat.

For each of the five resources in Part One, careful distinction must be made between amounts *in place* and amounts *thought to be recoverable*. The chapters differ in that respect because information on coal resources is for tonnage in place, whereas crude oil and nuclear fuels are represented in official documents as amounts recoverable under certain assumptions.

In each of the five chapters the units employed are peculiar to the resource, such as tons for coal, barrels for crude oil, and cubic feet for natural gas. Comparisons among resources are not possible unless the measurements are translated into common-denominator energy units, such as *calories* or *British Thermal Units*. That is done in the *Overview* in Part Three.

1 Coal

Although coal is thought to have been used as an energy source in China as long ago as 1000 B.C., it was not until the thirteenth century that notable coal mining operations appeared in England, Scotland, and on the continent of Europe. During this period, coal was not regarded as an important energy resource because of both the noxious fumes it emitted on combustion and the abundant supply of timber available at the time in Europe. With the development of coal-fired brick kilns in the fifteenth and sixteenth centuries, demand for coal began to increase. The establishment of iron smelting works and other engineering and metallurgical developments in England during the seventeenth and eighteenth centuries brought about a dramatic increase in demand for coal and ushered in an era in which coal was to be the dominant source of energy.

In the United States, coal appears to have been first mined in the early 1700s along the James River in Virginia. Almost 150 years elapsed before the discovery of the rich and expansive Appalachian bituminous coal field which was to become the focus of coal commercialization in the United States. In these early years, movement of coal overland was difficult, but the advent of the steam locomotive solved the land transport problem and opened huge new markets. In 1900, coal fulfilled about 90 percent of the national energy needs despite the development of oil and gas reserves some years earlier. From 1900 to 1920, however, oil began to compete with coal as an important source of energy (Fig. 1-1). From 1920 to the mid-1970's, coal's share of the United States' energy market declined rather steadily in face of competition from the cleaner-burning oil and gas. By 1976, coal's share of the nation's energy market had declined to less than 20 percent of the total. Despite its diminished role in the United States, the volume of coal production has significantly increased in the past two decades due to a large increase in the country's demand (Fig. 1-2).

Since 1976, however, coal's share of domestic energy consumption has risen because of its low cost, relative to oil, and natural gas. By 1983, coal accounted for 28 percent of the United States energy production and 21 percent of the domestic consumption (Department of Energy, May 1984). In order to meet domestic and foreign demands, coal production is forecast to rise by an average of 4.2 percent per

←Surface mining is a very large open-pit operation at Kemmerer, in southwestern Wyoming. The coal is burned in a nearby power plant. Photo courtesy of the American Petroleum Institute and Union Oil of California.

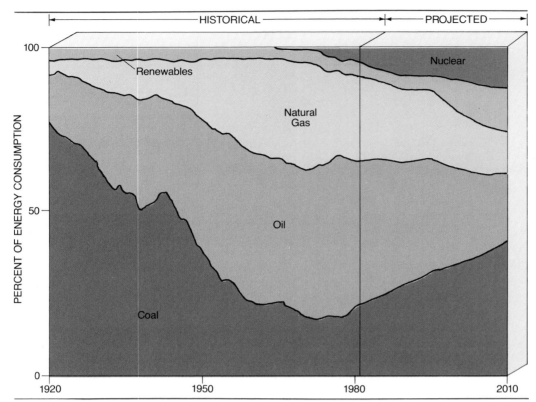

Fig. 1–1 Roles of various energy sources in the U.S., 1920–2010.
Source: United States Department of Energy, *Outlook for U.S. Coal*, August 1982.

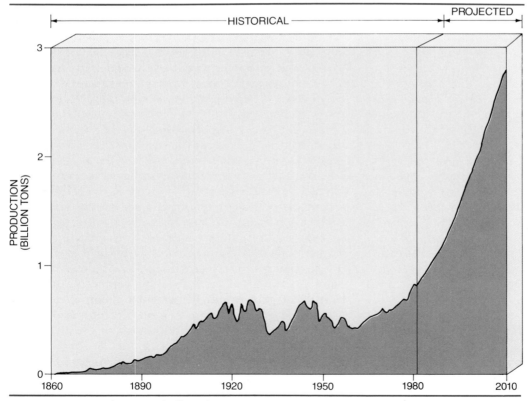

Fig. 1–2 U.S. Production of all coal, 1860–2010.
Source: United States Department of Energy, *Outlook for U.S. Coal*, August 1982.

year from 1983 to 1990 then slow to 2.7 percent per year from 1990 to 1995. Coal is expected to become the major domestically produced energy source by the late 1980s, contributing approximately 36 percent of the nation's energy production by 1995 (Department of Energy, May 1984).

Coal is used primarily as a boiler fuel to generate steam. About 85 percent of the domestic consumption is for electricity generation by utilities. In the near future, most coal will continue to be burned directly in electric generating plants with the use of devices to control stack emissions of sulfur oxides and particulates. If coal usage is to increase at the rate projected, more effective and economical methods of cleaning coal before combustion must be developed. In addition, present emission standards will have to be abided by and perhaps made even more rigorous. Long-term projections indicate that conversion of coal to synthetic crude and synthetic natural gas could provide substitutes for conventional oil and gas and would help to reduce air pollution from coal burning. Unfortunately, greatly increased coal mining could be accompanied by the environmental damage strip mining causes. Surface mining should be avoided where possible, and requirements for land restoration after surface mining must not be relaxed. As shown in this chapter, very substantial amounts of desirable low-sulfur coal are available through underground mining methods. Economics aside, this coal should be preferred to that obtainable by surface mining.

ORIGIN AND CHARACTER OF COAL

Coal was formed from vegetal matter that once thrived in some areas of the earth. The accumulation of coal-forming plant debris occurred millions of years ago, mainly during the Carboniferous period, and to a lesser degree during the Cretaceous and Tertiary periods. The earth's climate was particularly favorable for extravagant plant growth during this time. In addition, areas of level swamp land where plant life thrived were in abundance.

In the swamps, materials were preserved by virtue of the low-oxygen stagnant waters which arrested the bacterial action causing plant decomposition. Vegetal matter, accumulated in the shallow swamp waters, formed peat. As the land subsided or the sea rose (or both), the plant debris was covered by clays, sands, and lime muds, which were the basis of the shales, sandstones, and limestones found on top of the coal seams today. Over thousands of years the peat was compacted, became more dense, and was gradually transformed into coal.

Coal deposits are distributed worldwide although they are much more common in the Northern Hemisphere. Within the United States, coal-bearing strata underlie about 13 percent of the land area and are present in varying degrees in 37 states (Averitt, 1973). In many areas of the United States, coal seams continue over great distances and large areas as, for instance, the Pittsburgh Seam which underlies some 15,000 square miles of the Appalachian Highlands.

Coal seams in the United States range in thickness from a thin film to 100 feet or more, but most are between 2 and 10 feet thick (Department of Energy, October 1982).

RANK AND SULFUR CONTENT

Coal is classified on the basis of the amount of fixed carbon, moisture, and volatile matter present. Fixed carbon is the percentage of stable carbon found in coal. Volatile matter is the proportion of organic gases, such as oxygen and hydrogen. The rank of coal is generally based on the percentage of carbon, which increases through the ranks from lignite to subbituminous to bituminous to anthracite (Fig. 1-3). Figure 1-3 also shows that the highest heat content (energy content) of all four ranks is found in bituminous coal which has relatively few volatiles and low moisture content. The heat or energy content is lowest in lignite which has a high moisture content.

The changes in fixed carbon, moisture, and volatile matter content through the ranks are an expression of the progressive alteration of original peat materials and depend upon depth and heat of burial, compaction, time and structural deformation (Averitt, 1973). Energy content by rank ranges from 14×10^6 BTUs per short ton for lignite to 26.4×10^6 BTUs for anthracite (Bureau of Mines, 1976).

Figure 1-4 shows the distribution of United States coal deposits by rank. Lignite deposits, excepting those in Alabama, are found exclusively in young rocks west of the Mississippi River, the largest one being located in North Dakota. Subbituminous coals are found only in the western part of the United States, particularly in Montana, Wyoming, and New Mexico. Bituminous coals are found throughout the United States, but most are confined to Paleozoic rocks in the Appalachian and Mid-Continent regions. Anthracite deposits are scarce and are found only in Pennsylvania, Virginia, Arkansas, Colorado, Washington, and Alaska.

The quality or grade of coal is determined by the relative amounts of ash, sulfur, and certain other noxious constituents. Thus far, coal resources of the United States have been classified only according to sulfur content (Bureau of Mines, Information Circulars 8680 and 8693, 1975). Sulfur is a particularly undesirable constituent of coal because it lowers its quality, contributes to corrosion in boilers, creates air and water pollution problems, and inhibits plant growth in spoil banks. Although the amounts vary in both chemical composition and weight percentage, all coals contain some sulfur. According to Averitt, the sulfur content of coal in the United States ranges from 0.2 percent to about 7 percent, but the average in all coal is 1 to 2 percent. The bituminous coals of the Pennsylvanian Age in the Appalachian and Interior coal basins have the highest sulfur content, with over 40 percent of the identified resources containing at least 3 percent. The subbituminous coal and lignite of the Rocky Mountains and Northern Great Plains, however, generally contain less than 1 percent sulfur (Averitt, 1973).

In order to control sulfur emissions, the Environmental

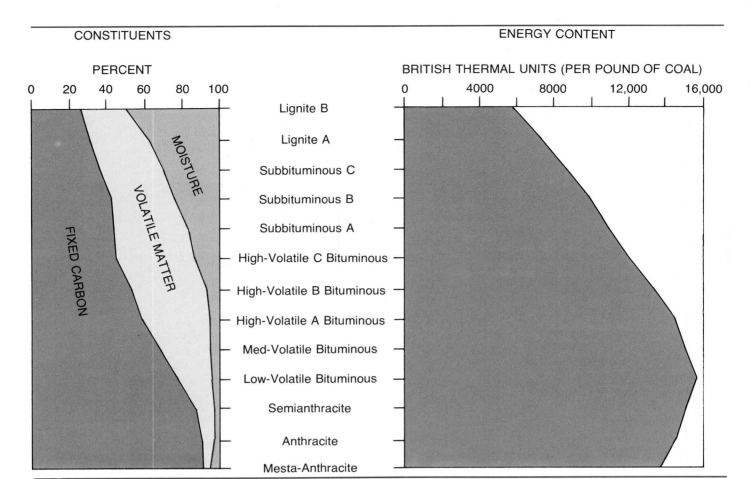

CONSTITUENTS

PERCENT

ENERGY CONTENT

BRITISH THERMAL UNITS (PER POUND OF COAL)

FIXED CARBON
VOLATILE MATTER
MOISTURE

Lignite B
Lignite A
Subbituminous C
Subbituminous B
Subbituminous A
High-Volatile C Bituminous
High-Volatile B Bituminous
High-Volatile A Bituminous
Med-Volatile Bituminous
Low-Volatile Bituminous
Semianthracite
Anthracite
Mesta-Anthracite

Fig. 1–3 Constituents and energy content of different ranks of coal.
Source: Averitt, 1975.

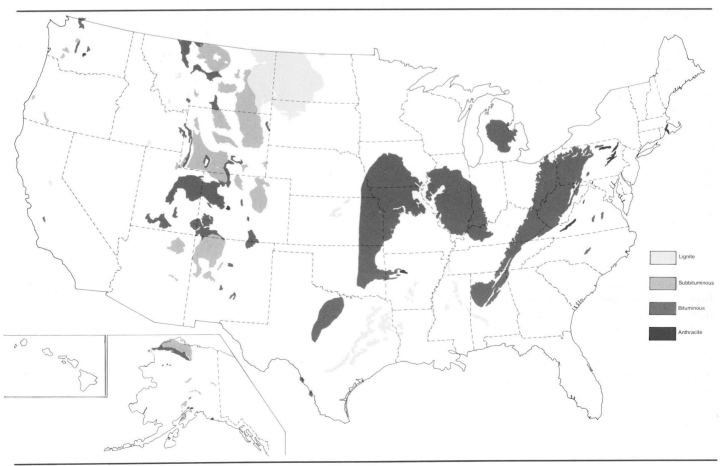

Lignite
Subbituminous
Bituminous
Anthracite

Fig. 1–4 Occurrence of coal of four ranks.
Source: Department of Energy, *Outlook for U.S. Coal*, 1982.

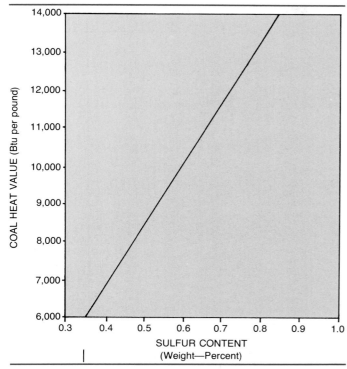

Fig. 1-5 Maximum permissible sulfur content versus energy content of coal commensurate with EPA standard for air quality.

Source: U.S. Bureau of Mines, "The Reserve Base Of U.S. Coals by Sulfur Content—The Western States," *Information Circular, 8693,* 1975.

Protection Agency has established national stack emission limits for new stationary sources with a capacity of greater than 250 million BTUs per hour heat input (Bureau of Mines, Information Circular 8680, 1975). The EPA regulates stack emissions, not sulfur content of coal burned, the limit being 1.2 pounds of sulfur dioxide per million BTUs of heat input. Because coals of different ranks have different heat values, the sulfur contents commensurate with allowable emissions will vary (Fig. 1-5). For example, a lignite with heat value of 7,000 BTUs per pound could hold no more than 0.4 percent sulfur, while bituminous coal of 13,000 BTUs per pound could hold up to 0.78 percent sulfur.[1]

MINING METHODS AND RECOVERY RATES

There are two basic methods used for extracting coal from the seam: underground and surface. In underground methods, coal is extracted without removal of overburden, that is, the soil deposits found on top of the coal. Strip mining involves removal of the overburden and extracting the coal from the exposed seam. Underground mining is used when the coal is buried too deeply in the ground to make surface mining feasible or possible. There are three different types of underground mines, as illustrated in Figure 1-6: shaft, drift, and slope types. The type of mining method used to recover

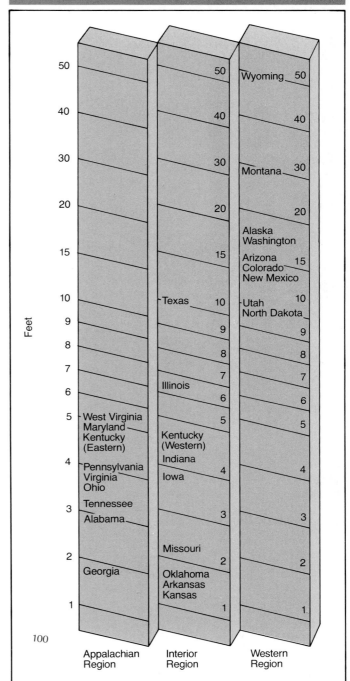

AVERAGE THICKNESS OF COALBEDS MINED, 1979

Coal seams in the United States range in thickness from a thin film to 50 feet or more. In the Appalachian Region the average thickness of coalbeds mined ranges from 2 feet in Georgia to 5 feet in West Virginia and Maryland. In the Interior Region the average thickness of the beds range from a foot in Kansas to 10 feet in Texas. The thickest coalbeds in the United States are in the West where they range from 10 feet in thickness in Utah and New Mexico to 50 feet in Wyoming. The average thickness of all coalbeds mined in the United States in 1979 was 10.7 feet.

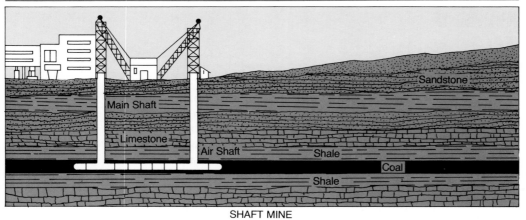

SHAFT MINE

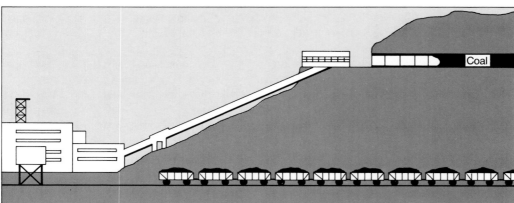

DRIFT MINE

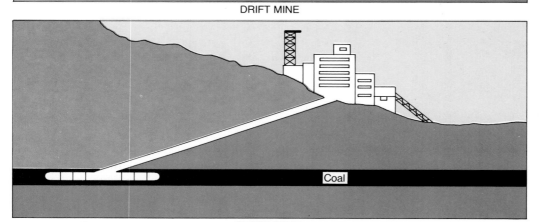

SLOPE MINE

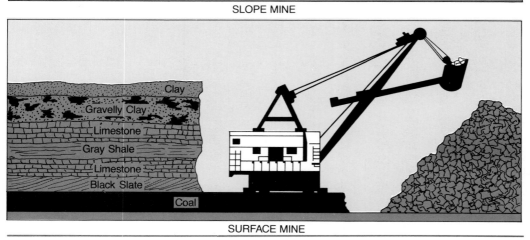

SURFACE MINE

Fig. 1–6 Four types of coal mines.
Source: National Coal Association. *Bituminous Coal Facts 1972*, 1972.

coal depends upon the total area of coal available, the thickness and inclination of the seam, the thickness of the overburden, the value of the surface land, as well as other economic factors. In 1982, almost all of the coal produced in the United States was recovered by either underground (40 percent) or strip mining (60 percent) methods (Department of Energy, September 1983).

There are two basic types of underground mining methods: room and pillar and longwall. In room and pillar mines, coal is removed by cutting rooms, or large tunnels, in the solid coal leaving pillars of coal for roof support. Longwall mining takes successive slices over the entire length of a long working face (Fig. 1-7). In the United States, almost all of the coal recovered by underground mining is by room and pillar method. In Great Britain, however, most of the coal is mined by longwall methods. Longwalling is advantageous because it can be carried out at depths greater than those at which the room and pillar system is used. This is simply because at great depths the pillars might break under the weight of overburden.

Up to now, longwalling has not been used extensively in the United States because most of the coal has been extracted from the thin and relatively shallow seams of the eastern part of the country. One would expect longwall methods to be used more as the demand increases for the thick and deep coal deposits of the western states.

The percentage of coal recovered from a mineable seam depends on a number of factors. These include: seam thickness, number of pillars needed to safely support the roof, and the degree to which the surface land is protected. Figure 1-8 compares the amount of coal recovered by room and pillar versus longwall recovery for beds of various thickness. Longwall mining can recover 85 percent of the coal in beds less than 10 feet thick whereas the recovery rate in these same beds using room and pillar techniques is only 57 percent. The recovery factor for both methods drops off very rapidly for beds over 10 feet thick.

As improved earth-moving equipment became available, strip mining developed very rapidly in the United States. Strip mining consists of removing the soil and rock above a coal seam and then removing the exposed coal. Modern strip mines employ equipment capable of removing overburden of more than 200 feet in thickness. Strip mining has surpassed underground techniques in the United States as the favored recovery method because it is cheaper and allows a higher recovery rate (85 to 90 percent). Strip mining, however, has been extremely destructive to the land surface, as it leaves huge scars and piles of broken rock, denudes the area of vegetation, and greatly promotes water pollution in adjacent streams. To counteract this destruction of the land, the state and federal governments now require the coal mining industry to restore stripped land to conditions favorable to future productivity. It is seldom possible, however, to restore a stripped area to its precise original condition.

RESOURCE TERMINOLOGY

The U.S. Geological Survey and the Bureau of Mines have classified coal resources to depths of 6,000 feet (Averitt, 1975 and Bureau of Mines, Information Circulars 8680, 8693, and 8678, 1975). As indicated in Figure 1-9, coal resources are designated as either *Identified* or *Hypothetical*. Identified resources, assessed to a depth of 3,000 feet, consist of mineral-bearing rock whose existence and location are known. The quantity of these deposits that meet specific depth and thickness criteria is termed the *Reserve Base*. The criteria used by the Bureau of Mines stipulate a thickness of 28 inches or more for bituminous and anthracite coals, and 60 inches for subbituminous and lignite. The

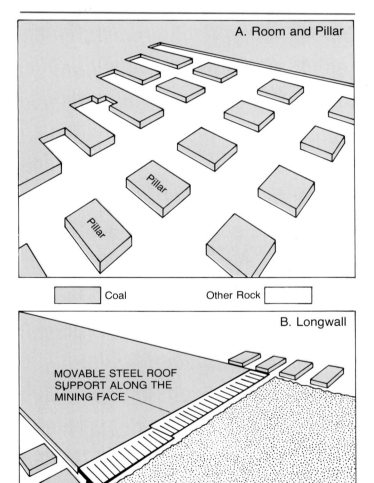

Fig. 1-7 Underground mining methods: Room and pillar versus longwall.
Source: U.S. Bureau of Mines, *Mining Technology Research*, 1976.

Conventional, or Room and Pillar Mining. A five-foot seam is slotted by this cutting machine in a Kentucky mine. The six-inch slot will allow the coal to expand and fall when it is blasted. Not shown are pillars of coal that are left to support the roof. (Courtesy of Westinghouse Air Brake.)

Longwall mining. Whirling cutters on the mining machine shave coal from the face on the left. The exposed coal face and the work area may be several hundred feet across with no coal pillars to support the roof. Support is provided by the steel roof and hydraulic jacks shown here. As the coal face retreats, the roof and supports are moved to the left and the roof behind (to the right) is allowed to fall. Because no coal is left as pillars, the recovery rate is much higher than in room and pillar mining. (Courtesy of National Coal Association.)

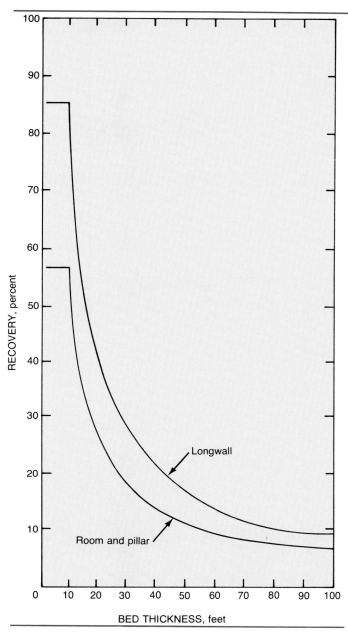

Fig. 1-8 Recovery ratios: Room and pillar versus longwall mining methods for beds of various thickness.
Source: U.S. Bureau of Mines, "The Reserve Base of Coal for Underground Mining in the Western United States," *Information Circular, 8678,* 1975.

Demonstrated Reserve Base refers to inplace coal, both Measured and Indicated, that is technically and economically mineable.

The part of the Reserve Base that can be recovered is termed the *Reserve.* Whether or not the coal will be recovered depends on a number of factors: the characteristics of the coal bed; mining methods used; and any legal constraints placed upon mining a deposit because of natural and cultural features. Although the recovery ratio may vary from 40 to 90 percent, it is thought that at least half of all in-place coals in the country can be recovered.

Hypothetical resources can be estimated on the basis of broad geologic knowledge and theory. They are defined as undiscovered deposits, whether of recoverable or subeconomic grade, whose existence in certain districts is geologically predictable (Averitt, 1975).

Giant shovel at work in surface mining operation near Percy Illinois. (Courtesy of American Petroleum Institute and Ashland Oil Company.)

depth criterion for all ranks, except lignite, is 1,000 feet. (Bureau of Mines, Information Circular 8678, 1975). In the case of lignite, only beds that can be mined by surface methods are included in the Reserve Base. These generally lie no more than 120 feet below the surface. In some instances, however, coal deposits which do not conform to the depth criteria are included in the Reserve Base if they are either presently being mined or could be mined commercially. The term "Demonstrated" refers to both Measured and Indicated categories of coal as defined by the Bureau of Mines and the U.S. Geological Survey. *In summary, the*

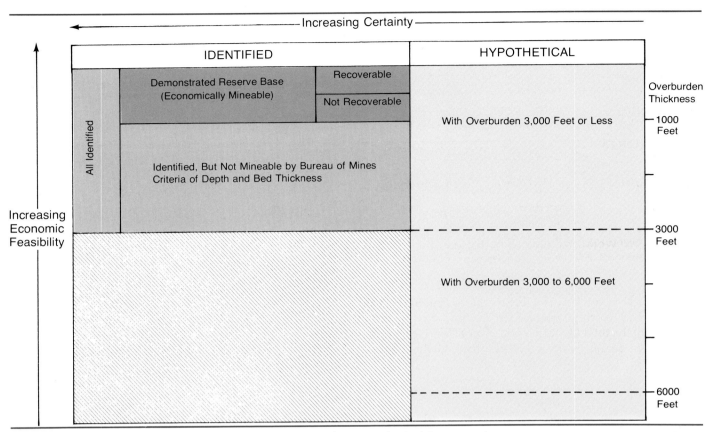

Fig. 1–9 Coal resource categories, combining concepts of U.S. Geological Survey and U.S. Bureau of Mines.

WORLD COAL RESOURCES

Coal resources of the world have been estimated at about 10.8 trillion short tons (*World Coal,* November 1975). The recoverable reserve portion of world coal resources amounts to approximately 986 billion short tons (World Energy Conference, 1983). Almost three-fourths of recoverable reserves are bituminous coals.

Eastern Europe and the Soviet Union account for one-third of the world's recoverable reserves while North America accounts for nearly that much (Table 1–1 and Figure 1–10). Two additional regions—the Middle and Far East and Oceania, and Western Europe—house 19 and 10 percent respectively of the world's recoverable reserves. Three countries, the United States, the Soviet Union, and The People's Republic of China have two-thirds of the recoverable reserves. The United States and the Soviet Union account for 29 and 27 percent respectively while The Peoples Republic of China has 11 percent.

Despite the phenomenal growth in the use of oil and gas in the past three decades, world coal production has increased 27 percent since 1973 (Department of Energy, September 1983). In 1982, three countries, the Soviet Union, the United States, and The Peoples Republic of China accounted for almost 54 percent of the world's coal production. The United States produced 19 percent of the total, and the Soviet Union and China produced 18 and 17 percent respectively. At the same time, over 50 percent of all energy consumed in the Soviet Union and China were derived from coal, whereas in the United States approximately 21 percent of total energy consumed came from coal.

UNITED STATES COAL RESOURCES

Identified and Hypothetical Resources

As of January 1974, the estimated remaining coal resources of the United States were 3.9 trillion short tons (Table 1–2). Of the total, 48 percent was classified as Identified and the remaining 52 percent as Hypothetical. Table 1–3 presents recent revisions of Averitt's coal resources estimates for six states. According to information published by the Department of Energy in 1978, total Identified and Hypothetical coal resources of the United States should be increased by 2,450 billion tons (Department of Energy, February, 1978). Alaska accounts for 72 percent of this increase. All of the additional resources in Alaska are Hypothetical. They are believed to be located in the Northern Alaska coal resource region. Montana's and Wyoming's total coal resources have been upwardly revised based on new estimates of Hypothetical coals in the Powder River Basin area.

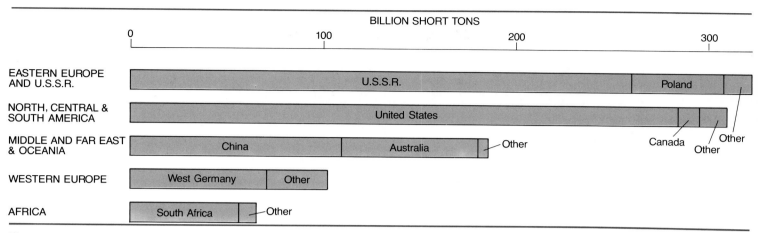

Fig. 1–10 1980 World coal reserves in selected countries divided by region.
Source: U.S. Department of the Interior, *Energy Resources on Federally Administered Lands*, 1981.

Almost 90 percent of the Identified coal is found at depths of 1,000 feet or less, and most is in thick beds. Therefore, of the total Identified coal, about 400 billion tons are in thick seams with no more than 1,000 feet of overburden. This is the portion of the coal resources that is either being mined at the present or is of current economic interest.

The Hypothetical amounts are estimates based on extrapolations of Identified resources in adjacent areas. Although Hypothetical resources in total are slightly larger than the Identified resources, they are considered inaccessible for mining at the present. About 20 percent of these resources are thought to lie at depths of 3,000 to 6,000 feet. Less than 200 billion tons of the Hypothetical resources are in seams at depths of less than 1,000 feet.

Figures 1–11 and 1–12 show Identified coal as a proportion of all coals, and as amounts in each state. Although coal deposits of the United States are widely distributed (being found in thirty-six states), five western states, Wyoming, North Dakota, Montana, Colorado, and Alaska, account for almost 67 percent of total estimated resources, 60 percent of the Identified resources, and 73 percent of the Hypothetical.[2] Montana and North Dakota alone house over 37 percent of the total Identified coal and 20 percent of the Hypothetical.[3]

In the eastern part of the country, Illinois, Kentucky, West Virginia, Pennsylvania, and Indiana house the largest share of the total remaining coal. These states account for over 17 percent of the total, 25 percent of the Identified, and 10 percent of the Hypothetical coal. Within these five states, 71 percent of the remaining coal deposits are classified as Identified and 29 percent Hypothetical.

Figure 1–13 shows the differing amounts of Identified coals by rank. In the United States, 28 percent are lignites, 28 percent subbituminous, 43 percent bituminous, and 1 percent is anthracite. Bituminous coal predominates in the eastern United States whereas subbituminous coals are found exclusively in the West. Over 80 percent of all Identified subbituminous coals are located in Alaska, Montana, and Wyoming. North Dakota contains only lignite deposits accounting for 70 percent of the total lignite deposits in the United States. Ninety-five percent of the country's anthracite deposits are located in Pennsylvania.

TABLE 1–1 Word Recoverable Reserves of Coal, 1980			
REGIONS	**RECOVERABLE RESERVES (BILLION SHORT TONS)**		
	Anthracite and Bituminous	**Lignite**	**Total**
Eastern Europe and U.S.S.R.			
U.S.S.R.	166.67	98.20	264.89
Poland	30.00	13.20	43.20
Other	3.26	11.55	14.81
Total	199.93	122.95	322.90
North, Central, and South America			
United States	248.16	35.25	283.41
Canada	4.18	2.33	6.51
Other	19.25	0.02	19.28
Total	271.60	37.60	309.20
Middle and Far East and Oceania			
China	108.90	—	108.90
Australia	32.52	39.90	72.42
Other	3.02	2.33	5.35
Total	144.44	42.23	186.67
Western Europe			
West Germany	32.98	38.67	71.65
Other	10.20	20.65	30.85
Total	43.18	59.32	102.50
Africa			
South Africa	57.04	—	57.04
Other	8.17	—	8.17
Total	65.21		65.21
World Total	724.36	262.10	986.46

Source: World Energy Conference, *World Energy Conference on Energy Resources, 1983*, London.

TABLE 1–2
All U.S. Coal Resources Identified and Hypothetical, to Depths of 3,000 and 6,000 Feet as of February 1978, Coal In-Place
(in millions of short tons)

| STATE | OVERBURDEN 0–3,000 FEET | | | | | Estimated Hypothetical Resources in Unmapped and Unexplored Areas | Estimated Total Identified and Hypothetical Resources | OVERBURDEN 3,000–6,000 FEET Estimated Additional Hypothetical Resources in Deeper Structural Basins | OVERBURDEN 0–6,000 FEET Estimated Total Identified and Hypothetical Resources |
| | Remaining Identified Resources, February 1978 | | | | | | | | |
	Anthracite and Semi-Anthracite[1]	Bituminous Coal[1]	Subbituminous Coal	Lignite	Total				
Alabama	0	13,262	0	2,000	15,262	20,000	35,262	6,000	41,262
Alaska	0	19,413	110,666	0	130,079	130,000[1]	260,079	5,000[1]	2,029,000[1]
Arizona	0	21,234	0	0	21,234	0	21,234	0	21,231
Arkansas	428	1,638	0	250	2,416	4,000	6,416	0	6,416
Colorado	78	109,117	19,733	20	128,948	161,272	290,220	143,991	434,211
Georgia	0	24	0	0	24	60	81	0	84
Illinois	0	146,001	0	0	146,001[1]	100,000[1]	246,001	0[1]	331,600[1]
Indiana	0	32,868	0	0	32,868	22,000	54,868	0	54,868
Iowa	0	6,505	0	0	6,505	14,000	20,505	0	20,505
Kansas	0	18,668	0	0	18,668	4,000	22,668	0	22,668
Kentucky									
Eastern	0	28,226	0	0	28,226	24,000	52,226	0	52,226
Western	0	36,120	0	0	36,120	28,000	64,120	0	64,120
Maryland	0	1,152	0	0	1,152	400	1,552	0	1,552
Michigan	0	205	0	0	205	500	705	0	705
Missouri	0	34,184	0	0	31,184	17,489	18,673	0	48,673
Montana	0	2,299	176,819	112,521	291,639[1]	180,000[1]	471,639	0[1]	670,000[1]
New Mexico	4	10,748	50,639	0	61,391[1]	65,556[1]	126,947	74,000[1]	283,000[1]
North Carolina	0	110	0	0	110	20	130	5	135
North Dakota	0	0	0	350,602	350,602	180,000	530,602	0	530,602
Ohio	0	41,166	0	0	41,166	6,152	47,318	0	47,318
Oklahoma	0	7,117	0	0	7,117	15,000	22,117	5,000	27,117
Oregon	0	50	284	0	334	100	434	0	434
Pennsylvania	18,812	63,940	0	0	82,752	4,000	86,752	3,600	90,352
South Dakota	0	0	0	2,185	2,185	1,000	3,185	0	3,185
Tennessee	0	2,530	0	0	2,530	2,000	4,530	0	4,530
Texas	0	6,048	0	10,293	16,341	112,100	128,441	0	128,441
Utah	0	23,186	173	0	23,359	22,000	45,359	35,000	80,359
Virginia	335	9,216	0	0	9,551	5,000	14,551	100	14,651
Washington	5	1,867	4,180	117	6,169	30,000	36,169	15,000	51,169
West Virginia	0	100,150	0	0	100,150[1]	0[1]	100,150	0[1]	116,600[1]
Wyoming	0	12,703	123,240	0	135,943[1]	700,000[1]	835,943	100,000[1]	1,240,000[1]
Other States	0	610	32	46	688	1,000	1,688	0	1,688
TOTAL	19,662	747,357	485,766	478,134	1,730,919	1,849,649	3,580,649	387,696	5,418,684[1]

[1]These numbers have been revised by the U.S. Department of Energy. The total identified and hypothetical resources to 6,000 feet reflects revisions to Averitt's 1974 figures.

Sources: Averitt, Paul. Coal Resources of the United States, January 1, 1974, *U.S. Geological Survey Bulletin 1412*, Washington, 1975 and U.S. Department of Energy. *Underground Coal Conversion Program*, Volume III, *Resources*, Washington, February 1978.

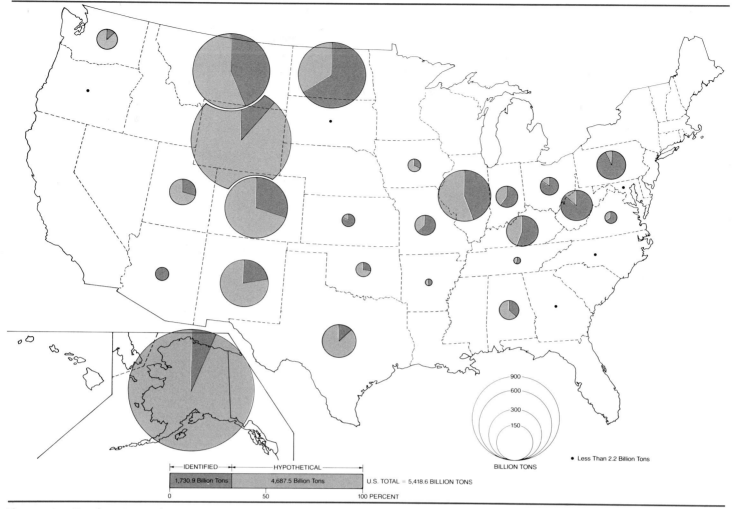

Fig. 1–11 Total U.S. coal resources (1978) showing proportions Identified and Hypothetical.

Sources: Averitt, 1975; and Department of Energy. *Underground Coal Conversion Program*, Volume III, *Resources*, February 1978.

TABLE 1–3
Recent Revisions to Averitt's Total United States Coal Resources, 1978 (in billions of tons)

| STATE | TOTAL IDENTIFIED AND HYPOTHETICAL (OVERBURDEN TO 6,000 FEET) | | |
	Averitt's Estimate	D.O.E. Revision	Change
Alaska	265.10	2,029.00	+ 1,763.90
Illinois	246.00	331.60	+ 85.60
Montana	471.64	670.00	+ 198.36
New Mexico	200.95	283.00	+ 82.05
West Virginia	100.15	116.60	+ 16.45
Wyoming	935.94	1,240.00	+ 304.06
TOTAL	2,219.78	4,670.20	+2,450.42

Source: U.S.G.S. *Bulletin 1412*, and Department of Energy, *Underground Coal Conversion Program*, Volume III, *Resources*, February 1978.

Note: In Alaska, most of the additional coal recognized is classified as Hypothetical. In western states, some is identified, and in the Powder River Basin (Wyoming and Montana) some is at depths less than 3,000 feet.

As ranks of coal differ in their BTU values, the tonnage does not represent a clear indication of how much usable energy is contained in the coal. Bituminous coals, due to their high BTU values, account for 53 percent of the total energy contained in Identified coals, but only 43 percent of the tonnage. On the other hand, lignites account for only 19 percent of the total energy content, but 28 percent of the tonnage. In practical terms, this means that the energy content of each ton of bituminous coal is equivalent to that of 1.3 tons of subbituminous coal and 1.9 tons of lignite. The difference in the energy content of the various ranks of coal prejudices the uses of the abundant but low-energy western coals because approximately one-third more western sub-bituminous coal than eastern bituminous coal would have to be transported to meet the same energy need.

Demonstrated Reserve Base

The Demonstrated Reserve Base refers to in-place coal deposits that are presently mineable both technically and economically (Department of Energy, October 1982).

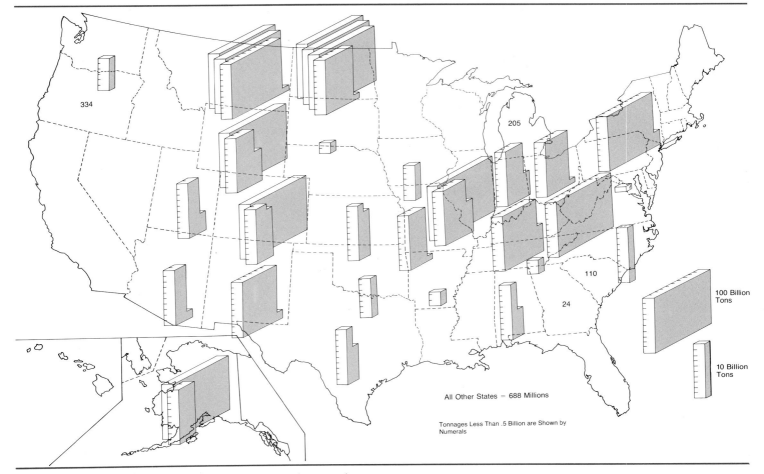

Fig. 1–12 Identified coal amounts to 3000 feet as of 1978.
Sources: Averitt, 1975; U.S. Department of Energy. *Underground Coal Conversion Program*, Volume III, *Resources*, Washington, February 1978.

Essentially, they represent the mineable portion of Averitt's Identified resources. Although the Reserve Base is referred to as mineable throughout the following pages, it is mineable only on the basis of geologic critieria. Other factors, such as competing land use or legal and social considerations could make it unfeasible to recover the coal. In 1982, the Demonstrated Reserve Base of coal in the United States was estimated at almost 483 billion short tons.

Of the total Identified coal in the United States, almost 28 percent of the tonnage is considered Reserve Base. As shown in Figure 1–14, the mineable portion of the Identified coal resources varies widely from state to state. For example, 54 percent of the in-place coal in Illinois is considered mineable as compared to less than 2 percent in Arizona. In general, since the coals in the eastern United States are more accessible and better defined, a larger percentage is mineable. The depth and thickness of seams in the western states accounts for the considerably smaller portions of the in-place coals considered mineable. The most extreme example is found in North Dakota where less than 3 percent of more than 350 billions tons of identified lignite can be mined by present methods.

RANK Of the total Demonstrated Reserve Base of Coal, 483 billion short tons, 23 percent is located in the Appalachian coal fields, 28 percent in the Interior fields, and 49 percent in Western fields (Table 1–4). Figure 1–15 shows the total tonnage of mineable coals (Demonstrated Reserve Base) by rank. The breakdown for the entire country is 9 percent lignite; 37 percent for subbituminous; 52 percent for bituminous; and 2 percent for anthracite. The distribution of coal deposits by rank reveals the dominance of bituminous coals in the eastern states, and subbituminous and lignite in the West.

MINING METHODS The mining method used to extract coal determines, in part, the amount of different ranks of coal that are useable. The coal Reserve Base is presented according to mining method in Table 1–4 and Figure 1–16. Some coals are mined only by one of the two methods; other coals can be mined by both methods. About 67 percent of the United States Reserve Base is extractable by underground mining methods, while 31 percent can be mined by surface methods. This ratio varies from region to region and from state to state. In the Appalachian Region 81 percent of the coal is underground mineable while only 19 percent can be mined by surface methods. In the Interior Region, 69 percent of the coal is recoverable through underground mining and 31 percent through surface methods. In the Western Region, the ratio is 59 percent to 41 percent. In eastern states, such as Pennsylvania and West Virginia, 96 and 87 percent respectively of the coals can be mined by underground methods. On the other hand, western states such as Wyoming and Montana extract about 40 percent of their coals through surface methods. In states like Texas and North Dakota where all the coals are lignites, 100 percent are mineable by surface methods.

MINING METHOD AND RANK The recovery of coal by surface mining depends primarily on the ratio of thickness of the overburden to that of the coal bed. Basically, a limit of 15 feet of overburden per foot of coal is used to calculate the surface mineable Reserve Base. Available machinery currently limits surface mining to depths of less than 180 feet (Bureau of Mines, *Information Circular, 8680,* 1975).

Table 1-4 shows amounts of coal mineable by surface and underground methods for each rank. For the nation as a whole, 98 percent of the anthracite, 80 percent of the bituminous, 65 percent of the subbituminous, and none of the lignite are extractable by underground mining methods. As reflected in the national summary, the *lignite* located in Montana, North Dakota, Texas, and Alabama is all surface mineable. For the more deeply buried *subbituminous* coals of the western states, only 35 percent are surface mineable. Nevertheless, there is a considerable difference from state to state in the amounts of subbituminous coals that can be mined by surface methods. For example, in New Mexico 66 percent of these coals are surface mineable, but in Colorado only 3 percent are. About 20 percent of the *bituminous* coals are surface mineable, a percentage that holds true for states in both the Appalachian and Interior regions. However, in the Western Region, only 11 percent of the bituminous coals

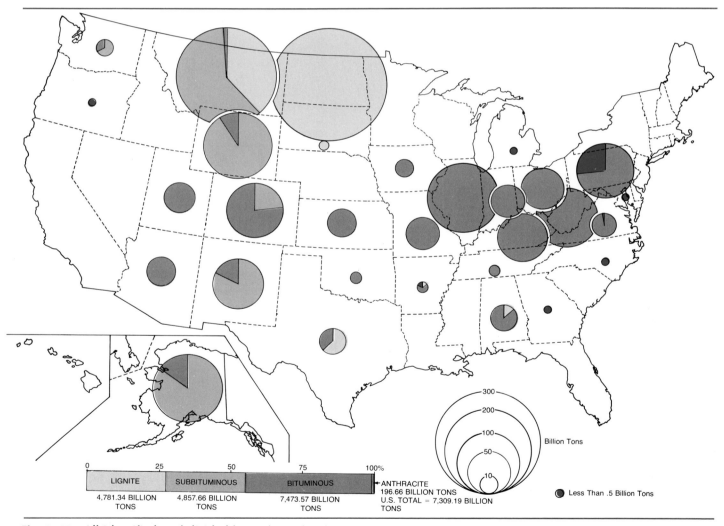

Fig. 1-13   All Identified coal divided by rank as of 1978.
Source: Averitt, 1975; and Department of Energy. *Underground Coal Conversion Program,* Volume III, *Resources,* February 1978.

Fig. 1–14 Demonstrated Reserve Base as a proportion of Identified coal resources.
Source: Department of Energy, *Coal Production 1982*, September 1983.

are surface mineable. Only 2 percent of the *anthracite* coal, found almost exclusively in Pennsylvania, is amenable to surface mining.

SULFUR CONTENT Stimulated by the need to identify clean-burning coals, the Bureau of Mines has delineated the 1974 coal Reserve Base on the basis of sulfur content, using four major categories: (1) *low-sulfur,* less than or equal to 1.0 percent sulfur; (2) *medium-sulfur,* 1.1 to 3.0 percent sulfur; (3) *high-sulfur,* greater than 3.0 percent sulfur; and (3) *unknown* sulfur content. As was discussed earlier in the section on rank and sulfur content, sulfur is an undesirable constituent of coal as it lowers the quality of the coal, corrodes boilers, and creates pollution problems such as sulfur dioxide smog and acid rain. In Table 1–5 and Figure 1–17 the state amounts by sulfur category were derived by applying the 1974 sulfur proportions to the 1982 Demonstrated Reserve Base.

Of the total United States Coal Reserve Base in 1982, 46 percent is low-sulfur, 21 percent medium-sulfur, 22 percent high-sulfur, and 11 percent of unknown sulfur content. As with the location of different ranks of coal, the amounts of sulfur in the east and west varies widely. In the Appalachian Region, 25 percent of the coals have a high-sulfur content and 26 percent are low-sulfur coals. Over two-thirds of the coals in the Appalachian region have less than 3 percent sulfur. The coals of the Interior Region have the highest sulfur content with almost two-thirds having a sulfur content of over 3 percent. In contrast, 78 percent of the coals of the Westen Region are low in sulfur content with 93 percent containing less than 3 percent sulfur.

Montana and Illinois demonstrate the variations in sulfur content between eastern and western coal Reserves. In Illinois, with about 16 percent of all the mineable coals in the United States, few low-sulfur coals are found and over 70 percent is classified as high-sulfur. On the other hand, in Montana, which has almost 25 percent of the nation's Reserve Base of coal, about 94 percent is low-sulfur. In the Appalachian Region of the east, West Virginia has the largest amount of low-sulfur coal—over 14 billion short tons.

SULFUR CONTENT AND RANK To a large extent sulfur content varies with the rank of coal. For example, low-sulfur coal is most often subbituminous and high-sulfur is usually bituminous. Most anthracites are low in sulfur content. Table 1–6 shows the amount of *low-sulfur* coal by rank. Of the more than 219 billions tons of low-sulfur coal, 5 percent is lignite, 74 percent subbituminous, 18 percent bituminous, and 3 percent anthracite. All of the low-sulfur lignite and subbituminous coal is in the western states. The states with the largest amounts of low-sulfur subbituminous coal are Montana and Wyoming, with a combined total of more than 150 billion tons. Montana and North Dakota house most of the 10.5 billion tons of low-sulfur lignite. Slightly more than 60 percent of the low-sulfur bituminous coal is located in the Appalachian Region of the East, with more than 2.2 billion tons in West Virginia and Kentucky. In the West, Colorado, Utah, New Mexico,

Wyoming, and Alaska have over 90 percent of the low-sulfur bituminous. More than 6 billion tons of low-sulfur anthracite is found in five states: Pennsylvania, Virginia, Colorado, Arkansas, and New Mexico. Over 98 percent of all low-sulfur anthracite is in Pennsylvania. Interestingly, 85 percent of all low-sulfur coal in Pennsylvania is anthracite.

As Table 1–7 shows, a total of 101 billion tons of *medium-sulfur* coal Reserves exist in the United States. By rank, the division is 16 percent lignite, 20 percent subbituminous, 64 percent bituminous, and less than 1 percent anthracite. Medium-sulfur lignite and subbituminous coals are again confined to the western states. Wyoming houses over 16 billions tons of medium-sulfur subbituminous, and North Dakota approximately 6.3 billion tons of medium-sulfur lignite. Of the roughly 64 billion tons of medium-sulfur bituminous coal, almost 90 percent is located in the eastern United States. Most of the eastern

TABLE 1–4
U.S. Demonstrated Coal Reserve Base, January 1, 1982 (million tons)

STATE AND REGION	ANTHRACITE Under-ground	ANTHRACITE Surface	BITUMINOUS Under-ground	BITUMINOUS Surface	SUBBITUMINOUS Under-ground	SUBBITUMINOUS Surface	LIGNITE Surface	TOTAL Under-ground	TOTAL Surface	TOTAL
Appalachian	7,086.3	118.3	83,061.9	19,813.1	—	—	1,083.0	90,148.2	21,014.5	111,162.7
Alabama	—	—	1,740.0	2,378.9	—	—	1,083.0	1,740.0	3,361.9	5,201.9
Georgia	—	—	1.9	1.7	—	—	—	1.9	1.7	3.6
Kentucky, Eastern	—	—	8,492.7	4,072.1	—	—	—	8,492.7	4,072.1	12,564.8
Maryland	—	—	706.1	103.8	—	—	—	706.1	103.8	809.9
North Carolina	—	—	10.7	—	—	—	—	10.7	—	10.7
Ohio	—	—	13,019.5	5,925.3	—	—	—	13,019.5	5,925.3	18,944.8
Pennsylvania	6,960.8	118.3	21,870.8	1,055.3	—	—	—	28,831.6	1,173.6	30,005.2
Tennessee	—	—	636.7	316.1	—	—	—	636.7	316.1	952.8
Virginia	125.5	—	2,372.3	824.4	—	—	—	2,497.8	824.4	3,322.2
West Virginia	—	—	34,211.2	5,135.6	—	—	—	34,211.2	5,135.6	39,346.8
Interior	88.6	15.5	93,726.2	27,782.9	—	—	13,505.9	93,726.2	41,304.4	135,119.2
Arkansas	88.6	15.5	183.9	104.4	—	—	25.7	272.5	145.6	418.1
Illinois	—	—	63,112.2	15,657.1	—	—	—	63,112.2	15,657.1	78,769.3
Indiana	—	—	8,934.8	1,575.3	—	—	—	8,934.8	1,575.3	10,510.0
Iowa	—	—	1,733.6	462.3	—	—	—	1,733.6	462.3	2,195.9
Kansas	—	—	—	990.9	—	—	—	—	990.9	990.9
Michigan	—	—	123.1	4.6	—	—	—	123.1	4.6	127.7
Kentucky, Western	—	—	16,921.1	4,025.9	—	—	—	16,921.1	4,025.9	20,947.0
Missouri	—	—	1,479.1	4,577.1	—	—	—	1,479.1	4,577.1	6,056.2
Oklahoma	—	—	1,238.4	385.4	—	—	—	1,238.4	385.4	1,623.8
Texas	—	—	—	—	—	—	13,480.3	—	13,480.3	13,480.3
Western	27.8	—	22,113.1	2,627.7	118,813.4	62,839.0	30,251.1	140,954.3	95,717.8	236,672.1
Alaska	—	—	617.0	80.5	4,805.9	635.1	14.0	5,423.0	729.6	6,152.6
Arizona	—	—	101.6	280.3	—	—	—	101.6	280.3	381.9
Colorado	25.5	—	8,408.6	636.1	3,837.9	124.8	4,189.9	12,272.0	4,950.8	17,222.8
Idaho	—	—	4.4	—	—	—	—	4.4	—	4.4
Montana	—	—	1,385.4	—	69,573.3	33,625.4	15,764.5	70,958.7	49,389.9	120,348.6
New Mexico	2.3	—	1,238.1	815.0	889.0	1,758.1	—	2,129.4	2,573.1	4,702.5
North Dakota	—	—	—	—	—	—	9,908.5	—	9,908.5	9,908.5
Oregon	—	—	—	—	14.5	2.9	—	14.5	2.9	17.5
South Dakota	—	—	—	—	—	—	366.1	—	366.1	366.1
Utah	—	—	6,154.7	267.9	1.1	—	—	6,155.8	267.9	6,423.7
Washington	—	—	303.7	—	1,028.6	128.6	8.1	1,332.3	136.7	1,469.0
Wyoming	—	—	3,899.6	547.9	38,663.0	25,564.1	—	42,462.7	27,112.0	69,674.6
U.S. TOTAL	7,202.7	133.9	198,901.3	50,223.8	118,813.4	62,839.0	44,840.0	324,917.3	158,036.6	482,954.0

Source: Department of Energy, *Coal Production 1982,* Washington, September 1983.

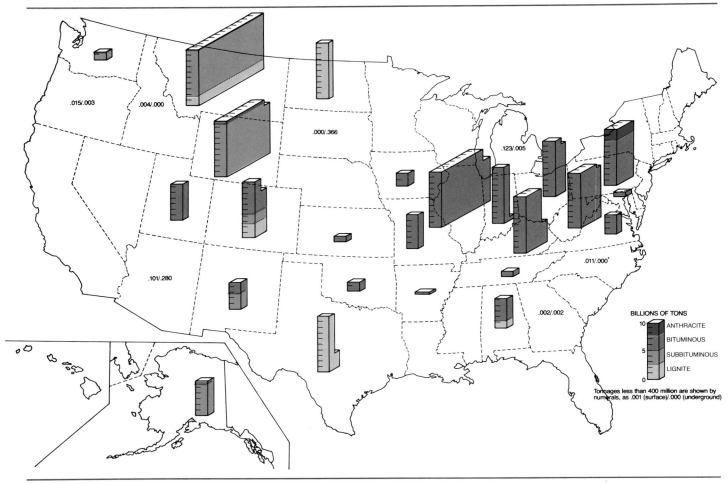

Fig. 1–15 Total tonnage of mineable coals by rank.
Source: Department of Energy, *Coal Production, 1982*, 1983.

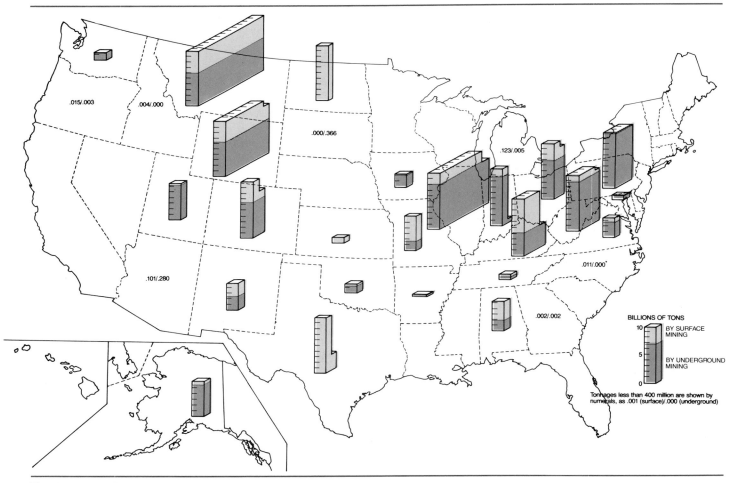

Fig. 1–16 Total tonnage mineable by surface and underground methods.
Source: Department of Energy, *Coal Production, 1982*, 1983.

TABLE 1-5
Mineable Amounts of Coal Showing Four Sulfur Categories, 1982
(in millions of short tons)

STATE	DEMONSTRATED RESERVE BASE	AMOUNTS BY SULFUR CATEGORY			
		Up to 1%	1.1–3.0%	>3.1%	Unknown
Alabama	5,201.9	1,090.8	1,919.5	27.9	2,161.9
Alaska	6,152.6	6,055.4	97.2	0.0	0.0
Arizona	381.9	188.8	193.1	0.0	0.0
Arkansas	418.1	51.0	291.5	29.0	46.6
Colorado	17,222.8	8,673.4	910.3	54.4	7,583.3
Georgia	3.6	—	—	—	—
Idaho	4.4	—	—	—	—
Illinois	78,769.3	147.3	8,360.6	55,626.1	14,634.5
Indiana	10,510.0	97.0	2,581.3	6,555.5	1,276.2
Iowa	2,195.9	1.0	172.9	1,604.1	418.2
Kansas	990.9	0.0	220.6	496.9	273.4
Kentucky	33,511.8	8,608.5	5,100.5	12,525.0	7,277.8
Maryland	809.9	104.3	534.0	144.5	27.1
Michigan	127.7	5.4	91.9	22.7	7.6
Missouri	6,056.2	0.0	116.2	3,335.8	2,604.3
Montana	120,348.6	112,818.4	4,567.2	558.4	2,405.8
New Mexico	4,702.5	3,802.6	870.1	1.1	29.8
North Carolina	10.7	—	—	—	—
North Dakota	9,908.5	3,337.7	6,394.8	166.6	9.3
Ohio	18,944.8	120.5	5,788.4	11,361.8	1,682.3
Oklahoma	1,623.8	345.3	410.7	302.7	565.1
Oregon	17.5	—	—	—	—
Pennsylvania	30,005.2	7,086.3	16,378.6	3,679.8	2,860.4
South Dakota	366.1	88.1	246.3	30.8	1.0
Tennessee	952.8	198.7	516.6	152.2	85.3
Texas	13,480.3	2,719.4	7,766.0	1,170.1	1,829.3
Utah	6,423.7	3,128.4	2,457.9	77.9	759.5
Virginia	3,322.2	1,949.4	1,059.4	12.8	300.6
Washington	1,469.0	451.1	955.9	29.5	32.5
West Virginia	39,346.8	14,011.0	13,925.6	6,783.8	4,626.4
Wyoming	69,674.6	44,304.7	19,148.7	2,222.6	3,999.3
U.S. Total	482,954.0	219,383.7	101,075.8	106,972.0	55,497.5

Sources: U.S. Department of Energy. *Coal Production, 1982*, Washington, 1983; U.S. Bureau of Mines. The Reserve Base of U.S. Coals by Sulfur Content—The Eastern United States, *Information Circular 8680*, Washington, 1975; and U.S. Bureau of Mines. The Reserve Base of U.S. Coal by Sulfur Content—The Western States, *Information Circular 8693*, Washington, 1975.

Note: Amounts of coal by sulfur content were derived by using the proportions in each sulfur category reported by the U.S. Bureau of Mines in the references listed above.

Reserves are found in West Virginia, Pennsylvania, Illinois, Ohio, and Kentucky. Four western states, Montana, Wyoming, Utah, and Colorado have a combined total of 7 billions tons of medium-sulfur bituminous coal. In addition, there are small amounts of medium-sulfur anthracite coal in Pennsylvania and Arkansas.

High-sulfur coals account for more than 106 billion tons of the coal Reserve Base. Over 98 percent of all of the high-sulfur coal is bituminous (Table 1-8). Only 2.3 billion tons of the lignite and subbituminous coal of the western United States has a high sulfur content. However, approximately 90 percent of the high-sulfur bituminous coals occur in the Interior and Appalachian regions of the eastern United States: Over one-half of it is found in Illinois. None of the Pennsylvania anthracite is classified as high-sulfur.

A total of 55.5 billion tons of the coal Reserve Base is of *unknown sulfur* content because the sulfur has not yet been measured (Table 1-9). Over three-fourths of these coals are bituminous, while subbituminous and lignite account for 12 and 10 percent respectively. Anthracite accounts for 1 percent of the total. Almost three-fourths of the bituminous coals of unknown sulfur content are located in the eastern United States, Illinois having the largest amount—approximately 14 billion tons. Figure 1-18 presents a national

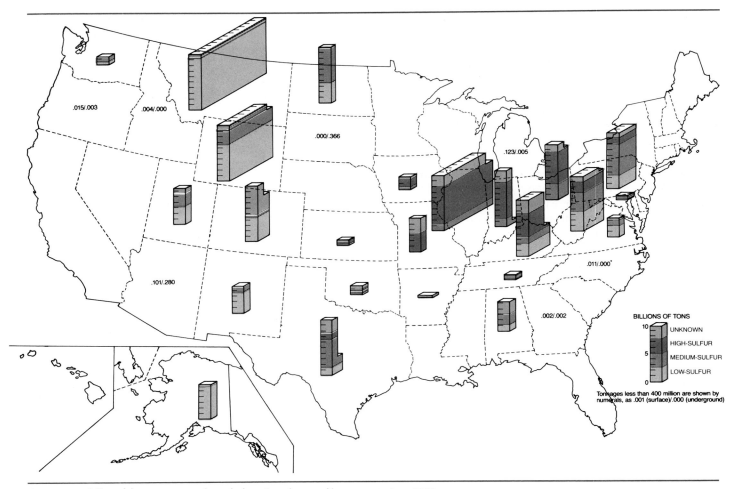

Fig. 1–17 Mineable amounts of coal showing four sulfur categories, 1982.
Sources: Department of Energy, *Coal Production, 1982*, 1983; U.S. Bureau of Mines. "The Reserve Base of U.S. Coals by Sulfur Content-Eastern States," *Information Circular, 8680*, 1975; and U.S. Bureau of Mines. "The Reserve Base of U.S. Coals by Sulfur Content-Western States," *Information Circular, 8698*, 1975.

summary of mineable coals showing which ranks account for the low-, medium-, and high-sulfur coals.

SULFUR CONTENT, RANK, AND MINING METHOD
In order to assess the overall quality and ease of recovery of the coal Reserve Base, it is helpful to look at the rank, sulfur content, and mining method together.

Tables 1–6 through 1–9 show, for the four sulfur levels in turn, the amounts of coal of each rank that are considered surface mineable and underground mineable. For *low-sulfur coals,* 36 percent is surface mineable, and 64 percent underground mineable. Over 75 percent of the coal mineable by surface methods is subbituminous, 13 percent lignite, 10 percent bituminous, and a very small proportion is anthracite. For the portion mineable by underground methods, 72 percent is subbituminous, 23 percent bituminous, and 5 percent is anthracite. Lignite is found at shallow depths and is, therefore, mined only by surface methods. Over one-half of the low-sulfur coal is in Montana. Most of it is subbituminous and underground mineable. In

the East, West Virginia, Pennsylvania, and Kentucky house approximately 14 percent of the low-sulfur coal. Excepting the more than 6 billion tons of low-sulfur anthracite, low-sulfur coal in the East is bituminous and over 80 percent is recoverable by underground mining.

Predominantly bituminous, *medium-sulfur coals* are usually recoverable by underground mining. Surface mineable medium-sulfur coals account for 38 percent of the total, with 43 percent being lignite, 38 percent subbituminous, and 19 percent bituminous. Underground mineable medium-sulfur coals amount to more than 62.7 billion tons, roughly 90 percent of which are bituminous. Three states, Pennsylvania, Wyoming, and West Virginia, contain almost one-half of the total medium-sulfur coal. In Pennsylvania, more than 95 percent of the medium-sulfur coal is underground mineable bituminous. Two-thirds of the Wyoming medium-sulfur coal is subbituminous and recoverable by surface methods.

High-sulfur coal, characteristically bituminous, is recovered by underground mining. Illinois has more than one-half of the high-sulfur coal with an additional 38 percent being found in Ohio, Kentucky, West Virginia, Indiana, and Missouri. Overall 79 percent of the high-sulfur coals are underground mineable and almost 98 percent are bituminous.

For those coals with *unknown* sulfur content, over three-fourths are recoverable by underground mining methods. Both the underground and surface mineable coals of unknown sulfur content are predominantly bituminous.

SUMMARY OF DEMONSTRATED RESERVE BASE
Much of the information presented earlier in tables and maps is summarized here in a series of bar graphs which provide a quick reference to relative amounts. In addition, these graphs emphasize the geographic dimension first by distinguishing Appalachian, Interior, and Western states, and subsequently by treating each state separately.

Figure 1-19 shows the underground and surface mineable coals in the United States divided by rank and by sulfur content. For each of these two characteristics the relative amounts are measured in tonnage.

In Figures 1-20 and 1-21, the prevalence of four ranks and four sulfur classes in surface mineable and underground mineable coals for Appalachian, Interior, and Western states is shown. Figure 1-21 compares Appalachian, Interior, and Western states in regard to prevalence of four ranks, and four sulfur classes, and the portions mineable by surface and underground mining methods.

Figures 1-22, 1-23, and 1-24 totals the amounts of surface and underground mineable coals for each state and, within each of those categories, the proportions in the four ranks and four sulfur classes.

Table 1-10 adds a new perspective to the Demonstrated Reserve Base by showing the proportion of all *Identified*

TABLE 1-6
Amounts of *Low-Sulfur* Coal by Rank and Mining Method, 1982 (in millions of short tons)

STATE	SURFACE MINEABLE[1]				UNDERGROUND MINEABLE[1]				TOTAL MINEABLE
	L	SB	B	A	L	SB	B	A	
Alabama	—	—	61.0	—	—	—	1,029.6	—	1,090.8
Alaska	145.3	3,118.8	634.7	—	—	2,156.6	—	—	6,055.4
Arizona	—	188.8	—	—	—	—	—	—	188.8
Arkansas	—	—	23.9	—	—	—	22.0	5.0	51.0
Colorado	—	—	838.7	—	—	3,510.1	4,293.2	31.3	8,673.4
Illinois	—	—	4.8	—	—	—	142.5	—	147.3
Indiana	—	—	17.0	—	—	—	80.0	—	97.0
Iowa	—	—	—	—	—	—	1.0	—	1.0
Kentucky	—	—	1,989.6	—	—	—	6,618.8	—	8,608.5
Maryland	—	—	22.4	—	—	—	81.9	—	104.3
Michigan	—	—	—	—	—	—	5.4	—	5.4
Montana	4,212.6	38,166.5	—	—	—	70,264.4	174.9	—	112,818.4
New Mexico	—	1,544.4	243.6	—	—	457.4	1556.1	1.1	3,802.ϵ
North Dakota	3,337.7	—	—	—	—	—	—	—	3.337.7
Ohio	—	—	17.1	—	—	—	103.4	—	120.5
Oklahoma	—	—	151.2	—	—	—	194.1	—	345.3
Pennsylvania	—	—	53.3	80.4	—	—	949.9	6,002.7	7,086.3
South Dakota	88.1	—	—	—	—	—	—	—	88.1
Tennessee	—	—	64.0	—	—	—	134.7	—	198.7
Texas	2,719.1	—	—	—	—	—	—	—	2,719,1
Utah	—	—	82.7	—	—	—	3,045.7	—	3,128.4
Virginia	—	—	375.3	—	—	—	1,526.7	47.4	1,949.4
Washington	—	126.2	—	—	—	190.4	134.5	—	451.1
West Virginia	—	—	2,987.7	—	—	—	11,023.3	—	14,011.0
Wyoming	—	17,235.4	—	—	—	25,459.7	1,609.6	—	44,304.7
U.S. Total	10,502.8	60,380.1	7,567.0	80.4	0.0	102,038.6	32,727.3	6,087.5	219,383.7

[1]L = lignite; SB = subbituminous; B = bituminous; and A = anthracite

Sources: Department of Energy, *Coal Production, 1982*, Washington, 1983; U.S. Bureau of Mines. The Reverse Base of U.S. Coals by Sulfur Content—Eastern States, *Information Circular 8680*, Washington, 1975; and U.S. Bureau of Mines. The Reverse Base of U.S. Coals by Sulfur Content—Western States, *Information Circular, 8693*, Washington, 1975.

Note: Amounts of coal by sulfur content were derived by using the 1982 Demonstrated Reserve Base and the proportions in each sulfur category reported by the Bureau of Mines in the above references.

TABLE 1-7
Amounts of *Medium-Sulfur* Coal by Rank and Mining Method, 1982 (in millions of short tons)

STATE	SURFACE MINEABLE[1]				UNDERGROUND MINEABLE				TOTAL MINEABLE
	L	SB	B	A	L	SB	B	A	
Alabama	—	—	144.8	—	—	—	1,774.7	—	1,919.5
Alaska	11.1	—	—	—	—	86.1	—	—	97.2
Arizona	—	193.1	—	—	—	—	—	—	193.1
Arkansas	—	—	96.3	—	—	—	146.7	48.5	291.5
Colorado	—	—	169.1	—	—	115.8	625.4	—	910.3
Illinois	—	—	828.9	—	—	—	7,531.7	—	8,360.6
Indiana	—	—	543.2	—	—	—	2,038.1	—	2,581.3
Iowa	—	—	—	—	—	—	172.9	—	172.9
Kansas	—	—	220.6	—	—	—	—	—	220.6
Kentucky	—	—	1,454.3	—	—	—	3,646.2	—	5,100.5
Maryland	—	—	51.2	—	—	—	482.2	—	534.0
Michigan	—	—	—	—	—	—	91.9	—	91.9
Missouri	—	—	30.6	—	—	—	85.6	—	116.2
Montana	1,987.8	426.2	—	—	—	1,208.6	854.6	—	4,567.2
New Mexico	—	620.1	22.3	—	—	188.3	38.3	1.1	870.1
N. Dakota	6,394.8	—	—	—	—	—	—	—	6,394.8
Ohio	—	—	890.6	—	—	—	4,897.8	—	5,788.4
Oklahoma	—	—	110.9	—	—	—	299.8	—	410.7
Pennsylvania	—	—	694.3	1.0	—	—	15,507.1	176.2	16,378.6
S. Dakota	246.3	—	—	—	—	—	—	—	246.3
Tennessee	—	—	158.0	—	—	—	358.6	—	516.6
Texas	7,766.0	—	—	—	—	—	—	—	7,766.0
Utah	—	—	236.7	—	—	—	2,221.2	—	2,457.9
Virginia	—	—	198.6	—	—	—	860.8	—	1,059.4
Washington	—	232.6	—	—	—	685.6	37.7	—	955.9
W. Virginia	—	—	1,414.8	—	—	—	12,510.8	—	13,925.6
Wyoming	—	13,223.9	—	—	—	3,012.7	2,912.1	—	19,148.7
TOTAL	16,406.0	14,695.9	7,265.3	1.0	0.0	5,387.1	57,094.2	225.8	101,075.8

[1]L =lignite; SB = subbituminous; B = bituminous; and A = anthracite

Source: Department of Energy. *Coal Production, 1982,* Washington, 1983; U.S. Bureau of Mines. The Reserve Base of U.S. Coals by Sulfur Content—Eastern States, *Information Circular 8680,* Washington, 1975; and U.S. Bureau of Mines. The Reserve Base of U.S. Coals by Sulfur Content—Western States, *Information Circular 8693,* Washington, 1975.

Note: Amounts of coal by sulfur content were derived by using the 1982 Demonstrated Reserve Base and the proportions in each sulfur category reported by the Bureau of Mines in the above references.

coals assigned to the Reserve Base. As noted before, the coals not assigned to the Reserve Base are excluded because of depth, thickness of bed, or conflicts with land use. When viewed according to all identified amounts, the states assume an order that is different from a ranking based on Reserve Base. North Dakota becomes the leading state on the basis of tonnage of coal located in the state. North Dakota coal consists of more than 350 billion tons of lignite, only a small portion of which—3 percent—is considered Reserve Base. Similarly, Colorado shows large amounts of *Identified* coal, much of it bituminous, which is not part of the Reserve Base.

Recoverable Coal Reserves

Coal Reserves are defined as the quantity of the Demonstrated Reserve Base that can actually be recovered, given present technology, and any economic or legal constraints (Bureau of Mines, *Information Circular 8680,* 1975). Specific factors influencing the amounts of coal that can be recovered are the mining method, environmental aspects, and the quality of the coal.

Initially, the quantity of the Demonstrated Reserve Base that can be recovered depends on whether or not the coal bed is suitable for underground or surface mining. As stated, in surface mining the recovery of coal depends primarily on the ratio of the thickness of the overburden to that of the coal bed. Normally, this ratio should be around 15:1. In addition, the local topography determines whether the coal can be recovered through contour stripping or area stripping. On the average, the recovery ratio for strip mining ranges from 80 to over 90 percent. For underground mining, Bureau of Mines studies indicates that average recovery is about 50 percent. The lower recovery ratio for underground mining is primarily due to the coal left unmined in order to support the roof.

Conflicting land use is another factor which may inhibit the mining of certain coal beds. When coal lies beneath urban areas, public facilities, such as parks and airports, or waterways, it is clearly impossible for the coal to be mined. In addition, coal beds overlying or underlying other worked-out beds can be hazardous and expensive to mine (Bureau of Mines, *Information Circular 8680, 1975*).

In the West, the underground mining of coals is often constrained by the bed thickness. Although approximately 35 percent of the total underground Reserve Base of the western states is found in coal beds greater than 10 feet thick (Bureau of Mines, *Information Circular 8678, 1975*), much of it is considered unavailable. Domestic underground

mining technology and equipment are generally limited to face heights of 10 feet or less. In a 20-foot seam, for example, only a 10-foot high slice is mined, while the remaining 10 feet is not touched. Assuming that room and pillar mining can recover 57 percent of the 10-foot slice, less than 30 percent of the 20-foot seam is mined. If the longwall mining method is used (see earlier section), 85 percent of the 10-foot slice is recovered, and therefore 43 percent of the 20-foot bed.

Environmental factors can seriously inhibit the mining of certain coal beds. Land disturbance due to surface mining, subsidence or sinking of parts of the earth due to underground mining, and the effect of mine drainage on water quality can all prevent mining. Furthermore, the chemical and physical properties of coal such as excessive sulfur, ash, and volatile materials may restrict the use of certain coals and reduce the amounts actually recovered.

TABLE 1-8
Amounts of *High-Sulfur* Coal by Rank and Mining Method, 1982
(in millions of short tons)

State	SURFACE MINEABLE[1]				UNDERGROUND MINEABLE[1]				Total Mineable
	L	SB	B	A	L	SB	B	A	
Alabama	—	—	3.3	—	—	—	24.6	—	27.9
Arkansas	—	—	10.7	—	—	—	11.3	6.9	29.0
Colorado	—	—	—	—	—	—	54.4	—	54.4
Illinois	—	—	12,242.7	—	—	—	43,383.4	—	55,626.1
Indiana	—	—	986.5	—	—	—	5,569.0	—	6,555.5
Iowa	—	—	—	—	—	—	1,604.1	—	1,604.1
Kansas	—	—	496.9	—	—	—	—	—	469.9
Kentucky	—	—	2,761.5	—	—	—	9,763.5	—	12,525.0
Maryland	—	—	12.4	—	—	—	132.1	—	144.5
Michigan	—	—	—	—	—	—	22.7	—	22.7
Missouri	—	—	1,044.3	—	—	—	2,291.5	—	3,335.8
Montana	51.2	—	—	—	—	—	507.2	—	558.4
New Mexico	—	—	—	—	—	—	—	1.1	1.1
North Dakota	166.6	—	—	—	—	—	—	—	166.6
Ohio	—	—	2,270.8	—	—	—	9,091.3	—	11,361.8
Oklahoma	—	—	48.8	—	—	—	253.9	—	302.7
Pennsylvania	—	—	224.7	—	—	—	3,455.1	—	3,679.8
South Dakota	30.8	—	—	—	—	—	—	—	30.8
Tennessee	—	—	53.3	—	—	—	98.9	—	152.2
Texas	1,170.1	—	—	—	—	—	—	—	1,170.1
Utah	—	—	67.0	—	—	—	10.9	—	77.9
Virginia	—	—	1.8	—	—	—	11.0	—	12.8
Washington	—	19.7	—	—	—	9.8	—	—	29.5
West Virginia	—	—	268.4	—	—	—	6,515.4	—	6,783.8
Wyoming	—	555.7	—	—	—	—	1,359.7	—	2,222.6
TOTAL	1,418.7	575.4	20,493.1	0.0	0.0	317.1	84,160.0	11.3	106,972.0

[1] L = lignite; SB = subbituminous; B = bituminous; and A = anthracite

Sources: Department of Energy. *Coal Production, 1982*, Washington 1983; U.S. Bureau of Mines. The Reserve Base of U.S. Coals by Sulfur Content—Eastern States, *Information Circular, 8680*, Washington, 1975; and U.S. Bureau of Mines. The Reserve Base of U.S. Coals by Sulfur Content—Western States, *Information Circular, 8693*, Washington, 1975.

Note: Amounts of coal by sulfur content were derived by using the 1982 Demonstrated Reserve Base and the proportions in each sulfur category reported by the Bureau of Mines in the above references.

TABLE 1–9
Amounts of Coal of *Unknown* Sulfur Content by Rank and Mining Method, 1982
(millions of short tons)

STATE	SURFACE MINEABLE[1]				UNDERGROUND MINEABLE[1]				TOTAL MINEABLE
	L	SB	B	A	L	SB	B	A	
Alabama	1,790.6	—	64.5	—	—	—	306.9	—	2,161.9
Arkansas	20.2	—	14.5	—	—	—	12.0	—	46.6
Colorado	—	—	—	—	—	1,871.8	5,711.5	—	7,583.3
Illinois	—	—	1,586.8	—	—	—	13,047.7	—	14,634.5
Indiana	—	—	108.8	—	—	—	1,167.4	—	1,276.2
Iowa	—	—	—	—	—	—	418.2	—	418.2
Kansas	—	—	273.4	—	—	—	—	—	273.4
Kentucky	—	—	3,444.0	—	—	—	3,833.9	—	7,277.8
Maryland	—	—	27.1	—	—	—	—	—	27.1
Michigan	—	—	—	—	—	—	7.6	—	7.6
Missouri	—	—	1,104.3	—	—	—	1,500.0	—	2,604.3
Montana	1,626.1	779.7	—	—	—	—	—	—	2,405.8
New Mexico	—	—	—	—	—	—	29.8	—	29.8
N. Dakota	9.3	—	—	—	—	—	—	—	9.3
Ohio	—	—	106.0	—	—	—	1,576.3	—	1,682.3
Oklahoma	—	—	233.6	—	—	—	331.5	—	565.1
Pennsylvania	—	—	81.3	5.8	—	—	2,144.8	628.4	2,860.4
S. Dakota	1.0	—	—	—	—	—	—	—	1.0
Tennessee	—	—	33.0	—	—	—	52.3	—	85.3
Texas	1,829.3	—	—	—	—	—	—	—	1,829.3
Utah	—	—	28.6	—	—	—	730.9	—	759.5
Virginia	—	—	42.8	—	—	—	183.9	77.4	306.6
Washington	—	—	—	—	—	15.9	16.6	—	32.5
W. Virginia	—	—	507.1	—	—	—	4,199.3	—	4,626.4
Wyoming	—	137.2	—	—	—	3,833.3	28.8	—	3999.3
TOTAL	5,276.5	916.9	7,655.8	5.8	0.0	5,721.0	35,219.4	705.8	55,497.5

[1]L = lignite; SB = subbituminous; B = bituminous; and A = anthracite

Sources: Department of Energy. *Coal Production, 1982*, Washington, 1983: U.S. Bureau of Mines. The Reserve Base of U.S. Coals by Sulfur Content—Eastern States, *Information Circular, 8680*, Washington, 1975; and U.S. Bureau of Mines. The Reserve Base of U.S. Coals by Sulfur Content—Western States, *Information Circular, 8693*, Washington, 1975.

Note: Amounts of coal by sulfur content were derived by using the 1982 Demonstrated Reserve Base and the proportions in each sulfur category reported by the Bureau of Mines in the above references.

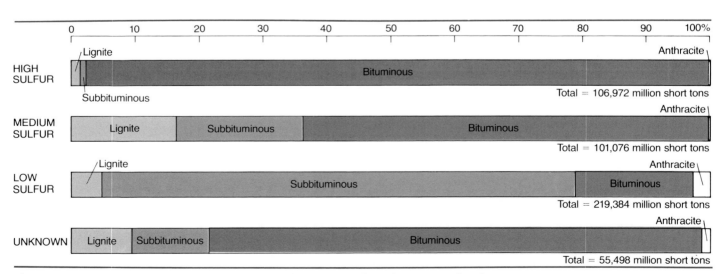

Fig. 1–18 National summary of mineable coals showing which ranks account for the low-, medium-, and high-sulfur coals, 1982.

Sources: Department of Energy, *Coal Production, 1982*, 1983; U.S. Bureau of Mines, "The Reserve Base of U.S. Coals by Sulfur Content—Eastern States," *Information Circular, 8680*, 1983; and U.S. Bureau of Mines, "The Reserve Base of U.S. Coals by Sulfur Content—Western States," *Information Circular, 8693*, 1975.

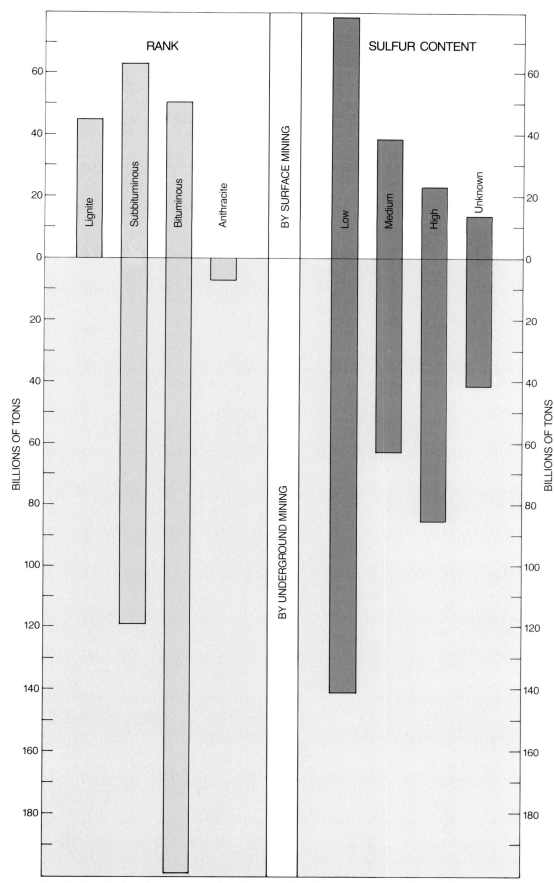

Fig. 1-19 National summary of surface and underground mineable coals, according to rank and sulfur content.

Sources: Department of Energy, *Coal Production, 1982*, 1983; U.S. Bureau of Mines, "The Reserve Base of U.S. Coals by Sulfur Content—Eastern States," *Information Circular, 8680*, 1975; and U.S. Bureau of Mines. "The Reserve Base of U.S. Coals by Sulfur Content—Western States," *Information Circular, 8693*, 1975.

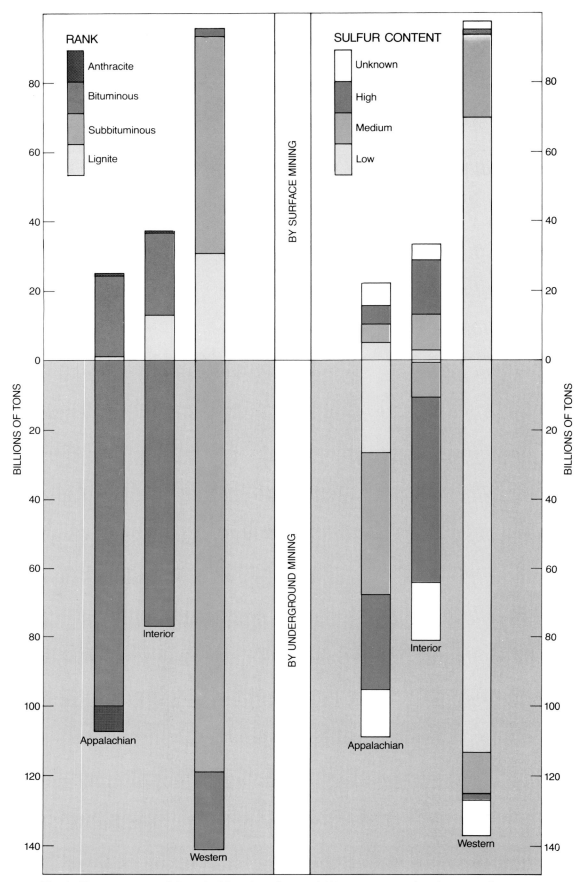

Fig. 1-20 Summary of Appalachian, Interior, and Western states showing prevalence of various ranks and sulfur contents in coals mineable by surface and underground mining methods.

Sources: Department of Energy, *Coal Production, 1982*, 1983; U.S. Bureau of Mines, "The Reserve Base of U.S. Coals by Sulfur Content—Eastern States," *Information Circular, 8680*, 1975; and U.S. Bureau of Mines, "The Reserve Base of U.S. Coals by Sulfur Content—Western States," *Information Circular, 8693*, 1975).

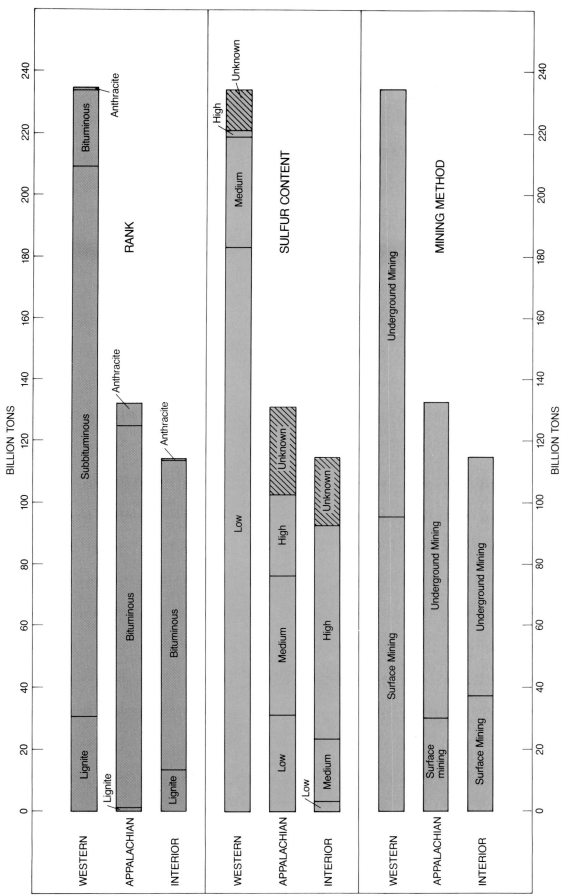

Fig. 1–21 Summary of Appalachian, Interior, and Western states showing mineable coals according to rank, sulfur content, and mining method.

Sources: Department of Energy, *Coal Production, 1982;* U.S. Bureau of Mines, "The Reserve Base of U.S. Coals by Sulfur Content—Eastern States," *Information Circular, 8680;* and U.S. Bureau of Mines, "The Reserve Base of U.S. Coals by Sulfur Content—Western States," *Information Circular, 8693,* 1975.

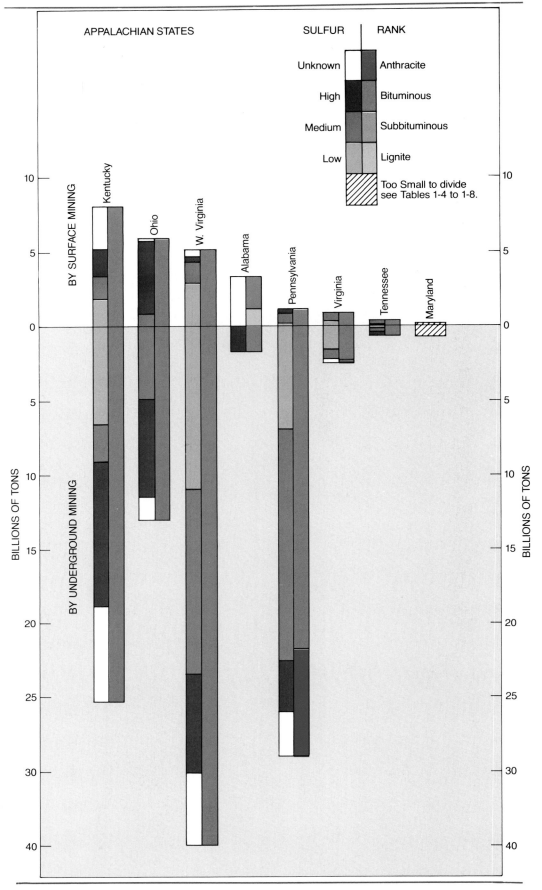

Fig. 1-22 Summary of mineable coals in Appalachian states showing rank, sulfur content, and mining method.

Sources: Department of Energy, *Coal Production, 1982*, 1983; U.S. Bureau of Mines, "The Reserve Base of U.S. Coals by Sulfur Content," *Information Circular, 8680*, 1975; and U.S. Bureau of Mines, "The Reserve Base of U.S. Coals by Sulfur Content—Eastern States," *Information Circular, 8693*, 1975.

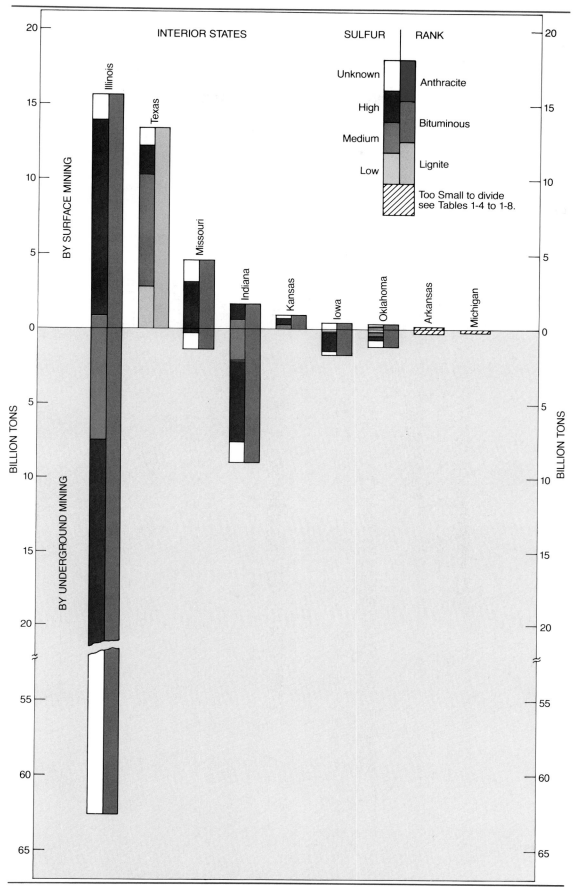

Fig. 1-23 Summary of mineable coals in interior states showing rank, sulfur content, and mining method.

Sources: Department of Energy, *Coal Production, 1982,* 1983; U.S Bureau of Mines, "The Reserve Base of U.S. Coals by Sulfur Content—Eastern States," *Information Circular, 8680,* 1975; and U.S. Bureau of Mines, "The Reserve Base of U.S. Coals by Sulfur Content—Western States," *Information Circular, 8693,* 1975.

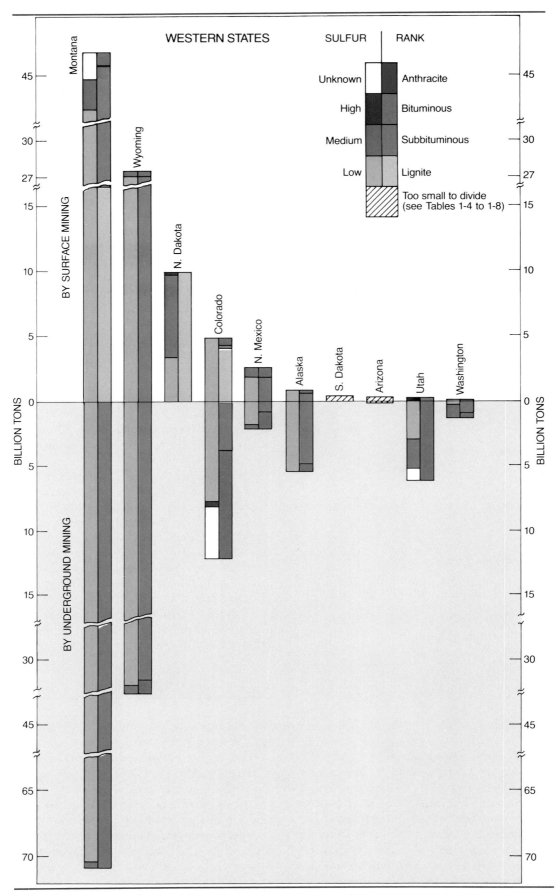

Fig. 1-24 Summary of mineable coals in Western states, 1982, showing rank, sulfur content, and mining method.

Sources: Department of Energy, *Coal Production, 1982*, 1983; U.S. Bureau of Mines, "The Reserve Base of U.S. Coals by Sulfur Content—Eastern States," *Information Circular, 8680*, 1975; and U.S. Bureau of Mines, "The Reserve Base of U.S. Coals by Sulfur Content—Western States," *Information Circular, 8696*, 1975.

TABLE 1-10
Total Identified Coals Showing Demonstrated Reserve
Base Portion (millions of short tons)

STATE	TOTAL IDENTIFIED	DEMONSTRATED RESERVE BASE	DRB PORTION OF IDENTIFIED (PERCENT) %
Alabama	15,262	5,202	34
Alaska	130,079	6,153	5
Arizona	21,234	383	2
Arkansas	2,416	146	6
Colorado	128,948	17,223	13
Georgia	24	4	17
Illinois	146,001	78,769	54
Indiana	32,868	10,510	32
Iowa	6,505	2,196	34
Kansas	18,668	991	5
Kentucky			
Eastern	28,226	12,565	45
Western	36,120	20,947	58
Maryland	1,152	810	70
Michigan	205	128	62
Missouri	31,184	6,056	19
Montana	291,639	120,349	41
New Mexico	61,391	4,703	8
North Carolina	110	11	10
North Dakota	350,602	9,909	3
Ohio	41,166	18,945	46
Oklahoma	7,117	1,624	23
Oregon	334	18	5
Pennsylvania	82,752	30,005	36
South Dakota	2,185	366	17
Tennessee	2,530	953	38
Texas	16,341	13,480	82
Utah	23,359	6,424	28
Virginia	9,551	3,322	35
Washington	6,169	1,469	24
West Virginia	100,150	39,347	39
Wyoming	135,943	69,674	51
Other	688	—	—
U.S. TOTAL	1,730,919	482,954	28

Sources: Averitt, Paul. Coal Resources of the United States, January 1, 1974. *U.S. Geological Survey Bulletin 1412*, Washington, 1975 and U.S. Department of Energy. *Coal Production, 1982*, Washington, 1983.

ESTIMATES OF RECOVERABLE COAL Due to conflicting land use, bed thickness, depth, and environmental factors, extreme caution should be exercised in interpreting coal Reserves. The estimates in Figure 1-25 and Table 1-11 can, therefore, be considered rough approximations. In these estimates, a 50 percent recovery factor was assumed for underground mining and 90 percent for surface mining.[4] Based on these assumptions, the recoverable portion of the Demonstrated Reserve Base may be as high as 304 billion tons. Surface mining would account for 47 percent of the total and underground mining for 53 percent. Almost two-thirds of the Reserves lie west of the Mississippi River.[5] Approximately 60 percent of the total underground

mineable reserves are located in the eastern states while three-fourths of the surface mineable reserves are in western states.

In the East, Illinois, West Virginia, Pennsylvania, and Kentucky contain 33 percent of the total reserves, 47 percent of the underground reserves, and 19 percent of the surface reserves. In the West, Montana and Wyoming have 41 percent of the total reserves, 35 percent of the underground reserves, and 48 percent of the surface reserves.

Furthermore, the Demonstrated Reserve Base, on which these estimates are based, is itself a conservative estimate of which beds are mineable according to geologic criteria of bed thickness and depth. Advances in mining technology may make possible the mining of Identified coals now *excluded* from the Reserve Base—especially those at depths greater than 1,000 feet. While it is impossible to make a careful estimate of additional amounts that may be recoverable, one approach is simply to estimate that perhaps 50 percent of all Identified coals in the nation may someday be recovered. The resulting amount, 865 billion tons, is shown in Table 1-11 for comparison with the 304 billion thought to be recoverable from the Reserve Base alone.

Figure 1-26 shows that the 304 billion tons recoverable from the Demonstrated Reserve Base, and the 865 possibly recoverable, together exhaust two-thirds of all Identified coals.

There is, of course, the possibility that improved mining techniques will allow for higher recovery rates than 50 percent. Coals with overburden greater than 3,000 feet are classified as Hypothetical: with overburden 3,000 feet or less are 1,850 billion tons, roughly the same amount as all Identified; and with overburden ranging from 3,000 to 6,000 feet, only 388 billion tons. Recent revision of these 1975 estimates by the Department of Energy (see Table 1-2) has doubled the total resource amount. Although most of the additional coal is in Alaska, substantial amounts are in Montana and Wyoming and are being considered for in-place coal gasification (see section on coal gasification).

COAL UTILIZATION

Although coal comprises an estimated 90 percent of the total United States' fossil fuel reserves, it meets less than 22 percent of the Nation's energy needs. Coal has not been used extensively because until recently oil and natural gas were both cheaper and cleaner sources of energy. Domestic shortages of oil, however, and the nearly ten-fold increase in the cost of a barrel of imported crude oil, have made coal a competitive source of energy. Faced with a certain shortage of domestic oil, President Carter called for a doubling of coal production by 1985 to meet the country's future energy needs. Most of the increased amounts of coal would be used in electric power plants and as liquid or gaseous fuels. The degree to which coal can meet future energy needs depends on the coal industry's ability to increase production, the cost of delivering the coal to the major markets of the East and

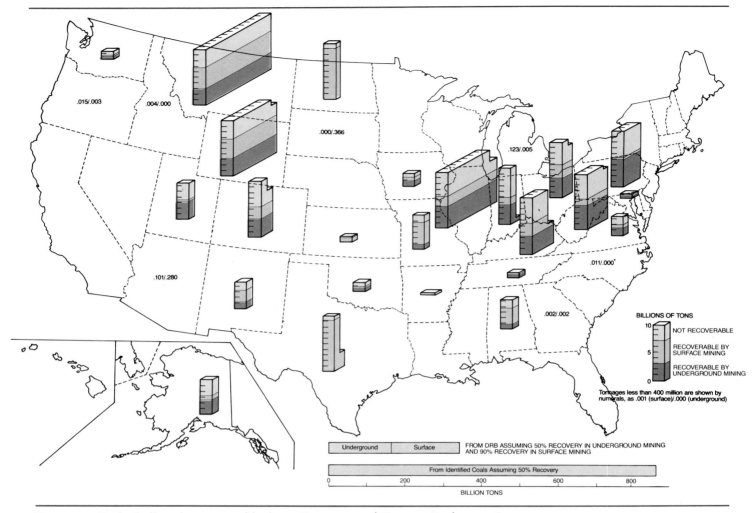

Fig. 1–25 Coal of all ranks recoverable from Demonstrated Reserve Bank assuming 50 percent recovery by underground mining and 90 percent recovery by surface mining.
Source: Department of Energy, *Coal Productions, 1982,* September 1983.

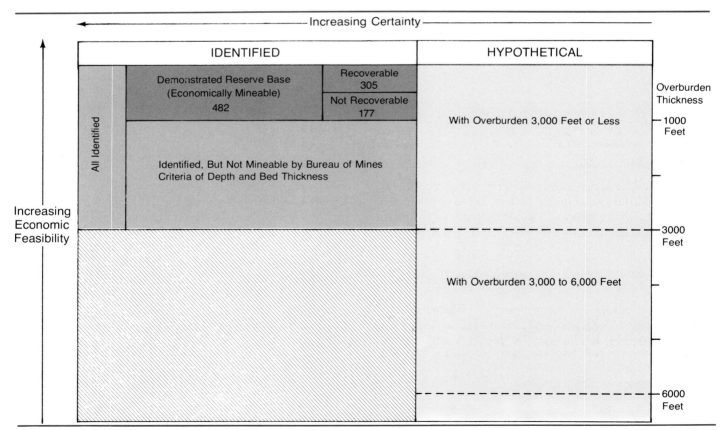

Fig. 1–26 Coal resource amounts, according to Averitt, in framework of varying geologic certainty and economic feasibility.
Sources: Averitt, 1975 and Department of Energy. *Coal Production, 1982,* September 1983.

TABLE 1-11

Coal of all ranks recoverable from Demonstrated Reserve Base assuming 50 percent recovery in underground mining 90 percent recovery in surface mining compared with the estimated coal recoverable from all Identified coals assuming a 50 percent recovery rate regardless of mining method (in millions of short tons)

| STATE | SURFACE MINEABLE | | UNDERGROUND MINEABLE | | TOTAL RECOVERABLE FROM DRB | TOTAL RECOVERABLE FROM ALL IDENTIFIED ASSUMING 50% RECOVERY |
	DRB Amounts	Recoverable Assuming 90% Recovery	DRB Amounts	Recoverable Assuming 50% Recovery		
Alabama	3,362	3,025	1,740	870	3,895	7,361
Alaska	730	657	5,423	2,712	3,369	65,040
Arizona	280	252	102	51	303	10,617
Arkansas	146	131	273	137	268	1,208
Colorado	4,951	4,456	12,951	6,476	10,932	64,474
Georgia	2	2	2	1	3	12
Illinois	15,657	14,091	63,112	31,556	45,647	73,001
Indiana	1,575	1,418	8,935	4,468	5,886	16,434
Iowa	462	416	1,734	867	1,283	3,253
Kansas	991	892	—	—	892	9,334
Kentucky						
Eastern	4,072	3,665	8,493	4,247	7,912	14,113
Western	4,026	3,623	16,921	8,461	12,084	18,060
Maryland	104	94	706	353	447	576
Michigan	5	5	123	62	67	103
Missouri	4,571	4,114	1,479	740	4,854	15,592
Montana	49,390	44,451	70,592	35,296	79,747	145,820
New Mexico	2,573	2,226	2,129	1,065	3,291	30,696
North Carolina	—	—	11	6	6	55
North Dakota	9,909	8,918	—	—	8,918	175,301
Ohio	5,925	5,333	13,020	6,510	11,843	20,583
Oklahoma	385	347	1,238	619	966	3,559
Oregon	3	3	15	8	11	167
Pennsylvania	1,174	1,057	28,832	14,416	15,473	41,376
South Dakota	366	329	—	—	329	1,093
Tennessee	316	284	637	319	603	1,265
Texas	13,480	12,132	—	—	12,132	8,171
Utah	268	241	6,156	3,078	3,319	11,680
Virginia	824	742	2,498	1,249	1,991	4,776
Washington	137	123	1,332	666	789	3,085
West Virginia	5,136	4,622	34,211	17,106	21,728	50,075
Wyoming	27,112	24,400	42,463	21,232	45,632	67,972
U.S. TOTAL	158,037	142,233	324,917	162,459	304,692	865,460

Sources: Averitt, Paul. "Coal Resources of the United States, January 1, 1974." *U.S. Geological Survey Bulletin 1412*, Washington, 1975 and U.S. Department of Energy. *Coal Production, 1982*, Washington, 1983.

Midwest, and technological developments that would allow coal to be burned without undue cost to the environment. The following discussion will focus on current and future production, and coal movement patterns.

Current Production

In 1982, total United States coal production was approximately 833 million tons (Table 1-12), a two percent increase from 1981 production. The production by rank was 74 percent bituminous, 19 percent subbituminous, 6 percent lignite, and 1 percent anthracite (Table 1-13 and Fig. 1-27). Approximately 67 percent of the production took place in eastern states and 33 percent in western. In the East, bituminous coals accounted for 99 percent of the production with the remainder consisting of Pennsylvania anthracite. The division by rank in the West was 77 percent subbituminous and lignite, and 23 percent bituminous.

Figure 1-27 shows the extent to which surface and underground mining have played a part in recent production. In 1982, surface mining accounted for 60 percent of all production.

Three eastern states, Kentucky, West Virginia, and

TABLE 1–12
United States' Coal Production for the Years 1980, 1981, and 1982 (thousands of tons)

STATE	PRODUCTION		
	1980	1981	1982
Alabama	26,403	24,363	26,226
Alaska	791	808	833
Arizona	10,905	11,609	12,364
Arkansas	319	218	138
Colorado	18,846	19,865	18,307
Georgia	3	5	—
Illinois	62,543	51,851	60,259
Indiana	30,873	29,124	31,722
Iowa	559	717	564
Kansas	842	1,361	1,401
Kentucky	150,144	154,760	147,930
Eastern	109,186	115,413	109,030
Western	40,958	39,347	38,900
Maryland	3,760	4,240	3,764
Missouri	5,503	4,882	5,336
Montana	29,872	33,545	27,882
New Mexico	18,425	18,685	19,940
North Dakota	16,975	18,122	17,848
Ohio	39,394	35,704	36,337
Oklahoma	5,358	5,785	4,770
Pennsylvania	87,069	82,986	78,279
Tennessee	9,850	9,706	7,287
Texas	29,354	34,818	32,814
Utah	13,236	13,706	17,029
Virginia	41,009	40,577	39,068
Washington	5,140	4,635	4,161
West Virginia	121,584	112,211	127,899
Wyoming	94,887	102,969	108,360
U.S. TOTALS	823,644	815,244	832,524

Sources: Department of Energy, *Coal Data: A Reference*, Washington, October 1982 and U.S. Department of Energy. *Coal Production, 1982*, Washington, September 1983.

Railroad cars crawl through coal-storage silos picking up coal from the Jacob Ranch mine southeast of Gillette, in the Powder River Basin, Wyoming.

coal is strip mined, thereby lowering the cost and minimizing safety problems; and finally most western mines are not unionized thereby cutting down on labor disputes.

IMPACT OF PAST, PRESENT, AND FUTURE PRODUCTION ON RECOVERABLE RESOURCES

With the rising interest in increased coal production to meet energy needs, it is important to examine the impact of past, present, and future production on the nation's recoverable coal supply. Table 1–14 is a very important summary that offers: (1) a comparison of cumulative production through 1982 with remaining reserves; (2) estimates of production capacity in 1982; and (3) a forecast of the life of remaining coal reserves. Through 1982, cumulative production of coal in the United States amounted to 43.1 billion tons. This past production has consumed approximately 12 percent of the estimated original coal reserves of the nation.

Figure 1–28 shows the impact of past production on the estimated original reserves of each producing state. As would be expected, more coal in eastern regions has been produced; from 6 to 48 percent of the original reserves have been used in these states. In Tennessee, over 48 percent of the estimated original reserves have been mined and 26 to 44 percent in the

Pennsylvania, produced 42 percent of the nation's coal and 63 percent of the total mined in eastern and interior coal fields. Among the western states, Montana, Texas, and Wyoming led in production, accounting for roughly 20 percent of the national total and 62 percent of the coal produced in western coal fields.

Since 1973, coal production in the states west of the Mississippi has been increasing at a substantial rate. In 1970, roughly 75 million tons of coal, or 11 percent of the national total, was produced in the West. In 1982, this increased to 274 million tons, or 33 percent of the national total. Some analysts have predicted that by 1990 western states will be supplying 50 percent of the nation's coal.

Several factors account for the rapid rise in production of western coals: They are low in sulfur and produce a minimum amount of sulfur dioxide pollution; much of the

TABLE 1–13
U.S. Coal Production during 1982 Showing Tonnages by Rank and Mining Method (thousands of tons)

STATE	TOTAL PRODUCTION	PRODUCTION BY RANK[1]				PRODUCTION BY MINING METHOD	
		L	SB	B	A	Underground	Surface
Alabama	26,226	—	—	26,226	—	11,291	14,936
Alaska	833	—	833	—	—	—	833
Arizona	12,364	—	—	12,364	—	—	12,364
Arkansas	138	11	—	99	29	—	138
Colorado	18,307	—	6,120	12,180	—	6,613	11,694
Illinois	60,259	—	—	60,259	—	34,660	25,599
Indiana	31,722	—	—	31,722	—	1,570	30,152
Iowa	564	—	—	564	—	—	564
Kansas	1,401	—	—	1,401	—	—	1,401
Kentucky	147,930	—	—	147,930	—	74,784	73,146
Eastern	109,030	—	—	109,030	—	57,069	51,961
Western	38,900	—	—	38,900	—	17,715	21,186
Maryland	3,764	—	—	3,764	—	1,899	1,865
Missouri	5,336	—	—	5,336	—	—	5,336
Montana	27,882	174	27,708	—	—	—	27,882
New Mexico	19,940	—	12,410	7,530	—	708	19,233
North Dakota	17,848	17,848	—	—	—	—	17,848
Ohio	36,337	—	—	36,337	—	12,219	24,117
Oklahoma	4,770	—	—	4,770	—	57	4,712
Pennsylvania	78,279	—	—	74,066	4,213	35,864	42,416
Tennessee	7,287	—	—	7,287	—	4,518	2,769
Texas	34,818	34,379	—	440	—	—	34,818
Utah	17,029	—	—	17,029	—	17,029	—
Virginia	39,068	—	—	39,068	—	30,992	8,076
Washington	4,161	—	4,161	—	—	—	4,161
West Virginia	127,899	—	—	127,899	—	103,531	24,368
Wyoming	108,360	—	107,135	1,225	—	1,276	107,084
U.S. Totals	832,524	52,411	158,367	617,504	4,242	337,010	495,514
Percent	100	6	19	74	1	40	60

[1]L = Lignite; SB = Subbituminous; B = Bituminous; and A = Anthracite

Source: Department of Energy. *Coal Production, 1982*, Washington, September, 1983.

states of Alabama, Maryland, Pennsylvania, and Virginia. By contrast, no state in the West has consumed as much as 26 percent of its estimated original reserves, and most have consumed less than 10 percent. In Montana, less than one percent of the original reserves have been mined and in Wyoming only 2.2 percent.

Coal Mine Utilization in 1982

In 1982 coal mines in the United States operated at approximately 89 percent of their total capacity of 931 million tons (Department of Energy, *Coal Production, 1982* September 1983). According to a survey of the coal industry in 1977, planned expansion and development of new coal mines was to have increased coal production capacity to over 1.4 billion tons by 1985 (Nielsen, 1977). This increase would amount to 50 percent over the 1982 capacity. Although it is

not likely that a 50 percent increase in mining capacity can be achieved by the end of 1985, it seems, however, feasible that U.S. coal production could reach a billion tons annually by that date. Most of the new capacity would be achieved through surface mining in the western coal fields. (Figure 1–29).

ESTIMATE OF LIFE OF REMAINING RESERVES To provide some insight into the impact of future coal usage of remaining reserves, an estimated life was computed on the basis of 1982 estimated production capacity (Table 1–14 and Fig. 1–30). The estimated life of the remaining reserves for the country as a whole is 327 years. This estimate represents a fairly optimistic future for remaining reserves because production capacity is expected to rise substantially over the next two or more decades. For example, if the national production capacity were 1.4 billion

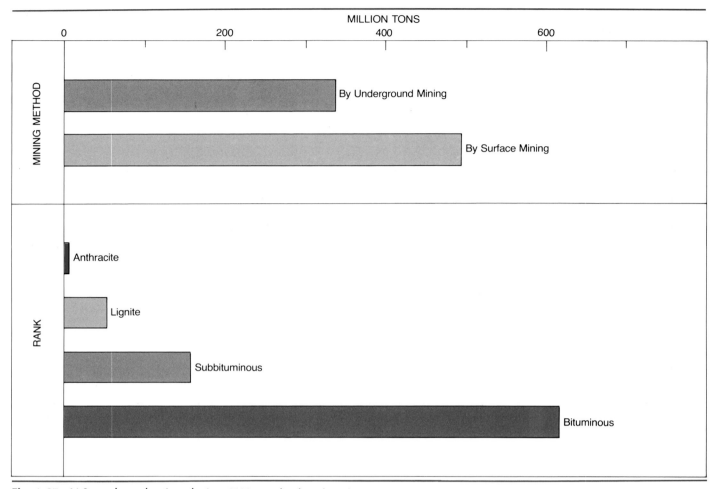

Fig. 1-27 U.S. coal production during 1982, method and rank.
Department of Energy. *Coal Production, 1982,* September 1983.

tons per year, the life of the remaining reserves would decrease from the 327 years to 187 years based on 1982 production capacity.

However, for the purpose of this analysis we will use the more optimistic projection. The estimated life among the states ranged from 24 years in Arizona to over 4,000 years in Alaska. For the major producing states in the East, West Virginia, Kentucky and Pennsylvania, the estimated life of the remaining reserves is 138, 161, and 87 years respectively. In the West the estimated life in Montana is 2,394 years and 144 years in Wyoming.

Coal Movement and Consumption Patterns

Although it is very important to have a careful estimate of the remaining coal reserves, and to determine how much can be mined, it is equally important to understand the cost of using coal to meet energy needs.

Since the energy crisis of 1973, and with the encouragement to use coal, the question has been raised of whether western coal can be transported to midwestern and eastern markets at a cost competitive with coals produced in these regions. The major reason for the keen interest in coals in the West, as noted before, is their low sulfur content. Nonetheless, the lower energy content of Western coals and the expense of transportation to market could offset the advantage of burning low-sulfur coals.

On the average, the heating value of western coal is about 9,000 BTUs per pound compared with about 12,000 BTUs per pound for eastern coal (Bureau of Mines, *Information Circular 8690,* 1975). A major reason for the low heat value of western subbituminous coal is its high inherent moisture content. However, Texaco and Bechtel have explored a hydrothermal coal-treating process to upgrade the BTU content of these coals. In a laboratory test, Western subbituminous coal was ground, slurried in water, and then heated to between 500 and 650 degrees Fahrenheit in heat exchangers pressurized to keep the slurry liquid. The BTU content of the resultant dewatered coal was raised by 25 percent. It is believed that this process could be easily adapted for minehead coal preparation plants or dewatering terminals on the slurry systems. If this proves to be feasible, the low heat content disadvantage of western subbituminous coal would, to a large degree, be eliminated (Wasp, 1983).

In addition to the low heat content, western coals must often be transported 1,200 to 1,400 miles to the East and South before reaching major market areas. On first examination, the low energy content and distance handicaps of western coals should favor sales of coal from eastern mines; western coals do have other important features which help them compete. In many areas of the West, such as the Powder River Basin in Wyoming, seams 100 feet thick lie just below the surface and favor cheap strip mining methods. In 1982 the average mine price for a ton of surface mined coal in the western states was $13.61. This compares to a $21.46 average mine price for a ton of coal surface mined in the Appalachian states (Department of Energy, *Coal Production,* 1982 September 1983). It would appear that

lower mining costs and a greater output per man-day are significant factors in allowing western coals to reach eastern markets.

Even though eastern coals are much closer to major markets, environmental restrictions imposed on electrical generating stations have made it more costly to burn the medium- and high-sulfur coals found in the Appalachian and Interior coal fields. These costs have helped to nullify the extra expense of long-distance transport of western coals. The development of economical ways to transport large volumes of coal to the East and South would also allow western coals to compete more effectively with coals mined in the Applachian and Interior coal fields.

MOVEMENT PATTERNS Figures 1-31 and 1-32 show the movement pattern of coal in the United States. The principal mode of coal transportation is by rail. In 1983 in the United States, almost 60 percent of all coal shipments to

TABLE 1-14
Past, Present, and Future Coal Production (millions of tons)

STATE	CUMULATIVE PRODUCTION THROUGH 1982	REMAINING RESERVES IN 1982[1]	ESTIMATED ORIGINAL RESERVES[2]	CUMULATIVE PRODUCTION AS A PROPORTION OF ESTIMATED ORIGINAL RESERVES	PRODUCTION CAPACITY IN 1982[3]	LIFE REMAINING AT 1982 MINING CAPACITY (YEARS)
Alabama	1,376	3,895	5,271	.261	30.2	129
Alaska	10	3,369	3,379	.003	0.8	4,211
Arizona	98	303	401	.244	12.4	24
Arkansas	108	268	376	.281	0.2	1,340
Colorado	702	10,932	11,634	.060	21.2	516
Illinois	4,929	45,647	50,576	.097	74.1	616
Indiana	1,656	5,886	7,542	.219	36.2	163
Iowa	370	2,601	2,971	.125	0.7	3,716
Kansas	298	892	1,190	.250	1.5	595
Kentucky	5,288	19,996	25,284	.209	160.7	124
Maryland	312	447	759	.411	3.9	115
Missouri	386	4,854	5,240	.073	5.7	852
Montana	445	79,747	80,192	.006	33.3	2,394
New Mexico	306	3,291	3,597	.085	25.8	128
North Dakota	273	8,918	9,191	.030	21.4	417
Ohio	3,048	11,843	14,891	.207	38.9	304
Oklahoma	241	966	1,207	.200	5.0	193
Pennsylvania	10,038	15,473	25,511	.393	86.9	178
Tennessee	571	603	1,174	.486	7.6	79
Texas	205	12,132	12,337	.017	33.4	363
Utah	418	3,319	3,737	.112	21.2	157
Virginia	1,584	1,991	3,575	.443	41.1	48
Washington	197	789	986	.200	4.2	188
West Virginia	9,192	21,728	30,920	.297	137.9	156
Wyoming	1,033	45,632	46,665	.022	126.4	361
U.S. TOTALS	43,084	304,692	347,776	.124	930.6	327

[1]Reserves are estimated tonnages recoverable from the Demonstrated Reserve Base, 1982 assuming 90 percent recovery in surface mining and 50 percent recovery in underground mining.

[2]Estimated original reserves are cumulative production plus remaining reserves.

[3]Based on the 1982 utilization reported by the Department of Energy.

Sources: Department of Energy, *Coal Production, 1982,* Washington, September 1983 and U.S. Department of Energy. *Coal Data: A Reference,* Washington, October 1982.

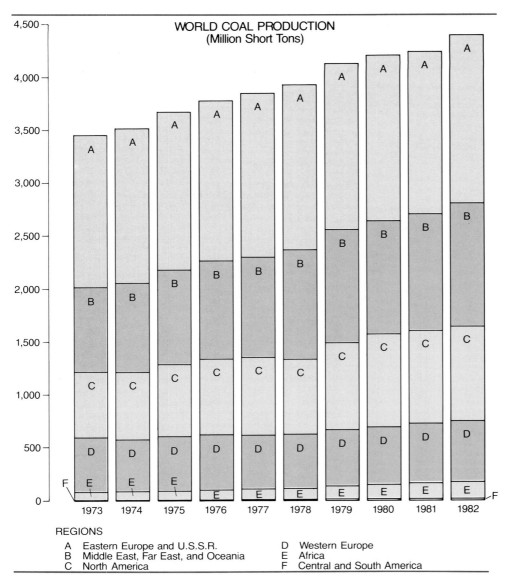

WORLD COAL PRODUCTION
(Million Short Tons)

REGIONS

A   Eastern Europe and U.S.S.R.   D   Western Europe
B   Middle East, Far East, and Oceania E   Africa
C   North America   F   Central and South America

World coal production rose from 3.4 billion short tons in 1973 to 4.4 billion short tons in 1982. Most dramatic increases were manifested in countries where formerly marginal coal deposits became viable alternatives to costly petroleum. Production in Africa increased by 123 percent between 1973 and 1982, principally as a result of mine expansion in South Africa. The Far East and Oceania region increased its output by 45 percent during this same period. The United States, the Soviet Union, and the Peoples Republic of China accounted for 54 percent of world coal production in 1982.

Source: Department of Energy. *1982 International Energy Annual*, Washington, September 1983.

consumers were rail. Water transportation ranked second with 16 percent, followed by truck with 14 percent and tramway/conveyor and slurry pipeline with 11 percent (Department of Energy, April 1984). The greatest amount of coal traffic was generated in the East. Coal moves by rail, water, and highway from the major mining districts in the East to interstate and foreign markets. The largest volume of coal moves by rail from mining districts in West Virginia, Virginia, and Kentucky to markets throughout the East and South as well as to foreign markets in Europe, Asia, and South America from the port of Norfolk, Virginia. A substantial volume of eastern coal is moved by water through the Ohio and Mississippi River systems and the Great Lakes. Coal movement along these waterways is destined for markets in the South, Midwest, and foreign countries.

In the West, significant amounts of coal are moved by rail from mining districts in Wyoming and Montana to destinations in the upper and lower Midwest. Smaller amounts are transported by rail from Montana and Wyoming to destinations in the western states. In addition, up to 5.5 million tons of coal are moved each year by way of the Black Mesa Slurry Pipeline from the Kayenta Mine in northeastern Arizona to the Mojave Power Plant in

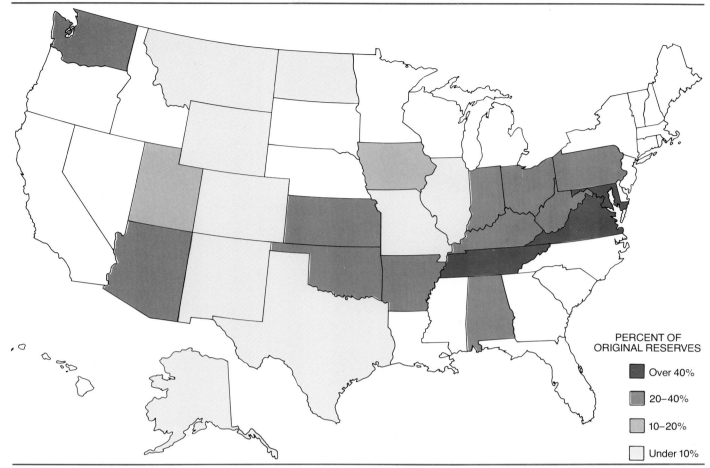

Fig. 1-28 Cumulative coal production in producing states through 1982 as a proportion of estimated original reserves.

Note: Excludes mines producing less than 10,000 short tons of coal during the year.

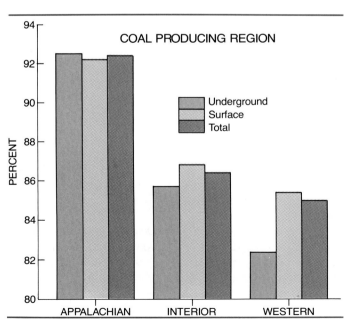

Fig. 1-29 Percentage utilization of U.S. coal mines, 1982.

Source: Department of Energy, *Coal Production, 1982*, September 1983.

southeastern Nevada, a distance of 273 miles (see slurry pipeline illustration and essay).

Presently, there are seven coal slurry pipeline projects that, if constructed, would connect the western coal mines to distant markets in Texas, Louisiana, Minnesota, Wisconsin, and California. In addition, the Coal Stream Slurry Pipeline Project would connect coal mines in West Virginia and Southern Illinois to power plants in Florida (Wasp, 1983). Up to this time, slurry pipeline projects have been slowed by the tedious task of acquiring rights of way needed for the pipeline and easements needed for the water-supply line. Moreover, opposition from railroad companies and the securing of a stable water supply has created substantial problems.

CONSUMPTION PATTERNS Figure 1-31 shows the end-use coal distribution by supplying area. Approximately 80 percent of coal consumed in recent years has been used for the generation of electricity (Department of Energy, October 1982), and each ton of coal consumed generates about 2,000 kilowatt hours of electricity. Coal-fired electrical generating plants account for about 50 percent of the total

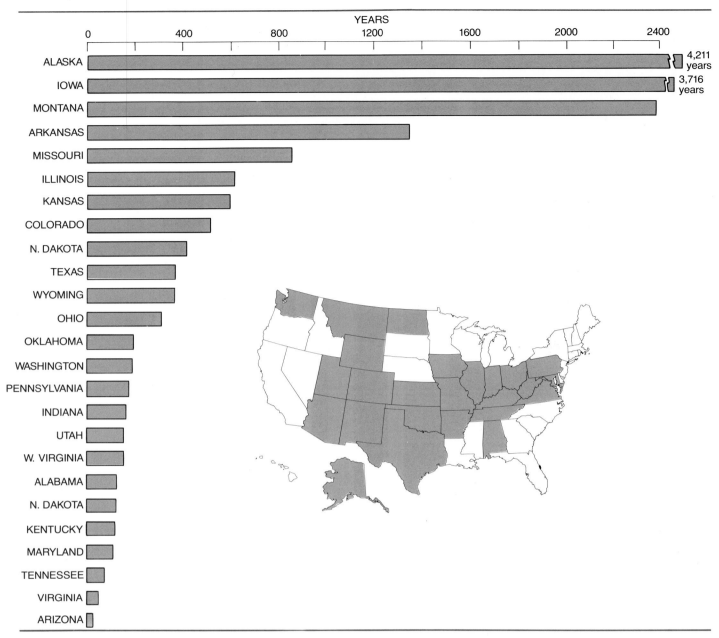

Fig. 1-30 Estimated life of remaining coal reserves in producing states based on 1982 mining capacities.

Sources: Department of Energy, *Coal Production, 1982*, September 1983; Department of Energy, *Coal Data: A Reference*, October 1982.

Notes: 1. Estimates assumed that mines in producing state are operating at 100 percent of their 1982 capacity.
2. The average life of all U.S. coal reserves is estimated to be 327 years.

electricity generated in the United States. There are about 500 coal-burning power plants in the country, about one-sixth of the total number of power plants. A very large plant can burn over 5 million tons of coal per year (Department of Energy, October 1982).

Another end-use of coal is coke plants, which consume approximately 8 percent of the annual coal production. Coking coal (a premium coal that must meet certain specifications) is used for smelting iron ore in blast furnaces. About 1,100 pounds of coke are consumed per ton of pig iron produced. There are about 55 coke plants located primarily in the eastern United States, and these plants use coal mined principally in the Appalachian region of the country.

Coal is also used in the manufacturing of calcium carbide, carbon, and graphite electrodes. Four states, Ohio, Pennsylvania, Indiana, and Illinois account for almost one-third of the coal consumed in the United States.

About 10 percent of the annual coal production is exported. Metallurgical and steam grade coals are the principal ones exported. The chief recipients of United States coal are Japan, Canada, and Western Europe. Most of the exported coal moves through the ports of Hampton Roads, Virginia; Baltimore, Maryland; New Orleans, Louisiana; Mobile, Alabama; and Los Angeles-Long Beach, California (Department of Energy, October 1982).

COAL LIQUEFACTION AND GASIFICATION

In July 1979, President Carter called for the establishment of a government-run Synthetic Fuel Corporation that would supervise the spending of 88 billion dollars over a

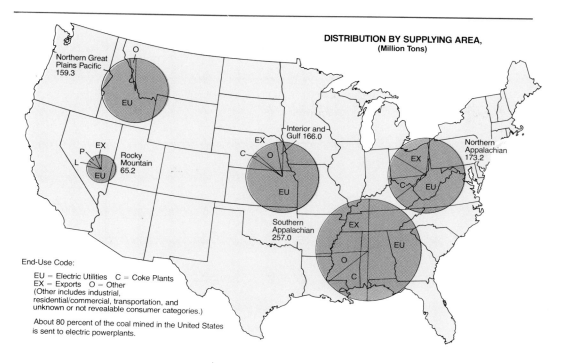

DISTRIBUTION BY SUPPLYING AREA,
(Million Tons)

Northern Great Plains Pacific 159.3

Interior and Gulf 166.0

Northern Appalachian 173.2

Rocky Mountain 65.2

Southern Appalachian 257.0

End-Use Code:

EU = Electric Utilities C = Coke Plants
EX = Exports O = Other
(Other includes industrial, residential/commercial, transportation, and unknown or not revealable consumer categories.)

About 80 percent of the coal mined in the United States is sent to electric powerplants.

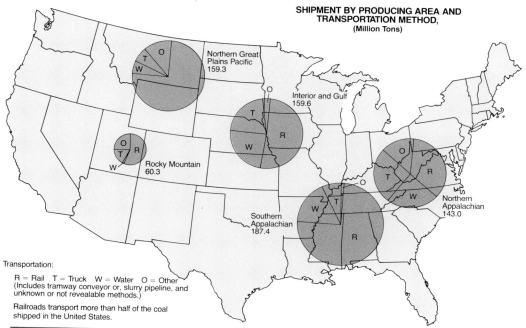

SHIPMENT BY PRODUCING AREA AND TRANSPORTATION METHOD,
(Million Tons)

Northern Great Plains Pacific 159.3

Interior and Gulf 159.6

Rocky Mountain 60.3

Southern Appalachian 187.4

Northern Appalachian 143.0

Transportation:

R = Rail T = Truck W = Water O = Other
(Includes tramway conveyor or, slurry pipeline, and unknown or not revealable methods.)

Railroads transport more than half of the coal shipped in the United States.

Fig. 1–31 1981 end-use coal distribution by supplying area (top) and coal shipments for U.S. consumption by producing area and transportation method (bottom).
Source: Department of Energy. *Coal Data: A Reference,* October, 1982.

Loaded coal barges on the Monongahela River near Morgantown, West Virginia. Such barges move coal from the fields of West Virginia downstream on the Monongahela to Pittsburgh, and beyond Pittsburgh on the Ohio River. (Courtesy of Consolidation Coal Company.)

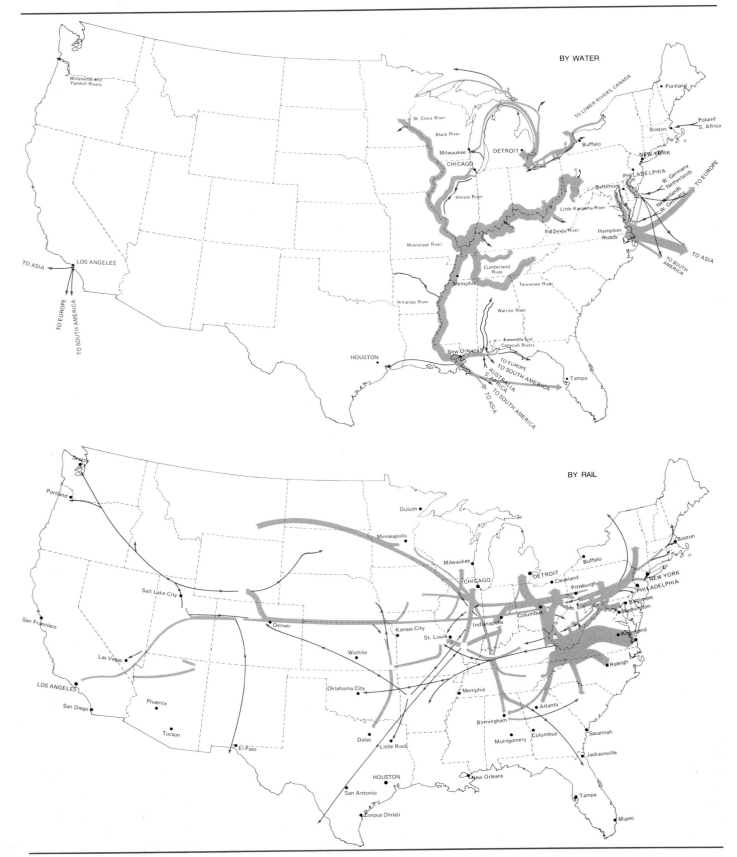

Fig. 1–32 Interstate coal movement patterns by water (top) and rail (bottom).
Source: Jimison, 1978.

COAL SLURRY PIPELINES

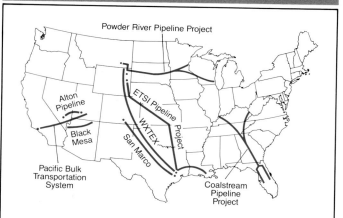

Reserves of coals in Western states are enormous and they have, in general, a low sulfur content. However, because the coals are located at great distances from most electric-power generating stations, the cost of transport is high. Coal slurry pipelines may be the answer to cheap, efficient transport of large volumes of Western coals to distant markets. Slurry transportation consists of pipelines through which coal is pumped suspended in water. At the mine the coal is crushed and formed into a slurry and set in motion by giant pumps. At the destination, the power plant site, the slurry is fed into centrifuges that remove the water and produce a material that looks like damp black beach sand which is ready for burning. Currently the longest slurry pipeline in the United States is the Black Mesa line, which has the capacity to transport 5.5 million tons of coal per year 273 miles from the Kayenta coal mine in northeastern Arizona to the Mojave Power Plant in southeastern Nevada. At present there are seven proposed slurry pipeline projects in the United States (see above map). Of these projects, the ETSI project (Energy Transportation Incorporated) is the one closest to construction. The ETSI pipeline would carry coal from Gillette, Wyoming to power plant destinations more than 1400 miles away in Texas and Louisiana. It would have the capacity to move 30 million tons of coal per year, about 3.6 percent of all U.S. coal produced in 1982.

advanced than other synthetic fuel technologies. The general objectives of the coal conversion proposal are to develop and demonstrate the technologies necessary to allow coal and coal-derived fuels to be substituted for oil and gas. Moreover, these objectives would be pursued in ways and at rates that are economically, environmentally, and socially acceptable (Department of Energy, March 1978).

The specific products that can be derived from coal conversion processes include: (1) crude oil, (2) fuel oil, (3) distillates, (4) chemical feedstock, and (5) pipeline quality gas (high-BTU) and fuel gas (low- and intermediate-BTU). Figure 1–33 depicts the various coal conversion processes. Basically, liquid or gaseous fuels are produced by decreasing the carbon to hydrogen ratio of solid coal. Coal has a carbon to hydrogen (C/H) weight ratio ranging from 12 for lignite to 20 for bituminous grades (Corey, 1976). Either by addition of hydrogen or by rejection of carbon, the C/H ratio can be lowered to 10 to produce, by molecular weight, synthetic crude oil. If the C/H ratio is further decreased to 3, methane gas can be produced.

Liquefaction

The history of coal liquefaction began in 1913 when work on the Bergius concept of direct hydrogenation of coal was undertaken in Germany (Office of Fossil Energy, 1976). During World War II, the Germans produced a major part of their aviation gasoline using liquefaction technology. Today there are several liquefaction plants in operation around the world. South Africa, for instance, produces 20,000 barrels a day of synthetic oil called *Sasol*.

Although there are no commercial coal liquefaction plants presently operating in the United States, several

The Coalstrip power plant near Billings, Montana, burns coal mined from the plains of southern Montana. Its 2,000 Megawatt generating capacity is matched by very few thermal-electric plants. (Courtesy of American Petroleum Institute and Bechtel Corporation.)

12-year period. The main purpose of the corporation would be to bring about, by 1992, the production of fuels from domestic resources in amounts sufficient to replace 2 million barrels per day of imported petroleum (*Chemical and Engineering News,* 1979). Much of the 2 million barrels per day of replacement fuels would be synthetics produced from coals, oil shales, peat, biomass (the organic matter in animal wastes and plants), and organic waste. The United States Synthetic Fuels Corporation was officially established with the signing by President Carter of the Energy Security Act. The Synthetic Fuels Corporation was directed to send to Congress a comprehensive strategy for the development of the synthetic fuels industry in 1984. If the strategy meets congressional approval, it may get as much as $68 billion to spend through 1992 (*Oil and Gas Journal,* 1981). Because of delays caused by the Reagan Administration, the strategy has yet to be finalized.

Up to now, coal conversion projects have received much attention and most of the federal dollars. Not only is coal the United States' most abundant fossil fuel resource, coal conversion processes are, in addition, generally more

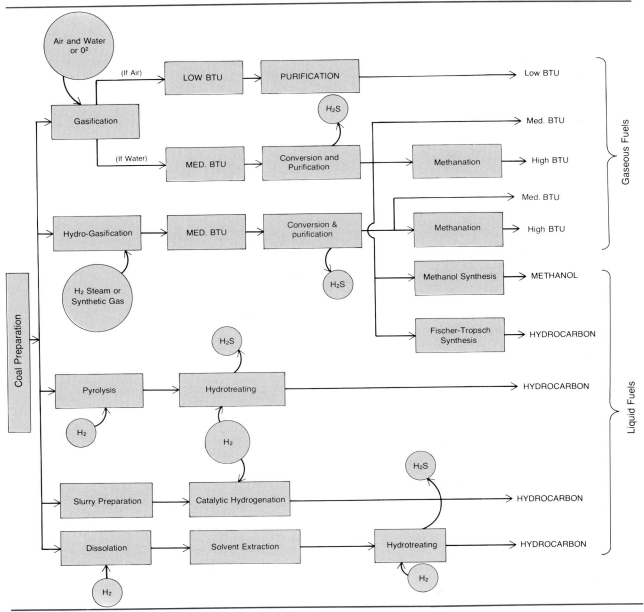

Fig. 1–33 Coal gasification and liquefaction processes at surface installations.
Source: Energy Resource Development Agency, *Energy from Coal,* 1976

projects have been initiated (Table 1-15 and Fig. 1-34). Among the coal liquefaction projects proposed, the W.R. Grace Project is the largest. If constructed the plant would be located at Baskett, Kentucky. It would cost an estimated $3 billion dollars and convert 29,000 tons per day of high-sulfur coal to synthetic gas to produce 50,000 barrels per day of high-octane unleaded gas (*Oil and Gas Journal,* 1981). Ashland Oil, Mapco, and Tennessee Eastman Company also have plans to build methanol plants at their respective sites in Ohio, Texas, and Tennessee.

Some feel that coal is a poor choice for producing synthetic crude oil. The molecular structure of crude oil typically contains 12 to 14 percent hydrogen. However, coal contains only about 5 percent hydrogen and upgrading the

liquids produced from it requires the addition of costly hydrogen molecules (Derbyshire, 1983). Shale oil and tar sand bitumen appear to be better choices for producing "syncrude" since they contain 9 to 12 percent hydrogen.

Coal liquefaction is an enormous undertaking. Each full-scale plant producing 50,000 barrels per day of boiler fuel or syncrude per day would require 15,000 to 30,000 tons per day of bituminous coal. The assumed fuel efficiency of the plant is 60 to 75 percent and water requirements are estimated at 10,000 acre feet per year (Corey, 1976).

Up to now, coal liquefaction research indicates that high yields of syncrude can be achieved from bituminous, subbituminous, and lignite coals. Most of the research in progress appears to use high-sulfur eastern coals as feedstock.

TABLE 1-15
Coal Liquefaction Projects in the United States as of June 1981

COMPANY	PROJECT LOCATION	PLANT CAPACITY (BARRELS/DAY)	COAL REQUIRED (TONS/DAY)	START-UP DATE	PRODUCTS
SRC Internation	Morgantown, W. Virginia	20,000	6,000	UK[1]	Syncrude
EXXON	Baytown, Texas	600	250	pilot unit on stream	Syncrude
Ashland Synthetic Fuels	Cattletsburg, Kentucky	600–1,800	200–600	pilot unit on stream	Syncrude
Texas Eastern Synfuels	Henderson, Kentucky	50,000[2]	28,000	UK[1]	Syncrude
W.R. Grace	Baskett, Kentucky	50,000	29,000	1986	Gasoline
Tennessee Eastman	Kingsport, Tennessee	—	1,600	1984	Methanol
Ashland Oil	South Point, Ohio	4,000	—	UK[1]	Methanol
Mapco Inc.	Moore County, Texas	1,419	—	1982	Methanol
Mobil R and D Corporation	Buffalo, Wyoming	40,000	—	UK[1]	Gasoline

[1]Unknown
[2]plus 145 million cubic feet/day of high BTU gas and chemicals
Source: *Oil and Gas Journal*, "Synthetic Fuels Report," June 29, 1981.

Fig. 1–34 Coal gasification and liquefaction projects in the United States as of June 1981.
Source: *Oil and Gas Journal*, "Synthetic Fuels Report," June 29, 1981.

Coal liquefaction pilot plant near Baytown, Texas. (Courtesy of U.S. Department of Energy.)

However, before commercialization of coal liquefaction, several environmental concerns must be dealt with. Beyond the heavy burden placed on the environment by the coal and water demand, liquefaction plants produce carcinogenic organic compounds in the coal residue and products and in trace metal pollutants.

Gasification

Gas made from coal had its first commercial success in the early 1800s when it was used in the cities of London and Baltimore in street lights. With the rapid exploitation of natural gas since World War II, gas derived from coal has played a rather minor role in commerce and industry. Although coal gasification processes have been available for many years, present technology is expensive, thermally inefficient, and, in many ways limited in the kinds and sizes of processable coal (Department of Energy, March, 1978).

Coal gasification research and development in the United States is focused on three main activities: (1) high-BTU gasification for clean pipeline quality gas, (2) low-BTU gasification for clean industrial and utility fuel gas, and (3) underground *in situ* coal gasification technology for on-site conversion of coal to gas in unmineable seams.

SURFACE GASIFICATION As indicated in the upper half of Figure 1–33 the initial step in the conversion of coal to gas may be either simple gasification or hydrogasification. The former primarily involves the direct reaction of steam with coal and the latter brings hydrogen produced elsewhere into contact with coal. The resultant products, low-BTU gas, medium-BTU gas, and high-BTU gas, are then purified and ready to be used as fuel. Low-BTU gas, with a heating value of 100 to 500 BTUs per cubic foot, is

suitable for use as a fuel feedstock or for power generation in combined gas-steam turbine power cycles. It is believed that low-BTU gas can be produced at a competitive cost if the gasifier is built on the premises of a power generating station, thereby eliminating long-distance pumping costs. Medium-BTU gas, with heating value of 500 to 950 BTUs per cubic foot, is usually a feed gas for production of high-BTU gas. High-BTU gas, with a heating value of 750 to 1000 BTUs per cubic foot can be used in the same way as natural gas (Corey, 1976).

Presently, the favored technology for producing high-BTU gas from coal is the German-developed Lurgi Gasifier. This process is already widely used in many countries to produce synthetic gas and is used in most commercial gasification projects throughout the world. The huge *Sasol* complex at Secunda, Republic of South Africa is the best example of a coal gasification facility. The *Sasol* complex handles up to 80,000 tons per day of coal and produces approximately 70,000 barrels per day of liquid fuels by gasifying the coal in 72 Lurgi gasifiers (Electric Power Research Institute, 1983). Many other gasification approaches are being tested, especially, where medium-BTU gas is needed.

Figure 1–34 and Table 1–16 show the location of a number of proposed coal gasification projects in the United States. In 1983, there were four coal gasification facilities under construction:

1. The Great Plains Synthetic Natural Gas Plant at Beulah, North Dakota which will use 14,000 tons per day of lignite to produce 137 million standard cubic feet of synthetic natural gas (SNG).

2. The Cool Water Gasification Project at Daggett, California uses 1,000 tons per day of Utah bituminous

TABLE 1–16
Coal Gasification Projects in the United States as of June 1981

COMPANY	PROJECT LOCATION	PLANT CAPACITY (MILLION CU. FT./DAY)	COAL REQUIRED (TONS/DAY)	START-UP DATE	GAS QUALITY
Great Plains Gasification	Mercer County, Dakota	125	14,000[1]	1984	High-BTU
Conoco Coal	Noble County, Ohio	19	1,250	UK[2]	High-BTU
Tenneco Coal	Wibaux, Montana	280	40,000[1]	UK[2]	High-BTU
Cool Water Coal Gasification	Dagget, California	67	1,000	UK[2]	Medium- or Low-BTU
Memphis Light, Gas and Water	Memphis, Tennessee	150	3,100	UK[2]	Medium- or Low-BTU
Gulf States Utilities	West Lake, Louisiana	225	1,400	UK[2]	Medium- or Low-BTU
TVA	Murphy Hill, Alabama	300[3]	20,000	1985–87	Medium- or Low-BTU
Mountain Fuel Resources	Salt Lake City, Utah	—	30	1982	Medium- or Low-BTU
Allis Chalmers	Wood River, Illinois	—	600	1983	Medium- or Low-BTU

[1]Lignite
[2]Unknown
[3]Plus liquid fuels
Source: *Oil and Gas Journal.* "Synthetic Fuels Report," June 29, 1981.

coal to produce 67.2 million standard cubic feet per day of medium-BTU gas to produce 100 megawatts of electric power.

3. The Tennessee Eastman Project at Kingsport, Tennessee is designed to use 900 tons per day of Tennessee bituminous coal to produce 50 million standard cubic feet per day of synthetic gas to be converted to methanol.

4. The KILNGAS Commercial Module at Wood River, Illonois is designed to use 600 tons per day of Illinois bituminous coal to produce 62.4 million standard cubic feet per day of low-BTU gas to be used for electrical generation (Electric Power Research Institute, 1983).

The Lurgi process works well with non-caking coals (lignite and subbituminous) found in the Western states. Eastern bituminous coals have a tendency to cake and clog gasifiers. Although caking tendencies of Eastern coals can be lessened by pretreating the coal, this additional step reduces the thermal efficiency of the process (*Chemical and Engineering News,* 1979).

The Electric Power Research Institute feels that converting the energy in coal to gas is a more environmentally acceptable way to produce electric power. They claim that sulfur and nitrogen oxide emissions would be lower using coal gasification and that there would be no scrubber sludge accumulation problem since the sulfur in a coal gasification plant would be removed in a highly pure form and sold (Electric Power Research Institute, 1983). In addition, they feel that plants can be designed for zero discharge of waste water.

Like coal liquefaction plants, coal gasification plants make heavy demands on coal resources and water. A supply of 250 million standard cubic feet per day of synthetic natural gas would require 14,700 to 17,900 tons per day of bituminous coal or 19,600 to 23,800 tons per day of lignite. Water requirement estimates range from 10,000 to 45,000 acre feet per year. The assumed fuel efficiency of such a plant is 56 to 58 percent, somewhat less than the fuel efficiency of a coal liquefaction plant.

UNDERGROUND COAL GASIFICATION As has already been indicated, coal is the nation's most abundant energy resource, with an estimated 6.4 trillion tons within 6,000 feet of the earth's surface, including Alaska's estimated coal resources (Department of Energy, February 1978). Because of the depth and thickness of the beds, however, more than 85 percent of these coal resources are not recoverable by conventional underground or strip mining methods. In order to recover the energy in the deep and thick coal seams of the United States, the Department of Energy, in partnership with private industry, is working to develop new recovery techniques. One of the more promising is underground coal gasification (UCG). This

Cool Water Gasification Plant near Daggett in the Mojave desert, 75 miles northeast of Los Angeles. Named for a nearby stream, not for a process, the Cool Water plant is the nation's first integrated coal gasification power plant. Placed in service in June, 1984, the plant, rated at 120 Megawatts, consumes roughly 1,000 tons of Utah coal per day and produces a medium-BTU gas, using a gasification process that is the property of Texaco Inc. Electric power is fed into the grid of Southern California Edison Company, whose solar-electric generating plant, *Solar One,* is adjacent to this site (see Chapter 6). (Courtesy of American Petroleum Institute and Texaco Inc.)

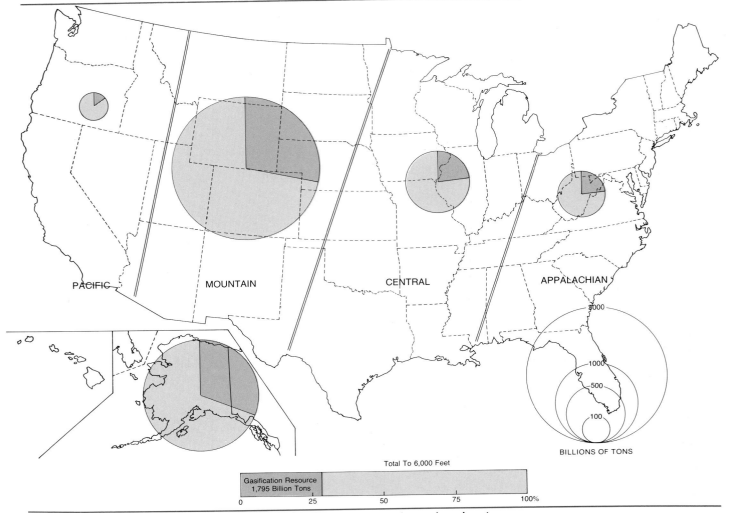

Total To 6,000 Feet

Gasification Resource 1,795 Billion Tons	

0 25 50 75 100%

Fig. 1–35 Estimates of all Identified and Hypothetical coals to depth of 6,000 feet showing the portion of beds suitable for underground gasification.

Source: Department of Energy, *Underground Coal Conversion Program*, Volume III, *Resources*, 1978.

technique involves the injection of air at a rapid rate into holes drilled into the coal bed. The bed is then set afire and the fire reacts with unburned coal and water to produce methane, carbon dioxide, and hydrogen. These gases are removed through a second hole called a production well. This process is called the Linked Vertical Wells (Fig. 1–35). Using this process, a coal bed covering a four square mile area and averaging 30 feet in thickness could supply enough energy to meet the electrical needs of a city of one million people for 20 years (*UA Journal*, 1979).

With such techniques, the Department of Energy believes that *as much as 1.8 trillion tons of unmineable coal* could be utilized by converting it in-place to clean-burning gases (Fig. 1–36 and Table 1–17). Comparing this figure with the 484 billion tons of Demonstrated Reserve Base that can be mined economically using current technology, underground coal gasification could potentially triple the recoverable energy in coal for the 48 contiguous states (Fig. 1–37). Using another comparison, the energy potentially recoverable by UCG is slightly less than that recoverable from all Identified coal to a depth of 3,000 feet assuming a 50 percent recovery ratio.

The products from underground coal gasification process are low-BTU gas suitable for local (on-site) use in

TABLE 1–17
Coal Resource for Underground Gasification in Five Regions of the Country, Showing Total of Identified and Hypothetical Coals to Depth of 6,000 Feet, Coal Suitable for Underground Gasification, and Estimated Recoverable Energy

REGION (see map)	TOTAL OF IDENTIFIED AND HYPOTHETICAL COALS TO 6,000 FEET DEPTH (in Billions of tons)	TONNAGE CONSIDERED RESOURCE FOR UNDERGROUND GASIFICATION[1] (in Billions of tons)	ESTIMATED ENERGY RECOVERABLE[2] (in Quads)
Alaska	2,029	640	6,400
Pacific	103	15	250
Mountain	3,263	920	9,200
Central	620	140	1,820
Appalachian	352	80	1,040
TOTAL	6,367	1,795	18,610

[1] According to Department of Energy, February, 1978 (Vol. III: *Resources*).

[2] On the basis of one-half of the coal energy in-place. For each region half the in-place tonnage was multiplied by an energy value dictated by rank of coal in the UCG resource: for the three western regions a value of 20 million BTUs per ton; for Central and Appalachian regions, a value of 26 million BTUs per ton.

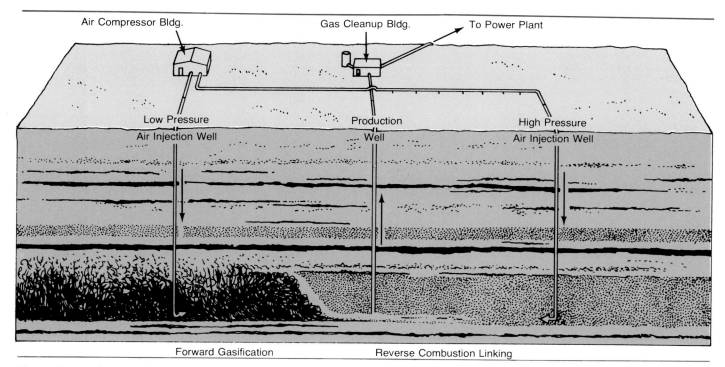

Fig. 1-36 Underground coal gasification by linked vertical walls process.
Source: Deparatment of Energy, *Fossil Fuel Energy Research and Development Program,* March 1978.

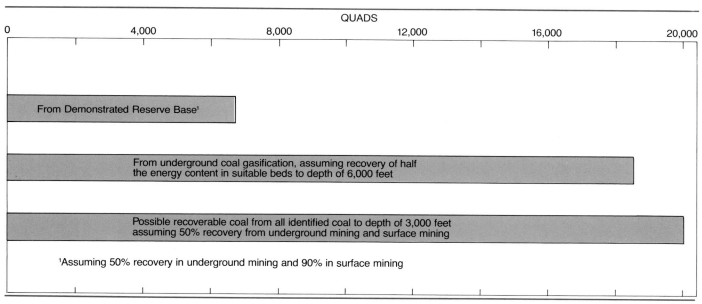

Fig. 1-37 Estimated energy Recoverable from underground coal gasification resource base (to depth of 6,000 feet) compared with energy Recoverable by two approaches to mining of identified coal to depth of 3,000 feet.

electrical power generation; medium-BTU gas suitable for chemical feedstock; and high-BTU gas that can be transported through pipelines directly into existing natural gas transmission systems.

The Department of Energy lists a number of advantages that underground coal gasification has over conventional coal mining techniques. In addition to tripling the recoverable energy from coal, it is expected to minimize health and safety problems associated with conventional coal mining; produce less surface disruption than strip mining; consume less water; generate less air pollution; and reduce both the capital investment and gas cost by at least 25 percent.

Suitable Coals. There is some overlap between coals

TABLE 1-18
Assumptions Used in Calculating the Volume of Coal Required to Support a UCG Product Fired Power Plant

1. POWER PLANT

Capacity (Megawatts)	500
Operating Life (years)	20
Operating Factor (percent)	70
Power Plant Efficiency (Kilojoules/Kilowatt hour)	10,551
Power Plant Efficiency (BTU/Kilowatt hour)	10,000

2. UCG

Overall Conversion Efficiency (Percent)	50
(net Megajoules of gas produced/Megajoules available in coal)	

3. COAL

	Bituminous (45° Dip)	Subbituminous (Horizontal)
Heating Value In-Place (Kilojoules/kg)	30,211	20,125
Heating Value In-Place (BTU/lb.)	13,000	8,860
Density In-Place (kg./cu. meter)	1,411	1,313
Density In-Place (lb./cu. ft.)	87	81

Source: Department of Energy, *Underground Coal Conversion Program,* Volume III, *Resources,* Washington: Government Printing Office, February, 1978.

suitable for underground gasification and those suitable for conventional mining. An important exception is the coal found in steeply dipping beds. These deposits are difficult to mine through conventional techniques but can be gasified successfully. In fact, they appear to have advantages over flat-lying seams. In any case, the total amount of recoverable coal appears to be so large that underground coal gasification and conventional mining should not have to compete for resources. For example, based on assumptions in Table 1-18, an underground coal gasification-fired 500 Megawatt electric power plant would require only 50 million tons of bituminous coal and 70 million tons of subbituminous coal over a 20 year period. The volume of coal required is inversely proportional to the overall UCG efficiency.

In general, western coals of low rank (lignite and subbituminous) are the easiest to gasify underground (Fig. 1-36). Bituminous coal is more difficult to gasify in-place because, when heated, it swells and produces viscous tars which plug natural or induced permeability escape channels (Department of Energy, February 1978). Because of the proximity of Appalachian and midwestern bituminous coals to market, a large effort is being made to overcome these difficulties.

The best prospects for underground coal gasification east of the Rocky Mountains are the bituminous coals in the Illinois Basin and the lignites of the Gulf Coast. The coal seams of the Illinois Basin are consistently thicker than 5 feet and swell less than Appalachian coals when heated. Gulf Coast lignites are the thickest coals east of the Rockies, highly reactive, and shrink upon heating, making them easier to gasify than bituminous.

With their vast resources of thick bituminous, subbituminous, and lignite coal, the Mountain states and those of the Northern Great Plains contain over one-half the coal of potential interest for near-term underground gasification. Markets will, undoubtedly, determine which of these deposits will be exploited earliest (Fig. 1-36).

Washington contains lignite, subbituminous, and bituminous coal of interest for underground conversion to gas. Much of the coal located there is in steeply dipping beds. Although Alaska contains huge coal resources of all ranks which would be suitable for underground gasification, its remoteness from the major markets makes exploitation unlikely.

Department of Energy Test Sites. Technical feasibility of underground coal gasification has been proven by the British and, most notably, by the Russians, who have had commercial underground coal gasification plants in operation for 20 years (Department of Energy, March 1978). Proving economic feasibility is the key to commercialization of underground coal gasification in the United States.

At the present, several major underground coal gasification process options are being developed by the Department of Energy and its subcontractors. At Hanna, Wyoming, tests have already been conducted using the Linked Vertical Wells process to produce a low-BTU gas for utility power generation or utility use from a seam of subbituminous coal 30 feet thick (Fig. 1-35). A second test at Hanna conducted in 1976 produced an average of 8.5 million standard cubic feet (SCF) per day of 172 BTU/SCF gas, the amounts equivalent to electrical needs of a town of 6,000 people. Present plans at Hanna call for a pilot plant to be built in the mid to late 1980s. The plant would have a capacity of 50 to 60 Megawatts, large enough to supply a city of 50 to 60 thousand people with its electrical needs (*UA Journal,* 1979).

A second process has been tested by Lawrence Livermore Laboratory at Hoe Creek, Wyoming. The Hoe Creek project uses underground explosions to fracture the coal and create permeable zones. Up to now, these tests have produced medium-BTU gas suitable for upgrading to methanol, gasoline, ammonia, or pipeline quality gas. Following further tests, the Hoe Creek project may move to a pilot-testing stage.

A third underground coal gasification process is in the very early stages of development at Princetown, West Virginia. At this site the objective is to develop a method of gasification for eastern coals that until now have been difficult to gasify because of their thin seams, low permeability, and high-swelling characteristics.

Based on the work conducted to date, it would appear that underground coal gasification has the potential of being an important energy option for the future. One future coal-based energy scenario puts the potential contribution of underground coal gasification in perspective by suggesting that electrical and synthetic natural gas mine-mouth UCG plants alone could comprise about 15 percent of the total coal usage by the year 2000 and 35 percent by 2050 (Department of Energy, March 1978).

NOTES

1. The above assumes that all sulfur is converted to sulfur dioxide (SO_2); but, while 95 to 100 percent of the sulfur in bituminous coal is converted to that gas, only 72 percent of sulfur in subbituminous coal becomes SO_2, which suggests the requirement in Fig. 1–5 may be modified or suggests that subbituminous coal can be used with less expensive SO_2 removal devices than would be required if all its sulfur were converted to SO_2.
2. Statement is based on Averitt's 1974 estimate of United States coal resources.
3. Amounts are not mapped for California, Idaho, Nebraska, Nevada, and Louisiana because the data were not available. The total tonnage of Identified coal for these states is estimated at 688 million tons.
4. The Bureau of Mines usually bases its estimates for surface mineable coal on an 80 percent recoverability factor.
5. Estimate may be high because of the thick seams of subbituminous coal in the West.

References

Averitt, Paul. "Coal." *U.S. Geological Professional Paper 820,* Washington, DC, Government Printing Office, 1973.

Chemical and Engineering News. August 27, 1979.

Corey, Richard. *Elements of Coal Science and Utilization Technology, and Some Energy Facts and Figures.* Washington, DC: Office of Fossil Energy, 1976.

Derbyshire, Frank. "Coal: Phoenix of the Twentieth Century." *Earth and Mineral Sciences,* 52, 3, Spring 1983. (Pennsylvania State University, University Park, Pennsylvania), pp. 1, 32–34.

Electric Power Research Institute. *Coal Gasification Systems: A Guide to Status, Applications and Economics.* Palo Alto, Cal., June 1983.

National Coal Association. *Bituminous Coal Facts, 1972.* Washington, 1972.

Nielsen, George F. "Coal Mine Development and Expansion Survey—617.3 Million Tons of New Capacity through 1985." *Coal Age,* February 1977, pp. 1–37.

Oil and Gas Journal. "Synthetic Fuel Report." June 29, 1981, pp. 71–87.

UA Journal. "Underground Coal Gasification Expands U.S. Energy Options." 41:3, March 1979.

U.S. Bureau of Mines. "Long-Distance Coal Transport: Unit Trains or Slurry Pipelines." *Information Circular,* 8690, Washington, DC, 1975.

U.S. Bureau of Mines. *Mining Technology Research.* Washington, DC, 1976.

U.S. Bureau of Mines. "The Reserve Base of U.S. Coals by Sulfur Content—The Eastern United States." *Information Circular,* 8680, Washington, DC, 1975.

U.S. Bureau of Mines. "The Reserve Base of U.S. Coals by Sulfur Content—The Western States." *Information Circular,* 8693, Washington, DC, 1975.

U.S. Bureau of Mines. "The Reserve Base of Coal for Underground Mining in the Western United States." *Information Circular,* 8678, Washington, DC, 1975.

U.S. Department of Energy. *Annual Energy Outlook, 1983.* Washington, DC: Government Printing Office, May 1984.

U.S. Department of Energy. *Coal Data: A Reference.* Washington, DC: Government Printing Office, October 1982.

U.S. Department of Energy. *Coal Distribution—January to December, 1983.* Washington, DC: Government Printing Office, April 1984.

U.S. Department of Energy. *Coal Production, 1982.* Washington, DC: Government Printing Office, September 1983.

U.S. Department of Energy. *Demonstrated Reserve Base of Coal in the United States on January 1, 1979.* Washington, DC: Government Printing Office, May 1981.

U.S. Department of Energy. *Fossil Energy Research and Development Program.* Washington, DC: Government Printing Office, March 1978.

U.S. Department of Energy. *1982 International Energy Annual.* Washington, DC: Government Printing Office, September 1983.

U.S. Department of Energy. *Outlook for U.S. Coal.* Washington, DC: Government Printing Office, August 1982.

U.S. Department of Energy. *Projected Costs of Electricity from Nuclear and Coal-fired Power Plants.* Washington, DC: Government Printing Office, August 1982.

U.S. Department of Energy. *Underground Coal Conversion Program,* Vol. III, *Resources.* Washington, DC: Government Printing Office, February 1978.

U.S. Department of the Interior. *Coal Resource Classification System of the U.S. Geological Survey.* Washington, DC: Government Printing Office, 1983.

U.S. Department of the Interior. *Energy Resources on Federally Administered Land.* Washington, DC: Government Printing Office, November 1981.

Wasp, Edward J. "Slurry Pipelines." *Scientific American,* November 1983. pp. 48–54.

World Coal. "A Look at Global Coal Resources," November 1975, pp. 1–11.

World Energy Conference. *World Energy Conference of Energy Resources, 1983.* London; 1983.

2 Crude Oil and Natural Gas

Crude oil, or petroleum, is a complex mixture of hydrocarbons (compounds of hydrogen and carbon) which, under normal conditions, exist as a liquid in the earth's crust. In chemical composition, an "average" crude oil is roughly 83 percent carbon and 12 percent hydrogen, with sulfur, oxygen, and nitrogen accounting for the remainder. No two crude oils ever contain the same mix of molecules. Some are far richer in carbon than others; some contain more sulfur than others. In physical composition, an important factor is the oil's viscosity. Some crude oils are black, thick, and tarry (asphaltic) while others are lighter in color, thinner, and more volatile. The asphaltic crudes are referred to as "heavy"; but, in fact, their specific gravity is lower than that of the less viscous "lighter" crude oils.

In refining, the first step is distillation—the vaporization and condensation which separates the lighter components, such as gaseous hydrocarbons, gasoline, and kerosene, from the heavier components, such as lubricating oils and residual fuel oil. The crude oils that are considered light will yield more gasoline per barrel than the heavy crudes which will yield greater amounts of lubricants (Fig. 2–1). In subsequent processes, some of the heavier and less volatile products, such as fuel oil, may be subjected to *catalytic cracking*, a process that alters their molecular structure and yields some lighter products such as gasoline. Similarly, some of the very light products may be treated by *polymerization* in order to produce gasoline. Thus, the ultimate output of a refinery may be adjusted according to season or the demands of local markets to produce more or less of products such as gasoline and heating oil.

About 6 percent of the output from refineries is in the form of petroleum gases. When combined with natural gas, these products provide the petrochemical industry with raw materials, such as methane, ethane, propane, butane, xylene, benzene, and toluene which are used in the production of fertilizers, antifreeze, plastics, solvents, synthetic rubber, polyester fibers, nylon, and a host of other products. Considering the output from refineries and from the petrochemical industry, roughly 15 percent of all the crude oil consumed in the United States goes into products that are not used for their energy content.

← A gusher at Midway-Lakeview, California, around 1910. Photo courtesy of the American Petroleum Institute and Union Oil of California.

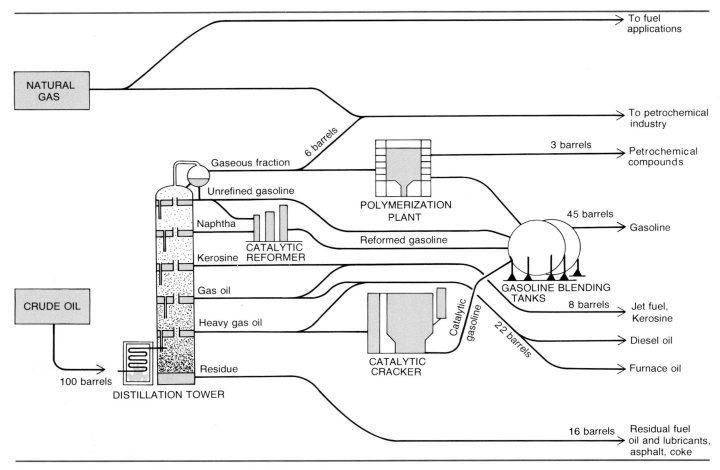

Fig. 2-1 Products from refining of crude oil. Volumes of output are a rough average for all U.S. refineries and assume 100 barrels of oil as input.
Source: Cook, 1976.

THE NATURE OF OIL AND GAS ACCUMULATIONS

The formation of crude oil and natural gas is not fully understood. One important clue to the process is the existence today of dark organic-rich marine shales, that is sedimentary rocks formed by the slow accumulation of clay materials on an ocean bottom (see Fig. 2-2). The organic material, thought to be derived from the remains of microscopic plant and animal life, can be converted to petroleum by high temperatures in the laboratory, so chemists deduce that a similar process took place in the high temperatures and high pressures that occur in a sedimentary rock that is deeply buried.

Squeezed out of the marine shale source rock, the liquid petroleum finds its way into adjacent porous and permeable rocks, such as sandstones or limestones, which allow the liquid to migrate upward as buoyancy and differential pressure dictate. It appears that migration is necessary to produce accumulations because nearby source beds may be absent or inadequate to account for the volumes of oil or gas found in a given reservoir.

The petroleum liquids accumulate in a *reservoir rock*, often the same rock unit through which migration has taken place. Sandstone, and carbonate rocks such as limestone and dolomite, frequently serve as reservoir rocks because they have pore spaces among grains (porosity) and the ability to transmit fluids (permeability).

If the upward migration of crude oil or natural gas were not interrupted somehow, the fluids would escape to the surface and be lost. However, with a barrier to the migration, the oil or gas can accumulate to form a reservoir that will yield substantial amounts when penetrated by the drill.

Some leakages, such as the burning gas at the Eternal Fires of Iraq, or the La Brea tar pits in California, are indications that gas and oil accumulated in large amounts below the surface, and then began to leak out when the reservoir rock was disturbed.

In oil and gas fields of today, some condition always deforms or otherwise interrupts the reservoir rock holding the oil or gas—so it always is apparent that the migrating fluid met a barrier and accumulated there. In general, the arrangement of rocks that interrupts migration is called a

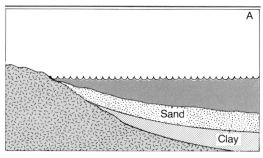

A. Sediments derived from the continent are deposited in ocean water near shore. Included in the clay are remains of microscopic animal and plant life.

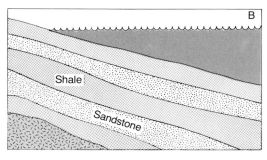

B. Buried by overlying younger sediments, the sand and clay become sandstone and shale which is rich in organic matter.

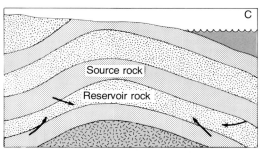

C. Heat and pressure transform the organic material into oil or gas which are squeezed out of source rock into adjacent sandstone, and migrate upward. If a trap is present the oil or gas will accumulate.

Fig. 2-2 Origin of oil and gas accumulations.

trap; and, whatever the kind of trap, there always is an impermeable *cap rock*, such as shale, salt, or dense limestone, overlying the *reservoir rock* and sealing in the oil or gas (Fig. 2-3).

There are two forms of traps: structural and stratigraphic. *Structural traps*, which account for roughly 70 percent of the oil in United States reservoirs, involve deformation or breaking of the reservoir rock, due sometimes to large movements within the crust and sometimes to nothing more than settling and compaction of sediments. The best-known structural traps occur when reservoir rock is warped into a dome or elongated domal form (anticline) as in Figure 2-3A. Traps of this sort are responsible for most of the world's largest oil fields, such as Prudhoe Bay,

in Alaska, and Burgan, in Kuwait. Another structural trap occurs with the faulting or displacement of a rock series as in Figure 2-3B. This often occurs in combination with anticlinal uplifts, as in fields of the Los Angeles Basin. A large faulted block by itself is responsible for the Officina field in Venezuela.

Stratigraphic traps are the result of a change in permeability or a discontinuity of the reservoir rock, rather than structural deformation. This form of trapping accounts for 30 percent of United States crude oil accumulation. A body of sand, such as an ancient sandbar, that gives way to a less permeable rock (Fig. 2-3C) is exemplified in the shoestring sands of eastern Kansas, and in the Pembina field in the Canadian province of Alberta. A buried coral reef, can create a trap if enclosed by less permeable rock (Fig. 2-3D). Reefs are responsible for some important reservoirs, such as the Leduc chain in Alberta, the Sirte Basin fields of Libya, and—in a more widespread form—some large fields in West Texas.

It appears there are five factors necessary for the occur-

Some crude oil is thick, viscous, and difficult to recover from the reservoir rock that holds it. Such oil may be referred to as *heavy crude,* but, in fact, its specific gravity is relatively low. (Courtesy of American Petroleum Institute and Standard Oil Company of California.)

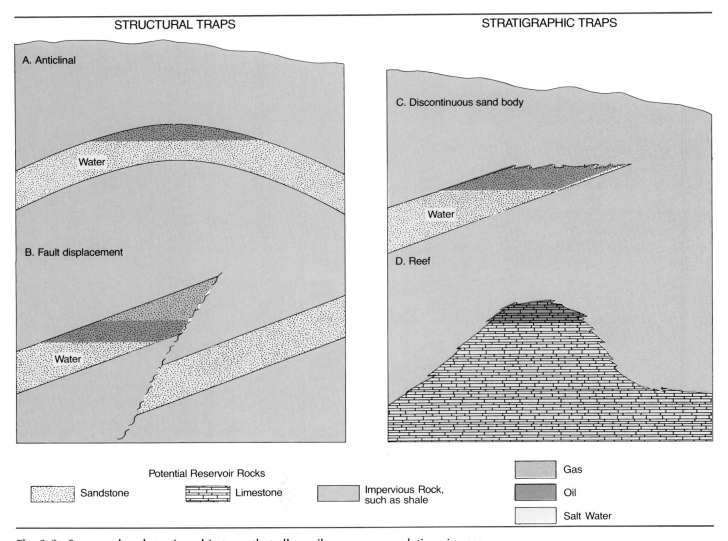

STRUCTURAL TRAPS   STRATIGRAPHIC TRAPS

A. Anticlinal

Water

B. Fault displacement

Water

C. Discontinuous sand body

Water

D. Reef

Potential Reservoir Rocks

Sandstone   Limestone   Impervious Rock,
                                      such as shale

Gas

Oil

Salt Water

Fig. 2-3 Structural and stratigraphic traps that allow oil or gas accumulations in reservoir rocks.

rence of crude oil in quantities sufficiently large to warrant exploitation. The first requirement is the existence of a source rock which is sufficiently rich in organic remains—preferably remains of marine origin and not continental. Deep burial of these organisms must occur to provide the temperatures and pressures needed to change the organic material into liquid hydrocarbons. In addition, porous and permeable rocks must exist for the migration and accumulation of the liquids. A suitable trap must be formed to cause the accumulation. And finally, the timing must be right: If the trap is formed *after* the period of migration, then it will be barren.

Exploration for oil or gas begins by identifying regions that will meet these five requirements. More specific identification of drilling sites is based largely on geologic or geophysical evidence that suggests a trap of some kind is likely in a certain area. Whether the trap exists, and whether it holds hydrocarbons or just salt water, can be determined only by drilling.

An oil or gas field is essentially a porous and permeable rock saturated with oil, gas, or both. The dimensions of the field and the amount of oil or gas in it can be estimated, as shown in Figure 2-4. The portion of the reservoir rock that is saturated with oil is termed the *pay zone*. Its thickness, measured in drill holes, often will vary across the field because of the inclination of the reservoir rock. The bottom of the pay zone is defined by the top of the water-saturated zone. Areal extent of the pay zone will sometimes depend on changes in permeability. More often, the pay zone simply gives way to a reservoir rock that is too low in elevation to be oil-saturated. Drill holes that are too far from the crest of the structure (Fig. 2-4) will encounter just salt water and will indicate the limits of the field. From its estimated thickness and area, the volume of saturated rock can be calculated. From core samples, the percentage of pore space can be measured. From these two numbers, the volume of *oil in-place* or *gas in-place* can be calculated.

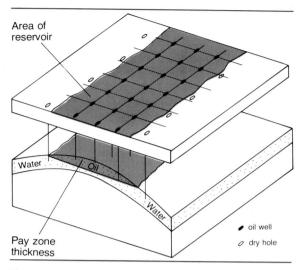

Fig. 2–4 Dimensions of a crude oil accumulation.

Around 85 percent of the natural gas in a reservoir can be easily recovered. Its extremely low viscosity allows it to flow through the reservoir to lower pressures in the borehole. Crude oil, and especially the more viscous "heavy" oil, adheres to the many surfaces in the reservoir rock and is not so easily recovered. In the United States an average of only 35 percent of the crude oil in-place has been recovered. More can be recovered by using special techniques. The amounts of oil that may be recovered from existing and future oil fields depends therefore upon whether prices justify the application of these special techniques. This factor is taken into account in the following review of resource terminology.

RESOURCE TERMS APPLICABLE TO CRUDE OIL AND NATURAL GAS

A number of special terms are used by the oil and gas industry and by the federal government to refer to: amounts that have been produced; resource amounts remaining; and techniques for recovery of crude oil. The terms that deal with the resource amounts appear in Figure 2–5, which is a variation on the general scheme for understanding amounts of mineral resources (see Introduction). The resource amounts that are more certain are placed toward the left of that diagram, and the amounts more easily attained are placed toward the top.

ORIGINAL OIL (OR GAS) IN-PLACE This is the amount of oil in known reservoirs prior to production. The term may be applied to reservoirs that currently are productive, to those that are known but not yet exploited, or to those that have been depleted. The estimation of

original oil in-place is made from calculations of the volume of saturated reservoir rock and the proportion of it that is pore space.

ALL REMAINING RECOVERABLE In this atlas, this term is used for amounts of oil or gas that are recoverable, given present economic limitations, from reservoirs that are either Identified or Undiscovered. Varying degrees of certainty are expressed by the categories below.

PROVED, OR MEASURED, RESERVES These are amounts shown by geologic and engineering data to be *recoverable from Identified reservoirs under existing economic and operating conditions*. Because both the reservoir's existence and the economic feasibility of recovery are certain, these amounts occupy the extreme upper-left corner of the resource diagram (Fig. 2–5).

INDICATED (OR INDICATED ADDITIONAL) RESERVES Applicable to crude oil, not natural gas, these are amounts (beyond the Proved reserves) that may be recovered through application of modest improved recovery techniques such as injection of water or gas to maintain reservoir pressures. For some fields these secondary

A new tri-cone rotary drilling bit, ready to be lowered into the drillhole. Above the bit will be two or three lengths of the heavy pipe (drill collars) visible in the photo. Attached to that will be 30-foot lengths of 4-inch drill pipe—all joined to reach depths of three or four miles. This sinuous drillstring is rotated by a diesel-powered device on the floor of the drilling rig. (Courtesy of American Petroleum Institute and British Petroleum Company.)

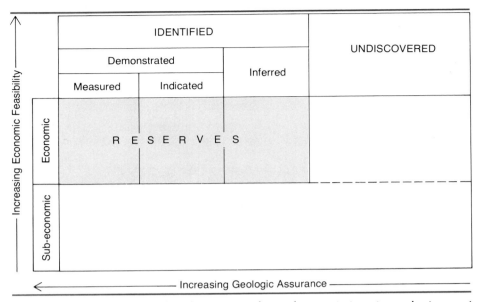

Fig. 2–5 Categories of crude oil resources depend on variations in geologic certainty and economic feasibility.

recovery techniques are likely to be successful and economical, but since they have not yet been applied, the anticipated additional crude is not considered Proved. These amounts tend to be *relatively* large in old productive regions or states whose proved reserves have been severely depleted, leaving substantial amounts of crude in the reservoirs.

DEMONSTRATED RESERVES This term refers to the sum of Proved (Measured) and Indicated reserves. It is used largely by the U.S. Geological Survey.

INFERRED RESERVES Used by the U.S. Geological Survey, this category refers to amounts of oil or gas in identified reservoirs near the edges of fields that have not yet been fully defined by drill holes. It is known that young fields tend to be underestimated by Proved reserves which consider only the amounts in reservoir rock that has been drilled. Inferred amounts thus anticipate what is reported annually by the Department of Energy as *extensions* and *revisions* (though some revisions are downward). Since the amounts are expected in identified reservoirs but are not yet measured, the inferred category occupies a position on the resources chart (Fig. 2–5) that is intermediate between Demonstrated and Underdiscovered.

UNDISCOVERED RECOVERABLE AMOUNTS Applicable to both oil and gas, these are amounts thought to be recoverable, usually under today's economic conditions, from reservoirs not yet discovered. The existence of additional reservoir rocks and traps is postulated on the knowledge that sedimentary rock sequences similar to those already productive do exist and are either unexplored or inadequately explored. Frontier areas untouched

by the drill are prime targets for such speculation. Also considered are remote or deep parts of sedimentary regions whose nearby and shallower beds have been explored. Amounts may be estimated on the basis of historic recovery rates, and are represented by position X in Figure 2–5.

CUMULATIVE PRODUCTION As the name suggests, this refers to amounts of oil or gas (or natural gas liquids) that have been produced to date in a given field, state, or region for which statistics are gathered. It is an indication of the richness of an area—or rather of its past richness—and is, when combined with information on amounts remaining, a valuable guide to the status of an area.

ULTIMATE RECOVERY As used by the American Petroleum Institute, for either oil or gas, this is simply the sum of Cumulative Production to date and current Proved reserves. It expresses, therefore, "ultimate recovery" *only from identified reservoirs* and only as determined by current technology and economics. It should not be construed as the ultimate ever to be recovered for an area or region, since it does not consider either undiscovered reservoirs or additional amounts that may be obtained from identified reservoirs by enhanced recovery techniques.

WORLD RESOURCES OF CRUDE OIL

Since the time of early civilizations crude oil from natural seepages has been used for fuels, mortar, and medicines. Some oil was distilled from shales in Scotland in the seventeenth century, but it was not until 1857 in Rumania that a successful hole was drilled expressly to seek petroleum underground. Two years later, one of the first American oil wells was completed at Titusville in Northwestern Pennsylvania. These discoveries of liquid petroleum, a versatile and transportable fuel and raw material, stored naturally at depth, unaltered by surface oxidation, and under pressures that eased its recovery through a drillhole, was a revelation that began a remarkable exploitation of world resources (Fig. 2–6).

The growth in production for different regions illuminates the present status of the United States in relation to other countries. For the first 100 years of the oil industry's existence, North America (dominated by the United States) was the world's leading producer. In fact, the United States accounted for 60 to 70 percent of world production through 1950, being surpassed only in the postwar years as the fields in the Middle East, Eastern Europe, and the Soviet Union expanded rapidly.

The differing rates for development among countries have led to extremely interesting differences in amounts of oil produced to date, that is, Cumulative Production for world regions, as shown on the *left side* of Figure 2–7.

The Commodities: Crude Oil, Natural Gas, and Natural Gas Liquids

Crude Oil A complex mixture of hydrocarbons, varying in composition from one field to another, existing as a liquid underground and remaining liquid at the surface.

Conventional Crude Oil This term distinguishes oil extracted as a liquid (through a borehole) from the crude that can be obtained by processing oil shales or tar sands. Both of these alternatives are considered here to be unconventional crude or synthetic crude from mineral sources. (They are dealt with in Chapter 3.)

Natural Gas A mixture of hydrocarbons, dominantly methane, which may exist in the reservoir in the gaseous phase or in solution with oil, but at surface pressure is in the gaseous state.

Natural Gas Liquids These consist of propane and similar heavier hydrocarbons which separate from natural gas through condensation in the reservoir or are separated from gas at the surface through condensation, absorption, or related methods in separators at or near the well-head. Since these liquids are produced only when the host gas is produced, their amounts produced and remaining are directly tied to the production and the reserves of natural gas. The amounts are expressed in barrels, and are often added to crude oil amounts to arrive at Total Petroleum Liquids.

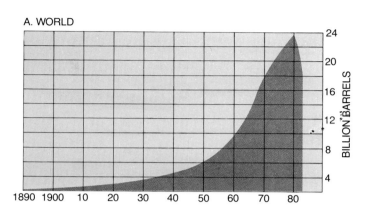

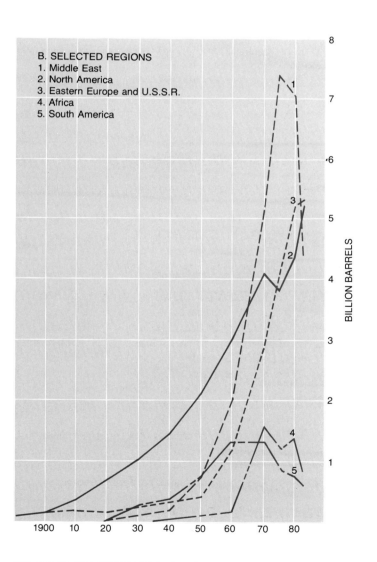

Fig. 2–6 Crude oil production for world and selected regions 1890–1983.
Sources: 1890–1970 from Tiratsoo, 1973; later years from *Oil and Gas Journal*, March issues.

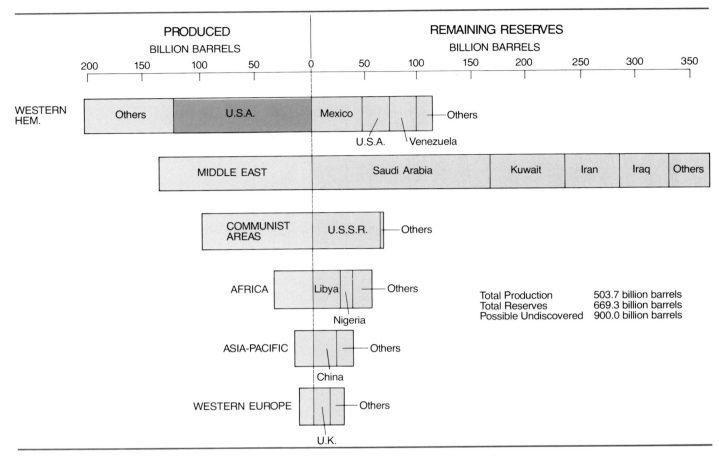

Fig. 2–7 World regions. Crude oil produced, and proved reserves remaining as of Dec. 31, 1983.
Sources: Cumulative production from National Academy of Sciences, 1975. Later production and current reserves from *Oil and Gas Journal*, March issues.

Clearly the Western Hemisphere has produced more than any other region; and the United States alone has produced almost as much as the entire Middle East.

On the right side of the graph and in Table 2–1 are the latest estimates of Proved Reserves remaining in each region. On this basis, the Middle East holds the majority of world oil: in fact, Saudi Arabia has larger reserves than the entire Western Hemisphere. Not shown in Figure 2–7 are estimates of *undiscovered* crude oil, which in both Western Hemisphere and Communist areas may be twice as large as undiscovered amounts in the Middle East. Relatively recent are North Sea reserves in the United Kingdom and Norway, and the large reserves in Mexico, due to development of the Chiapas–Tobasco area near the Guatemala Border.

The amounts of crude oil in the world may suggest the supply is quite comfortable: Only about 500 billion barrels have been produced, while 670 remain as Proved reserves and another 900 as undiscovered reserves. But, at the current world production rate of near 20 billion barrels annually, the Proved reserves would last only 33 years, and the total, including undiscovered, around 75 years.

Despite the certainty of an ultimate shortage, there are periods when reduced demand causes an oversupply. In 1978 and again in 1984, the combined effects of recession and conservation led to a market glut and a reduction in world oil prices.

A most significant feature of the world oil scene is the discrepancy between the patterns of oil supply and oil demand. As Figure 2–8 shows, the greatest share of demand is concentrated in North America and Western Europe, while the supply is largely in the Middle East, Eastern bloc nations, and Africa. This unbalance has led to an uneasy interdependence between Western industrial importers and less-developed exporters in the Middle East, Africa, and Latin America.

UNITED STATES CRUDE OIL RESOURCES

Conventional crude oil resources in the United States are reviewed in two parts. First, a broad view takes into account both identified and undiscovered reservoirs and makes use of past production and resource amounts

TABLE 2–1
World Crude Oil, Proved Reserves as of December 31, 1983.
Amounts Tabulated are Billions of Barrels (barrels $\times 10^9$)

North America		Middle East	
Canada	6.7	Bahrain	0.2
Mexico	48.0	Iran	51.0
United States	27.3	Iraq	43.0
Total	82.0	Kuwait	66.7
Central and South America		Oman	2.8
Argentina	2.4	Qatar	3.3
Bolivia	0.2	Saudi Arabia	168.8
Brazil	1.8	Syria	1.5
Chile	0.7	United Arab emirates	32.3
Colombia	0.6	Other	*
Ecuador	1.7	Total	369.7
Peru	0.8	Africa	
Trinidad and Tobago	0.6	Algeria	9.2
Venezuela	24.9	Angola	1.7
Other	*	Cameroon	0.5
Total	33.7	Congo	0.4
Western Europe		Egypt	3.5
Denmark	0.3	Gabon	0.5
Germany, West	0.3	Libya	21.3
Italy	0.8	Nigeria	16.6
Netherlands	0.3	Tunisia	1.8
Norway	7.7	Other	1.5
United Kingdom	13.2	Total	56.9
Other	0.8	Far East and Oceania	
Total	23.4	Australia	1.6
Eastern Europe and U.S.S.R.		Bangladesh	0
U.S.S.R.	63.0	Brunei	1.4
Other	2.5	China	19.1
Total	65.5	India	3.5
		Indonesia	9.1
		Malaysia	3.0
		New Zealand	0.2
		Pakistan	0.1
		Thailand	*
		Other	0.1
		Total	38.1
		World Total	669.3

*less than 0.05 billion barrels
Source: *Oil and Gas Journal*, Dec. 26, 1983.

assigned to 15 regions of the country. A later section focuses upon identified reservoirs only, and uses current production and reserve data for states.

Resources in Identified and Undiscovered Reservoirs

This section is based almost entirely on one report by the U.S. Geological Survey. It provides a unique overview of oil and gas amounts of varying degrees of certainty for onshore and offshore regions of the lower 48 states and Alaska (U.S.G.S. *Circular 860*, 1981). This report is a sequel to *Circular 725* which made comparable resource assessments for the year 1974. The latest report contains estimates dated to the end of 1979. Although the amounts of cumulative production and proved reserves are now six years old, the estimates of inferred and undiscovered amounts are invaluable because, when combined with past production and proved reserves, they convey a comprehensive picture of crude oil resources in the nation and in 15 geologic regions.

Figure 2-9 shows the 15 regions, ranked according to crude oil produced, that is, cumulative production through the year 1979. On the right side of the graph are *all amounts* of crude oil thought to remain, in each region, in both identified and undiscovered reservoirs.

PROVED CRUDE OIL RESERVES AS OF DEC. 16, 1983

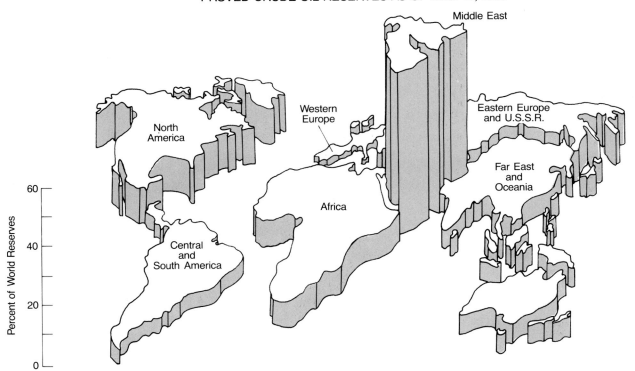

APPARENT CONSUMPTION OF INTERNATIONAL PETROLEUM SUPPLY, 1981

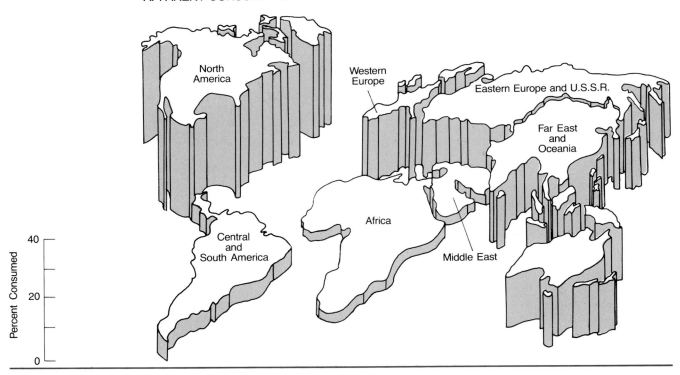

Fig. 2-8 Regional differences in shares of crude oil reserves and petroleum consumption.

Sources: Proved reserves as of Dec. 21, 1983 from *Oil and Gas Journal*, Dec. 26, 1983. Petroleum consumption as of 1981 from DOE, *Annual Energy Review*, 1983. Drawn by Yvonne Keck Holman, Alternative Productions, Paoli, Pennsylvania.

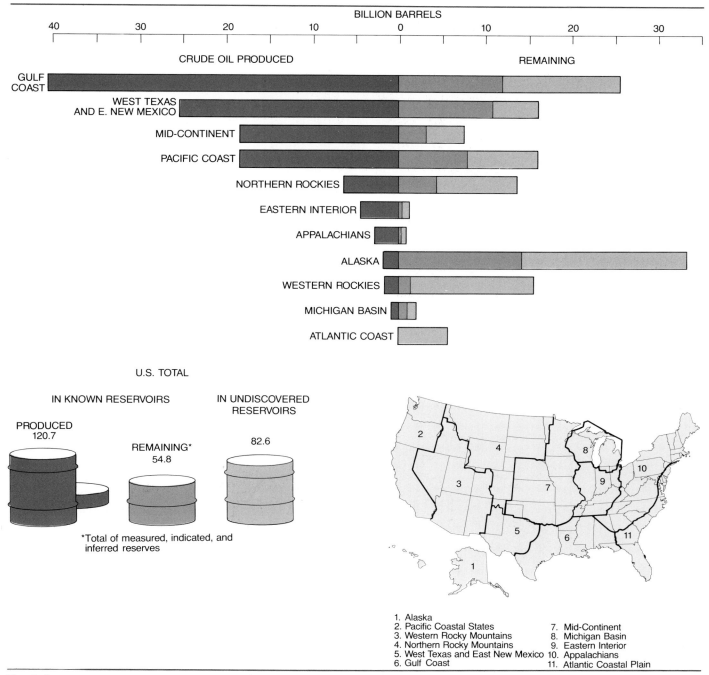

Fig. 2–9 Crude oil produced and remaining in 15 regions of the United States as of December 31, 1979.
Source: U.S.G.S. *Circular 860*, 1981.

Four regions account for around 80 percent of the nation's oil production to 1979: Gulf Coast, West Texas and New Mexico, Pacific Coast, and Mid-Continent. If the estimates of remaining oil are accurate, then most regions are more than half spent. The exceptions are Alaska, Western Rockies, and the Atlantic Coast where substantial amounts await discovery and little or no production has taken place. A large portion of the oil thought to remain in identified reservoirs is the *inferred* amounts

(see Table 2–2). These amounts anticipate the growth of young fields through development drilling. There are no such resources postulated for the Atlantic offshore area, because no fields exist there. But, in more mature areas such as Alaska, West Texas, and the Gulf Coast, inferred amounts are roughly equal to proved reserves.

To put these crude oil amounts in perspective, realize that the U.S. annual production from all fields is around 3 billion barrels per year. On that basis, all the oil in Alas-

TABLE 2-2
Summary of Crude Oil Produced and Remaining in the United States as of
December 31, 1979, Showing 15 Regions (billions of barrels)

| | | | CRUDE OIL REMAINING | | |
| | | | In Known Reservoirs | | |
REGION		CUMULATIVE PRODUCTION TO DEC. 31, 1979	Measured & Indicated	Inferred	In Undiscovered Reservoirs (mean estimate)
1	Alaska	1.2	8.7	5.0	6.9
1A	Alaska Offshore	0.7	0.2	0.1	12.2
2	Pacific	16.5	4.8	1.2	4.4
2A	Pacific Offshore	1.9	1.2	0.5	3.8
3	Western Rockies	1.7	0.3	1.0	14.2
4	Northern Rockies	6.9	1.3	2.9	9.4
5	West Texas & E. N. Mexico	25.2	6.7	4.0	5.4
6	Gulf Coast	34.9	4.0	5.3	7.1
6A	Gulf Coast Offshore	5.6	1.7	1.0	6.5
7	Mid-continent	18.2	1.7	1.4	4.4
8	Michigan Basin	0.8	0.2	0.8	1.1
9	Eastern Interior	4.3	0.2	0.1	0.9
10	Appalachians	2.8	0.2	0.1	0.6
11	Atlantic Coast	0.1	—	0.1	0.3
11A	Atlantic Offshore	—	—	—	5.4
	U.S. Total	120.7	31.4	23.4	82.6

Source: U.S.G.S. *Circular 860*, 1981.

ka's identified reservoirs would supply the nation's production for a little over 4 years; and the nation's total oil in known reservoirs would last for 18 years.

UNDISCOVERED RECOVERABLE AMOUNTS, AND FRONTIER AREAS

As suggested earlier, there are amounts thought to be recoverable, according to present-day economics, from reservoir rocks not yet discovered. The basis for all such speculation is analogy: Where sedimentary sequences exist and are similar to known productive rocks, additional fields may be expected. In making estimates, basic information, such as the total *volume* of unexplored sedimentary rocks, is augmented by clues to the presence of source rocks and the existence of structures or discontinuities that might lead to traps.

For undiscovered amounts reported here, volumes of sedimentary rock were estimated from geophysical and drilling data that indicated depth to the crystalline basement (see Introduction). Excluded are any rocks thought to be severely deformed or metamorphosed. Also excluded as inaccessible are rocks deeper than 30,000 feet, and any offshore areas under ocean water whose depth exceeds 2,500 meters (8,200 feet).

For each of the geologic provinces making up a region, committees of government geologists agreed on a *low* estimate, consistent with a 95 percent probability that a certain amount of oil or gas exists, a *high* estimate, consistent with only a 5 percent probability, and a third estimate considered the most probable amount. The three were added, and divided by three to yield the *statistical mean for the province.*

Table 2-3 lists the low, high, and statistical mean for each region, while Figure 2-10 maps the mean estimates. The western Rockies region, according to this estimate, holds the largest undiscovered amount, with almost as large an amount in offshore Alaska. The generous assessment of the western Rockies region depends on the Western Overthrust Belt in Wyoming, Utah, and Idaho which is thought to be one of the most promising regions for exploration (see later section on exploration).

The mean total of undiscovered crude, 82.6 billion barrels, is practically identical to the 82 billion estimated

TABLE 2-3
Estimated Crude Oil Recoverable from Undiscovered Reservoirs as of 1980

| REGION | 1981 ESTIMATE (BILLIONS OF BARRELS) | | |
	Low (1)	High (2)	Mean
1	2.5	14.6	6.9
1A	4.6	24.2	12.2
2	2.1	7.9	4.4
2A	1.7	7.9	4.4
3	6.9	25.9	14.2
4	6.0	14.0	9.4
5	2.7	9.4	5.4
6	3.6	12.6	7.1
6A	3.1	11.1	6.5
7	2.3	7.7	4.4
8	0.3	2.7	1.1
9	0.3	1.9	0.9
10	0.1	1.5	0.6
11	0.1	0.8	0.3
11A	1.1	12.9	5.4
Total Onshore	41.7	71.0	54.6
Total Offshore	16.9	43.5	28.0
U.S. Total	64.3	105.1	82.6

(1) Low estimate has probability of 95 percent.
(2) High estimate has probability of 5 percent.
Source: U.S.G.S. Circular 860, 1981.

An exploratory well probes the overthrust belt in Arizona. (Courtesy of American Petroleum Institute and Phillips Petroleum Company.)

in the previous U.S. Geological Survey report (*Circular 725,* 1974). This latest estimate, though, considers offshore areas that extend beyond the continental shelf to include the continental slope, whereas the previous report excluded the continental slope, and bounded offshore areas by the 200-meter depth line. Figure 2-11A shows the mean total is clearly in line with estimates made by other authorities before and after the 1975 U.S.G.S. report (Gillette, 1974; Habbert, 1974; Nehring, 1981; Theobald et al., 1972).

The Survey deals with the uncertainty of these crude oil amounts by providing a probability distribution for each of the 15 regions. Here we present probabilities for only the national total (Fig. 2-11B). From the curve it is possible to derive the probability for any selected amount between the high and the low extremes. For example, the chance of their being over 100 billion barrels is roughly 10 percent.

SUMMARY OF AMOUNTS REMAINING, AS OF 1980 All conventional crude oil known and thought to remain as of the end of 1979 is shown in Figure 2-12 and is contrasted with the total amount produced to that time. The amount produced, 120.7 billion barrels, is nearly as large as the total amount remaining, 137.4 billion, which

is a 45-year supply at the current production rate of 3 billion per year. When remaining amounts are divided by certainty, it is clear that by far the largest part of oil thought to remain is either inferred, or undiscovered. The division by location shows how the state of Alaska holds a disproportionate share of all remaining crude. In both the lower 48 and Alaska, onshore amounts account for roughly 75 to 80 percent of the totals. Just as locations are masked in the upper bar, the degree of certainty is masked in the lower bar, so any observations about the richness of various parts of the nation should be made with caution.

Crude Oil in Identified Reservoirs

A state-by-state dissection of crude oil resources is made possible by *annual* data on reserves and production published now by the U.S. Department of Energy and informally referred to here as annual reserve reports. Until 1979, this information was provided by an annual published jointly by the American Petroleum Institute (API),

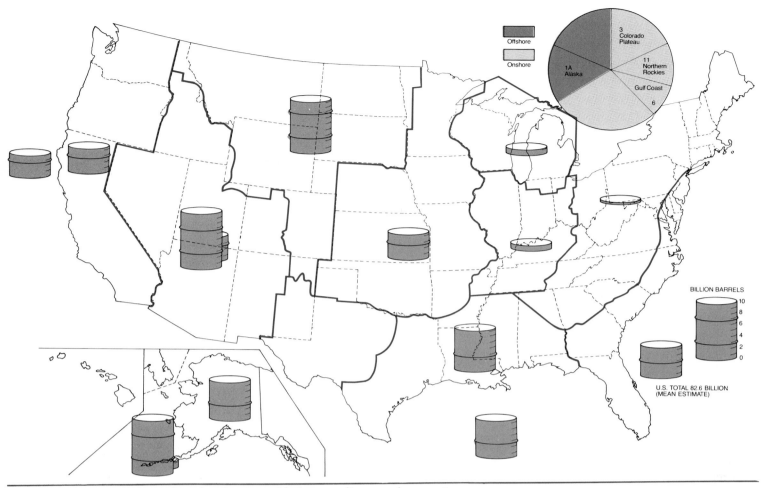

Fig. 2–10 Estimated crude oil recoverable from undiscovered reservoirs.
Source: U.S.G.S. *Circular 860,* 1981.

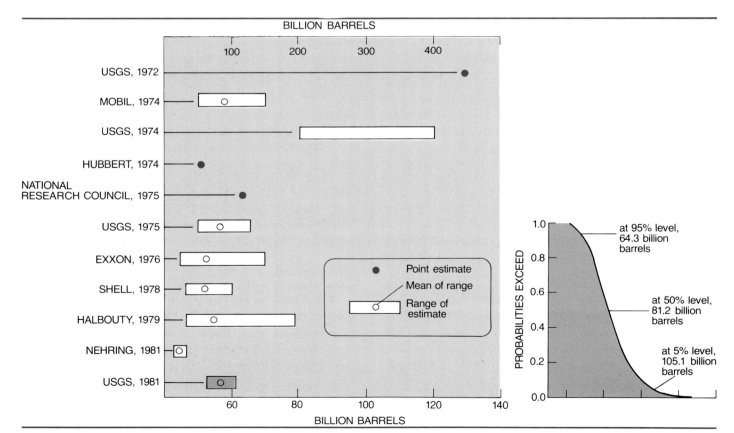

Fig. 2–11 The uncertainty of undiscovered crude oil. A. Various estimates of undis-
covered crude since 1971. B. Probability distribution for the amounts estimated by the
U.S. Geological Survey in 1981.
Source: U.S.G.S. *Circular 860,* 1981.

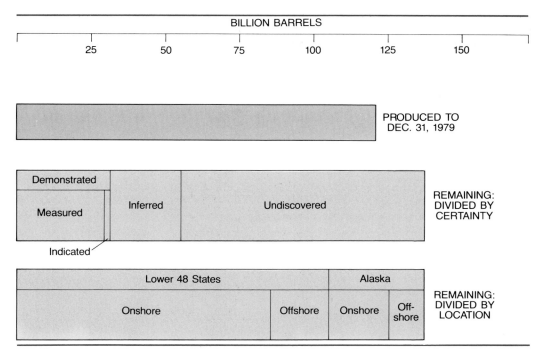

Fig. 2-12 Summary of crude oil produced and remaining as of December 21, 1979, showing certainty and location.

the American Gas Association, and the Canadian Petroleum Association. Both of these series are very thorough with regard to reserves and production, but lack estimates of inferred or undiscovered amounts, making the foregoing section and this one complementary. This section begins with the character of the oil accumulations in each state, then proceeds to the amounts produced and remaining, and ends with a discussion of special recovery techniques that could lead to expanded reserve estimates.

ESTIMATED ORIGINAL OIL IN-PLACE This measure is the summation of all crude oil that has been discovered in each state: Much of it has been produced now from old fields, but some is in fields just recently discovered. The most recent of the API annual reports includes this measure, but the Department of Energy reserve reports do not. The measure does offer a useful view of crude oil in various states, because it includes oil that was found and produced years ago and not apparent in present day reserves. Figures 2-13 and 2-14 show, for instance, that very large amounts have been discovered, through time, in California—over twice the amount that has been found in Louisiana.

Circles are divided in the first two maps to show the nature of the reservoir rocks and traps in which the oil occurs. While the overall national proportion of sandstone as the reservoir rock (71 percent) is reflected in many states, there are some where either sandstones or carbonates are absolutely dominant (Fig. 2-13). California's fields, through the San Joaquin and Los Angeles Basins, are virtually all in sandstones, as is the Prudhoe Bay field

in Alaska. The many fields of Texas, however, show a near-even split betwen sandstones and carbonates. Carbonates appear to dominate only in states whose total oil in-place is relatively small. The national average of 71 percent shows an importance of sandstones in United States reservoirs that is not found in all nations; in fields in Canada and in the Middle East for instance, carbonates play a much larger role.

The importance of structural rather than stratigraphic trapping mechanisms in past and present fields is apparent in most states (Fig. 2-14) and is striking in Alaska, where one field of the structural type accounts for most of the oil in-place. The states with smaller amounts of oil most strongly demonstrate the opposite character, for example, West Virginia, New York, Pennsylvania, and Indiana.

ESTIMATED ULTIMATE RECOVERY Estimated Ultimate Recovery is the total of Cumulative Production and Proved reserves, or the extent to which oil in-place in identified reservoirs has been, and will be, recovered. It reflects, therefore, the all-important factor of recovery rates. In 1978, the national average of 32 percent recovery was exceeded in states such as Louisiana, which have either exceptionally permeable reservoir rocks, crude that is "light" (less viscous), or a combination of the two. Conversely, relatively "heavy" crude leads to lower recovery rates in Ohio, Pennsylvania, and New York (see Fig. 2-15). State averages, of course, do not reflect the much greater extremes of recovery which occur among actual reservoirs. The fact that 60 to 75 percent of the original oil in-place *remains in-place* in many reservoirs, means that large

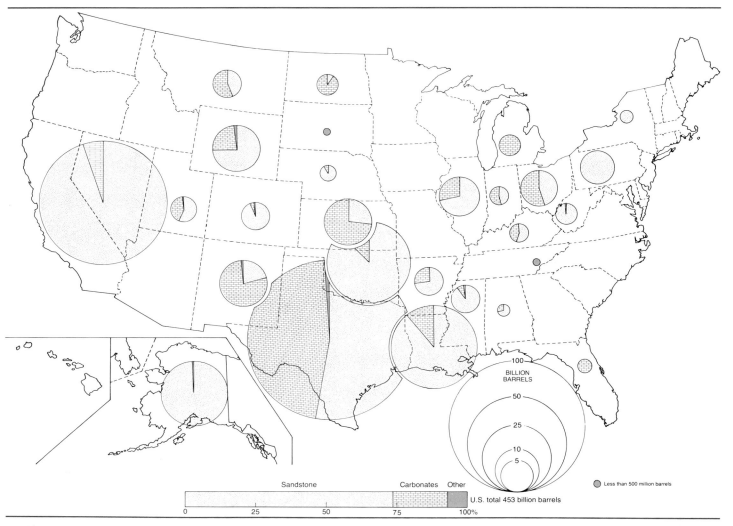

Fig. 2–13 Original oil in place, by state, showing roles of three classes of reservior rock.
Source: American Petroleum Institute, 1980.

volumes might be recovered now from old fields, and also suggests that present and future fields might yield far more than 35 percent of their oil in-place if special techniques are employed (The possible amounts are discussed in a later section on higher recovery rates).

When the two components of *ultimate recovery*, that is *cumulative production* and *proved reserves*, are studied separately and compared, they throw new light on the history of a state and its present status (ignoring inferred and undiscovered amounts). Table 2–4 lists, for each state, cumulative production and proved reserves updated to the end of 1983. The two measures are mapped in Figures 2–16 and 2–17. In most states, the past production is far greater than proved reserves, which is not surprising, since proved reserves are only the currently economic amounts of oil and will be enlarged in most states by some future discoveries. Alaska is unusual, because its development is so recent that reserves still exceed cumulative production.

Cumulative production (Fig. 2–16) is very large in Texas. In fact, the amount produced through time in that state is comparable to the amount produced to date by Saudi Arabia. California, by virtue of its early and productive fields, has produced 18.45 billion barrels, an amount comparable to the cumulative production of Kuwait. Louisiana and Oklahoma rank next among the states; and they are followed by a longer list of states whose cumulative production is in the 1 to 5 billion barrel range.

PROVED RESERVES, CURRENT AND THROUGH TIME Figure 2–17 maps the current proved reserves, which are concentrated largely in the same states as cumulative production. Exceptions are California and Oklahoma, which are less conspicuous in reserves than in past production, and Alaska, as noted previously. While proved reserves are only a part of the resource picture, it is useful nevertheless to realize that these amounts, the most certain, will last only 9 years at the current national rate of production.

Altogether, the 27.7 billion barrels of reserves are located in thousands of fields scattered throughout the country. But a surprising proportion of the oil is in a few large fields (Fig. 2-18). As of the end of 1979, around 30 fields, with reserves exceeding 0.1 billion barrels accounted for over 50 percent of the national total. Moreover, three giant fields with reserves exceeding 1 billion barrels—Prudhoe Bay in Alaska and the Yates and East Texas fields in Texas—hold over 35 percent.

The history of proved reserves shows the United States in a circumstance that had never occurred until the late 1960s: Its reserves are not steadily increasing. Figure 2-19, plotting the long-term history, shows reserves increasing throughout the twentieth Century until a plateau was reached in the 1960s. Then, a decline—unprecedented in U.S. history—began in 1968. It was arrested by the giant

Prudhoe Bay, Alaska, discovery in 1970. Since then, reserves have continued to decline and now have levelled off around 27 billion barrels, slightly below the 1968 amount.

In Figure 2-20, which focuses on the postwar period, annual changes in reserves are seen as the result of two factors—*new oil added* during the year, and *oil produced* during the year. In any year with production less than new oil added, reserves will increase. If production exceeds new oil added, reserves will fall that year. In recent years production has been around 3 billion barrels per year, but new oil added has averaged closer to 2 billion barrels, therefore reserves have decreased.

INDICATED RESERVES These additional reserves are amounts of crude oil, relatively small for the nation as a whole, that can be recovered economically from certain fields by applying *secondary* recovery techniques, such as

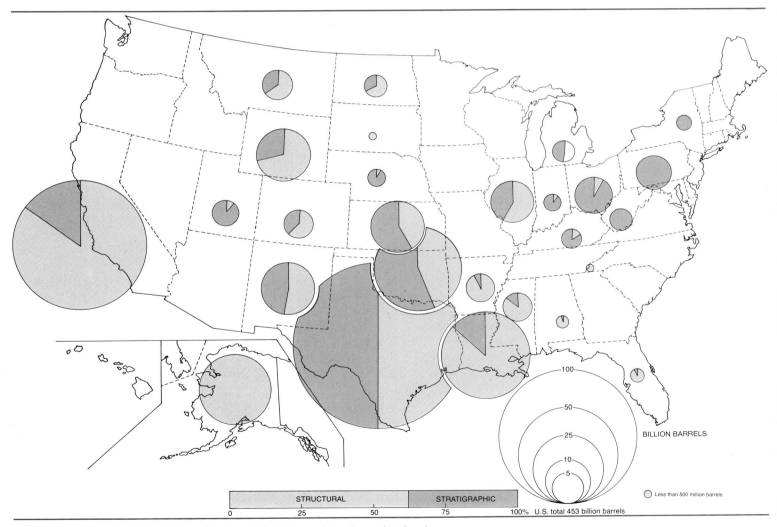

Fig. 2-14 Original oil in-place, by state, showing roles of two kinds of trap.
Source: American Petroleum Institute, 1980.

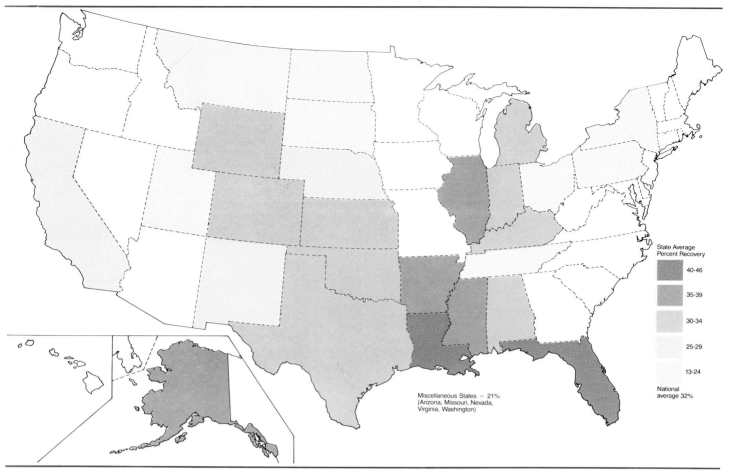

Fig. 2–15 Historic recovery rates from known reservoirs: Estimated ultimate recovery as proportion of original oil In-place.

Source: American Petroleum Institute, 1980.

State Average Percent Recovery

40-46

35-39

30-34

25-29

13-24

National average 32%

Miscellaneous States = 21% (Arizona, Missouri, Nevada, Virginia, Washington)

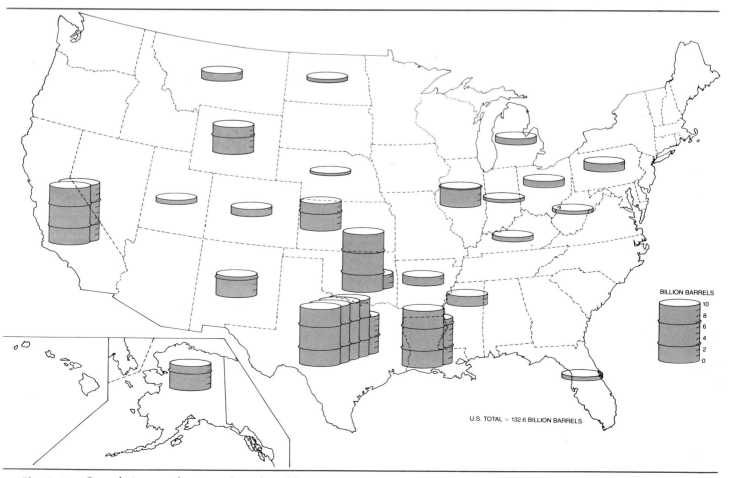

BILLION BARRELS

10
8
6
4
2
0

U.S. TOTAL = 132.6 BILLION BARRELS

Fig. 2–16 Cumulative production of crude oil by state, to Dec. 31, 1983.

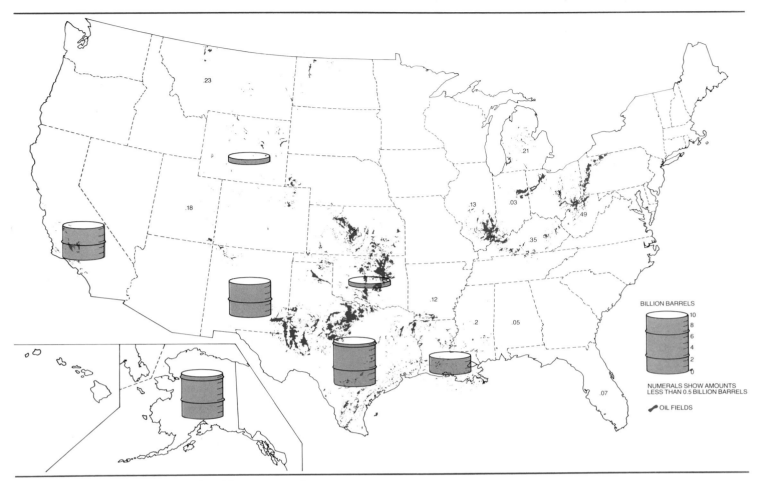

Fig. 2–17 Proved reserves of crude oil by state, as of Dec. 31, 1983.
Source: Department of Energy, *Annual Reserves Report, 1983*, 1984.

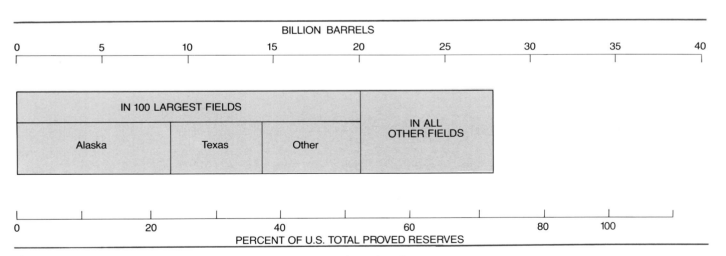

Fig. 2–18 Role of 100 largest fields in proved reserves of crude oil in 1979.

injection of water to maintain pressure in a reservoir so the oil will flow into the borehole more readily. These amounts, listed in the last column of Table 2–4, are too small to show by barrels as scaled in Figures 2–16 and 2–17, so the amounts are mapped in Figure 2–21 as proportions of the total of *indicated* and *proved* reserves (a quantity called *demonstrated* reserves by the U.S. Geological Survey). The proportions tend to be large in states

TABLE 2-4
Crude Oil from Known Reservoirs: Amounts Produced through
Time, and Amounts Remaining (Thousands of barrels)

STATE	CUMULATIVE PRODUCTION THROUGH DEC. 31, 1983	PRODUCTION DURING 1983	PROVED RESERVES AS OF DEC. 31, 1983	INDICATED ADDITIONAL RESERVES IN 1982
Alabama	211,328	7,000	51,000	— (1)
Alaska	1,871,937	556,000	7,307,000	—
Arkansas	1,871,937	18,000	120,000	4,000
California	18,450,114	400,000	5,348,000	615,804
Colorado	1,262,409	30,000	186,000	100,460
Florida	333,839	21,000	78,000	1,000
Illinois	3,121,158	19,000	135,000	882
Indiana	469,327	5,000	34,000	1,492
Kansas	4,843,432	62,000	344,000	12,859
Kentucky	651,867	6,000	35,000	13,010
Louisiana	16,266,253	417,000	2,707,000	48,993
Michigan	803,616	31,000	209,000	34,197
Mississippi	1,671,604	29,000	205,000	84,535
Montana	1,017,446	28,000	234,000	6,235
Nebraska	376,991	5,000	44,000	—
New Mexico	3,470,557	70,000	576,000	76,135
New York	229,947	—	—	—
North Dakota	554,205	51,000	258,000	8,083
Ohio	856,301	18,000	130,000	—
Oklahoma	11,728,045	140,000	931,000	25,389
Pennsylvania	1,297,755	3,000	41,000	—
South Dakota	7,127	—	—	—
Tennessee	6,213	—	—	—
Texas	43,977,440	855,000	7,539,000	376,556
Utah	677,826	22,000	187,000	— (1)
W. Virginia	522,529	3,000	49,000	8,376
Wyoming	4,551,050	111,000	957,000	58,244
Miscellaneous	26,536	4,000	30,000	—
US Total	120,730,299	3,020,000	27,735,000	1,477,732

(1) data not disclosed

Sources: Cumulative Production to end 1979 from API, June, 1980, revised with data
from DOE, *U.S. Crude Oil, Natural Gas, and Natural Gas Liquids Reserves,* annual
reports 1980 through 1983. Proved reserves from DOE, *1983 Annual Report.* Indicated
Additional Reserves from DOE *1982 Annual Report.*

such as California and New Mexico where past production has reduced proved reserves and left many old reservoirs holding oil that was only partially recovered. If, however, the secondary techniques have been applied in recent years, then the amounts of such potential oil will be sharply reduced.

THE IMPACT OF IMPROVED RECOVERY RATES ON CRUDE OIL ESTIMATES

Improved recovery, often called *tertiary,* or *enhanced* recovery could greatly affect the amounts of oil recovered from identified and undiscovered reservoirs. A number of different techniques have been researched and are now being applied to fields in various states (see essay).

The potential for this additional oil in the country is the subject of estimates that range from 11 billion to a maximum possible (at any price) of 51 billion barrels (see *Oil and Gas Journal,* July 25, 1977). The target for these techniques is the large volume of oil in-place that is left behind after primary and secondary methods have been applied—and have led, on the average, to recovery of only about 35 percent of the original oil in-place. Because of their large fields and long production histories, Texas, Louisiana, and California account for well over half of such oil (see Fig. 2-22). In the past few years the federal government has encouraged enhanced recovery by allowing higher prices for crude produced that way. The industry has responded by applying the methods—mostly steam injection and miscible gas injection—at a large number of fields.

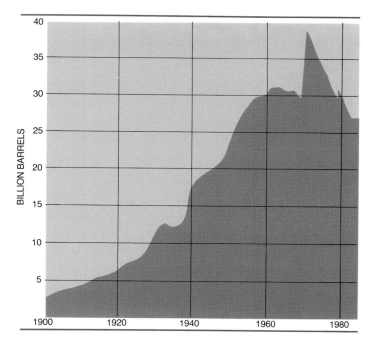

Fig. 2–19 U.S. crude oil reserves, 1900–1983.
Sources: 1900–1960 from Hubbert, 1969; later years from API *Annual Reserve Reports* and DOE, *Annual Reserve Reports*.

Summary of Oil in Identified and Undiscovered Reservoirs

Figure 2–23 brings together the key resource amounts from the foregoing sections and places them in a framework showing the two dimensions that are so important in understanding resource estimates: degree of certainty and economic feasibility. The measured and indicated reserves are from the most recent Department of Energy reserves report. Inferred and Undiscovered amounts are from U.S.G.S. *Circular 860*, and are, therefore, less recent. All those amounts, including the Undiscovered, assume re-

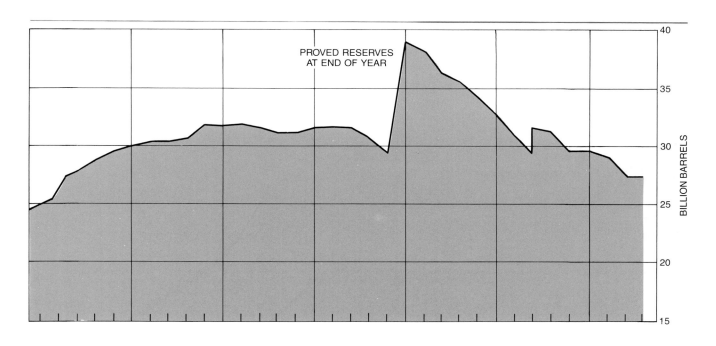

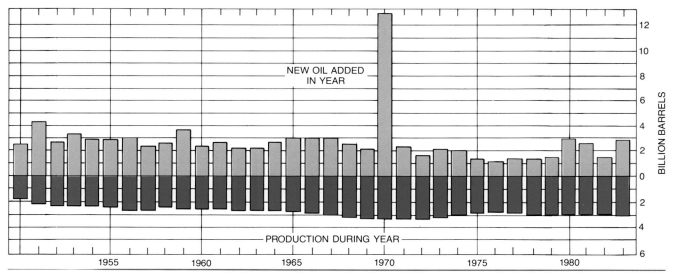

Fig. 2–20 U.S. crude oil reserves, 1950–1983, showing effects of annual production and annual additions to reserves.

Sources: 1950–1979, API, *Annual Reserve Reports;* later years from DOE, *Annual Reserve Reports.*

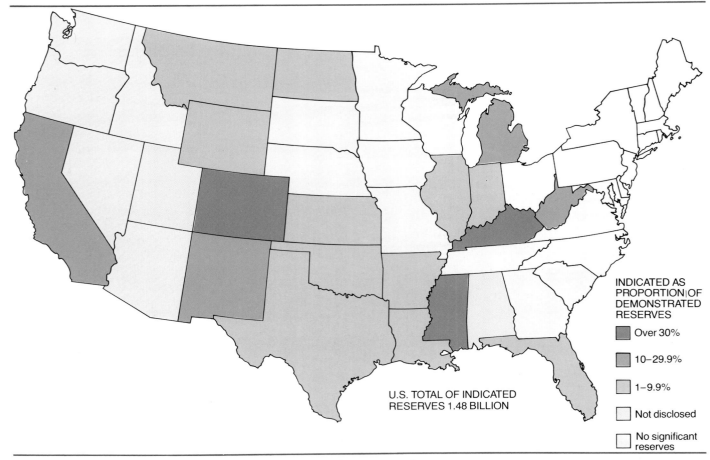

INDICATED AS PROPORTION|OF DEMONSTRATED RESERVES

- Over 30%
- 10–29.9%
- 1–9.9%
- Not disclosed
- No significant reserves

U.S. TOTAL OF INDICATED RESERVES 1.48 BILLION

Fig. 2-21 Indicated reserves in relation to demonstrated reserves—showing states where secondary recovery methods will return relatively large amounts of crude.

Sources: Indicated reserves as of end 1982, from DOE, *Reserves Report*, 1983; Proved reserves as of end 1983, from DOE, *Reserves Report*, 1983.

THE STRATEGIC PETROLEUM RESERVE

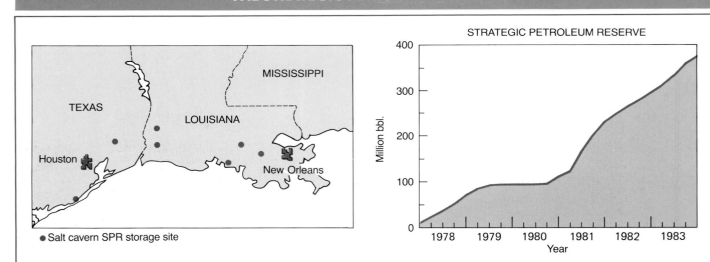

● Salt cavern SPR storage site

In response to the Arab oil embargo of 1973-1974, Congress passed the Energy Policy and Conservation Act. Included in this legislation was the creation of the Strategic Petroleum Reserve program, designed to set aside a reserve of up to one billion barrels of crude oil and/or petroleum products to reduce the impact of any supply disruption caused by international discord. The reserves can be withdrawn only after the President has determined such an action is necessary. Procedures for using the oil have not been completely specified, but they may include auctioning the oil to refining companies.

Oil for the SPR has come largely from foreign sources, Mexico being the most prominent. Through the year 1980, the growth of SPR was slow, but since 1981 it has filled rapidly to the 1983 level of near 400 million barrels. The goal is 750 million barrels to be reached sometime in 1990. That amount, *three-quarters of a billion barrels*, will make the

SPR comparable in size to the West Texas field, one of the five largest oil fields in the nation.

SPR's resemblance to reserves in an oil field lies only in the fact that the SPR crude is a sure thing and can be produced readily. The way it is stored is quite unnatural. Unlike natural gas, which is pumped into the pores of exhausted natural gas reservoir rock (such as sandstones or limestones) to provide storage for gas pipeline companies, the SPR crude is pumped into giant *caverns* created in salt beds by leaching, i.e., solution-mining. There are seven sites in coastal Texas and Louisiana (see map).

Source: Sites from *Oil and Gas Journal*, Jan. 10, 1983. History of reserve amounts from DOE/EIA, *Annual Energy Review, 1983*, and *Petroleum Supply Annual, 1983*.

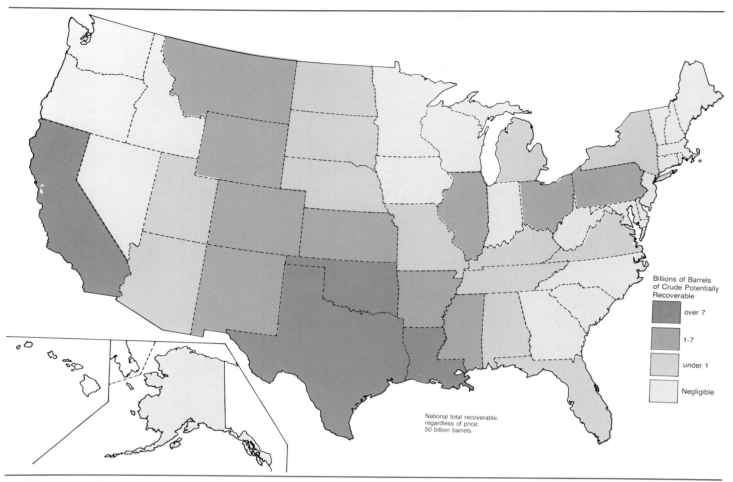

Billions of Barrels
of Crude Potentially
Recoverable

over 7

1-7

under 1

Negligible

National total recoverable,
regardless of price:
50 billion barrels.

Fig. 2–22 Additional crude oil attainable through enhanced (tertiary) recovery methods, showing states with greatest potential.
Source: U.S. Energy Research and Development Administration, *Fossil Energy Research Meeting*, reported in *Oil and Gas Journal*, July 25, 1977.

Waste steam is vented from a steam-flood operation in which thick, low-gravity crude is recovered from a field in San Ardo, California. (Courtesy of American Petroleum Institute and Texaco Inc.)

ENHANCED OIL RECOVERY

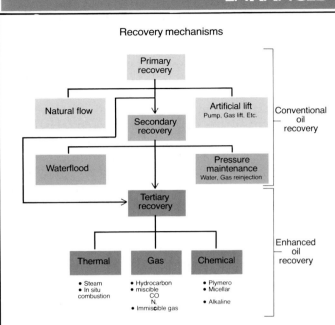

Recovery mechanisms

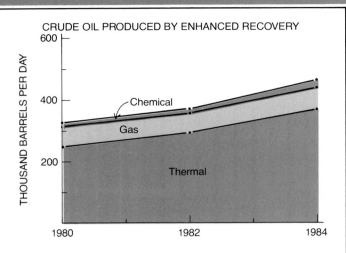

CRUDE OIL PRODUCED BY ENHANCED RECOVERY

Production by enhanced recovery is growing slowly. In 1984 the rate of production from all methods is 461,000 barrels per day. On an annual basis, that is equivalent to 0.17 billion barrels, or roughly 5.5 percent of all U.S. production. In 1984, 79 percent of the production came from steam injection, while 18 percent came from gas injection— mostly carbon dioxide. Some industry experts foresee spending twice as much on enhanced recovery in the near future because it will be more attractive than exploring frontier areas. Production may rise by 1990 to 0.36 barrels per year, or roughly 12 percent of all production.

Primary Recovery. Oil flows through the reservoir rock into the borehole, driven by natural pressure and aided by pumping.

Secondary Recovery. As formation pressure drops, the operator may inject fluid to maintain pressure: Water will be injected at a level below the oil-saturated zone; if natural gas is used it will be injected above that zone.

Tertiary Recovery. As the arrows suggest, these methods may be used after secondary recovery or immediately after primary recovery has failed. In fact, many operators are returning to fields abandoned years ago, when primary production reached rates that were unsatisfactory. Depending on the nature of the oil and the kind of reservoir rock, one of the following three major tertiary methods are used:

1. *Thermal.* This raises the temperature of sluggish oil, lowers its viscosity, and makes it flow more readily. Injection of steam into the rock is most common. Controlled burning of the oil itself is an alternative for very thick (heavy) crude.

2. *Gas.* The gases used, such as carbon dioxide, nitrogen, and certain *immiscible* gases, are injected directly into the oil-saturated zone, where they unite with the oil and free it from the reservoir rock.

3. *Chemical.* Polymer, micellar, and alkaline chemicals are used in place of gases to dislodge molecules of oil from the reservoir rock. The chemicals must be separated from the produced oil at the surface.

Sources: *Oil and Gas Journal*, Mar. 31, 1980, April 2, 1984, April 23, 1984

PROJECTS, ACTIVE AND PLANNED IN 1980

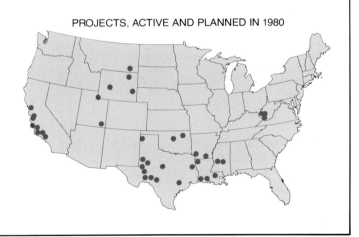

covery rates that are not significantly greater than the historic rates. In the lower part of the diagram are additional amounts that reflect the application of enhanced recovery methods. For Identified reservoirs, a value of 30 billion barrels was selected arbitrarily, as roughly midway between a low (11 billion) and a high (51 billion) estimate. To be complete, the diagram should also show how the oil to be recovered from now-undiscovered reservoirs would be augmented if enhanced recovery techniques were applied. The incremental amount would be 35 billion barrels. All these amounts are shown by the bar graph against cumulative production in the United States. The grand total of conventional crude oil resources, using these assumptions, is around 200 billion barrels, against cumulative produc-

tion of 121.7 billion. Oil resources *excluded* are the *unconventional*, that is, the oil that may be recovered from oil shales, heavy crude, and tar sands. The potential for such oil is very great (see Chapter 3).

PRODUCTION, TRANSPORTATION, AND IMPORTS OF CRUDE OIL

As reserves grew through the twentieth century, productive capacity grew, and production climbed to a peak in 1972. Then, as reserves declined, production fell to the

	IDENTIFIED			UNDISCOVERED
	Demonstrated		Inferred	
	Measured	Indicated		(64.3 to 105.1) 82.6[2]
ECONOMIC	27.7[1]	1.5[1]	23.4[2]	
SUB-ECONOMIC	Through Enhanced Recovery Methods: 30 Billion (mean of estimates)			Assuming 50% Enhanced Recovery: 35 Billion

[1]As of 12/31/83
[2]As of 12/31/80

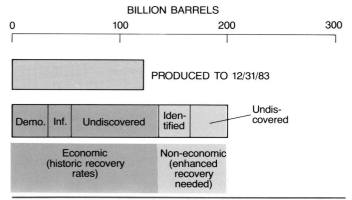

Fig. 2–23 Summary of conventional crude oil resources in the United States.
Sources: 1983 data from DOE, *Annual Reserve Report, 1983*, 1984; 1980 data from U.S.G.S. *Circular 860*, 1981; Identified subeconomic from Energy Research and Development Administration, 1977.

3 billion barrel level where it has remained fairly steady since 1975 (Fig. 2–24).

In 1983, Texas led in production, with 0.85 billion barrels, while Alaska, Louisiana, and California ranged from 0.4 to 0.55 billion (see Table 2–4). These four states together accounted for 74 percent of the nation's production. Alaska's production has increased by more than three times since 1977, as fields in the Prudhoe Bay area have been developed.

Imports and Exports of Petroleum

For the United States as for other industrial Western nations, the international aspects of petroleum supply have been vital. Like the countries of Western Europe, the United States' demand for petroleum exceeds its capacity to produce; so, since the 1940s, foreign crude has played a role in meeting the U.S. demand. In 1947, imports were only 7 percent of demand. But as productive capacity fell in the early 1970s and demand for petroleum remained high, the United States developed a growing dependence

on imported supplies, until, in 1977, imports were 46 percent of the total supply.

The growing imports are shown in Figure 2–25, which also reveals how the source nations have changed. In the 1960s, Venezuela was the largest supplier, but through the 1970s Middle Eastern nations supplied more and more, until, at the peak of imports in 1977 and 1978, Algeria and Saudi Arabia were the dominant sources. Since that time, economic recession and energy conservation have reduced demand sharply in the United States, and there has been a deliberate policy of reducing dependence upon the unstable nations of the Middle East and North Africa. The result is not only a dramatic drop in imports, but also a reduction in the supply roles of Saudi Arabia and Algeria.

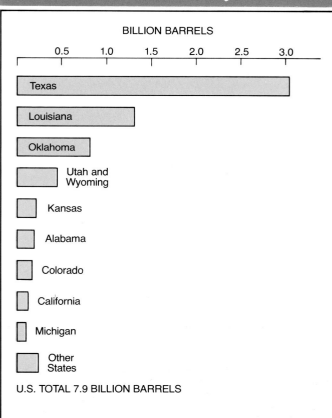

PROVED RESERVES OF NATURAL GAS LIQUIDS

U.S. TOTAL 7.9 BILLION BARRELS

Although derived from natural gas, not from crude oil, these hydrocarbons are liquid, measured in barrels, and therefore are associated with crude oil in any tabulations of *total petroleum liquids*. The liquids are mostly *condensate*, obtained at the well head during production of wet gas. The largest proved reserves, therefore, are in those states with large reserves of natural gas. The national total of 7.9 billion barrels is roughly 22 percent of total petroleum liquids as of the end of 1983. The complete picture of NGL resources must include *undiscovered* amounts as estimated by the U.S. Geological Survey in 1981. The total for onshore and offshore regions is 17.7 billion barrels.

Source: DOE, *Annual Reserves Report for 1983*

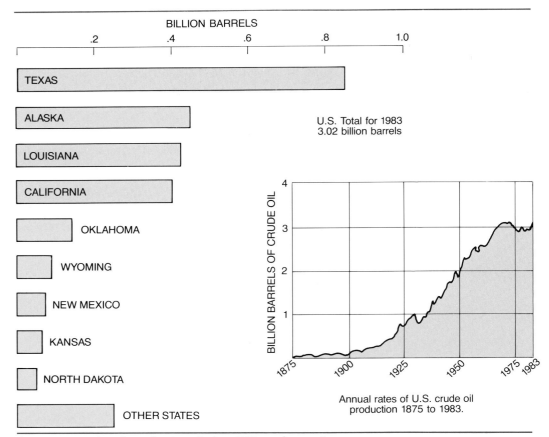

BILLION BARRELS

.2   .4   .6   .8   1.0

TEXAS

ALASKA

U.S. Total for 1983
3.02 billion barrels

LOUISIANA

CALIFORNIA

OKLAHOMA

WYOMING

NEW MEXICO

KANSAS

NORTH DAKOTA

OTHER STATES

Annual rates of U.S. crude oil
production 1875 to 1983.

Fig. 2–24 Crude oil production during 1983, and over time.

Sources: 1983 state production from DOE, *Annual Reserve Report, 1983*, 1984; history from Hubbert, 1969, updated with data from API *Annual Reserves Reports* and DOE *Annual Reserves Reports.*

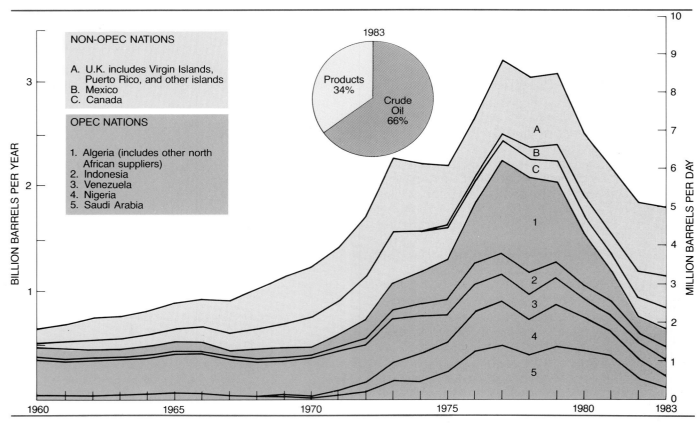

NON-OPEC NATIONS

A. U.K. includes Virgin Islands,
   Puerto Rico, and other islands
B. Mexico
C. Canada

OPEC NATIONS

1. Algeria (includes other north
   African suppliers)
2. Indonesia
3. Venezuela
4. Nigeria
5. Saudi Arabia

1983

Products
34%

Crude
Oil
66%

Fig. 2–25 Petroleum liquids imported to the United States, 1960–1983, showing proportions in crude oil and refined products.

Source: Department of Energy, *Annual Energy Review*, 1983.

TABLE 2-5
Leading Nations Supplying Crude Oil to the
United States during 1983

NATIONS	CRUDE OIL IMPORTED BY THE UNITED STATES (THOUSANDS OF BARRELS)
Arab OPEC:	total 194,639
Algeria	64,274
Saudio Arabia	117,325
Other OPEC:	total 344,401
Indonesia	115,156
Nigeria	109,807
Venezuela	59,854
Non-OPEC:	total 676,185
Canada	100,142
Mexico	279,703
Norway	23,836
Trinidad and Tobago	30,207
United Kingdom	133,142
Total Crude Oil Imports	1,215,225

Source: DOE, *Petroleum Supply Annual, 1983*, 1984.

In 1983, Arab OPEC nations supplied 16 percent of the *crude oil*, other OPEC 28 percent, the United Kingdom 11 percent, and Mexico 23 percent (Table 2-5). A large part of *petroleum products* were shipped from Caribbean nations, simply because a number of major oil companies have located refineries there.

The contrast in foreign suppliers of crude oil between 1978 and 1983 is developed in Figure 2-26, which shows how the proportions from Mexico, Canada, and the United Kingdom increased as most OPEC nations supplied less. Indonesia and Venezuela are exceptions, with percentage contributions that increased slightly between 1978 and 1983.

As the sources of crude oil changed, so did the character of the crude arriving in the United States (see Fig. 2-27). Part A of the illustration shows that in the 1983 imports there was a higher percentage of *low gravity* crude, that is, thicker, more viscous oil. Part B shows an increase in the percentage of crude that is *high in sulphur*. Both these trends are results of less Saudi Arabian and Nigerian crude in the supply. Their crudes are high gravity ("light") low-sulphur oil. Because that oil carries a premium price, refiners have modified their equipment to accommodate the heavier, high-sulphur crude which is cheaper and available from other nations. In 1984, the general oversupply of crude oil was compounded for Saudi Arabia and Nigeria by the reluctance of buyers to take this premium oil which, traditionally, had been sought after.

In the United States, the destinations of foreign crude oil can be read in Figure 2-28, outlining for each of five Petroleum Administration for Defense Districts what proportion of refinery inputs were foreign and what domestic

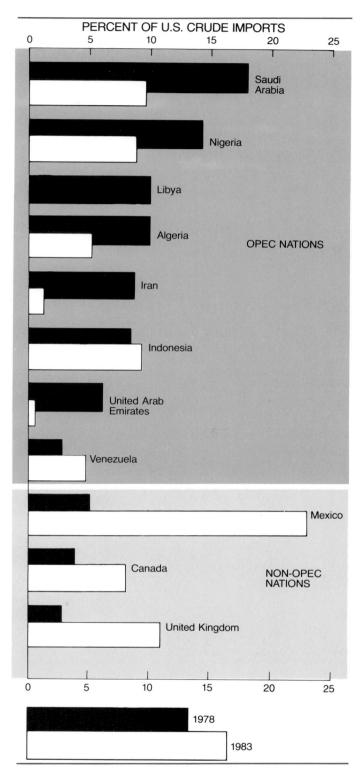

Fig. 2-26 Changing sources of crude oil imports, 1978 versus 1983.
Source: Department of Energy, *Petroleum Supply Monthly*, Feb. 1984.

in 1983. The dominance of domestic crude in the nation as a whole (over 70 percent) is borne out in most of the five districts. District I, holding East coast refineries, is the

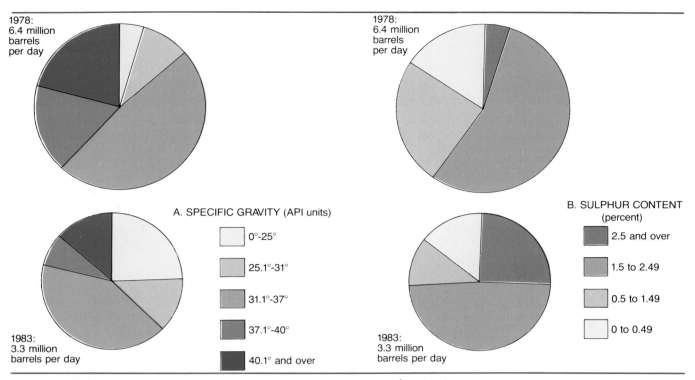

1978:
6.4 million
barrels
per day

1978:
6.4 million
barrels
per day

A. SPECIFIC GRAVITY (API units)

- ☐ 0°–25°
- ☐ 25.1°–31°
- ☐ 31.1°–37°
- ☐ 37.1°–40°
- ☐ 40.1° and over

B. SULPHUR CONTENT
(percent)

- ☐ 2.5 and over
- ☐ 1.5 to 2.49
- ☐ 0.5 to 1.49
- ☐ 0 to 0.49

1983:
3.3 million
barrels per day

1983:
3.3 million
barrels per day

Fig. 2–27 Changing character of crude oil imports, 1978 versus 1983, showing increasing proportions of lower-gravity crude and high-sulphur crude.
Source: Department of Energy, *Petroleum Supply Monthly*, Feb. 1984.

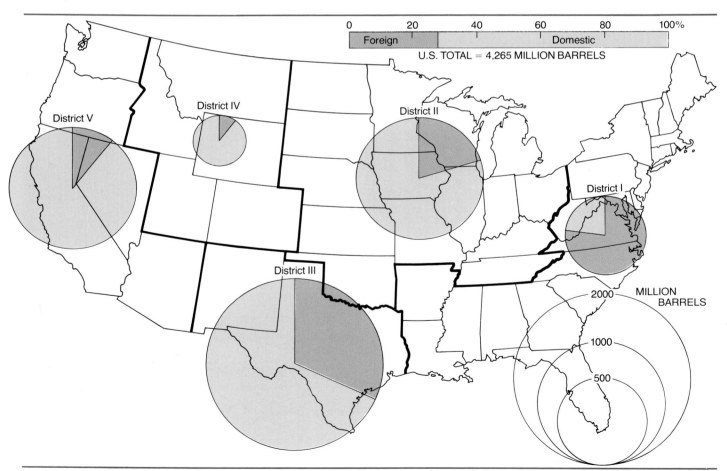

0   20   40   60   80   100%

Foreign   Domestic

U.S. TOTAL = 4,265 MILLION BARRELS

District V

District IV

District II

District I

District III

MILLION
BARRELS

2000

1000

500

Fig. 2–28 Refinery receipts in 1983, showing proportions foreign and domestic crude oil in each of five PAD districts.
Source: Department of Energy, *Annual Energy Review*, 1983.

exception: Here, more than 75 percent of the refinery input is foreign. Much of the domestic crude refined in Districts II and V, incidentally, is not from the lower 48 states, but is shipped from Alaska.

Petroleum exports over time are shown by Figure 2-29, whose vertical scale has only *one-tenth* the magnitude of that on Figure 2-25. Exports, 78 percent refined products, not crude oil, increased sharply in 1977 to a peak in 1982, dropping slightly in 1983. Carribbean nations, Japan, and Canada are the prominent destinations of these exports.

Crude Oil Pipelines and Movements

Crude oil was transported from the country's earliest producing areas in Pennsylvania by means of barrels on wagons and flatboats. A railroad line to the fields was completed in 1862 and carried the crude at first in barrels, then, after 1865, in tankcars which consisted of two large upright barrels on a flatcar.

Pipeline technology was developing at the same time, and by the 1870s a 6-inch line was moving oil from the fields about 130 miles to Williamsport from which tankcars carried it to refineries. By 1880, pipelines were

carrying the crude from Pennsylvania fields to Pittsburgh, Cleveland, Buffalo, and New York City. As more oil fields were discovered in Ohio, Indiana, Illinois, Oklahoma, Kansas, and Texas a continental network of pipelines developed.

In 1974, when the most recent comprehensive study of energy transport was produced (Jimison, 1977) around 75 percent of crude movements were by pipeline, and the balance by water and truck. The situation has changed markedly since that time (see Fig. 2-30). With the increase of tanker shipments from Alaska, the role of tankers and barges has grown so much that in 1982 water transport accounted for over 50 percent of all crude movements.

There is a significant difference now between how crude moves within regions of the nation and how it moves from region to region. Within regions, pipelines are the main carriers—taking crude from wells to tank farms and to terminals at rivers or the ocean. Between regions, tanker and barge are the main carriers. The flows for 1983 by tanker and barge are shown in Figure 2-30. The very large flows originating in PAD District V are due to tanker shipments from Alaska, which is included in that district. While some Alaskan crude is unloaded at California refineries, most continues down the West Coast, through

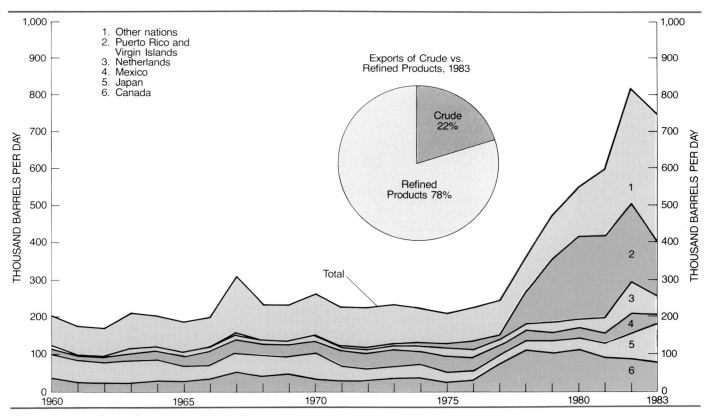

Fig. 2-29 Petroleum liquids exported from the United States, 1960–1983, showing proportions in crude oil and refined products.
Source: Department of Energy, *Annual Energy Review*, 1983.

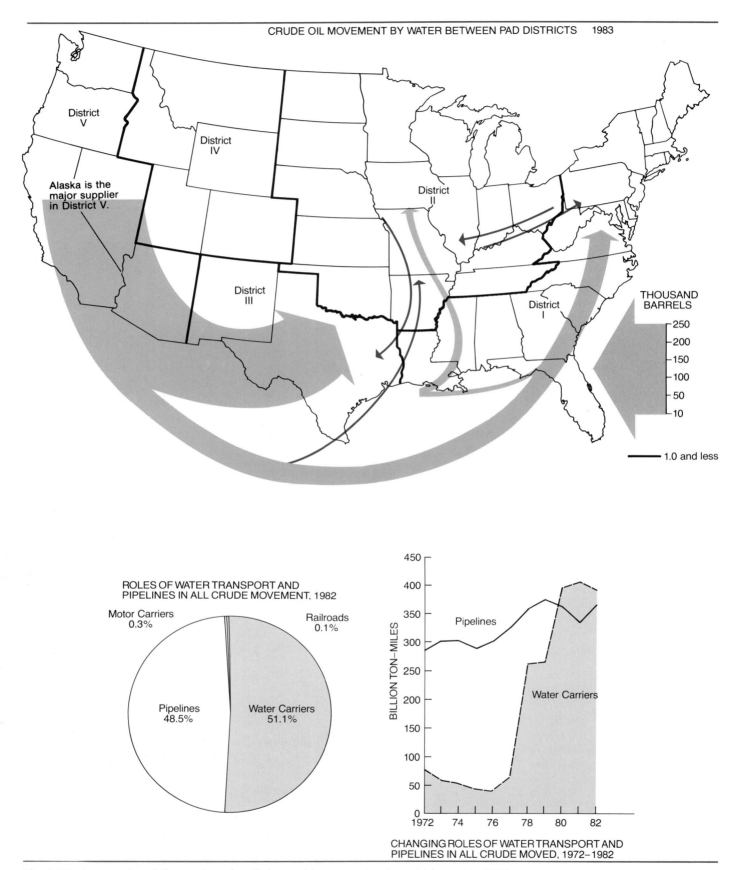

CRUDE OIL MOVEMENT BY WATER BETWEEN PAD DISTRICTS   1983

District V

District IV

Alaska is the major supplier in District V.

District II

District III

District I

THOUSAND BARRELS

— 250
— 200
— 150
— 100
— 50
— 10

—— 1.0 and less

ROLES OF WATER TRANSPORT AND
PIPELINES IN ALL CRUDE MOVEMENT, 1982

Motor Carriers
0.3%

Railroads
0.1%

Pipelines
48.5%

Water Carriers
51.1%

CHANGING ROLES OF WATER TRANSPORT AND
PIPELINES IN ALL CRUDE MOVED, 1972–1982

BILLION TON-MILES

Pipelines

Water Carriers

Fig. 2–30   Interregional flows of crude oil shipped by water (tanker and barge) in 1983.
Source: Department of Energy, *Petroleum Supply Monthly*, March 1983.

Storage tanks at Valdez, on Alaska's south coast hold crude oil from North Slope oil fields, brought by pipeline for shipment by tanker from this terminal.

the Panama Canal, and into Texas and Louisiana ports in District III, though some follows the Atlantic coast into Region I. Shipments from Region III to Region II are by barge, up the Mississippi, while much smaller amounts go from the Gulf along the Atlantic coast to central Atlantic ports such as Baltimore, Philadelphia, and New York. Flows from Region I and II are very small, and they make use of the Mississippi and Ohio rivers.

NATURAL GAS

Natural gas is a mixture of the lighter hydrocarbons that exist in a gaseous phase underground or may be in solution with crude oil at the pressures which occur in the reservoir. The principal hydrocarbons in the mixture are methane, ethane, propane, butanes, and pentanes. Some nonhydrocarbon gases that may occur in the mixture are carbon dioxide, helium, hydrogen sulfide, and nitrogen.

Natural gas is often found trapped in regions (and in rock sequences) where crude oil is not abundant. In some cases the gas is in deeply buried rocks where the high temperatures at depth may have carried the conversion of organic materials beyond the crude oil stage until only gas remained. In other cases, it appears that the presence of *terrestrial* rather than *marine* organic matter in source

The tanker, *Texaco Caribbean*, discharging foreign crude at a deep-water port off the Louisiana coast. Beneath this buoy are pipes that carry the crude oil to shore. (Courtesy of American Petroleum Institute and Texaco Inc.)

WHAT IS THE FUTURE FOR US OIL?

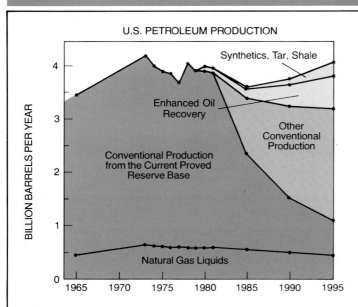

U.S. PETROLEUM PRODUCTION

If estimates of undiscovered crude that can be recovered economically are near the mark, then the USA has used up a little less than half the oil it was born with.

There is some debate, though, about how easily that remaining oil can be found and brought into production. The reservoirs already found and exploited were easily discovered, relatively shallow, and in areas that are accessible—mostly onshore in the lower 48 states. Much of the crude oil awaiting the drill is likely to be in smaller accumula- tions, deeply buried, and in frontier areas. In fact, about 35 percent of the undiscovered recoverable oil estimated by the US Geological Survey in 1981 is in offshore areas.

One approach to the future is to assess the possibilities for future *production*. In a 1984 report, Joseph Riva of the Library of Congress Research Service predicts a 17 percent and a possible 29 percent decline in production by the year 2000.[1] He reasons, on the basis of historic ratios between production rates and reserves, that to maintain the current rate of 3 billion barrels per year, it will be necessary to discover almost 45 billion barrels of oil between now and the year 2000. That 45 billion would be roughly 70 percent of the now undiscovered oil (using a figure lower than the USGS estimate of 82 billion barrels undiscovered). In preceding years, however, the industry has found on average only 44 percent of the oil thought to be remaining at the start of the period. Thus the chances are, Riva says, that the needed discovery rate will not be realized.

In the projected 17 percent decline, Riva sees production dropping in all areas except the Northern Rockies-Great Plains region: Production will drop 16 percent in Alaska, 40 percent in West Texas, and 44 percent in the Gulf Coast. If no oil is found off the Atlantic Coast or in the Oregon-Washington area, and the discovery rate does not improve in the Northern Rockies-Great Plains region, the national production decline will be 29 percent.

In a projection made in 1982, the Department of Energy[2] recognized how production from the currently proved reserves will decline (see graph) while production from newly discovered reserves ("Other Conventional Production") will largely fill the gap. In order to maintain the 1982 rate of production through 1995, it will be necessary for enhanced oil recovery and crude from oil shales and tar sands to grow appreciably.

[1]Congressional Research Service, Report No. 84–129 SPR. Reported in *Science*, Oct. 26, 1984

[2]U.S. Department of Energy, *Outlook for Oil Imports*, Sept., 1982

---

rock led to formation of gas, not crude oil. Some occurrences of natural gas without oil may be explained by migration processes that separated the two fluids.

In the past a great deal of natural gas has been burned at the well-head (flared) when it was produced at the same time as crude oil, and there were no facilities for storage and transportation. Now, welded pipelines that can effectively move gas, storage systems, and even methods for liquefying and shipping gas in tankers have given this fuel new status.

WORLD RESOURCES OF NATURAL GAS

The status of the world and its regions, including past production and current proved reserves, is shown by Figure 2-31. As with crude oil, the Western Hemisphere has produced most of the gas, with the United States responsible for the largest part of that production. The greatest reserves, though, are in the Soviet Union, which has such a dominant position in gas that it is not rivaled by even the entire Middle East. In crude oil (Fig. 2-7) no single nation has such a large share of the world's total. The gas reserve figure for the Soviet Union may, in fact, be not quite comparable to those for other nations because Soviet estimates usually include some anticipated reserves. Table 2-6 shows that the Soviet reserves, at 1,400 trillion cubic feet, are 44 percent of the world total.

In every region except the Western Hemisphere there appears to be a healthy balance between the amount produced to the end of 1983 and the proved reserves as of that same date. While figures for gas produced in the Western Hemisphere are probably realistic, the small amounts shown for the Middle East and Africa may be due to underrecording of gas released in association with crude oil production and then lost through flaring.

The world total of 3,200 trillion cubic feet as proved reserves would last nearly 60 years at the annual production rate of 50 trillion per year. In contrast, world proved reserves of conventional crude oil have a life of only 33 years, at current rates of production.

UNITED STATES RESOURCES OF NATURAL GAS

As in the foregoing section on United States crude oil, this review of natural gas in the country is in two parts. First, a broad survey considers both identified and undiscovered reservoirs and makes use of information assigned to 15 regions of the country. A second part considers only the identified reservoirs and uses current reserve and production data for states, not regions.

Gas in Identified and Undiscovered Reservoirs

As with the corresponding section on crude oil, this section depends on latest estimates by the U.S. Geological Survey (U.S.G.S. *Circular 860*, 1981).

Figure 2-32 shows the 15 regions, ranked according to natural gas produced, i.e. cumulative production, through the year 1979. On the right side of the graph are *all amounts* of conventional natural gas thought to remain in each region in both identified and undiscovered reservoirs.

Cumulative gas production is concentrated strongly in two or three regions: the Gulf Coast, Midcontinent, and West Texas–East New Mexico. Remaining gas is overwhelmingly concentrated in the Gulf Coast region. Most regions have produced roughly as much gas as remains,

but the Western Rockies, Northern Rockies, and Alaska all have more remaining than produced—if the undiscovered amounts are valid. The amounts shown as remaining in identified reservoirs are totals of proved and inferred reserves (see Table 2-7). The inferred component is especially large in the Gulf Coast and West Texas–East New Mexico regions where expansion of existing fields is expected to result in very significant additions to reserves.

If the 15 regions are compared on the basis of all past production, plus proved and inferred reserves, the Gulf Coast regions (onshore and offshore) together hold 50 percent of the nation's total. If estimated recoverable gas in undiscovered reservoirs is included, the Gulf Coast regions hold 43 percent of the total. This remarkable concentration of natural gas is held in relatively young sedimentary rocks, mostly of Tertiary age (see Geologic time scale in Appendix) which underly the Coastal Plain and continue offshore as the continental shelf adjacent to Texas and Louisiana.

The resource amounts in Table 2-7 assume greater meaning if compared to the 1983 annual production rate of 16 trillion cubic feet. At that rate, Proved reserves would

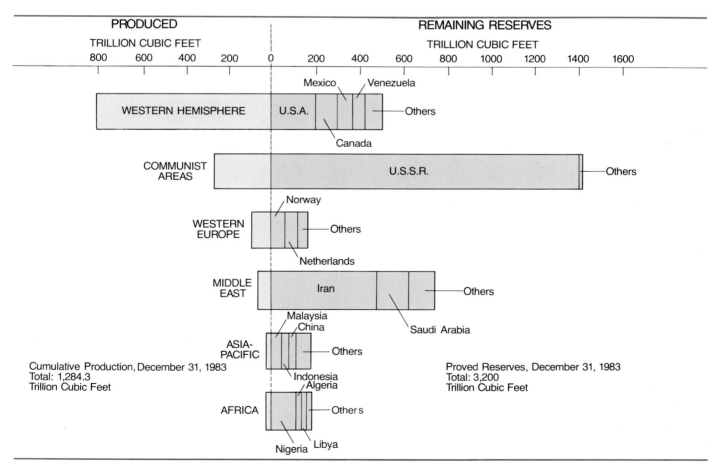

Fig. 2–31 World regions; Cumulative production of natural gas and proved reserves as of Dec. 31, 1984.

Sources: Cumulative production from National Academy of Sciences, 1975; later production and current reserve, *Oil and Gas Journal*, March issues.

TABLE 2-6
World Natural Gas, Proved Reserves as of December 31, 1983.
Amounts tabulated are trillions of cubic feet (cu. ft. $\times 10^{12}$)

North America		Middle East	
Canada	91	Bahrain	7
Mexico	75	Iran	480
United States	198	Iraq	29
Total	364	Kuwait	35
Central and South America		Oman	3
Argentina	24	Qatar	62
Bolivia	5	Saudi Arabia	125
Brazil	3	Syria	1
Chile	2	United Arab Emirates	31
Colombia	4	Other	(*)
Ecuador	4	Total	774
Peru	1	Africa	
Trinidad and Tobago	13	Algeria	110
Venezuela	55	Angola	2
Other	(*)	Cameroon	4
Total	111	Congo	2
Western Europe		Egypt	7
Denmark	3	Gabon	1
Germany, West	7	Libya	21
Italy	4	Nigeria	35
Netherlands	50	Tunisia	4
Norway	59	Other	4
United Kingdom	25	Total	190
Other	10	Far East and Oceania	
Total	158	Australia	18
Eastern Europe and U.S.S.R.		Bangladesh	7
U.S.S.R.	1,400	Brunei	7
Other	17	China	30
Total	1,417	India	15
		Indonesia	30
		Malaysia	48
		New Zealand	6
		Pakistan	16
		Thailand	9
		Other	2
		Total	187
		World Total	3,200

*less than 0.05 trillion cubic feet
Source: *Oil and Gas Journal,* Dec. 26, 1983.

last 12 years, proved plus inferred, 23 years, and all remaining gas, 60 years if the estimate of undiscovered is valid.

UNDISCOVERED RECOVERABLE AMOUNTS
These amounts are estimated through the same procedures used for crude oil and assume the same offshore limits. Table 2-8 lists the low, high, and mean estimate for each region, and Figure 2-33 maps the mean values. The Gulf Coast regions hold the most, but the Western Rockies, with 90.1 trillion cubic feet, is regarded as very promising by the Geological Survey. As with the estimates of crude oil, the Western overthrust belt is thought to hold much of this undiscovered gas.

The uncertainty of estimates can be evaluated by reference to Figure 2-34. This latest U.S.G.S. estimate is slightly higher than the one of 1975: The mean value for the nation as a whole is 593.8 trillion cubic feet versus 484 trillion in 1975. As with undiscovered crude oil, these estimates can be qualified by probability values for amounts between the low of 474.6 trillion and the high of 739.3 trillion. For instance, the chance of there being more than 665 trillion cubic feet is around 20 percent.

SUMMARY OF AMOUNTS REMAINING, AS OF 1980
All *conventional* natural gas known and thought to remain as of the end of 1979 is shown in Figure 2-35 (see also Table 2-7) and is contrasted there with cumulative

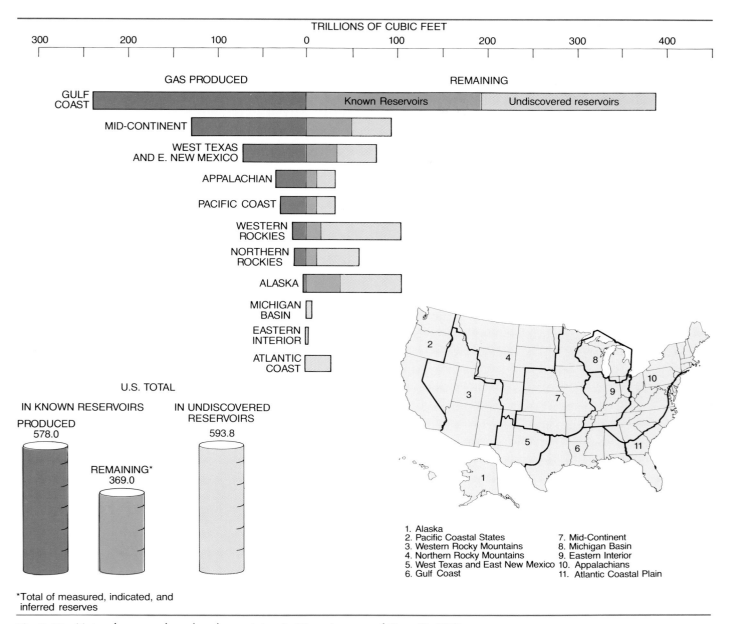

GAS PRODUCED   **REMAINING**

| TRILLIONS OF CUBIC FEET |
| 300   200   100   0   100   200   300   400 |

GULF COAST — Known Reservoirs — Undiscovered reservoirs
MID-CONTINENT
WEST TEXAS AND E. NEW MEXICO
APPALACHIAN
PACIFIC COAST
WESTERN ROCKIES
NORTHERN ROCKIES
ALASKA
MICHIGAN BASIN
EASTERN INTERIOR
ATLANTIC COAST

U.S. TOTAL

IN KNOWN RESERVOIRS
PRODUCED 578.0
REMAINING* 369.0

IN UNDISCOVERED RESERVOIRS
593.8

1. Alaska
2. Pacific Coastal States
3. Western Rocky Mountains
4. Northern Rocky Mountains
5. West Texas and East New Mexico
6. Gulf Coast
7. Mid-Continent
8. Michigan Basin
9. Eastern Interior
10. Appalachians
11. Atlantic Coastal Plain

*Total of measured, indicated, and inferred reserves

Fig. 2-32 Natural gas produced and remaining in 15 regions as of Dec. 31, 1979.
Source: U.S.G.S. *Circular 860,* 1981.

production of natural gas in the United States through 1979. The amount produced, 578 trillion cubic feet, is 60 percent of the total amount remaining in known and undiscovered reservoirs. Or, to put it another way, the cumulative production is 37 percent of the estimated *ultimate recovery*, the total of amounts recovered to date and expected to be recovered in the future. The division by certainty on the bar graph shows how large the undiscovered portion is for the whole nation. Divided by location, the lower bar reveals how important onshore areas will be in the lower 48 states—if the assessment and regional allocation of undiscovered resources are valid.

Natural Gas in Identified Reservoirs

Annual data on reserves and production, published now by the U.S. Department of Energy, makes possible a state-by-state current view of gas resources—without

benefit, though, of inferred or undiscovered estimates. It is necessary, as with crude oil, to integrate the current data on reserves and production by state with slightly older estimates of inferred and undiscovered amounts organized by the 15 regions in the previous section.

This section begins with natural gas reserves, their character and their history, then deals with unconventional (and presently uneconomic) sources of natural gas. After these resources are summarized, the matters of production and consumption are reviewed.

PROVED RESERVES These most certain amounts—economically recoverable from firmly defined reservoir rocks—are summarized in Figure 2-36 and Table 2-9 (see Fig. 2-49). Texas and Louisiana hold 47 percent of the nation's total. If Alaska's 34 trillion cubic feet are added, the three states account for 63 percent of the U.S. total. This concentration of reserves in a few states

101

TABLE 2-7
Summary of Natural Gas Produced and Remaining in the United States as of December 31, 1979, Showing 15 Regions (trillions of cubic feet)

REGION	CUMULATIVE PRODUCTION TO DEC. 31, 1979	TOTAL GAS REMAINING		In Undiscovered Reservoirs (mean estimate)
		In Known Reservoirs		
		Measured	Inferred	
1 Alaska	1.2	30.0	4.4	36.6
1A Alaska Offshore	0.6	2.0	1.2	64.6
2 Pacific	27.9	11.4	3.8	14.7
2A Pacific Offshore	1.5	1.9	0.4	6.9
3 Western Rockies	15.0	11.4	4.5	90.1
4 Northern Rockies	13.2	7.1	5.5	45.7
5 West Texas and East New Mexico	70.7	15.9	18.1	42.8
6 Gulf Coast	228.5	45.3	74.3	124.4
6A Gulf Coast Offshore	55.3	34.4	39.5	71.8
7 Midcontinent	127.8	32.0	18.8	44.5
8 Michigan Basin	1.3	1.1	1.1	5.1
9 Eastern Interior	1.8	—	0.1	2.7
10 Appalachians	34.2	6.2	5.9	20.1
11 Atlantic Coast	—	—	—	0.1
11A Atlantic Offshore	—	—	—	23.7
Total USA	578.0	191.5	177.5	593.8

Source: U.S.G.S. *Circular 860*, 1981.

TABLE 2-8
Estimated Natural Gas Recoverable from Undiscovered Reservoirs as of 1980

REGION	LOW[1]	HIGH[2]	MEAN
	(TRILLIONS OF CUBIC FEET)		
1	19.8	62.3	36.6
1A	33.3	109.6	64.6
2	8.2	24.9	14.7
2A	3.7	13.6	6.9
3	53.5	142.4	90.1
4	29.6	69.0	45.7
5	22.4	75.2	42.8
6	56.5	249.1	124.4
6A	41.7	114.2	71.8
7	22.9	80.8	44.5
8	1.8	10.9	5.1
9	1.2	5.0	2.7
10	6.4	45.8	20.1
11	—	0.4	0.1
11A	92	42.8	23.7
Total Onshore	322.5	567.9	426.8
Total Offshore	117.4	230.6	167.0
Total U.S.	474.6	739.3	593.8

[1]Low estimate has probability of 95 percent.
[2]High estimate has probability of 5 percent.
Source: U.S.G.S., *Circular 860*, 1981.

TABLE 2-9
Proved Reserves of Natural Gas, December 31, 1983, Showing Proportions Non-Associated and Associated-Dissolved

STATE	PROVED RESERVES 12/31/83 (BILLIONS OF CUBIC FEET)		PROPORTIONS (%)	
	Non-Associated	Total Gas	Non-Associated	Associated
Alabama	822	853	96	4
Alaska	3,883	34,291	11	89
Arkansas	1,887	2,081	90	10
California	2,000	5,965	33	67
Colorado	3,200	3,373	95	5
Florida	0	64	0	100
Kansas	10,051	10,183	99	1
Kentucky	581	600	97	3
Louisiana	36,061	43,663	83	17
Michigan	529	1,353	39	61
Mississippi	1,523	1,603	95	5
Montana	813	921	88	12
New Mexico	10,868	12,371	88	12
New York	295	295	100	0
North Dakota	416	673	62	38
Ohio	396	2,031	19	81
Oklahoma	14,992	17,261	87	13
Pennsylvania	1,861	1,882	99	1
Texas	42,830	53,803	80	20
Utah	2,112	2,472	85	15
Virginia	175	175	100	0
W. Virginia	2,238	2,324	96	4
Wyoming	9,769	10,728	91	9
Miscellaneous	50	81	62	38
U.S. Total	147,352	209,046	70	30

Source: DOE/EIA, *Crude Oil, Natural Gas, and Natural Gas Liquid Reserves, 1983 Annual Report*, 1984.

will be altered if the Geological Survey's estimates of undiscovered gas in the northern Rockies region prove valid.

Some natural gas occurs in the reservoir rock as free gas filling pore spaces of the rock, in a zone underlain by water-saturated rock. This type is called *non-associated* gas. In other reservoirs the natural gas occurs in a zone overlying an oil-saturated zone (*associated* gas) or it may be dissolved in crude oil at the pressures that exist in the reservoir rock (*dissolved* gas). The distinction is made in gas reserves, therefore, between *non-associated* and *associated-dissolved* gas. The first kind can be produced without regard for oil production, but production of associated gas from a "gas cap" will reduce formation pressure and affect the recovery of oil from the underlying zone, while production of gas-rich oil will inevitably release dissolved gas at the surface as a by-product of oil production.

The prevalence of one type of gas versus the other is shown in Table 2-9, and mapped in Figure 2-37. Nationally, 70 percent of gas reserves are non-associated. The states vary widely in this regard: In some states with relatively small reserves, the gas is almost entirely non-associated; and in those with the largest reserves, Louisiana and Texas, the gas is about 80 percent non-associated. Alaska, however, has large reserves which are mostly associated-dissolved.

The history of gas reserves is not unlike that for crude oil reserves in the country (Fig. 2-38). The reserve

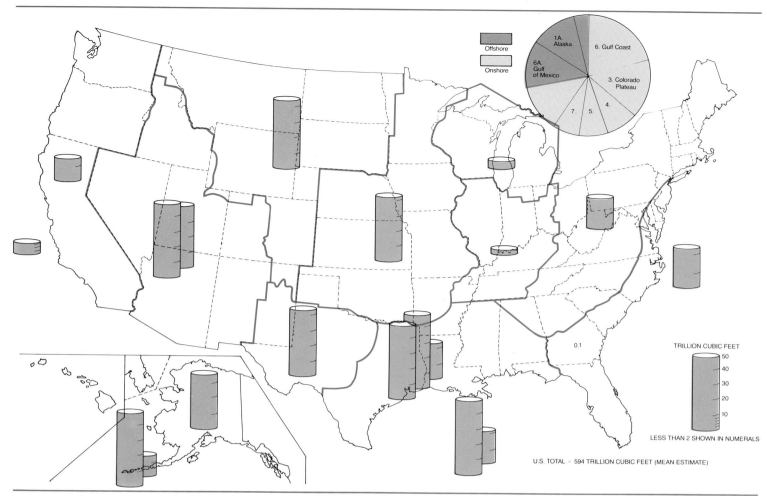

Fig. 2–33 Estimated natural gas recoverable from undiscovered reservoirs: mean estimate for the 15 regions.
Source: U.S.G.S. *Circular 860,* 1981.

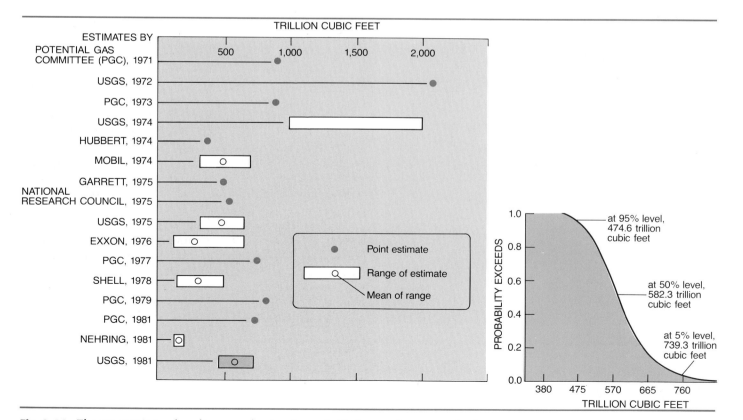

Fig. 2–34 The uncertainty of undiscovered natural gas. A. Various estimates made since 1961. B. Probability distribution for amounts estimated by U.S. Geological Survey in 1981.
Source: U.S.G.S. *Circular 860,* 1981.

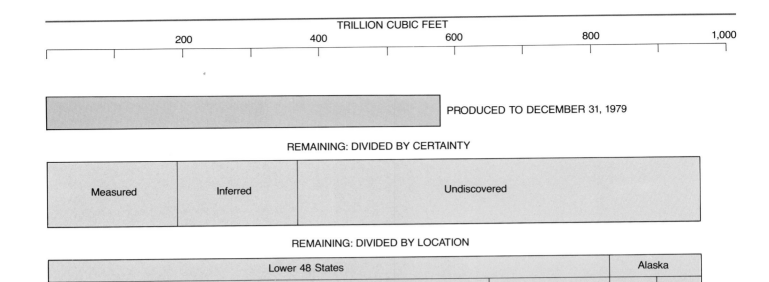

TRILLION CUBIC FEET

| 200 | 400 | 600 | 800 | 1,000 |

PRODUCED TO DECEMBER 31, 1979

REMAINING: DIVIDED BY CERTAINTY

| Measured | Inferred | Undiscovered |

REMAINING: DIVIDED BY LOCATION

Lower 48 States		Alaska	
Onshore	Offshore	On-shore	Off-shore

Fig. 2–35 Summary of natural gas produced and remaining as of Dec. 31, 1979, showing certainty and location.

A completed gas well in the marshlands of southern Louisiana. Production man checks a valve on the wellhead. (Courtesy of American Petroleum Institute and Tenneco Inc.)

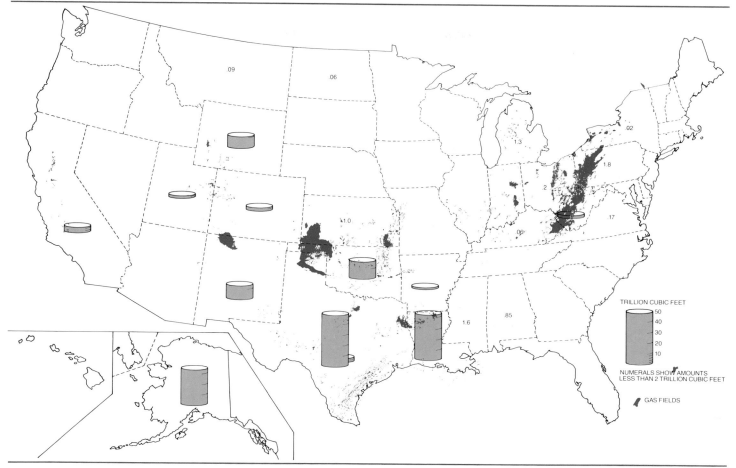

Fig. 2–36 Proved reserves by state as of December 31, 1983.
Source: Department of Energy, *Annual Reserves Report, 1983,* 1984.

amounts grew steadily until 1967 or 1968, when they, like oil reserves, levelled off and then fell for the first time in history. The relief provided by Alaskan discoveries was not significant for gas, and the reserves fell steadily to the present plateau near 200 trillion cubic feet. That level is now being maintained since new gas added is roughly the same as annual production.

Unconventional Natural Gas

Just as oil shales and oil sands offer potential for extraction of "unconventional" crude that will not flow into a borehole (see following chapter), there are possibilities for additional natural gas beyond the amounts considered Proved, Inferred, or Undiscovered in reservoirs of the usual kind.

The occurrences of unconventional natural gas are quite varied. Two occurrences entail natural gas of normal origin which does not flow easily into a drill hole because reservoir rocks are not sufficiently permeable. Another is

methane associated with coal beds. A fourth is gas of normal origin that is involved in a geothermal occurrence.

TIGHT GAS SANDS In the Rocky Mountain areas of Wyoming, Utah, Colorado, and New Mexico (Fig. 2–39) these sands occur in sequences of clay, chalk, and sandstone interbedded with shale. Commercial flows of gas cannot be obtained unless the sandstone is fractured to open communication routes to the borehole. Experimental *nuclear* fracturing was tried in the Gasbuggy, Rulison, and Rio Blanco projects in 1967, 1969, and 1973, with disappointing results. Massive hydraulic fracturing of the sort often used to stimulate conventional oil or gas flows is being attempted in projects initiated in 1979 by the Department of Energy. These are a 7-well project in the Uintah basin of Utah, and a 1-well project in the Piceance Creek area of Colorado (DOE, March, 1978). Gas in-place in such reservoirs in the Rocky Mountain region may amount to 600 trillion cubic feet, while amounts recoverable may be 300 trillion cubic feet. A recent study, however, suggests gas in-place is only 400 trillion cubic

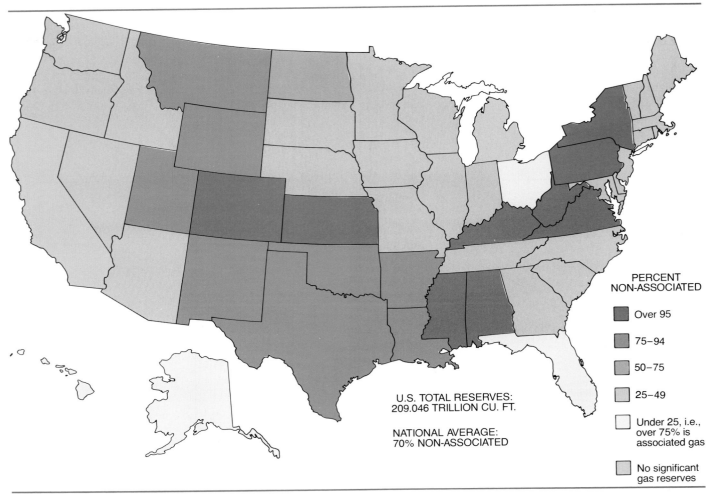

PERCENT
NON-ASSOCIATED

- Over 95
- 75–94
- 50–75
- 25–49
- Under 25, i.e., over 75% is associated gas
- No significant gas reserves

U.S. TOTAL RESERVES:
209.046 TRILLION CU. FT.

NATIONAL AVERAGE:
70% NON-ASSOCIATED

Fig. 2–37 Proportions of 1983 proved reserves that are non-associated (free) gas.
Source: Department of Energy, *Annual Reserves Report, 1983*, 1984.

feet, while recoverable amounts are 100 to 180 trillion cubic feet at gas prices of 3 dollars per thousand cubic feet (Kuuskraa *et al.*, 1978).

DEVONIAN SHALES OF THE APPALACHIANS

These shales contain natural gas in an occurrence similar to that in the Rockies, that is, in a rock which yields adequate flows only when fractured. Gas has already been produced from these formations with the aid of natural fractures, as in the Big Sandy Field of eastern Kentucky. Other areas of potential extraction are in the states of New York, Pennsylvania, Ohio, West Virginia, Kentucky, Michigan, Indiana, Illinois, and Tennessee. In 1979 the Federal Department of Energy began a demonstration project for hydraulic fracturing of Devonian age shales and associated sandstones in each of these states (DOE, March, 1978). Gas in-place in such occurrences through the Appalachians may amount to 490 trillion cubic feet, with 8 to 16 trillion cubic feet recoverable (Kuuskraa *et al.*, 1978).

Federal government estimates of gas *in-place* in West-

ern gas sands and Devonian shales are shown, along with locations in Figure 2–39.

METHANE FROM COAL BEDS

Methane occurs in coal deposits. While methane is known as the dominant constituent of conventional natural gas, it is derived in this case from peat materials as they matured and altered chemically to become coal. Because methane is a highly noxious gas, it must be vented from mines at some expense, nearly 0.1 trillion cubic feet being wasted each year. In untouched Identified coal deposits, some of which are too deep for mining, there is perhaps 300 trillion cubic feet *in-place* (Maurice Deul, quoted by *Science*, 1976). Recoverable amounts are quite uncertain, but may be as high as 20 to 25 trillion cubic feet (Kuuskraa *et al.*, 1978).

GAS IN GEOPRESSURED ZONES OF THE GULF COAST

The nature of geopressured zones is elaborated in the chapter on geothermal resources, but here it can be noted that natural gas dissolved in high temperature brines under great pressure at depths from 6,000 to 22,500 feet could be available at the surface if the heat and the hydrostatic pressure of those brines were ever exploited. The amounts of gas liberated would depend upon the specific plan for development of the geothermal resources. Two plans described by the U.S.G.S. (*Circular 790*) would be consistent with recovery of 97 or 617 trillion cubic feet.

Summary of All Natural Gas Resources

Figure 2-40 uses bars to compare conventional with unconventional gas resources. Recoverable amounts are emphasized, and show the amounts thought to be recoverable from tight sands and Devonian shales are relatively small when compared to the gas recoverable from

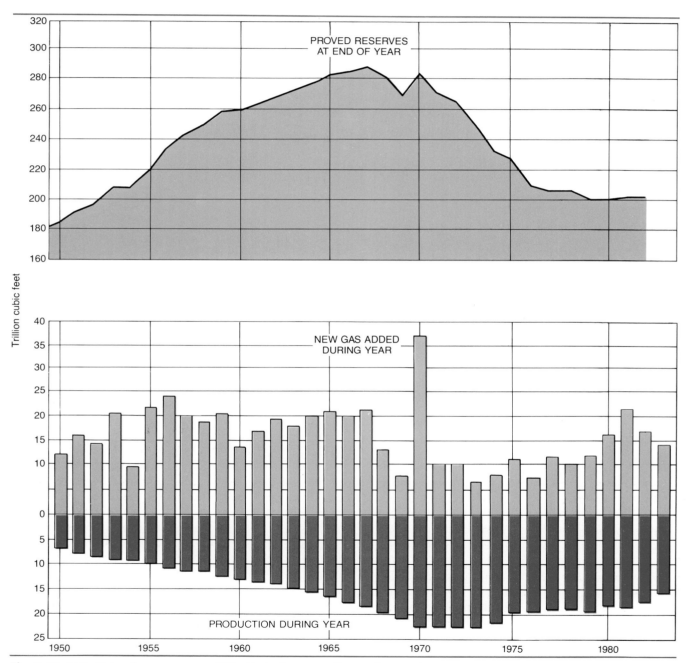

Fig. 2-38 U.S. Natural gas reserves, 1958–1983, showing effects of annual production and annual additions to reserves.
Sources: 1950–1978, API, *Annual Reserve Reports;* later years from DOE, *Annual Reserve Reports.*

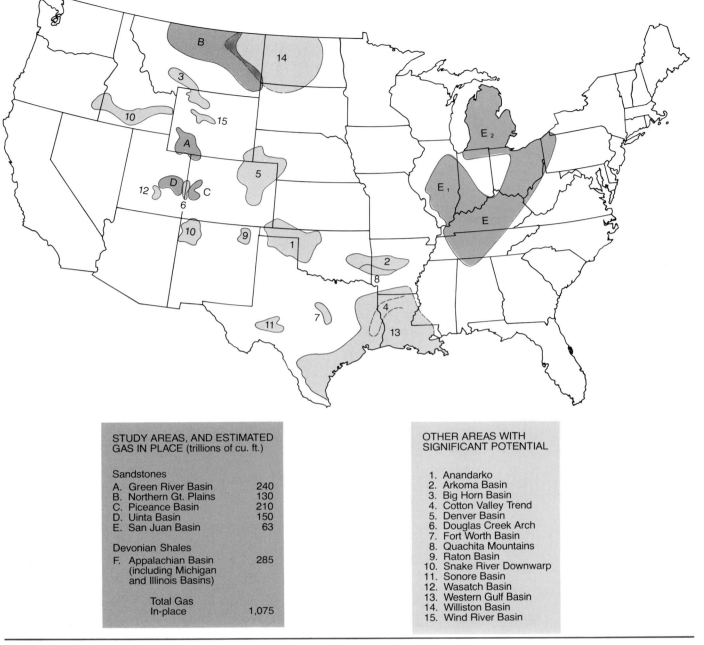

STUDY AREAS, AND ESTIMATED
GAS IN PLACE (trillions of cu. ft.)

Sandstones
A. Green River Basin	240
B. Northern Gt. Plains	130
C. Piceance Basin	210
D. Uinta Basin	150
E. San Juan Basin	63

Devonian Shales
F. Appalachian Basin (including Michigan and Illinois Basins)	285
Total Gas In-place	1,075

OTHER AREAS WITH
SIGNIFICANT POTENTIAL

1. Anandarko
2. Arkoma Basin
3. Big Horn Basin
4. Cotton Valley Trend
5. Denver Basin
6. Douglas Creek Arch
7. Fort Worth Basin
8. Quachita Mountains
9. Raton Basin
10. Snake River Downwarp
11. Sonore Basin
12. Wasatch Basin
13. Western Gulf Basin
14. Williston Basin
15. Wind River Basin

Fig. 2–39 Potential for unconventional natural gas from tight gas sands and devonian shales.
Source: U.S. Department of Interior, *Energy Resources on Federally Administered Lands,* 1981.

known and undiscovered conventional reservoirs. Gas in Gulf Coast geopressured reservoirs, however (see Geothermal chapter) is very significant. These various categories of gas resources are placed, in the lower part of Figure 2–40, in a framework that recognizes both certainty and economic feasibility.

Gas Production, Consumption, and Movements

State-by-state gas production during 1983 is plotted in Figure 2–41, which also classifies each state according to its share of the nation's gas production in that year. A

rough idea of the location of gas fields is also added. As might be expected, production is very large in Texas and Louisiana and roughly equal in those two states. Together, their 11.1 trillion cubic feet of production is 65 percent of the total.

Complementary to that illustration is Figure 2–42, mapping the *consumption of gas* by indicating state shares of the national total, rather than absolute amounts. Apparently the states that produce the lion's share also consume a large part of the total. Illinois is anomalous, though: It has no significant production, yet is ranked in the second group for proportion of the nation's consumption.

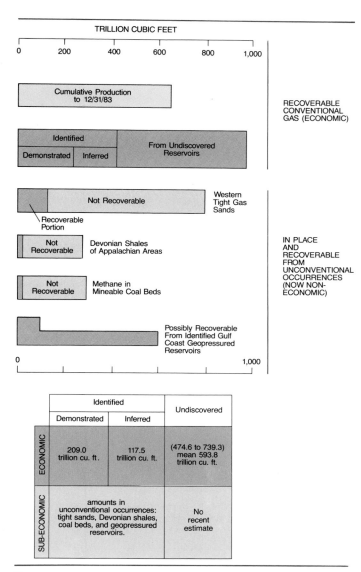

Fig. 2-40 Summary of conventional and unconventional natural gas resources.

Sources: For methane in coal beds, Duel, 1978; for geopressured reservoirs, U.S.G.S. *Circular 790* (see Chapter 5).

Conventional, unconventional, and synthetic gases, showing the unconventional gases reviewed in this chapter.

GASES	FREE GAS OR GAS IN SOLUTION IN VARIOUS KINDS OF ROCK	Conventional natural gas in porous and permeable reservoir rocks
		In tight gas sands of western states
		In Devonian shales of Appalachians
		In geopressured reservoirs
		In un-mined coal beds
	SYNTHETIC GASES	From coal gasification at mine-head or in place underground
		From biomass

To serve those states that have little gas production, pipeline companies buy gas from producers and sell to utility companies. To create supplies adequate for the winter season, these companies establish underground storage reservoirs near the markets. Figure 2-43 classifies states according to their shares of *gas storage capacity* as of spring 1983, and also plots the locations of storage reservoirs. North central and northeastern states hold largest shares of national capacity. These reservoirs, in the rock of exhausted gas fields, are filled during the summer and drawn upon in the winter season.

Shipments of gas between states occurs in a complex pattern that may be simplified by using the *net result* of shipments into and out of all states. States are classified, then, as those with net receipts overall and those that show net deliveries during the specified year. The extent of the receipts or deliveries is noted on each state. Only Vermont was uninvolved in this process in 1982 (Fig. 2-44).

Recent patterns of natural gas pipelines are the result of an evolution. The center of gas production and transport has not always been the southwestern states. The earliest development and transportation of natural gas was in Pennsylvania, Ohio, and West Virginia, and made use, incidentally, of gas lines that existed for coal gas used in lighting. Additional small gas production was developed later in Indiana, New York, Illinois, and Kansas. Gas discoveries that dwarfed any previous finds were made in Louisiana, Texas, Kansas, and Oklahoma during the first decade of this century—just as production capacity in the Northeast was beginning to dwindle. The need for a connection between the large fields of the Southwest and the markets in northern and northeastern states became apparent, but the southwestern and northeastern networks were not linked until 1930. By 1950, gas from the Southwest was being piped to 33 states, Canada, and Mexico.

IMPORTS OF NATURAL GAS Imports have never formed a large part of the U.S. gas supply. In 1982, the proportion was only 5 percent. The largest amounts entered by pipeline from Canada (783.4 billion cubic feet) and from Mexico (94.7 billion cubic feet). Gas from overseas arrived in the form of Liquefied Natural Gas (LNG) from Algeria: 55.1 billion barrels entered in 1982, considerably increased from the 15 billion typical of the late 1970s. Of the many terminal sites planned and built for LNG, only two were active in 1982, at Everett, Massachusetts and at Lake Charles, Louisiana (Fig. 2-45).

THE ALASKAN GAS SUPPLY A pipeline to bring gas from Alaska was approved in 1977, but has not yet been undertaken. The route (see Fig. 2-46) is to follow the trans-Alaska oil pipeline until Fairbanks, and then follow the Alaska Highway into British Columbia, and ultimately into Southern Alberta, where it can join existing pipelines. Now it appears that the Canadian route will be approved. The application has been filed with Canada's National

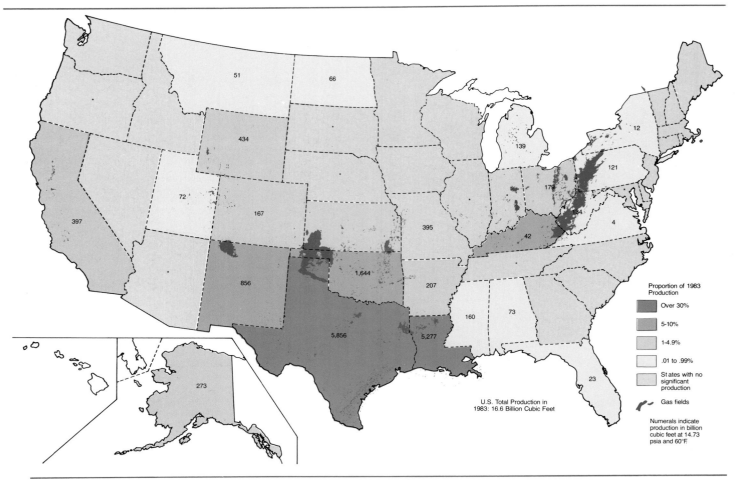

Fig. 2-41 State shares of natural gas production, 1983, showing gas fields.
Source: Department of Energy, *Annual Reserves Report, 1983*, 1984.

Proportion of 1983 Production

- Over 30%
- 5-10%
- 1-4.9%
- .01 to .99%
- States with no significant production
- Gas fields

U.S. Total Production in 1983: 16.6 Billion Cubic Feet

Numerals indicate production in billion cubic feet at 14.73 psia and 60°F.

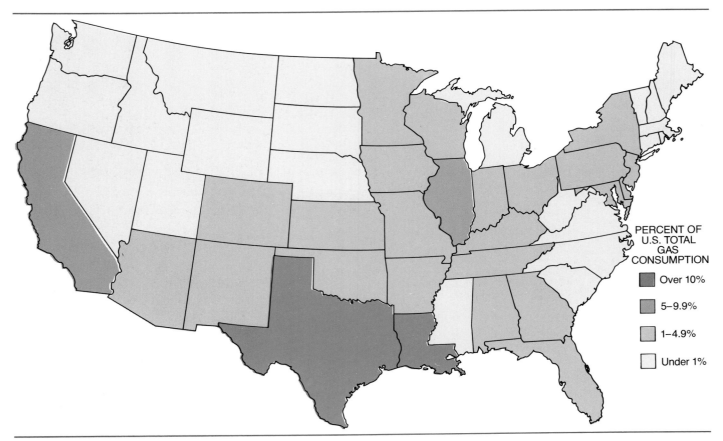

Fig. 2-42 State shares of natural gas consumption, 1983.
Source: Department of Energy, *Natural Gas Annual*, 1982.

PERCENT OF U.S. TOTAL GAS CONSUMPTION

- Over 10%
- 5-9.9%
- 1-4.9%
- Under 1%

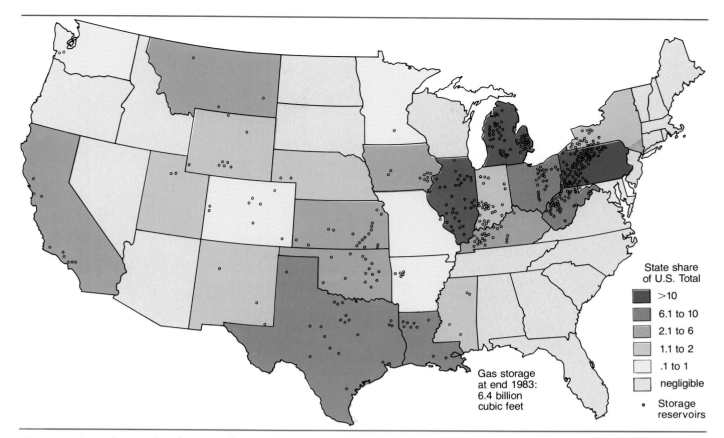

Fig. 2-43 State shares of underground natural gas storage capacity, Mar. 31, 1982.

Source: (Map) Department of Energy, *Natural Gas Annual,* 1982; (Graph) U.S. Department of Energy, *Annual Review,* 1983. Reservoir locations from American Gas Association, 1983.

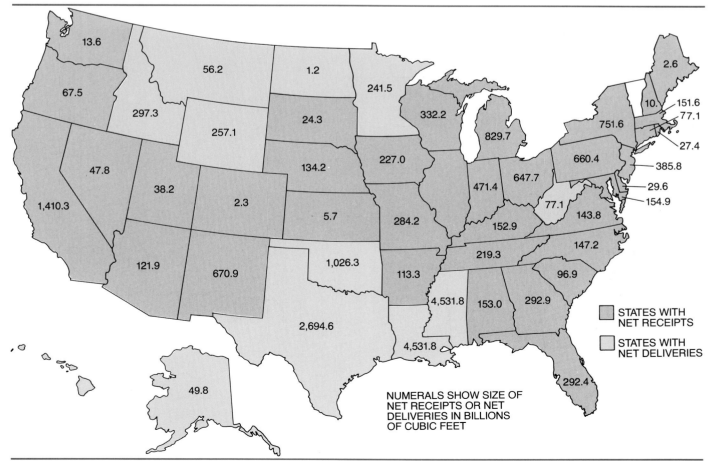

Fig. 2-44 Interstate shipments of natural gas during 1982, showing net receipts and net deliveries by state.

Source: Department of Energy, *Annual Energy Review, 1983,* 1984.

LNG tanker, *Lake Charles*, which carries liquefied natural gas from Algeria to a terminal near Lake Charles, Louisiana. (Courtesy of American Petroleum Institute and Panhandle Eastern Pipeline Company.)

Fig. 2-45 Terminals for liquified natural gas (LNG) imports.

Source: Various sources. Status in 1982 from Department of Energy, *Natural Gas Annual*, 1982.

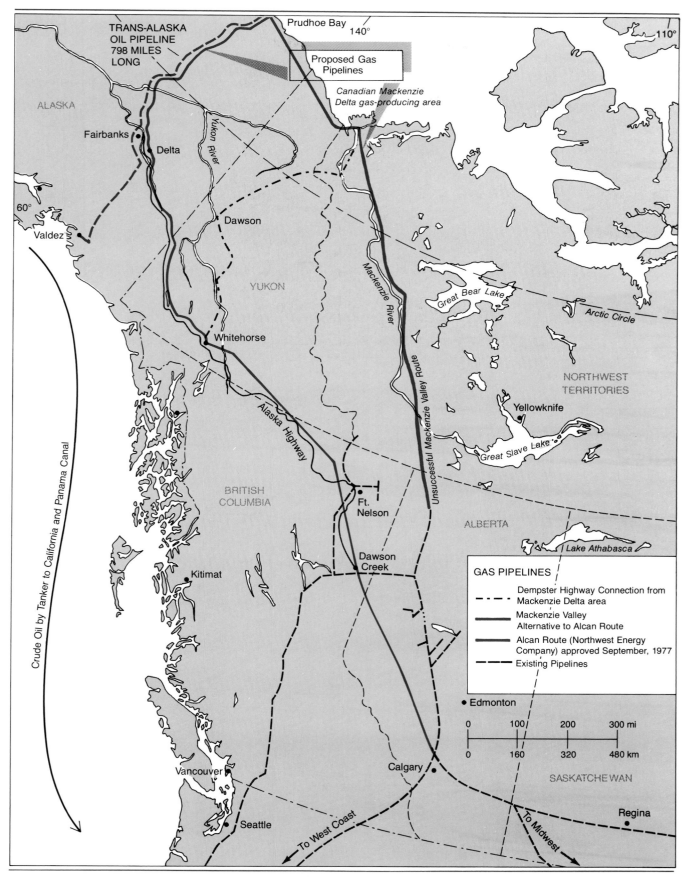

Fig. 2-46 Routes for Alaska's gas and crude oil.
Source: *Canada Today/D'Aujourd'hui,* 8:5, 1977.

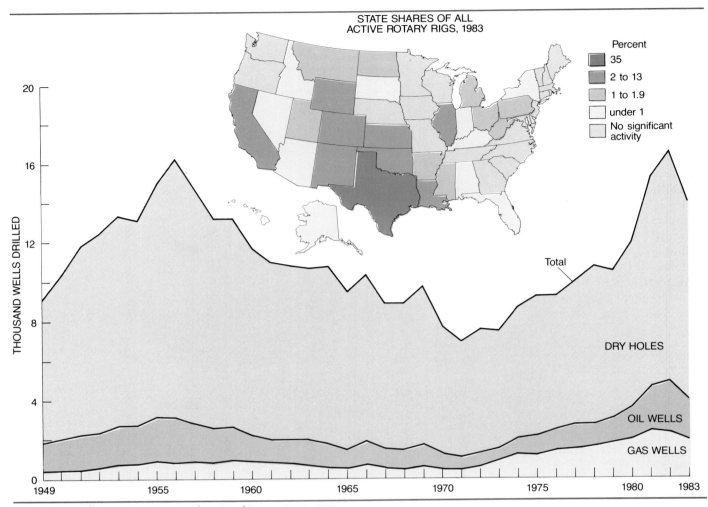

Fig. 2-47 Drilling activity, 1983, showing history 1949–1983.
Source: Rig numbers from *Oil and Gas Journal*, Jan. 30, 1984. History from DOE, *Annual Energy Review, 1983*, 1984.

Drilling from a man-made island in the Beaufort Sea near Prudhoe Bay on Alaska's north coast.
(Courtesy of American Petroleum Institute and Standard Oil Company of Ohio.)

Energy Board by Canada Polar Gas, and completion is expected in the early 1990s (*Oil and Gas Journal*, Oct. 29, 1984).

OIL AND GAS EXPLORATION, AND FRONTIER AREAS

This section offers some information on the pace and location of drilling activity and a close-up view of some frontier areas, especially those offshore.

Drilling activity, as represented by numbers of wells drilled (Fig. 2–47) fell in 1983 from its 1982 peak because of slackening demand and lower prices for oil. The graph shows *all holes* drilled, including, therefore, development drilling to prove and expand known fields, as well as true exploration. It is not appropriate to infer an *exploration success ratio* from the proportion of all holes that are shown to be oil or gas wells—roughly 28 percent.

Offshore exploration drilling from a specially-equipped drillship, shown here with accompanying workboat in the Santa Barbara Channel off California's coast. (Courtesy of American Petroleum Institute and Standard Oil Company of California.)

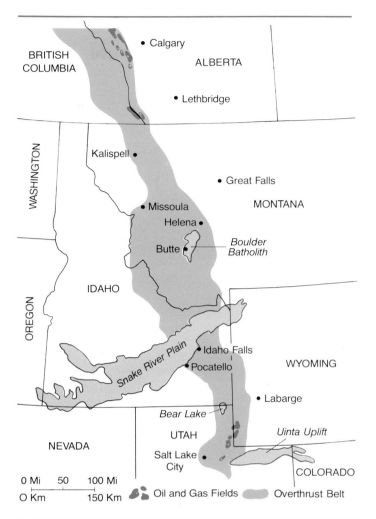

ESTIMATED UNDISCOVERED RECOVERABLE CRUDE OIL AND NATURAL GAS IN WESTERN OVERTHRUST BELT (MEAN ESTIMATES).		
	Crude Oil (billions of barrels)	Natural Gas (trillions of cubic ft.)
Wyoming-Utah-Idaho	6.7	57.4
Montana	0.6	9.3
Total	7.3	66.7

Fig. 2–48 **Western overthrust belt, and estimated oil and gas undiscovered resources.**
Source: U.S. Department of the Interior, *Energy Resources on Federally Administered Lands*, 1981.

The map in Figure 2–47 classifies states according to their shares of all *rotary* drilling rigs active in 1983. Excluded are the few old-style *cable-tool* rigs that are still active in Pennsylvania. Drilling activity in 1983 apparently concentrated mostly in Texas, with lesser proportions in New Mexico, Oklahoma, Wyoming, Colorado, Kansas, and California.

One of the frontier areas, and one responsible for much of the drilling in Wyoming and Utah, is the Western Overthrust Belt. It also is recognized in the Geological Survey's estimates of large amounts of undiscovered oil and natural gas in the Northern Rockies region. Figure 2–48 outlines this area at the front of the Rocky Moun-

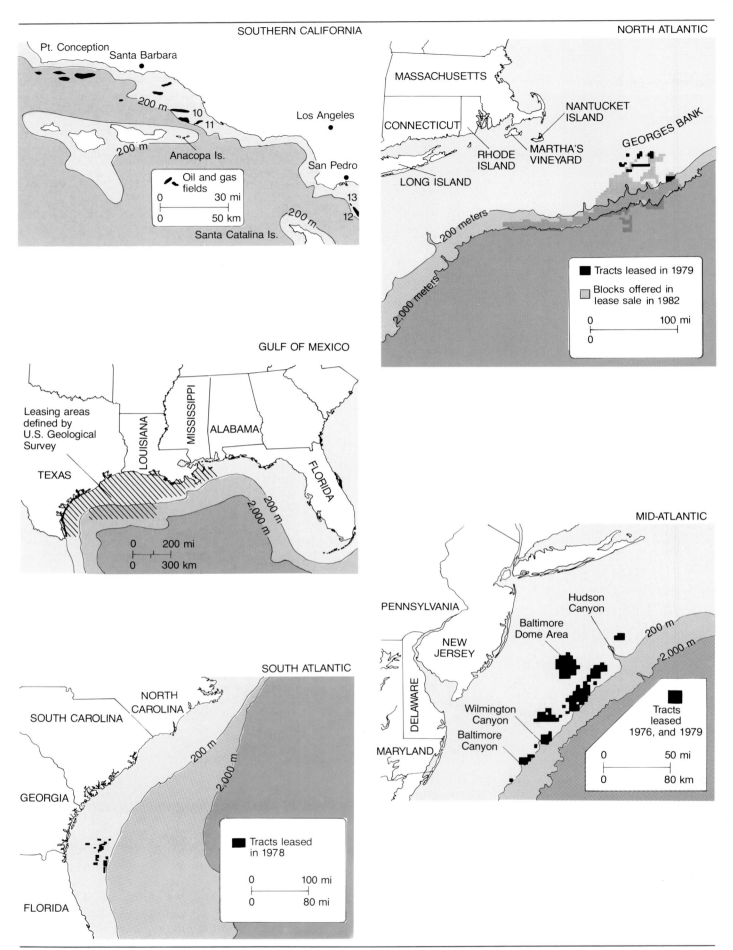

Fig. 2–49 Offshore exploration areas in lower 48 states.

Source: U.S. Department of the Interior, *Energy Resources on Federally Administered Lands*, 1981.

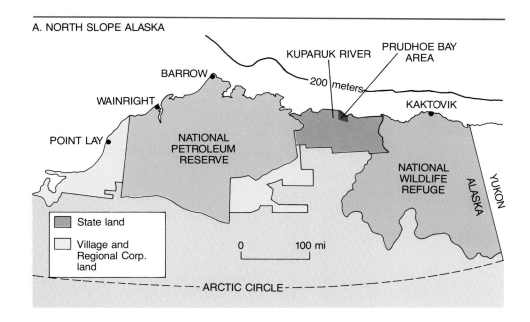

A. NORTH SLOPE ALASKA

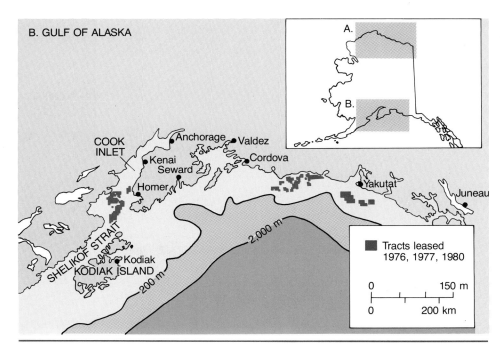

B. GULF OF ALASKA

Fig. 2-50 Alaska's north slope and the gulf of Alaska.
Source: U.S. Department of the Interior, *Energy Resources on Federally Administered Lands,* 1981.

tains where compressional forces associated with the mountain-building have strongly folded and faulted a series of sedimentary rocks and caused whole sheets of rock to override other sheets. In a similar complex structural setting in the Province of Alberta one of the earliest oil fields was discovered (Turner Valley) and some very large gas fields were discovered more recently (Pincher Creek and

Waterton). Along the same geologic trend, are fields in northern Utah. The belt between these successes is where the prospects are thought to be. U.S. Geological Survey estimates of undiscovered oil and gas are noted in Figure 2-48.

For reference, a series of five offshore areas in the lower 48 states are mapped in Figure 2-49, which shows water

TABLE 2–10
Oil and Gas Resources in Lower 48 Offshore Areas

| OFFSHORE AREA | DEMONSTRATED RESERVES | | UNDISCOVERED RECOVERABLE | |
	Crude Oil (billions of barrels)	Natural Gas (trillions of cubic feet)	Crude Oil (billions of barrels)	Natural Gas (trillions of cubic feet)
Southern California	0.787	1.752	3.200	3.400
Gulf of Mexico	3.052	40.16	6.600	71.900
North Atlantic	—	—	1.400	5.700
Middle Atlantic	—	—	3.100	14.200
South Atlantic	—	—	0.900	3.600

Source: Various reports of the U.S. Geological Survey, as reported in U.S. Department of the Interior, *Energy Resources on Federally Administered Lands,* Nov. 1981.

TABLE 2–11
Oil and Gas Resources in Alaska—North Slope and Gulf of Alaska

RESERVES	CRUDE OIL (BILLIONS OF BARRELS)	NATURAL GAS (TRILLIONS OF CUBIC FEET)	UNDISCOVERED RECOVERABLE RESOURCES	CRUDE OIL (BILLIONS OF BARRELS)	NATURAL GAS (TRILLIONS OF CUBIC FEET)
Beaufort Sea	—	—	Beaufort Sea	7.0	35.0
Chukchi Sea	—	—	Chukchi Sea	1.4	6.4
Prudhoe Bay	7.819	28.831	Arctic Coastal Plain	4.4	18.1
Kuparuk River	0.448	0.206	North Foothills	1.4	11.7
South Barrow	not available	0.240	Southern Foothills and Brooks Range	0.2	2.0
Gulf of Alaska	—	—	Gulf of Alaska	0.4	2.2

Source: Various reports of the U.S. Geological Survey, as reported in U.S. Department of the Interior, *Energy Resources, on Federally Administered Lands,* Nov. 1981.

depths, leasing activity, and *fields* in the case of California. U.S. Geological Survey estimates of undiscovered oil and gas are listed in Table 2–10.

In Figure 2–50, two portions of Alaska are mapped. The northern map shows the setting for Prudhoe Bay area, where the nation's largest single oil field was discovered in 1970. The southern map shows areas of leasing activity in the Gulf of Alaska, and also includes Valdez, the terminal where crude from the trans-Alaska pipeline is loaded on to tankers. Crude oil and natural gas resources in Alaska are summarized in Table 2–11.

References

American Gas Association. *The Underground Storage of Gas in the United States and Canada, Dec. 31, 1982.* Arlington, Va., 1983.

American Petroleum Institute. *Reserves of Crude Oil, Natural Gas Liquids, and Natural Gas in the United States and Canada as of Dec. 31, 1979.* Vol. 34, June 1980.

Canada Today/D'aujourd' hui 8:5, 1977. Published by Canadian Embassy, Washington, DC.

Cook, Earl. *Man, Energy, Society.* San Francisco: W. H. Freeman and Co., 1976.

Deul, Maurice. Quoted in "Natural Gas: United States Has It If Price Is Right." *Science,* Feb. 13, 1976.

Gillette, R. "Oil and gas resources—Did the USGS gush too high?" *Science,* 185:4146, 1974, 127–130.

Hubbert, King. "Energy Resources." In *Resources and Man.* Committee on Resources and Man, National Academy of Sciences and National Research Council. San Francisco: W. H. Freeman and Co., 1969.

Hubbert, King. "U.S. Energy Resources, A Review as of 1972, pt. 1." *In A National Fuels and Energy Policy Study: U.S.* 93rd. Congress, 2d. session, Senate Committee on Interior and Insular Affairs, Committee Print, Serial no. 93-40 (92-75), 1974.

Jimison, John W. *National Energy Transportation; Volume 1, Current Systems and Movements.* Prepared by Congressional Research Service, with maps jointly prepared by that service and the U.S. Geological Survey, Publication No. 95-115, May 1977.

Kuuskraa, V. E., et al. "Vast Potential Held by Four Unconventional Gas Sources." *Oil and Gas Journal,* June 12, 1978, 48-54.

Oil and Gas Journal. Published weekly by Petroleum Publishing Company, 1421-South Sheridan Rd., Tulsa, Oklahoma. Various issues, cited specifically in text.

National Academy of Sciences. *Mineral Resources and the Environment.* A report by Committee on Mineral Resources and the Environment, National Research Council, Washington, DC, 1975.

Nehring, Richard, (with Van Driest, E. R., II). *The Discovery of Significant Oil and Gas Fields in the United States.* The Rand Publication Series R-2654/1-USGS/DOE, (with Appendix v. 2, of 477 p.), 1981.

Riva, John, U.S. Congressional Research Service Report No. 84-129 SPR, Washington, D.C., 1984. Reported in *Science,* Oct. 26, 1984.

Theobald, P. K., Schweinfurth, S. P., and Duncan, D. C. "Energy Resources of the United States." U.S.G.S. Circular 650, 1972.

U.S. Department of Energy (DOA). *Annual Energy Review, 1983.*

DOE/EIA-0384, Washington, DC.: Government Printing Office, April 1984.

U.S. Department of Energy. *Fossil Energy Research and Development Program of the U.S. Department of Energy,* FY 1979. Assistant Secretary for Energy Technology, Washington, D.C. March, 1978.

———. *Natural Gas Annual, 1982.* DOE/EIA-0131 (82), Washington, DC.: Government Printing Office Oct. 1983.

———. *Outlook for Oil Imports.* Washington, D.C.: Government Printing Office Sept., 1982.

———. *Petroleum Supply Annual, 1983.* DOE/EIA-0340 (83), Volume 1 of 2. Washington, DC: Government Printing Office, June 1984.

———. *Petroleum Supply Monthly.* Various issues.

———. *U.S. Crude Oil, Natural Gas, and Natural Gas Liquid Reserves, 1983 Annual Report.* DOE/EIA-0216 (83), Advance Summary. Washington, DC: Government Printing Office, Sept. 1984.

U.S. Department of the Interior. *Energy on Federally Administered Lands.* Washington, DC: Government Printing Office, Nov. 1981.

U.S. Energy Research and Development Administration. *Fossil Energy Research Meeting.* Washington, DC, December 1977. CONF 7706100.

U.S. Geological Survey. *Estimates of Undiscovered Recoverable Conventional Resources of Oil and Gas in the United States.* U.S.G.S. *Circular 860,* 1981.

———. *Assessment of Geothermal Resources of the United States —1978.* U.S.G.S. Circular 790, 1979.

———. *Geological Estimates of Undiscovered Recoverable Oil and Gas Resources in the United States.* U.S.G.S. Circular 725, 1975.

3 Oil Shales and Tar Sands

←Green River oil shales near entrance to a mine at the Anvil Points facility of Paraho Development Corporation. Photo courtesy of U.S. Department of Energy.

OIL SHALES

Oil shale represents one of the largest untapped sources of hydrocarbon energy known to man. Deposits occur throughout the world, on every continent, and constitute a greater potential energy source than any other natural material with the exception of coal. The Rocky Mountain area of the United States alone contains oil shale representing billions of barrels of oil. This resource, however, has remained virtually undeveloped because supplies of conventional crude have been available at lower development costs.

Oil shales and tar sands potentially represent two of the major sources of synthetic crude oil in the United States. Next to coal, oil shales comprise the second largest fossil fuel resource in the country. The major oil shale deposits are found in western states: Colorado, Wyoming, and Utah. For more than 50 years, oil companies have been experimenting with extracting oil from shale, but until now, economic realities have prevented commercialization.

In Canada, the bitumen content of the well-known Athabasca Tar Sands is already being converted to syncrude on a commercial scale. Although the tar sands of the United States are not nearly as extensive or rich as those in Canada, scattered rich deposits do occur in the West, principally in Utah. However, none of these deposits are anywhere near commercial exploitation.

Technically, oil shale contains no oil; instead, it contains an organic-rich material known as *kerogen*. Because the organic material in these shales may be converted to liquid oil by heating, the shales may be thought of as *source rocks* for petroleum (see Chapter 2). Apparently these shales never were subjected to enough natural heat to accomplish the conversion process. In a mining and retorting operation, the shale is subjected to destructive distillation or "retorting" in a large pot (retort). High temperatures in that process convert the waxy organic *kerogen* to a liquid hydrocarbon, thereby finishing the evolution of oil that was begun by natural processes. During the distillation process, oil vapors begin to appear at

121

temperatures between 480 and 660 degrees Fahrenheit. Practically all of the *kerogen* is converted to oil when the temperature approximates 900 degrees Fahrenheit.

In comparison with conventional crude, shale oil exhibits a number of important physical differences. It usually has a high viscosity due to wax content and does not flow freely at room temperature. In order to facilitate movement through a pipeline it must be heated to a temperature over 90 degrees Fahrenheit. In addition, shale oil commonly contains excessive impurities, such as sulfurs, nitrogen, and ammonia. When these are removed, however, shale oil can be refined into a full range of petroleum products.

Oil shale deposits of various sizes and grades are found throughout the world in unmetamorphosed sedimentary rocks dating geologically from the Cambrian to Tertiary ages. The organic-rich material present in these rocks was derived from aquatic plants and animals deposited in water. The principal environments ranged from small bodies of water, such as lakes and lagoons near coal-forming swamps, to large marine and lake basins (Duncan and Swanson, 1965). Most of the world's oil shale resources are found in only a few locations—the Tertiary deposits of the western United States, the Permain deposits of southern Brazil, the Cambrian and Ordovician deposits in northern Europe and Asia, in the Jurassic rocks of western Europe and Russia, and the Triassic oil shale of central Africa.

Resource Terminology

As with crude oil and natural gas, some shale oil resources are geologically better known and others are less certain. Those shale oil amounts based on deposits whose magnitude and richness can be established within reasonable limits are considered *Identified*. In contrast, *Hypothetical* resources of shale oil, which may be loosely compared with undiscovered amounts of conventional oil and gas, are in deposits whose approximate dimensions can be predicted by geologic inference, but whose richness or yield is based on very scanty data. In assessing Identified resources, estimates can be made of oil (kerogen) in place in the deposits. Recoverable amounts of oil can then be derived by estimating what proportion of oil in-place can be extracted. This estimate must take into consideration the thickness and continuity or persistence of beds and how accessible the beds are to mining or other recovery methods.

In the following discussion of world shale oil are three types of *Identified* resources: (1) *Proved recoverable reserves,* (2) *Estimated additional resources,* and (3) *Potentially recoverable resources.*

Proved recoverable reserves are the fraction of proved reserves in-place that can be recovered economically. *Estimated additional resources* are all resources, in addition to proved reserves, that are of at least forseeable eco-

nomic interest. The estimates provided for additional resources reflect, if not certainty about existence of the entire quantities reported, at least a reasonable level of confidence. *Potentially recoverable resources* are all those identified resources whose economic extraction is potentially feasible.

WORLD SHALE OIL RESOURCES

The total *Identified* shale oil resources in the world are estimated at over 3 trillion barrels (Table 3–1 and Fig. 3–1). According to the World Energy Conference, only 5 percent of the total shale oil resources are *Proved recoverable resources* while 66 percent are *Estimated additional recoverable resources* (World Energy Conference, 1980).

Of the *Proved recoverable resources,* 35 percent are found in Morocco, 32 percent in the Soviet Union and 13 percent in the United States. On the other hand, the United States houses 83 percent of the *Estimated additional recoverable resources* while the Soviet Union has most of the remainder. Taking into account the total *Identified* resources reported by the World Energy Conference, the United States houses 68 percent and the Soviet Union 25 percent.

Although the world's shale oil resources are widely distributed over the earth, the deposits vary in thickness from a few feet to a few hundred feet, and the kerogen content varies from a few gallons per ton to over 100 gallons per ton in the Marahu Shale in Brazil.

The literature reveals that other estimates of world shale oil resources are similar to those reported by the World Energy Conference. Based on 10 to 25 gallons per ton of shale, the Exxon Corporation has estimated that the total shale oil resources have an oil content of approximately 3.2 trillion barrels. They estimate that only 180 billion barrels of this are recoverable. The Federal Institute for Geosciences and Natural Resources, Hanover, Germany, place total world resources at 3.4 trillion barrels of oil of which approximately 210 billion barrels are considered to be economically recoverable. (Grathwohl, 1982).

Presently, commercial shale oil is produced only in the Soviet Union and the Peoples Republic of China. Demonstration projects have been started in the United States and Brazil, and commercial oil shale production should begin in both of these countries by the end of this decade.

UNITED STATES SHALE OIL RESOURCES

Oil shale occurs in marine and lacustrine (sediments produced by lakes) deposits in three principal areas in the United States: (1) in the Chattanooga Shales and equivalent deposits in central and eastern states; (2) in marine

TABLE 3-1
World Distribution of Shale Oil Resources (billions barrels of oil)

COUNTRY	PROVED RECOVERABLE RESERVES[1]	ESTIMATED ADDITIONAL RESOURCES[2]	POTENTIALLY RECOVERABLE RESOURCES[3]
United States	20.1	1,694.0	2,104.0
Soviet Union	49.0	353.0	—
Brazil	0.6	—	768.3
Zaire	—	—	93.3
Morocco	53.1	—	—
Canada	—	—	50.3
Italy	—	—	35.9
China (Peoples Rep.)	—	—	28.7
Thailand	14.5	—	—
Sweden	6.3	—	—
Jordan	5.7	—	—
Australia	—	3.5	—
Germany, F.R.	1.8	—	—
Spain	0.1	—	—
Other Countries	—	—	14.4
Total	151.2	2,050.5	3,094.9

[1]*Proved recoverable reserves.* The fraction of proved reserves in place that can be economically recovered.

[2]*Estimated additional resources.* All resources, in addition to proved reserves, that are of at least forseeable economic interest. The estimates provided for additional resources reflect, if not certainty about existence of the entire quantities reported, at least a reasonable level of confidence. Speculative resources are not included.

[3]*Potentially recoverable resources.* A concentration of naturally occurring solid, liquid, or gaseous materials in or on the earth's crust in such form that economic extraction of a commodity is potentially feasible.

Notes: 1. One metric ton of oil is equivalent to 42.2×10^9 Joules or one metric ton of oil is equivalent to 7.18 barrels of oil.
2. For countries that list only potentially recoverable resources, economic extraction of some of those resources may currently be feasible.

Source: World Energy Conference. *Survey of Energy Resources, 1980.* Munich: September 1980.

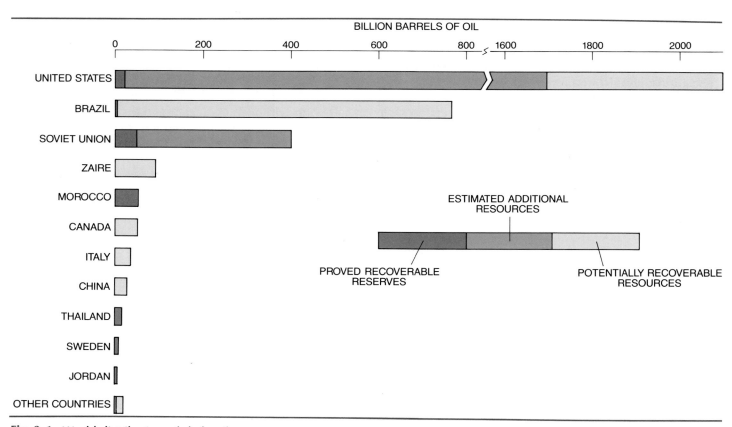

Fig. 3-1 World distribution of shale oil resources.
Source: World Energy Conference, 1980.

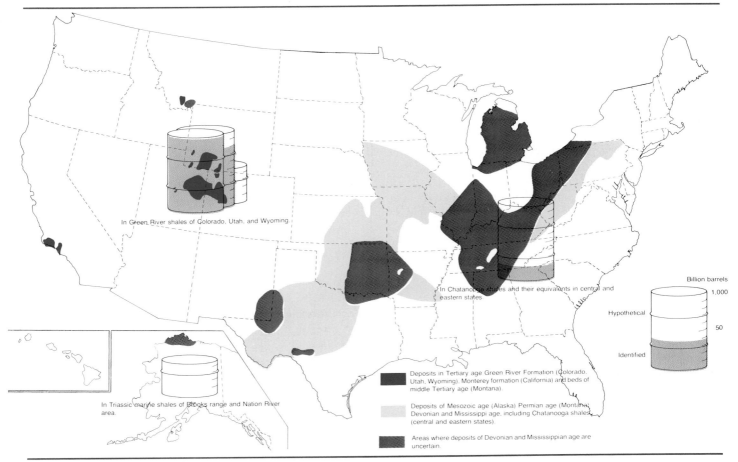

Fig. 3–2 Shale oil amounts in U.S. deposits of richness 15 gallons per ton and over.
Source: Culbertson and Pitman, U.S.G.S. *Professional Paper, 820, 1973.*

shales in Alaska; and (3) in the Green River formation in Colorado, Utah, and Wyoming.[1]

For these areas the approximate amounts of shale oil *in-place* in beds of richness over 15 gallons per ton are shown in Figure 3–2 according to their status as Identified or Hypothetical amounts. Table 3–2 roughly divides the in-place amounts according to richness of the deposits. From these two summaries it is apparent that the western states have the major portion, two-thirds of which is in the Identified category. The total deposits in central and eastern states are smaller; most are Hypothetical in character, and are spread over a very large area. Alaskan deposits, according to present knowledge, are relatively small and almost entirely Hypothetical.

Chattanooga Shales and Equivalent Formations

Black shale deposits of Mississippian and Devonian age underlie vast areas in the eastern-central United States between the Appalachian and Rocky Mountains (Fig. 3–2). They yield so little oil by conventional recovery methods, 1 to 15 gallons per ton, however, that it is unlikely that recovery of oil from these shales will be considered economically feasible any time in the forseeable future (Duncan and Swanson, 1965; Culbertson and Pit-

TABLE 3–2
United States Identified and Hypothetical Shale Oil Resources In-Place, Showing Richness Categories, as Estimated in 1973 (in billions of barrels in-place)

DEPOSIT	IDENTIFIED[1]		HYPOTHETICAL[2]		TOTAL
	25–100 Gal/Ton	10–25 Gal/Ton	25–100 Gal/Ton	10–25 Gal/Ton	
Green River Formation, Colorado, Utah, and Wyoming	418[3]	1,400	50	600	2,468
Chattanooga Shale and equivalent formations, Central and Eastern United States	0	200	0	800	1,000
Marine Shale, Alaska	small	small	250	200	450
Total	418	1,600	300	1,600	3,918

[1] Identified Resources: Specific, identified mineral deposits that may or may not be evaluated as to the extent and grade, and whose contained minerals may or may not be profitably recoverable with existing technology and economic conditions.

[2] Hypothetical Resources: Undiscovered mineral deposits, whether of recoverable or subeconomic grade, that are geologically predictable as existing in known districts.

[3] The 25–100 gallons per ton category is considered virtually equivalent to the category average of 30 or more gallons per ton.

Source: Culbertson, W., and Pitman, J. *U.S. Geological Professional Paper, 820,* 1973.

man, 1973). Using Duncan and Swanson's minimum thickness and grade criteria of 5 *feet of shale yielding 10 gallons of oil per ton,* the Identified resources of these deposits are estimated to be 200 billion barrels of oil and the Hypothetical resources 800 billion barrels. Two of the more interesting formations containing marine black shales are the Chattanooga Black Shales in the central part of the eastern Highland Rim of Tennessee, and the New Albany Shale in Indiana and Kentucky. The Chattanooga Shales are on the average 15 feet thick, and yield an average of 10 gallons of oil per ton. Some areas of the Albany Shales contain zones 20 to 100 feet thick and yield an average of 10 to 12 gallons of oil per ton of shale.

Alaskan Marine Shales

Oil shales of marine origin underlie substantial areas in both northern and eastern Alaska, in rock sequences ranging from Mississippian to Cretaceous in age. As of yet, none of the Alaskan deposits have been appraised carefully enough to allow estimates of Identified amounts. In eastern Alaska near the Nation River, however, local out-crops of Triassic oil shales reportedly are 200 feet thick and yield 30 gallons of oil per ton (Duncan and Swanson, 1965). Triassic and Cretaceous oil shales on the north slope of the Brooks Range are reported to contain thin zones, less than 5 feet thick, that yield as much as 160 gallons of oil per ton of rock. Other even thinner zones are known to yield 15 gallons of oil per ton. Duncan and Swanson estimated that Hypothetical resources of Alaska total 450 billion barrels of oil of which 200 billion barrels would be in high-grade deposits.

The Green River Formation

By far the richest and most extensive oil shale deposit in the United States, and the world, is in the Green River Formation, which covers roughly 17,000 square miles in northwestern Colorado and adjacent areas in Utah and Wyoming (Fig. 3–3). This formation was deposited about 50 million years ago in two large lakes of the Eocene age: Lake Uinta, which covered northwestern Colorado and part of Utah, and Lake Gosiute, which covered southern Wyoming.[2] The formation ranges in thickness from a few

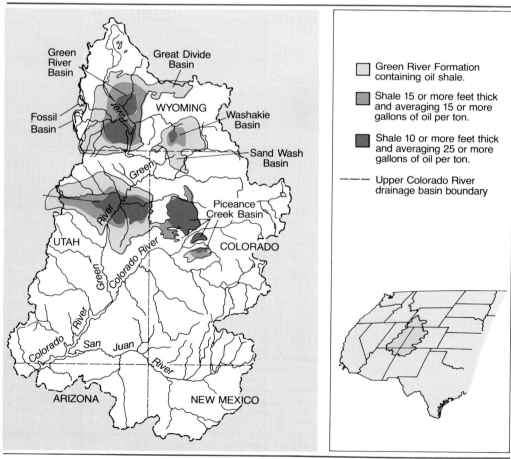

Fig. 3–3 Oil shale deposits in the Green River Formation.
Source: U.S.G.S. *Synthetic Fuels Development,* 1979.

This twenty-ton pile of rich oil shale can yield enough oil to fill twelve 55-gallon oil drums. (Courtesy of American Petroleum Institute and Standard Oil Company of California.)

TABLE 3–3
In-Place and Recoverable Shale Oil Amounts in Identified Deposits of the Green River Formation (in billions of barrels)

LOCATION	IN-PLACE	RECOVERABLE
Over 15 Gallons/Ton		
Colorado	1,200	400
Utah	321	105
Wyoming	321	105
TOTAL	1,842	610
Over 25 Gallons/Ton		
Colorado	607	202
Utah	64	21
Wyoming	60	20
TOTAL	731	243
Over 30 Gallons/Ton		
Colorado	365	118
Utah	50	17
Wyoming	13	4
TOTAL	418	139
Over 30 Gallons/Tons in a Selected Mineable Seam		
Colorado		48
Utah		6
Total		54

Note: Oil shale resources in the 15, 25, and 30 gallons per ton categories, estimated by John Donnell of the U.S.G.S. for *Project Independence*, appeared in the February 1976 issue of *Shale Country*. The resources with over 30 gallons per ton in a selected mineable seam of the Mahogany Zone were reported by the National Petroleum Council in *U.S. Energy Outlook–Oil Shale Availability, 1973.*

feet to several thousand feet, and consists of marlstone interbedded with various amounts of siltstone, sandstone, halite, nahcolite, and dawsonite.[3] If the estimated 1800 billion barrels of Identified oil in-place is to be properly assessed, its varying richnesses and the amounts recoverable must be understood.

Table 3–3 and Figures 3–4 and 3–5 show the locations of various components of the total according to both the oil yield per ton of rock, and the amounts recoverable, which in every case are simply *one-third* of the estimated amounts in-place. This one-third ratio, similar to the average primary recovery rate for conventional crude oil, is based on very limited experience in recovering shale oil, but is a reasonable approximation to apply to large areas with deposits of varying character (Donnell, 1976).

As with many mineral deposits, there are large amounts of lower-grade shale, and much smaller amounts of the higher-grade. If shales yielding only 15 gallons of oil per ton are considered, there is a total of slightly over 1,840 billion barrels of oil in-place and 610 billion barrels recoverable, two-thirds of which is in Colorado (see Figs. 3–4, 3–5). If a more rigorous criterion demanding over 25 gallons per ton is applied, there are 731 billion barrels in-place and 243 billion recoverable, about 90 percent of which is in Colorado. In the richest category, 30 gallons per ton or over, there are only 418 billion barrels of oil in-place, and 139 billion barrels recoverable. Of these high-grade resources, 85 percent are in Colorado, 12 percent in Utah, and 3 percent in Wyoming.

Colorado leads all other states in volume of shale oil because of the thick and rich deposits in the Piceance Basin identified in Figure 3–4. This geologic basin lies at the center of the ancient Lake Uinta, contains beds of high kerogen content, and varies from a few feet at its edges to 2,300 feet in the basin's center. The overburden is quite thick, about 1,000 feet (Welles, 1970).

Two principal layers of rich shale occur in the Piceance Basin: (1) the Mahogany Zone, which is 50 to 200 feet thick and averages 30 gallons per ton, and (2) a deeper zone, much thicker than the Mahogany, but less rich and restricted to the center of the basin. Considering both of the rich zones together, a single square mile at the center of the basin may contain as much as 2.5 billion barrels of shale oil in-place (Welles, 1970). Because the rich Mahogany Zone is exposed in the accessible cliffs of the Colo-

Rich beds of Green River Oil Shales exposed in Colorado.

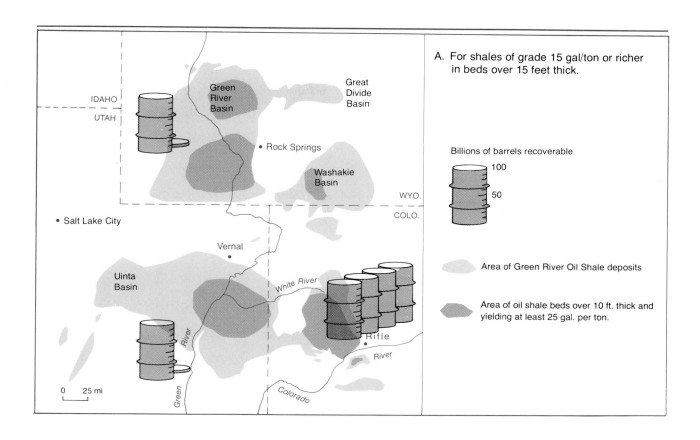

A. For shales of grade 15 gal/ton or richer in beds over 15 feet thick.

Billions of barrels recoverable

100
50

Area of Green River Oil Shale deposits

Area of oil shale beds over 10 ft. thick and yielding at least 25 gal. per ton.

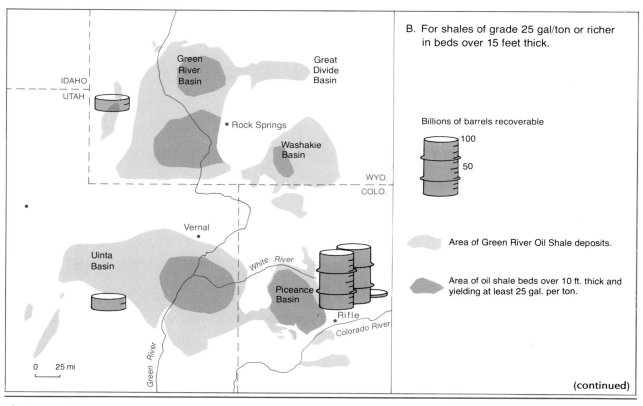

B. For shales of grade 25 gal/ton or richer in beds over 15 feet thick.

Billions of barrels recoverable

100
50

Area of Green River Oil Shale deposits.

Area of oil shale beds over 10 ft. thick and yielding at least 25 gal. per ton.

(continued)

Fig. 3–4 Shale oil amounts recoverable from various beds in the Green River Formation, showing areas of deposits and of richer beds.

Sources: Donnell, 1976 and National Petroleum Council, 1973.

Fig. 3–4 *(continued)*

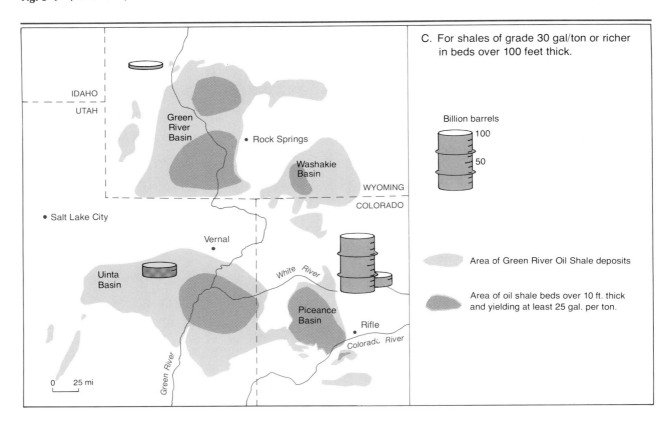

C. For shales of grade 30 gal/ton or richer in beds over 100 feet thick.

Billion barrels

Area of Green River Oil Shale deposits

Area of oil shale beds over 10 ft. thick and yielding at least 25 gal. per ton.

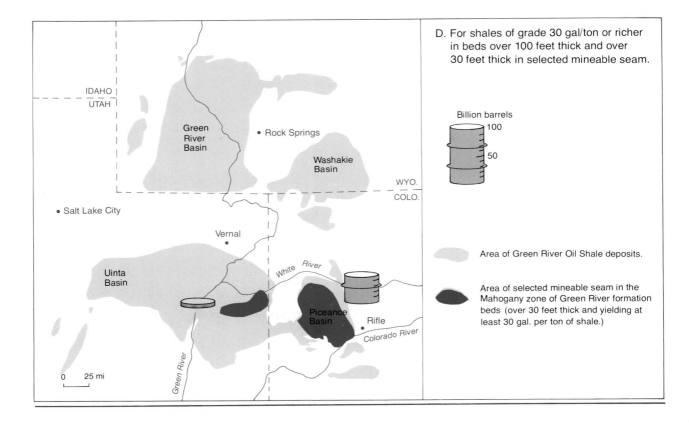

D. For shales of grade 30 gal/ton or richer in beds over 100 feet thick and over 30 feet thick in selected mineable seam.

Billion barrels

Area of Green River Oil Shale deposits.

Area of selected mineable seam in the Mahogany zone of Green River formation beds (over 30 feet thick and yielding at least 30 gal. per ton of shale.)

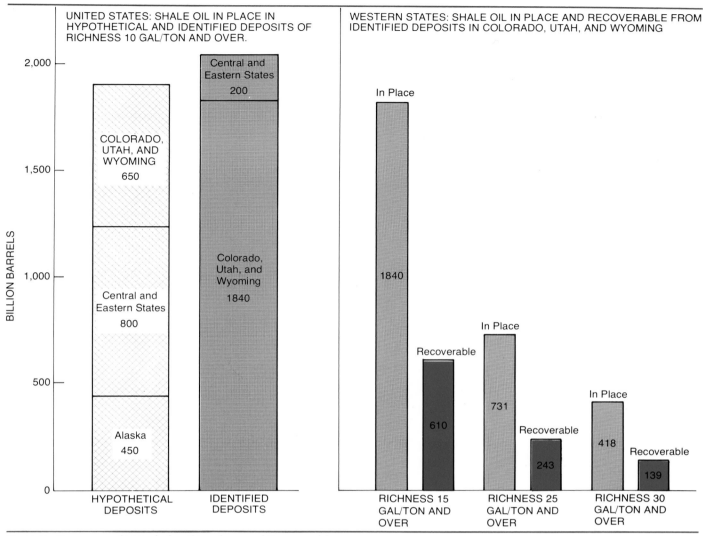

Fig. 3–5 Summary of U.S. shale oil in-place and Recoverable.
Sources: Culbertson and Pitman, U.S.G.S. *Paper 820,* 1973 and Donnell, 1976.

rado River and its tributaries, such as the one near Rifle, Colorado, oil shale research operations have been concentrated in such locations.

Table 3–3 and Figure 3–4 single out particular mineable beds that average over 30 gallons of oil per ton and are at least 30 feet thick everywhere. Therefore they are susceptible to about 60 percent recovery of the shale through underground room and pillar mining.[4] (See Chapter 2 for description of these mining methods.) Oil recoverable from these most attractive beds amounts to 48 billion barrels in Colorado and 6 billion barrels in Utah. So far, none have been identified in Wyoming because exploration has not been so extensive.

Assessment of Resources

Given the current status of shale oil development in the United States it is, perhaps, risky to assign recoverable amounts to *reserves.* Closest to the line of economic feasi-

Green River oil shale in cores cut by drilling rig in background. (Courtesy of American Petroleum Institute and Colony Development Corporation.)

bility are the amounts just discussed. The beds presently being developed are principally those in the thick and rich Mahogany Zone. Progressively more remote in economic feasibility are the leaner beds that do not yield as many gallons per ton. If *all* beds averaging over 15 gallons per ton were processed, 610 billion barrels would be recoverable. It is important to realize that the four recoverable estimates shown in Table 3–3 cannot be added. In each case, they indicate the total recoverable oil based on four different assumptions about which beds (rich or lean) will contribute to the total.

The recoverable amounts are, of course, only estimates derived from applying a "blanket" recovery ratio of one-third (with the exception of the higher recovery for 30-foot beds averaging over 30 gallons per ton). In specific locations this one-third ratio may not apply, as factors of bed thickness and continuity, overburden, or water in the deposit vary from place to place. If, for instance, the rich deposits in the Piceance Basin, which contribute heavily to the recoverable estimates, contain unmanageable amounts of water, then recovery will be less than estimated because certain beds will not be mined. If, on the other hand, *in situ* recovery processes are utilized, then oil may be recovered from deeply buried low-grade deposits not represented in the present estimates of amounts recoverable. Despite the uncertainty of the estimates, it is nevertheless interesting to note the following comparison: If high-grade beds averaging 30 gallons per ton are exploited, the total recoverable oil in them (139 billion barrels) is roughly the same as the total of *conventional* crude oil

recoverable from identified and undiscovered reservoirs on the basis of traditional recovery rates.

Recovery Processes

Two major options are being considered for oil shale recovery: (1) mining followed by surface processing, and (2) *in situ* processing (Fig. 3–6). Mining followed by surface processing is better understood, but economic and environmental factors associated with it have raised serious doubts as to its use in the early years of commercial-scale production. The *in situ* process is presently being developed, and the modified version successfully tested by Occidental oil appears to be the favored recovery process at this time.

MINING AND SURFACE PROCESSING The best-developed mining technique is the underground roof and pillar method developed by the Bureau of Mines between 1944 and 1956, and later improved by Union Oil, Colorado School of Mines, and the Colony Development Operation. Simply put, the technique involves cutting a series of room openings and roof support pillars, both 60 feet square, in a mineable seam of shale. The Bureau of Mines has achieved a shale extraction ratio of 75 percent using this technique, the supporting pillars being the remaining 25 percent.

Surface or open-pit mining has not been tested on oil shale to any large degree. Since the technique is highly developed for mining other ores, it may be practical in areas where overburden is thin enough.

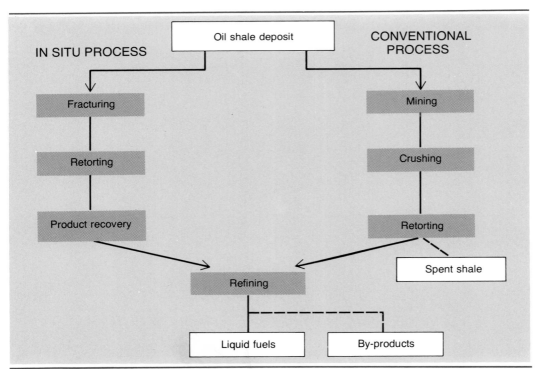

Fig. 3–6 Various operations required in *in-situ* and in conventional surface oil shale processing.

Inside an oil shale mine. (Courtesy of American Petroleum Institute and Colony Development Corporation.)

MODIFIED IN SITU PROCESSING One of the major goals of oil shale research has been to recover shale oil without mining. *In situ* processing, or processing the shale while it is still underground, is the method by which the shale does not have to be mined. The major problem with the use of *in situ* processing, however, is the fact that shale is not a permeable rock. As a result, it is difficult to circulate the fluids necessary to heat the rock in place.

The modified *in situ* process developed by Occidental Oil relieves some of the difficulties. In this process, about 20 percent of the volume of shale that is to be tapped is first mined. This mining serves to form chambers in which the rock is fragmented by drilling and blasting in a procedure that ultimately leaves a series of chambers separated by undisturbed rock (Fig. 3–8). The removed shale can then be retorted at the surface. Occidental has produced about 30,000 barrels of oil using their modified *in situ* process. Five thousand barrels have been marketed through a Michigan utility where it was successfully burned in boilers without further refining.

The use of *in situ* recovery processes substantially reduces the environmental impact of a shale oil plant. It requires only one-third of the work force needed for mining and surface retorting techniques, only 25 to 33 percent of the water, and it reduces spent shale disposal by at least 80 percent (*Science,* December 9, 1977). The modified *in situ* process recovers 60 to 70 percent of the organic material, about the same conversion efficiency as is achieved by surface retorting. This modified process may be practical only in thick beds close to the surface. In the long term, therefore, *in situ* conversion with no mining will be necessary.

After the shale is mined, it is crushed and then transported to surface retorts. These are of three types classified according to the method used to provide heat. The crushed shale may be heated by gas combustion, recycled gas, or recycled heat-carrying solids. The TOSCO II process developed by The Oil Shale Corporation supplies heat to the shale by direct contact with recycled, heat-carrying ceramic balls. This is one of the processes that would likely be used should surface techniques be used in recovering oil from shale (Fig. 3–7).

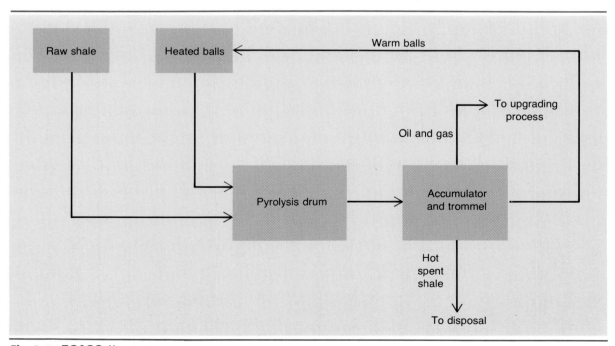

Fig. 3–7 TOSCO II retort.
Source: U.S. Department of Interior. *Final Environmental Statement for the Oil Shale Leasing Program* (Vol 1 of 6), *Regional Impacts of Oil Shale Development,* 1973.

A U.S. Department of Energy field site near Rocky Springs, Wyoming, where experiments are being conducted on in situ recovery of crude oil from the Green River oil shales. (Courtesy of U.S. Department of Energy.)

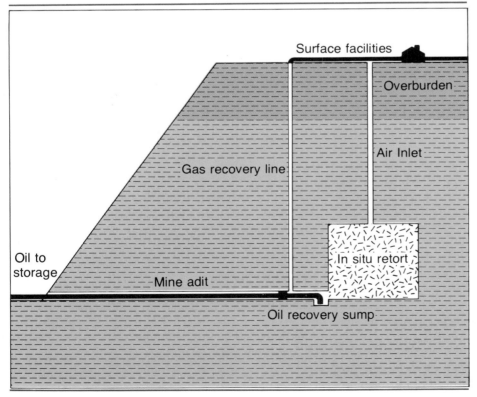

Fig. 3–8 *In-situ* shale oil recovery process.
Source: Occidental Petroleum, "Logan Wash: Where the Rocks Yield Oil," 1978.

History of Oil Shale Development

Extraction of oil from shale is not a recent event. In the United States, lamp oil, lubricants, and medicines were produced from lean oil shales in the eastern Appalachians during the early 1800s. The 1859 discovery of oil at Titusville, Pennsylvania, however, destroyed hopes for a commercial oil industry because it offered a cheaper source of crude oil. Since that time, several attempts have been made to revitalize interest in oil shale.

Shortly after World War I, the U.S. Bureau of Mines constructed an oil shale experimental plant near Rifle, Colorado which operated as a research facility until 1929 (East and Gardner, 1962). The plant was closed with the discovery of large oil fields in California, Oklahoma, and Texas. In the years after World War II new interest was kindled by the fear of near-term energy shortages. Many individuals began to predict that the oil supply might not fulfill the demand for it. As a result of these concerns, interest in oil shale development continued at a slightly accelerated pace throughout the 1950s and 1960s. During this period, the most publicized oil shale operation was one launched by a group of private companies in 1964 called the Colony Project.[5] This consortium began the construction of a pilot plant in western Colorado designed to establish the technical and economic feasibility of a commercial oil shale industry. After several years of successful work, a poor economic climate and the unwillingness of Congress to pass supportive legislation forced the suspension of the plan and operations by the Colony Group in 1974.

Although commercial oil shale technology was substantially developed during the 1960s, shale oil still could not be produced profitably at a price that would compete with domestic conventional crude oil and the still cheaper oil from the Middle East. But the 1973 Middle Eastern oil embargo brought on an instant collision of the supply and demand curves. Although the embargo lasted only a short time, it appears to have had a major impact on the United States' energy outlook: It clearly illustrated the United States' growing dependence on foreign oil and ended decades of cheap oil and cheap energy. With foreign oil rising to more than 11 dollars per barrel, it appeared that the economic barrier to commercial oil shale development had been removed, and shale oil's entry into the commercial market was inevitable.

THE PROTOTYPE OIL SHALE LEASING PROGRAM Approximately 72 percent of the land and 60 to 65 percent of the shale in the western United States is owned by the federal government. The government's renewed interest in shale development in the 1970s appeared to be the needed stimulus for establishment of a commercial oil shale industry. In 1971, President Nixon, in response to growing concerns regarding the adequacy of the nation's energy resources, requested that the Department of Interior initiate a leasing program to develop oil shale resources on public lands, provided that environmental considerations could be satisfactorily resolved (Department of Interior, *Vol. III*, 1973). Initially, the prototype program was formulated to make six tracts of land of not more than 5,120 acres each available for private development. After gathering resource information on various federal sites, the Department of Interior announced the selection of six tracts of about 5,000 acres each, two in the Piceance Creek Basin, two in the Uinta Basin, and two in the Washakie Basin (Table 3-4 and Fig. 3-9). In early 1974, the six tracts were offered to private companies in lease sales. The four in Colorado and Utah were bid upon and the highest bids were accepted by the Department of the Interior, but no bids were made on the two Wyoming tracts. The total value of the accepted bids paid to the federal government amounted to almost 450 million dollars for the four tracts.

Under the prototype leasing program, development has not progressed as smoothly as might have been expected. Following a year or more of gathering environmental baseline data, the companies submitted detailed development plans to the Department of Interior. But in the face of the rising cost of construction, lack of congressional support, environmental objections, and prohibitive environmental regulations, the companies elected not to proceed with development and requested a one-year suspension of their leases. Subsequently, their request was granted by the Department of Interior.

After the suspension of the leases, a number of events brightened the prospects for commercial oil shale development. Occidental Petroleum, which has had a great deal of success in developing underground conversion techniques, joined Ashland Oil as an equal partner in the Colorado Tract C-b venture. Shell Oil and the Oil Shale Corporation have dropped out of the lease. In 1978, Ashland Oil also dropped out of the C-b lease. Standard Oil of Indiana and Gulf Oil, the operators of Colorado tract C-a, also filed new development plans calling for *in situ* recovery of the shale oil.

In July 1979, President Carter called for the establishment of a government-run Energy Security Corporation that would supervise the spending of 88 billion dollars over a 12-year period. The major purpose of the corporation was to aid the development of synthetic fuels produced from oil shale, tar sands, coal, biomass, and organic waste. In 1980 the Energy Security Act was passed by Congress and gave the newly established Synthetic Fuels Corporation (SFC) a mandate to foster the development of a 500,000 barrels per day synthetic fuel industry by 1987. It authorized the SFC to commit as much as $14.9 billion in loan and price guarantees to synfuels projects. About 4.9 billion dollars of this money has been earmarked for oil shale projects.

TABLE 3-4
Summary Data: Oil Shale Prototype Tracts

TRACT	ACRES	ESTIMATED RECOVERABLE OIL SHALE (in billions of tons)		EXPECTED RECOVERY METHOD	COST OF LEASE (millions of dollars)	LEASING COMPANIES
		30 or More Gal/Ton	20 or More Gal/Ton			
Colorado C-a	5,089	1.857		underground	210.3	Standard Oil of Indiana and Gulf
Colorado C-b	5,093	1.012		underground	117.8	Atlantic Richfield, The Oil Shale Corp., Shell Oil and Ashland Oil[1]
Utah U-a	5,120	.342		underground	75.6	Phillips Petroleum and Sun Oil
Utah U-b	5,120	.372		underground	45.0	Standard of Ohio, Phillips Petroleum, and Sun Oil
Wyoming W-a	5,111		.354	in situ	no bid	
Wyoming W-b	5,083		.352	in situ	no bid	

[1] In the first year of the lease, Atlantic Richfield and the Oil Shale Corporation dropped out of tract C-b development. In the fall of 1976, Shell Oil dropped out of the lease and Occidental oil has become an equal partner with Ashland oil.

Sources: U.S. Department of Interior. *Final Environmental Statement for the Prototype Oil Shale Leasing Program*, Volume III of VI, *Specific Impacts of Prototype Oil Shale Development*, Washington: Government Printing Office, 1973 and *Department of Interior News Releases, 1974.*

Prospects for Commercialization

The basic problem that has long bothered the shale oil industry is that of economics. Can shale oil be produced at a cost competitive with oil imports? Because of the high rate of inflation during the past decade, private investors have been reluctant to invest the large amounts of capital necessary to foster the development of a commercial shale oil industry. Table 3-5 illustrates the effect of inflation on the prospects of commercial oil shale development. The data were derived for the Colony Project, until recently a private venture of Exxon and The Oil Shale Corporation. The Colony data were selected because they afford the best-developed base for which detailed cost estimates are available. As the data indicate, the necessary capital investment for a Colony-type plant increased nearly seven-fold between 1972 and 1981 and the required price for upgraded oil rose by the same ratio. The drop of world crude oil prices in recent years has compounded shale oil development problems. However, the price and loan guarantees offered by the Synthetic Fuels Corporation has moved a number of projects very near commercial status.

CURRENT DEVELOPMENT ACTIVITIES In 1984 there were five major commercial shale oil projects in various stages of completion: (1) *Rio Blanco,* (2) *Clear Creek,* (3) *Union Oil,* (4) *Logan Wash,* and (5) *Colony.* The following is a description of these projects and their current statuses:

Rio Blanco This project is a partnership of Gulf Oil and Standard Oil of Indiana located on Federal Tract C-a near Rangely, Colorado. The long-range plans of Rio Blanco are to produce 50,000 barrels per day of oil using both modified in-situ and surface retorts. The partnership presently has a 2,000 barrels per day surface retort demonstration plant. In 1981, 23,000 barrels of shale oil were produced from the *in situ* demonstration retort during a 6-month burn. In 1982, Rio Blanco asked the Department of Interior for more time to start commercial mining and retort operations. The companies claim that the success of the entire project depends on acquisition of a surface lease that is needed for plant siting and waste disposal. Without the lease, Rio Blanco claims they will not be able to begin commercial operation by 1995 as required by the oil shale

TABLE 3-5
Cost Estimates: Colony Shale Oil Plant of 55,000 Barrels per day Capacity

ESTIMATE DATE	INITIAL CAPITAL ($mm)	DIRECT OPERATING COSTS ($/bbl)	OIL PRICE FOR 10% DCF RATE OF RETURN ($/bbl)
1972	255	1.82	4.40
March 1974	425	3.13	8.00
August 1974	653	4.30	11.50
September 1975	750	4.55	12.70
June 1981	1,700	—	30.00

Sources: Whitecombe, John A. *Oil Shale Development: Status and Prospects*, prepared of the Society of Petroleum Engineers and AIME, 50th Annual Meeting. September 28–October 1, 1975, Dallas, Texas; *Oil and Gas Journal.* "Synfuels Offer Challenging Future." June 29, 1981.

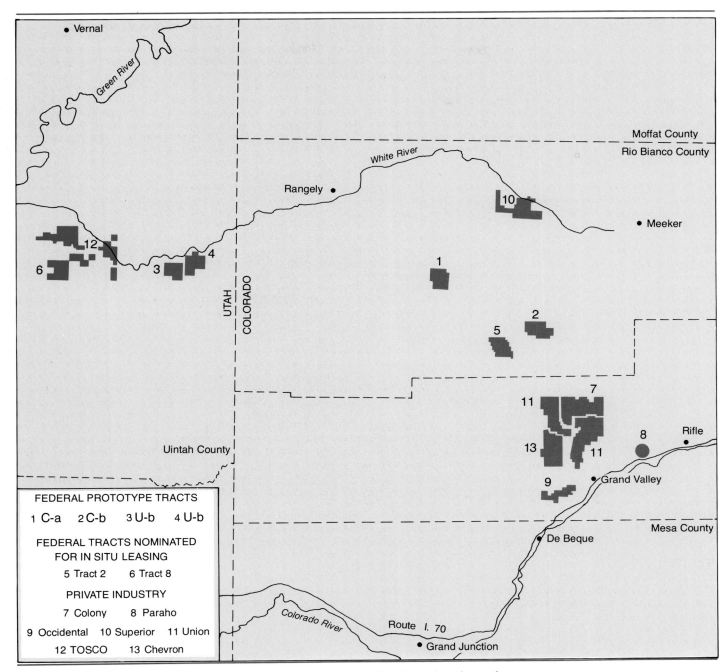

Fig. 3-9 Federal prototype and private oil shale tracts in Colorado and Utah as of 1984.

lease. Presently, the lease is in temporary suspension awaiting a decision on the surface lease.

Occidental–Logan Wash Occidental Petroleum and Tenneco Shale Oil Company's project is located in Rio Blanco County, Colorado. The project plans call for a 55,000 barrels per day plant using surface and modified *in situ* retorts to be operable by 1990. In 1981, Occidental operated a $75 million eight retort experimental plant which produced more than 100,000 barrels of oil. In 1982, the project was delayed awaiting a decision by the Syn-

thetic Fuels Corporation concerning loan and price guarantees. In 1983 the Synthetic Fuels Corporation agreed to provide the project with as much as 2.19 billion dollars in loan and price guarantees. The SFC has agreed to a $60 per barrel price guarantee. The present plan is to produce 14,100 barrels per day of synthetic crude starting in 1987. Occidental estimates the cost of full production to be around 3 billion dollars.

Clear Creek The Chevron Shale Oil Company's Clear Creek project is located in Garfield County, Colorado.

The present plans call for a 50,000 barrel per day plant to be operable in the early 1990s. Current development is progressing slowly.

Colony Until 1982, the Colony Project was thought to have the best chance to become the first commercial scale shale oil facility in the United States. The project, a partnership of Exxon and The Oil Shale Corporation, was projected to produce 47,000 barrels per day of oil at its surface retort facility in Rio Blanco, Colorado. However, in 1982 Exxon suddenly withdrew its 60 percent share from the project and halted construction. According to the partnership agreement, Exxon has since bought out TOSCO's 40 percent interest in the project. Exxon gave high plant costs as the reason for their abandonment of the Colony venture. The project is now temporarily suspended (*Oil and Gas Journal,* May 10 1982).

Union Oil The Union Oil of California facility is located at Parachute Creek, Colorado. It is designed to produce ultimately around 90,000 barrels of oil per day by means of surface retort. The oil shale tract that Union is developing contains an estimated 18 billion barrels of

Davis Gulch, Colorado, property that is part of the Colony Development Project. (Courtesy of American Petroleum Institute.)

shale oil in place. Potential production is estimated at 200,000 barrels per day of oil for 100 years (7.3 billion barrels) (*Oil and Gas Journal,* 1982). The first phase of the Union project was to have been completed by 1983 for a 10,000 barrels per day facility. For this phase Union has a contract with the Defense Department to produce 10,000 barrels per day of diesel and jet fuel. Under the contract, if the market price of the shale oil products is less than $42.50 per barrel, the government makes up the difference.

In 1984, Union received 2.7 billion dollars in price guarantees for an 80,000 barrels per day expansion of the project. Under the terms of the guarantees they will receive $60 per barrel for a period of 10 years. (Oil and Gas Journal, Aug. 8, 1983)

TAR SANDS

Another important source of crude is the *tar sands* of the world. On the one hand, they are similar to oil shales in the sense that their oil cannot be recovered through a drillhole by conventional petroleum production methods. On the other hand, they are quite different because hydrocarbons in oil sands are much more akin to conventional crude than is the kerogen in oil shales. In most cases, oil sands contain highly viscous asphaltic crude oil, often referred to as *bitumen,* in a sandstone that in some deposits is a consolidated rock, and in others is a relatively soft sand body held together by bitumen. In addition, asphaltic oil occurs in limestones, such as in the deposits in a number of European countries and in Oklahoma and Texas. The largest and most attractive occurrences in North America, however, are in sands.

Crude oil in oil sands is called *unconventional* crude oil because of the recovery techniques required to extract it. Any recoverable amounts are therefore considered additional to conventional crude. Nevertheless, in some deeply buried oil sands occurrences, there is no clear distinction between oil sands and reservoirs holding conventional crude of very high viscosity, that is, "heavy crude." As with shale oil extraction, recovery may be accomplished by mining, if deposits are accessible, or by *in situ* processes, if deposits are deeply buried. Some of the *in situ* processes for extraction of bitumen from oil sands make use of steam or combustion to heat the oil in-place, or solvents to flush it out, and are therefore quite similar to methods of enhanced recovery used to obtain oil left behind in conventional reservoirs (see Chapter 2).

World Tar Sand Resources

According to the World Energy Conference, 1980, the world's total tar sands resources amounted to over 2.5 trillion barrels. Only 11 percent of the total is *proved recoverable reserves* and 22 percent *estimated additional resources*

SHALE OIL'S FIVE MOST ACTIVE PROJECTS

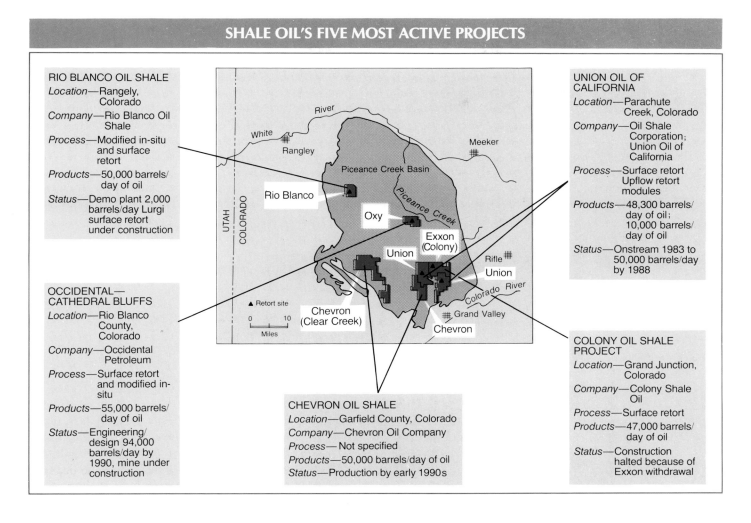

RIO BLANCO OIL SHALE

Location—Rangely, Colorado

Company—Rio Blanco Oil Shale

Process—Modified in-situ and surface retort

Products—50,000 barrels/ day of oil

Status—Demo plant 2,000 barrels/day Lurgi surface retort under construction

OCCIDENTAL— CATHEDRAL BLUFFS

Location—Rio Blanco County, Colorado

Company—Occidental Petroleum

Process—Surface retort and modified in-situ

Products—55,000 barrels/ day of oil

Status—Engineering/ design 94,000 barrels/day by 1990, mine under construction

UNION OIL OF CALIFORNIA

Location—Parachute Creek, Colorado

Company—Oil Shale Corporation; Union Oil of California

Process—Surface retort Upflow retort modules

Products—48,300 barrels/ day of oil; 10,000 barrels/ day of oil

Status—Onstream 1983 to 50,000 barrels/day by 1988

COLONY OIL SHALE PROJECT

Location—Grand Junction, Colorado

Company—Colony Shale Oil

Process—Surface retort

Products—47,000 barrels/ day of oil

Status—Construction halted because of Exxon withdrawal

CHEVRON OIL SHALE

Location—Garfield County, Colorado

Company—Chevron Oil Company

Process— Not specified

Products—50,000 barrels/day of oil

Status—Production by early 1990s

Map labels: River, White, Rangley, Meeker, Piceance Creek Basin, UTAH, COLORADO, Rio Blanco, Oxy, Piceance Creek, Exxon (Colony), Union, Rifle, Union, Colorado River, Chevron (Clear Creek), Grand Valley, Chevron, ▲ Retort site, 0 10 Miles

A chunk of tar sand at Vernal, Utah, showing mixture of sand and bituminous oil. (Courtesy of American Petroleum Institute.)

(Fig. 3–10). Canada, Venezuela, and Colombia house almost 96 percent of the total resources while Canada and Venezuela have 98 percent of the *proved recoverable reserves*. The United States has less than 1 percent of the *proved recoverable reserves* and just over 1 percent of the total resources.

United States Tar Sand Resources

In 1983 the Lewin Study conducted for the Interstate Oil Compact Commission of Oklahoma estimated hydrocarbons in U.S. tar sands at 53.7 billion barrels in place excluding off-shore California (*Oil and Gas Journal*, 1983). These estimates are considerably larger than the World Energy Conferences' estimate. The Lewin study classified tar sands into two types, *measured resources* and *speculative resources* (Fig. 3–11). Forty-one percent of U.S. tar sand resources is classified as *measured* and 59 percent *speculative*.

Of the total resources identified, 37 percent are found in Utah and 19 percent in Alaska. All the resources in Alaska are considered *speculative* while almost 60 percent

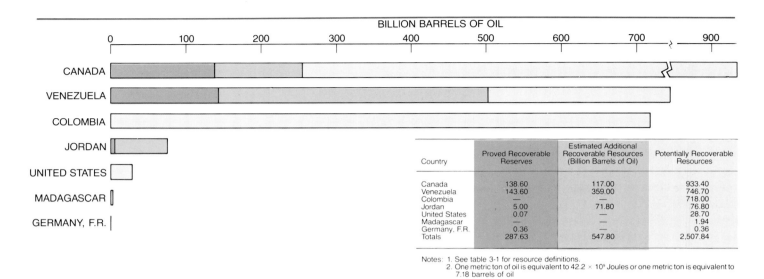

Country	Proved Recoverable Reserves	Estimated Additional Recoverable Resources (Billion Barrels of Oil)	Potentially Recoverable Resources
Canada	138.60	117.00	933.40
Venezuela	143.60	359.00	746.70
Colombia	—	—	718.00
Jordan	5.00	71.80	76.80
United States	0.07	—	28.70
Madagascar	—	—	1.94
Germany, F.R.	0.36	—	0.36
Totals	287.63	547.80	2,507.84

Notes: 1. See table 3-1 for resource definitions.
2. One metric ton of oil is equivalent to 42.2 × 10⁹ Joules or one metric ton is equivalent to 7.18 barrels of oil

Fig. 3–10 World distribution of tar sands and heavy oils.
Source: World Energy Conference, *Survey of Energy Resources 1980,* 1980.

of the Utah resource is *measured.* Utah houses almost 55 percent of the total *measured resource.* Figure 3–12 shows the distribution of major tar sands deposits in the United States.

If a one-third recovery ratio assumed for large deposits in Utah is applied to all bitumen in Figure 3–11 then about 18 billion barrels of oil may be recovered if all deposits were mined. Comparison with potential crude oil from Green River oil shales and estimated crude recoverable from identified and undiscovered conventional reservoirs, shows that tar sands potential in the United States is relatively small (Fig. 3–13).

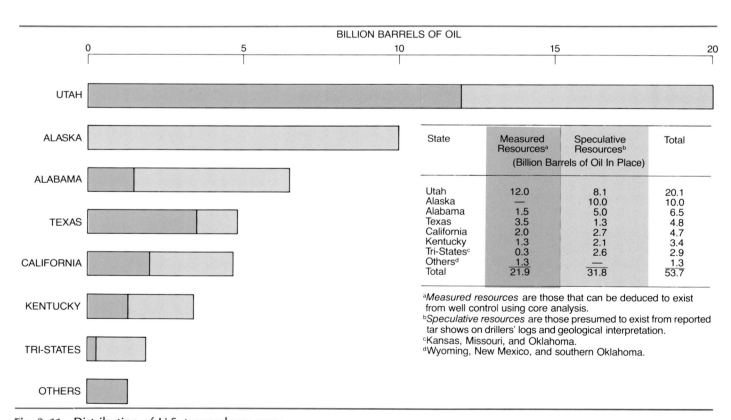

State	Measured Resources[a]	Speculative Resources[b]	Total
	(Billion Barrels of Oil In Place)		
Utah	12.0	8.1	20.1
Alaska	—	10.0	10.0
Alabama	1.5	5.0	6.5
Texas	3.5	1.3	4.8
California	2.0	2.7	4.7
Kentucky	1.3	2.1	3.4
Tri-States[c]	0.3	2.6	2.9
Others[d]	1.3	—	1.3
Total	21.9	31.8	53.7

[a]*Measured resources* are those that can be deduced to exist from well control using core analysis.
[b]*Speculative resources* are those presumed to exist from reported tar shows on drillers' logs and geological interpretation.
[c]Kansas, Missouri, and Oklahoma.
[d]Wyoming, New Mexico, and southern Oklahoma.

Fig. 3–11 Distribution of U.S. tar sand resources.
Source: *Oil and Gas Journal,* "Lewin Study Hikes Estimate of Tar Sand Resources," August 29, 1983.

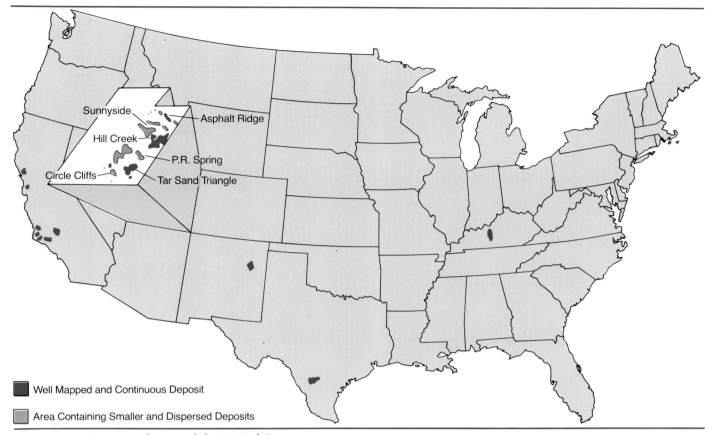

Fig. 3–12 Major tar sand areas of the United States.
Source: U.S. Department of Interior, *Energy Resources on Federally Administered Lands*, November 1981.

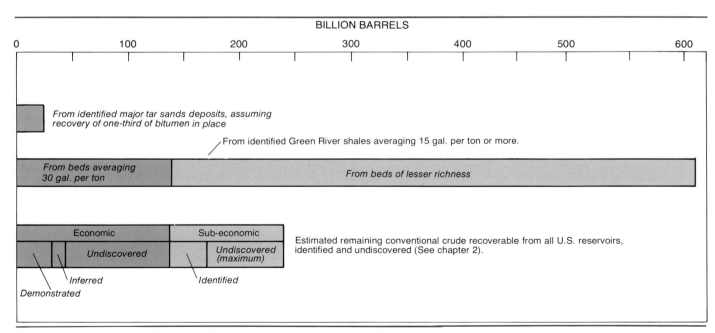

Fig. 3–13 Estimated synthetic crude oil Recoverable from Identified oil shale and tar sands deposits in the U.S. compared with all conventional crude oil thought to remain in Identified and Undiscovered reservoirs.

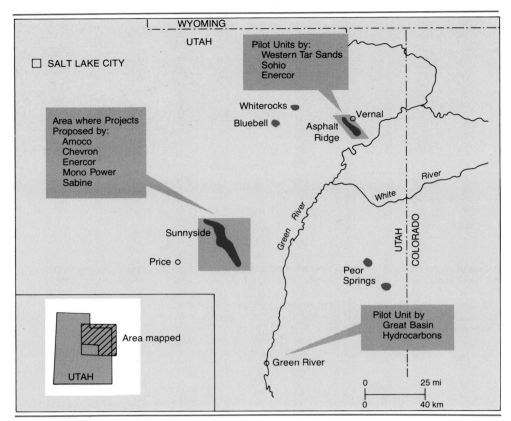

Fig. 3–14 Summary of major tar sand projects in Utah.

COMPANY	PROCESS	PRODUCTION	STATUS
	Pilot Units		
Great Basin Hydrocarbons	Koppers-Totzek	60,000 b/d	
Western Tar Sands		100 b/d	on-line
SOHIO	Solvent Extraction	24 b/d	1984
Enercor	Hot Water Extraction	50 b/d	on-line
	Proposed Projects		
Amoco	Solvent Extraction	50,000 b/d	1998
Chevron	Solvent Extraction	10,000 b/d	1997
Enercor	Hot Water Extraction	20,000 b/d	1991
Mono	Solvent Extraction	30,000 b/d	1990
Sabine	In situ	5,000 b/d	1989

Sources: *Oil and Gas Journal*, "Synfuels Offer Challenging Future," June 29, 1981; and *Oil and Gas Journal*, "Development of Utah Tar Sand Eyed," December 5, 1983.

CURRENT DEVELOPMENT ACTIVITIES There are a number of tar sands projects underway at the present time in the United States. Most of these projects are located in Utah as indicated in Figure 3–14. If the Utah projects (excluding pilot units) were completed, they would produce 115,000 barrels per day of oil. Outside of Utah, a 12,000 barrel per day demonstration plant is being planned in western Kentucky. In Maverick County, Texas a 20,000 barrel per day project is being planned by Texas Tar Sands Limited. At the Texas site a pilot recovery plant was placed on line in 1984. The pilot unit has produced an average of 200 to 300 barrels per day with peak production exceeding 500 barrels per day.

NOTES

1. The Chattanooga Shales are Devonian and Mississippian. The marine shales of Alaska are Triassic. The age of the Green River formation is Tertiary.

CRUDE OILS	CRUDE OIL AS SUCH IN VARIOUS RESERVOIR ROCKS	Conventional crude recovered by primary, secondary, or tertiary (enhanced) recovery techniques
	SYNTHETIC CRUDE OILS	From kerogen in oil shales
		From bitumen in tar sands
		From coal liquefaction
		From biomass

Conventional and unconventional (synthetic) oils, highlighting the synthetic crudes reviewed in this chapter.

Moving heap of tar sands mined near Vernal, Utah. (Courtesy of U.S. Department of Energy.)

2. Green River shales are preserved in seven geologic basins: the Green River, Great Divide, Washakie, and Fossil Basins in Wyoming; the Uinta in Utah; and the Piceance and Sand Wash Basins in Colorado.

3. Nahcolite is a sodium-bearing mineral, and dawsonite is an aluminum-bearing mineral that may be recovered in the future along with the shale oil.

4. If 90 percent of the oil in the mined shale is recovered, this implies a recovery of 54 percent of the shale oil in place.

5. Standard Oil of Ohio, Cleveland Cliffs Iron Company, and The Oil Shale Corporation were the original partners. Atlantic Richfield took over as operator in 1970, and in 1974 Shell Oil and Ashland Oil joined the venture. Exxon and The Oil Shale Corporation were the most recent operators.

References

Burwell, E.L. et al. "Shale Oil Recovery by In-Situ Retorting; A Pilot Study," *Journal of Petroleum Technology,* December 22, 1970.

Culbertson, William C. and Pitman, Janet K. "Oil Shale," U.S.G.S. *Professional Paper, 820.* Washington: U.S. Government Printing Office, 1973.

Donnell, John. "Oil Shale Resources: How Much?" *Shale Country,* February 1976.

Duncan, Ronald C. and Swanson, Vernon E. "Organic-Rich Shale of The United States and World Land Areas," *U.S. Geological Survey Circular, 532.* Washington, D.C.: Government Printing Office, 1965.

East, J.H. and Gardner, E.D. "Oil Shale Mining, Rifle, Colorado, 1944–56." *U.S. Bureau of Mines Bulletin, 611.* Washington: U.S. Government Printing Office 1904.

Oil and Gas Journal, "Synfuels Offer Challenging Future," June 29, 1981.

Oil and Gas Journal, "Exxon Halts Colony Oil Shale Project," May 10, 1982.

Oil and Gas Journal, "Union May Need Price Support for Full Oil Shale Development," August 30, 1982.

Oil and Gas Journal, "Shale Projects Get SFC Loan-Price Guarantees," August 8, 1983.

Oil and Gas Journal, "Lewin Study Hikes Estimate of Tar Sands Resources," August 29, 1983.

Oil and Gas Journal, "Controversies, Sluggish Markets Slow U.S. Sunthetic Fuels Corporation," April 23, 1984.

Science, 190 (4321). "Oil Shale Prospects on the Upswing," December 9, 1977.

Wells, Christopher. *The Elusive Bonanza—The Story of Oil Shale, America's Richest and Most Neglected National Resource.* New York: E.P. Dutton, 1970.

World Energy Conference. *Survey of Energy Resources,* Munich: World Energy Conference, September, 1980.

U.S. Department of Interior. *Final Environmental Statement for the Oil Shale Leasing Program,* Volume III, *Specific Imputs of Prototype Oil Shale Development.* Washington: U.S. Government Printing Office, 1973.

4 Nuclear Fuels

The use of nuclear materials for the purpose of generating electricity began in the 1950s as a by-product of the development of the atomic bomb. In 1954, the Atomic Energy Commission was given the right by Congress to encourage the development of nuclear power for electrical generation. By the 1960s, scores of plants were built, dozens planned, and, in 1968, the AEC predicted that one thousand plants would be operating by the year 2000.

Since these declarations were made, nuclear power has been surrounded by controversy. Well before the Three Mile Island accident in March, 1979, electric power companies had ceased requesting permits for nuclear power facilities. Between 1975 and 1979, less than half a dozen reactors were purchased. Since 1979 no new orders for nuclear reactors have been placed. Moreover, from 1979 through 1983 46 reactor projects were cancelled (Atomic Industrial Forum, 1984).

Critics of nuclear power have attacked the industry for not providing safe, economic means of disposing of the hazardous waste products. In addition, they have maintained that the availability of nuclear materials throughout the world could result in the proliferation of atomic bombs. Proponents see nuclear power as a cleaner energy source than coal and one that is essential if the nation is to reduce its dependence on foreign oil. They claim that the problems of toxic waste materials, leakages of contaminated gas, and technical and economic difficulties, can be solved. The outcome of this controversy will have a significant effect on the proportion of energy provided by nuclear power in the future.

THE NATURE OF THE RESOURCE

A nuclear reaction is produced when the nuclei of atoms are changed and produce an atom of a different element or new isotopes of the same element. (Isotopes are atoms of the same element differing only in the number of neutrons present in the atomic nucleus). Energy may be derived from nuclear materials by two different processes:

A bundle of fuel rods containing enriched uranium oxide. Approximately 600 bundles like this make up the fuel assembly in the core of a nuclear reactor. Photo courtesy of the American Petroleum Institute and Exxon Corporation.

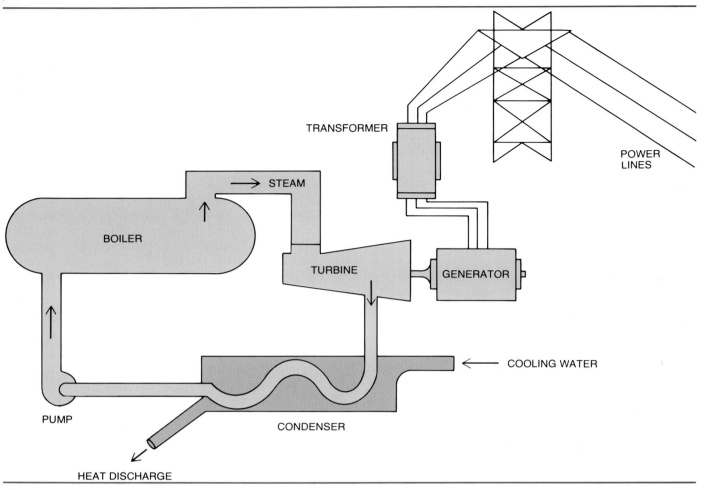

Fig. 4–1 Electrical power generation by steam turbine.

fission and *fusion*. Energy is released through the fission process by splitting certain atomically heavy radioactive elements into two or more lighter radioactive elements. In the fusion process, energy is released by melting together two or more very light elements to produce heavier elements. Fission, the atomic bomb reaction, is the one presently being used in nuclear power reactors for the generation of electricity. Fusion, or the hydrogen bomb reaction, has yet to be advanced technologically to the stage where it can be used as a source of energy. The production of energy from fusion may be possible in the early years of the twenty-first century.

The following discussion outlines the basic principles associated with the production of energy through the use of both fission and fusion, and presents a geographic evaluation of the United States' present nuclear fuel supplies as well as future requirements.

FISSION

At present, most of our electricity is produced in fossil-fueled plants in which coal, oil, or natural gas is used as the energy source to produce steam in a boiler. The steam is then used to spin a turbine which drives a generator and produces electricity (Fig. 4–1). The steam is then cooled, condensed, and pumped back to the boiler for re-use.

In a nuclear power plant, a fission reactor is substituted for the fuel-fired boiler. The fuels for the nuclear reactor are the nuclei of uranium, plutonium, and thorium, all of which occur in ores of various types. The only naturally occurring fissionable material used in nuclear reactors is Uranium-235. Two artificial isotopes, Uranium-233 and Plutonium-239, can be produced from naturally occurring Thorium-232 and Uranium-238. In a nuclear reactor, the energy used for the production of steam is released as shown in Figure 4–2. The Uranium-235 reaction, shown in Figure 4–2A, the one commonly used in nuclear power plants, releases energy when a heavy atom of Uranium-235 is split apart by a slow-moving neutron, producing lighter fission fragments plus two or three additional neutrons. This reaction also allows the release of a large amount of energy.

The newly created neutrons, in turn, split additional Uranium-235 atoms and produce more energy and more neutrons. When the splitting of Uranium-235 atoms is repeated again and again under controlled conditions, the result is a self-sustaining chain reaction releasing very large amounts of energy. When it is produced in an uncontrolled manner, the result is an atom bomb.

Reactions shown in Figures 4-2B and 4-2C illustrate how Uranium-233 and Plutonium-239 are "bred" from Thorium-232 and Uranium-238. Up to now, thorium has not been used extensively as a nuclear fuel; and Uranium-238, by far the largest component of natural uranium, has been considered a waste product. (An assessment of the potential role of Thorium-232 and Uranium-238 and how they may affect the United States' supply of fissionable fuels will be discussed in a later section).

THE NATURE AND OCCURRENCE OF URANIUM

Uranium, the basic raw material for nuclear energy, is a hard, nickel-white metal containing three radioactive isotopes: Uranium-238, Uranium-235, and Uranium-234.[1] Many nuclei, other than uranium, are radioactive, meaning that they are unstable and undergo changes, even when they are not being bombarded. Radioactive nuclei emit alpha particles (helium nuclei), beta particles (high energy electrons), or gamma rays (electromagnetic waves having greater energies than X-rays). When alpha and beta particles are emitted, a radioactive nucleus decays, becoming a different nucleus. Natural uranium contains only about 0.7 percent Uranium-235. Most of the remain-

ing 99.3 percent is Uranium-238, which, under neutron bombardment yields Plutonium-239, a fissionable material. Plutonium-239 would become the major nuclear fuel if the United States pursued the development of breeder reactors which produce more nuclear fuel than they consume by converting nonfissionable Uranium-238 into fissionable Plutonium-239.

Natural uranium is not an abundant metal. It makes up only about two parts per million of the earth's crust, occurring in combination with other elements in ores of two broad types—those in igneous and metamorphic rocks, which account for 20 percent of the world resources, and those in sedimentary rock, which account for roughly 80 percent (Woodmanse, 1975).

Ores of the first type are "original" occurrences of a uranium compound as they were deposited in veins or in a pegmatite (very coarsely crystalline igneous rock) by hot solutions associated with an intrusion of magma. Such deposits, which tend to be rich but small, are found at Yellowknife on Great Bear Lake in the Canadian Shield, in the shield area of Western Australia, and in the Congo at Shinkolobwe.

Ores of sedimentary type occur in various rocks, such as bituminous black shales, sandstones, conglomerates, and phosphate rocks usually associated with vanadium. The age of the host rock ranges from Pre-Cambrian, as in

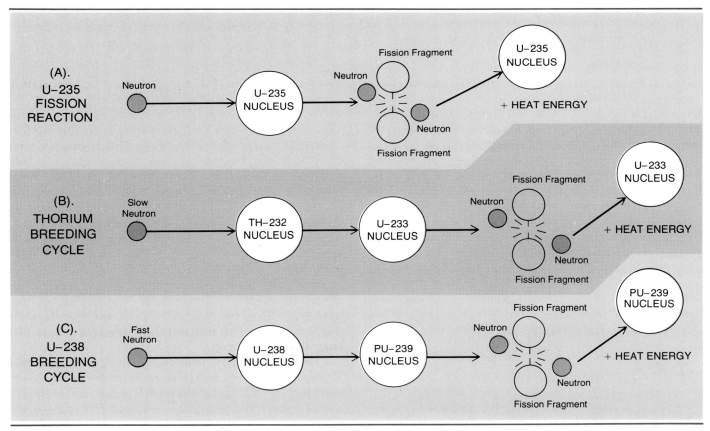

Fig. 4-2 Three nuclear reactions: one using naturally fissile U_{235} and two which breed fissile materials from U_{238} and Th_{232}.

URANIUM

Uranium is a hard grey radioactive metal that occurs in a number of isotopic forms. As it is found in nature, uranium consists of three isotopes of mass numbers 238, 235, and 234. The important isotopes of uranium for use in the production of nuclear energy are Uranium-238, Uranium-235, and Uranium-233. The complete fission of all the nuclei in one pound of uranium releases the same energy as burning 1,360 tons of bituminous coal.

Symbol	U
Atomic number	92
Atomic weight	238.0
Density	19.05 g/cm^3
Melting point	1135°C
Boiling point	4000°C

the conglomerates of Eliot Lake, Ontario, through Devonian, as in the Chattanooga shales of the eastern United States, to Tertiary, as in some sandstones in the western United States.

Uranium oxide is the major compound in which uranium occurs. Sandstones in western states such as New Mexico, Colorado, and Wyoming, contain the primary ore minerals, uraninite, and coffinite. In some areas, weathering processes have produced carnotite and other secondary minerals which are very rich in uranium oxide. The ore minerals occupy pore spaces of the sandstones, or in some occurrences replace the sand grains or replace plant remains. Many hypotheses have been proposed for the genesis of these sandstone deposits. Generally, it is thought that the uranium compounds were formed by the leaching of original igneous sources in veins, pegmatites, or volcanic ash. The compounds were transported in ground water and precipitated when they encountered organic materials and the associated low-oxygen (reducing) environments.

Less rich sedimentary ores occur in deposits that lack the enrichment afforded by weathering. The ancient conglomerates of Eliot Lake and Blind River in Canada contain over 0.10 percent U_3O_8. Black bituminous shales range from 0.03 percent in an exceptionally rich Swedish example to 0.007 percent in the Chattanooga Formation of the eastern United States; phosphate rocks in the Phosphoria Formation of Idaho and neighboring states contain around 0.012 percent (Finch et al., 1973).

Another type of low-grade deposit of uranium is found in the Conway granite in the White Mountains of New England which averages 0.001 to 0.003 percent. Whether these low-grade sedimentary or igneous deposits can be used as resources depends both on the future demand for

uranium and on whether the more abundant isotope, U-238, is used as a fuel supply.

Resource Terms Applicable to Uranium

The U.S. Department of Energy scheme for showing the relationship between uranium reserves and potential resources may be placed in the broad framework used by the U.S. Geological Survey (Fig. 4-3). *Reserves* are the most certain resources, constituting deposits that have been delineated by drilling or other sampling techniques. *Potential* resources are divided into three categories, *probable, possible,* and *speculative,* in declining order of certainty. These terms describe the uranium thought to be present in incompletely drilled or undiscovered deposits (National Uranium Resource Evaluation, 1976).

In addition to this classification, the Department of Energy also groups the resources into cost categories: under 30 dollars, 30 to 50 dollars, and 50 to 100 dollars per pound.

The three cost categories are the cost required *to produce* one pound of U_3O_8 concentrate from these resources. These costs are operating and capital costs (or *forward* costs) and do not reflect the market price. These include costs of labor, materials, electricity, royalties, payroll, insurance, and applicable general and administrative costs (U.S. Department of Energy, January, 1983). It is important to note that resources in each cost category embrace the resources of all lower-cost categories. For example, resources in the 50 dollars per pound category also include resources in the 15 and 30 dollars per pound categories.

World Resources

Figure 4-4 and Table 4-1 provide a summary of world uranium resources up to 50 dollars per pound U_3O_8. (The Soviet Union and China are excluded because data were not available.)

World resources of uranium, up to 50 dollars per pound of U_3O_8, amount to about 5 million metric tons (5.4 million short tons). Reserves (reasonably assured) account for 46 percent of the total. Most of the uranium reserves, approximately 85 percent, are found in seven countries: the United States, Canada, the Republic of South Africa, Niger, Australia, Brazil, and Namibia. The United States alone houses over one-fourth of the uranium Reserves. The same seven countries also account for over 90 percent of the estimated Additional resources.

Most uranium resources outside the United States, found largely in Pre-Cambrian conglomerates and vein deposits, average more than 0.10 percent uranium. In Sweden, however, uranium is recovered from black shales containing only 0.03 percent U_3O_8. Although uranium resource data were unavailable for the Soviet Union and China, it is believed that both countries have ample

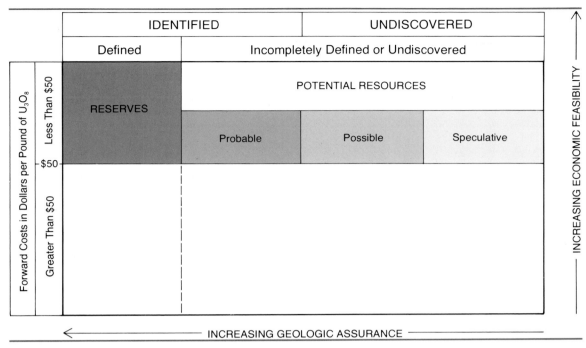

Fig. 4-3 Conceptual framework for organizing amounts of reserves and other resources of uranium oxide.

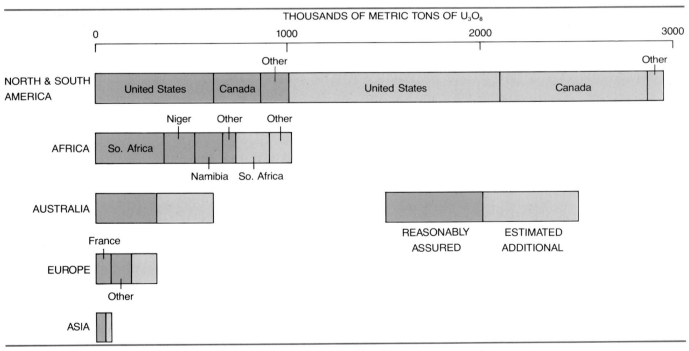

Fig. 4-4 World resources of uranium oxide showing regions and leading countries as of January 1, 1981.

Source: OECD Nuclear Energy Agency and The International Atomic Energy Agency, 1982.

Note: Includes resources with a forward cost of up to $50/lb. U_3O_8. Forward costs are the operating and capital costs, in present dollars, that will be incurred in the production of the uranium.

TABLE 4-1

World Uranium Resources as of January 1, 1981 excluding Communist Areas
(in metric tons U$_3$O$_8$)

COUNTRY/REGION	REASONABLY ASSURED[1]		ESTIMATED ADDITIONAL[2]		TOTAL
	<$30/lb	$30–$50/lb	<$30/lb	$30–$50/lb	<$50/lb
North and South America	739.0	276.3	1,127.5	830.2	2,979.7
Argentina	25.0	5.3	3.8	9.6	43.7
Brazil	119.1	—	81.2	—	200.3
Canada	230.0	28.0	358.0	402.0	1,108.0
Mexico	2.9	—	3.5	2.6	9.0
United States	362.0	243.0	681.0	416.0	1,702.0
Europe	82.4	98.2	41.4	92.8	314.8
Denmark	—	27.0	—	16.0	43.0
Finland	—	3.4	—	—	3.4
France	59.3	15.6	28.4	18.1	121.4
Greece	1.4	4.0	2.0	5.3	12.7
Italy	—	2.4	—	2.0	4.4
Portugal	6.7	1.5	2.5	—	10.7
Spain	12.5	3.9	8.5	—	24.9
Sweden	—	38.0	—	44.0	82.0
Turkey	2.5	2.1	—	—	4.6
United Kingdom	—	—	—	7.4	7.4
Africa	591.2	133.8	168.7	132.3	1,026.0
Algeria	26.0	—	—	—	26.0
Central African Rep.	18.0	—	—	—	18.0
Gabon	19.4	2.2	—	9.9	31.5
Namibia	119.0	16.0	30.0	23.0	188.0
Niger	160.0	—	53.0	—	213.0
Somalia	—	6.6	—	3.4	10.0
South Africa	247.0	109.0	84.0	91.0	531.0
Zaire	1.8	—	1.7	—	315
Australia	294.0	23.0	264.0	21.0	602.0
Asia	40.1	11.0	0.9	24.2	76.2
India	32.0	—	0.9	24.2	57.1
Japan	7.7	—	—	—	7.7
Korea	0.4	11.0	—	—	11.4
World	1,746.7	542.3	1,602.5	1,100.5	4,992.0

[1] Reasonably Assured Resources have a high assurance of existence and in the cost category below $30/lb U$_3O_8$ are considered as reserves.

[2] Estimated Additional Resources refers to uranium in addition to Reasonably Assured Resources, that is expected to occur, mostly on the basis of direct geological evidence, in: extensions of well explored deposits; little-explored deposits; and undiscovered deposits believed to exist along a well-defined geological trend with known deposits.

Source: OECD Nuclear Energy Agency and The International Atomic Energy Agency. *Uranium— Resources, Production, and Demand.* Paris: OECD, 1982.

resources to meet their needs (Nuclear Energy Policy Group Study, 1977).

United States Resources

The cost of future development of nuclear energy on the United States will, to a degree, depend on the availability of fuel supplies. It is important, therefore, to assess the nation's uranium resources as thoroughly as possible. A Department of Energy 1983 report estimates the United States' total uranium resources at more than 4.2 million tons of U$_3$O$_8$. In Figure 4–5 and Table 4–2, this total of 4.2 million tons of uranium is broken down according to certainty and cost. This breakdown reveals the following points:

1. In descending order of certainty, Reserves make up 21 percent, (889 thousand tons of U$_3$O$_8$) of the total

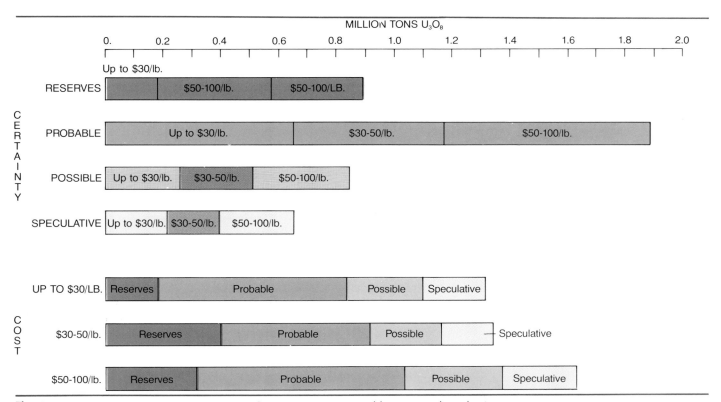

Fig. 4–5 National summary of uranium oxide resources arranged by cost and geologic certainty as of January 1, 1983.

Source: Department of Energy. *Statistical Data of the Uranium Industry,* January 1, 1983.

United States resources, and Potential resources 79 percent. Probable resources, the most certain of the Potential category, account for 44 percent of the total. Possible and Speculative resources account for 20 and 15 percent of the uranium supply respectively.

2. A breakdown of the uranium resources by cost category reveals that 31 percent of the total resources are available at a forward cost of 30 dollars per pound of U_3O_8, 31 percent at 30 to 50 dollars per pound, and 38 percent at 50 to 100 dollars per pound.[2]

3. Of the 1.3 million tons of uranium resources in the 30 dollar cost category, 86 percent are classified as Potential, and the bulk of these are assigned to the Probable and Possible categories. For uranium for which forward costs are 30 to 50 dollars per pound and 50 to 100 dollars per pound, 70 to 80 percent respectively are Potential resources. Together, these amount to almost 3 million tons of uranium, and account for 70 percent of the total uranium resources.

Considering that a large part of the total uranium

TABLE 4–2
United States' Uranium Resources as of January 1, 1983
(thousands of short tons U_3O_8)

U_3O_8 COST CATEGORIES	RESERVES	POTENTIAL RESOURCES			TOTAL
		Probable	Possible	Speculative	
$30/lb Average	180	654	257	216	1,307
$50/lb Average	576	1,167	508	391	2,642
$100/lb Average	889	1,887	842	652	4,270

Note: Columns should not be added because resources with costs averaging $100/lb *include* the two lower cost categories. The total in thousands of short tons = 4,270.

Source: U.S. Department of Energy. *Statistical Data of the Uranium Industry, January 1, 1983.* Washington: Government Printing Office, January 1, 1983.

An open pit uranium mine in the Gas Hills of Wyoming. (Courtesy of U.S. Department of Energy.)

resources of the United States is classified as either uncertain or low-grade (less than 0.10 percent U_3O_8), one might question the wisdom of recovering these ores in light of the high cost of the energy extracted from these materials. The fuel for nuclear power plants, however, is a relatively small element in the total cost of nuclear-generated electricity. One calculation of fuel costs for nuclear-generated electricity indicates the cost of uranium for fueling a light water reactor was 8.5 mills per Kilowatt hour at 1982 prices. For comparison, the cost of the fuel for a power plant using bituminous coal, in 1982, was estimated at 25.7 mills per Kilowatt hour in New England and 22.4 mills per Kilowatt hour in the Middle Atlantic states (Department of Energy, August 1983). The Office of Policy, Planning and Analysis, U.S. Department of Energy, concluded that assuming U_3O_8 costs of $43.60 per pound (1980 dollars) a reactor coming into operation in 1995 would use uranium fuel costing 8.49 mills per Kilowatt hour. These calculations further reinforce the argument that the cost of U_3O_8 should not be a critical factor in determining the cost of nuclear energy in this century.

An additional consideration is the environmental impact of mining uranium. At the present time, land disturbance resulting from uranium mining is low compared to coal mining. As lower-grade uranium ore begins to be mined in greater volume, the impact will increase. A 1,000-Megawatt coal-fired electrical generation plant consumes 3 million tons of coal per year, equivalent to 3.75 million tons of coal in-place. A nuclear power plant of the same capacity would not require this tonnage of ore to be mined until the U_3O_8 content declined from its present average level of 0.20 percent to a level of 0.006 to 0.007 percent, just about equal to that of Chattanooga Shales. In fact, presently defined resources do not include such low-grade ores (Nuclear Energy Policy Group, 1977). It should be noted, however, that the volume of solid wastes from coal is small when compared with uranium whose ore averages around 0.2 percent, so that 99.8 percent of the material mined is discarded as tailings (waste).

REGIONAL DISTRIBUTION OF UNITED STATES' RESOURCES *According to Certainty.* The Department of Energy now recognizes the regional distribution of uranium resources within the framework of 20 uranium resource regions (Fig. 4-6). Two regions, the Colorado Plateau and the Wyoming Basins, contain approximately

55 percent or 2.3 million tons, of the nation's 100 dollar per pound resources (Fig. 4–7). The Colorado Plateau contains 36 percent of the total resources followed by the Wyoming Basins with 19 percent. Most of these resources, found principally in sandstones, are located in areas of past and present production. In addition, these regions also contain proportionally more uranium Reserves, having more than 82 percent of the total 100 dollar per pound uranium (47 percent being on the Colorado Plateau and over 35 percent in the Wyoming Basins; Fig. 4–8). The division of the resources into varying degrees of certainty indicates that regional shares are similar to those for the country as a whole. The Wyoming Basins, which have not been mined as extensively as the Colorado Plateau, have a larger share of the total resources in the Reserve and Probable categories. The Coastal Plain and the Basin and Range regions have a substantial amount (24 percent) of the Potential resources.

As shown in Table 4–3, Potential resources are subject to considerable uncertainty. The amount used here to represent total Potential resources, 3.3 million tons, is the mean value. The table points out that there is a 95 percent probability that as little as 2.5 million tons exist as

TABLE 4–3
Probability Distribution Values for Potential Uranium Resources, January 1, 1983[1]

FORWARD-COST CATEGORY	THOUSAND TONS U_3O_8		
	Mean	95th Percentile	5th Percentile
$30/lb U_3O_8			
Probable	654	473	883
Possible	257	142	406
Speculative	216	118	411
Totals	1,127	791	1,556
$50/lb U_3O_8			
Probable	1,167	883	1,519
Possible	508	287	770
Speculative	391	222	700
Totals	2,066	1,502	2,748
$100/lb U_3O_8			
Probable	1,887	1,467	2,418
Possible	842	464	1,262
Speculative	652	379	1,103
Totals	3,381	2,502	4,403

[1]Estimated potential uranium resources include losses from mining; losses due to milling are not included and may range from 5 to 15 percent.

Source: U.S. Department of Energy. *Statistical Data of the Uranium Industry, January 1, 1983.* Washington: Government Printing Office, 1983.

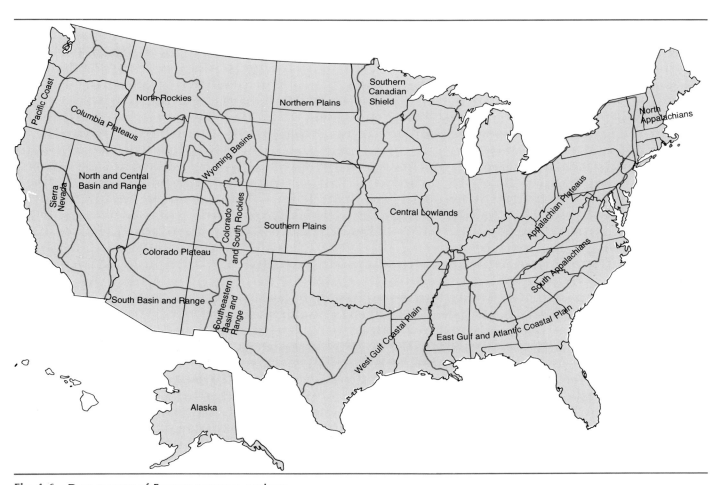

Fig. 4–6 Department of Energy resource regions.
Source: Department of Energy, *Statistical Data of the Uranium Industry,* January 1, 1983.

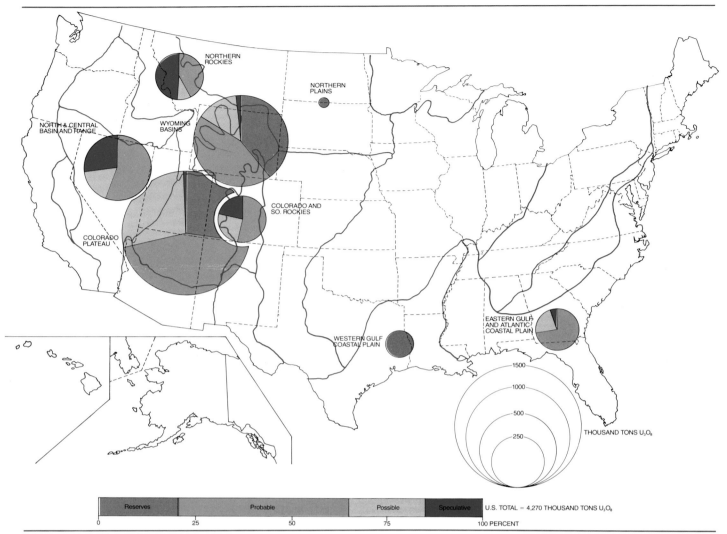

Reserves | Probable | Possible | Speculative | U.S. TOTAL = 4,270 THOUSAND TONS U₃O₈
0   25   50   75   100 PERCENT

Fig. 4-7 Geologic certainty of uranium oxide resources at costs up to $100 per pound by DOE resource region.

Source: Department of Energy. *Statistical Data of the Uranium Industry,* January 1, 1983.

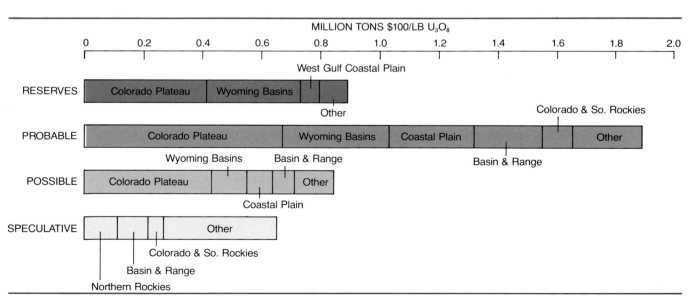

Fig. 4-8 National summary of uranium oxide resources by degree of certainty and by DOE resource region.

Source: Department of Energy, *Statistical Data of the Uranium Industry,* January 1, 1983.

Potential resources, but only a 5 percent chance that as much as 4.4 million tons of Potential resources exist. Figures 4–9 through 4–12 indicate the actual amounts of uranium oxide for each region according to the categories of Reserves or Probable, Possible, or Speculative resources.

According to Cost. The regional shares of the three cost categories generally reflect the national pattern of 31, 31, and 38 percent (Fig. 4–13 and Tables 4–4 through 4–6). Exceptions are found in the Colorado and Southern Rockies and Coastal Plain regions where a larger proportion of 30 dollar per pound uranium is found. Over 34 percent of lowest-cost (30 dollars per pound) uranium resources are located in the Colorado Plateau, with 17 percent in the Coastal Plain, and 13 percent in the Wyoming Basins (see bar graph, Fig. 4–14).

According to Certainty and Cost. Figures 4–15 and 4–16 bring together the information from individual maps and summarize all United States' resources by region and also by certainty and cost categories. The bar graph provides a comprehensive view of where the resources are located, their degree of certainty, and how much it is estimated to cost to produce one pound of U_3O_8 concentrate from these resources.

Application of Uranium Resources. Before considering the question of uranium, it is important to understand how uranium ores are used. There are two vital aspects of the use of uranium ores: the fuel cycle and the types of fission reactors.

THE NUCLEAR FUEL CYCLE Before the mined uranium can be used as a nuclear fuel it must undergo several stages of processing and conversion. The process begins at mills located near areas where uranium is mined (Fig. 4–17). At these mills, the ore is concentrated into U_3O_8, a uranium-uranyl oxide called "yellowcake." Most concentrates prepared contain an average of 80 to 85 percent U_3O_8, but must contain a minimum of 75 percent (Woodmanse, 1975).

After being transported to refineries, the concentrate mill product is processed to remove impurities. After purification, the U_3O_8 is converted into a hexafluoride product, UF_6, which is gasified for enrichment at federally operated gaseous diffusion plants at Oak Ridge, Tennessee, Paducah, Kentucky, or Portsmouth, Ohio (Fig. 4–17). The enrichment process increases the content of the Uranium-235 isotope from 0.7 percent to specific levels depending on the end-use of the enriched product. For the boiling water reactor, enrichment is to 2.03 percent Uranium-235 and for the pressurized water reactor 2.26 percent. For other reactors, such as the high-temperature gas-cooled reactor and for use in weapons, enrichment may be to over 90

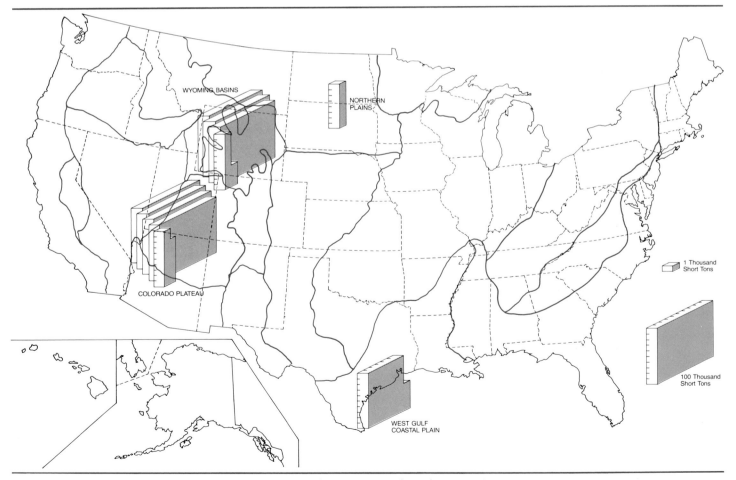

Fig. 4–9 Uranium oxide *reserve* tonnage at costs up to $100 per pound as of January 1, 1983 by DOE resource region.
Source: Department of Energy. *Statistical Data of the Uranium Industry*, January 1, 1983.

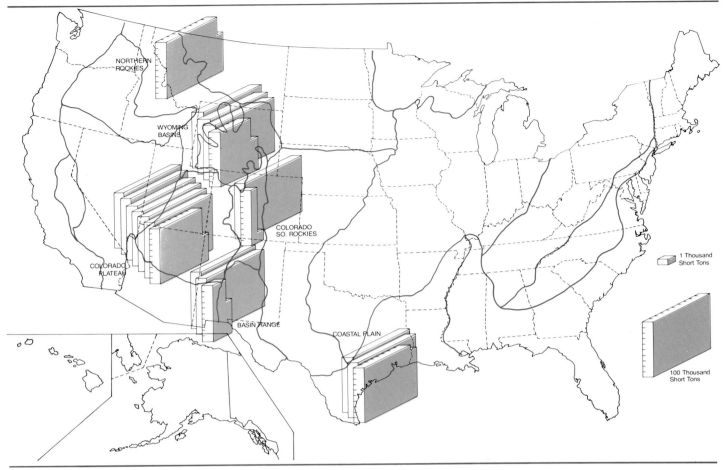

Fig. 4–10 Uranium oxide *probable* resource tonnage at costs up to $100 per pound as of January 1, 1983 by DOE resource region.

Source: Department of Energy, *Statistical Data of the Uranium Industry*, January 1, 1983.

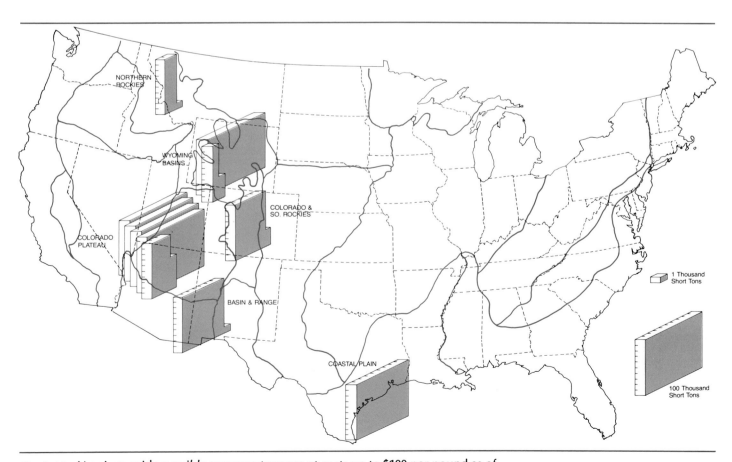

Fig. 4–11 Uranium oxide *possible* resource tonnage at costs up to $100 per pound as of January 1, 1983 by DOE resource region.

Source: Department of Energy. *Statistical Data of the Uranium Industry*, January 1, 1983.

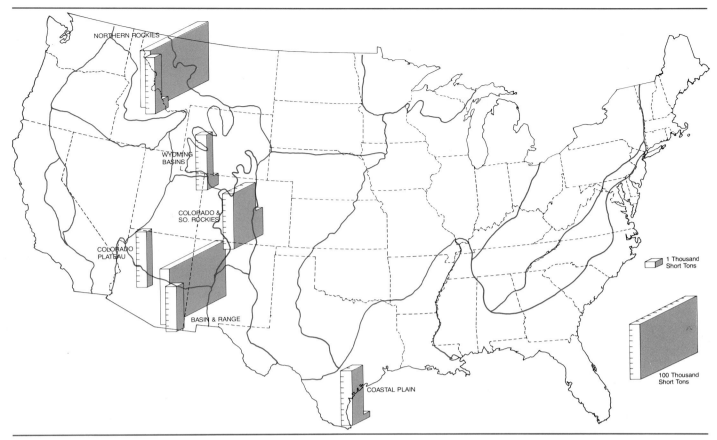

Fig. 4-12 Uranium oxide *speculative* resource tonnage at costs up to $100 per pound as of January 1, 1983 by DOE resource region.

Source: Department of Energy. *Statistical Data of the Uranium Industry*, January 1, 1983.

TABLE 4-4
Regional Summary of $30 per pound Uranium Reserves and Potential Resources as of January 1, 1983 (in thousands of tons U_3O_8

REGION	RESERVES	POTENTIAL RESOURCES Probable	Possible	Speculative	TOTAL
Colorado Plateau	117.2	229.9	97.7	1.7	446.5
Wyoming Basins	39.6	89.2	43.4	1.8	174.0
Western Gulf Coastal Plain	16.1	—	—	—	16.1
Coastal Plain	—	181.1[2]	53.6[2]	9.3[2]	244.0
Northern Rockies	—	10.8	9.5	30.4	50.7
Colorado and Southern Rockies	5.2	72.5	26.0	22.9	126.6
Basin and Range	—	34.4	11.9	43.9	90.2
Other	2.0[1]	35.6[3]	15.3[3]	105.5[3]	158.4
Total	180.0	653.5	257.4	215.5	1,306.5

[1] Includes Coastal United States, Colorado and Southern Rockies, Northern Rockies, Northern and Central Basin and Range, Southern Plains, Sierra Nevada, Pacific Coast, Southern Basin and Range, and Southeastern Basin and Range regions.

[2] Includes Western Gulf Coastal Region and Eastern Gulf and Atlantic Coastal Plains regions.

[3] Includes Appalachian Plateaus, Southern Appalachian, Northern Appalachian, Northern Plains, Southern Plains, Pacific Coast, Sierra Nevada, Central Lowlands, and Southern Canadian Shield regions.

Source: U.S. Department of Energy. *Statistical Data of the Uranium Industry, January 1, 1983.* Washington: Government Printing Office, 1983.

TABLE 4–5
Regional Summary of $50 per pound Uranium Reserves and Potential Resources as of January 1, 1983 (in thousands of tons U₃O₈)

| REGION | RESERVES | POTENTIAL RESOURCES | | | TOTAL |
		Probable	Possible	Speculative	
Colorado Plateau	284.6	433.9	239.7	5.0	963.2
Wyoming Basins	185.8	191.7	78.7	5.3	461.5
Western Gulf Coastal Plain	44.7	—	—	—	44.7
Coastal Plain	—	239.1[2]	72.5[2]	14.2[2]	325.8
Northern Plains	3.2	—	—	—	3.2
Northern Rockies	—	37.6	15.5	57.3	110.3
Colorado and Southern Rockies	—	93.0	39.1	34.3	166.4
Basin and Range	—	97.9	28.9	70.4	197.2
Other	57.7[1]	73.4[3]	33.4[3]	204.3[3]	368.8
Total	576.0	1,166.6	507.8	390.8	2,641.1

[1] Includes Coastal United States, Colorado and Southern Rockies, Northern Rockies, Northern and Central Basin and Range, Southern Plains, Sierra Nevada, Pacific Coast, Southern Basin and Range, and Southeastern Basin and Range regions.

[2] Includes Western Gulf Coastal Region and Eastern Gulf and Atlantic Coastal Plains regions.

[3] Includes Appalachian Plateaus, Southern Appalachian, Northern Appalachian, Northern Plains, Southern Plains, Pacific Coast, Sierra Nevada, Central Lowlands, and Southern Canadian Shield regions.

Source: U.S. Department of Energy. *Statistical Data of the Uranium Industry, January 1, 1983.* Washington: Government Printing Office, 1983.

TABLE 4–6
Regional Summary of $100 per pound Uranium Reserves and Potential Resources as of January 1, 1983 (in thousands of tons U₃O₈)

| REGION | RESERVES | POTENTIAL RESOURCES | | | TOTAL |
		Probable	Possible	Speculative	
Colorado Plateau	417.6	674.9	435.6	9.9	1,538.0
Wyoming Basins	313.4	356.2	114.4	11.7	795.7
Western Gulf Coastal Plain	63.7	—	—	—	63.7
Coastal Plain	—	289.3[2]	89.5[2]	20.7[2]	399.5
Northern Plains	7.5	—	—	—	7.5
Northern Rockies	—	96.8	21.4	111.8	230.0
Colorado and Southern Rockies	—	111.5	52.8	44.8	209.1
Basin and Range	—	226.4	71.1	107.9	405.4
Other	86.8[1]	132.3[3]	57.4[3]	344.9[3]	621.4
Total	889.0	1,887.4	842.2	651.7	4,270.3

[1] Includes Coastal United States, Colorado and Southern Rockies, Northern Rockies, Northern and Central Basin and Range, Southern Plains, Sierra Nevada, Pacific Coast, Southern Basin and Range, and Southeastern Basin and Range regions.

[2] Includes Western Gulf Coastal Region and Eastern Gulf and Atlantic Coastal Plains regions.

[3] Includes Appalachian Plateaus, Southern Appalachian, Northern Appalachian, Northern Plains, Southern Plains, Pacific Coast, Sierra Nevada, Central Lowlands, and Southern Canadian Shield regions.

Source: Department of Energy. *Statistical Data of the Uranium Industry, January 1, 1983.* Washington: Government Printing Office, 1983.

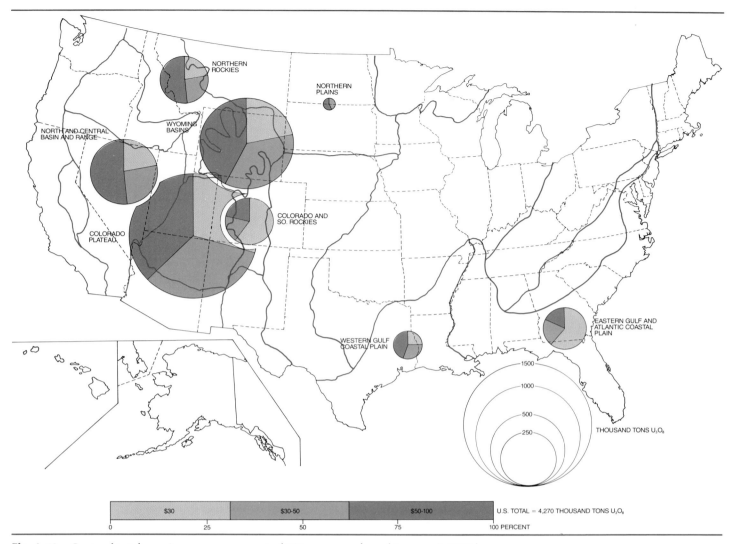

Fig. 4-13 Costs of total uranium resources up to $100 per pound as of January 1, 1983 by DOE resource region.

Source: Department of Energy, *Statistical Data of the Uranium Industry,* January 1, 1983.

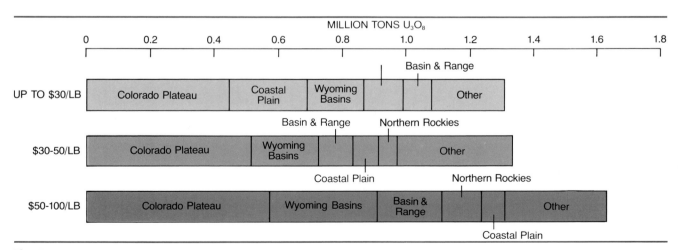

Fig. 4-14 National summary of uranium oxide resources by cost and by DOE resource region as of January 1, 1983.

Source: Department of Energy, *Statistical Data of the Uranium Industry,* January 1, 1983.

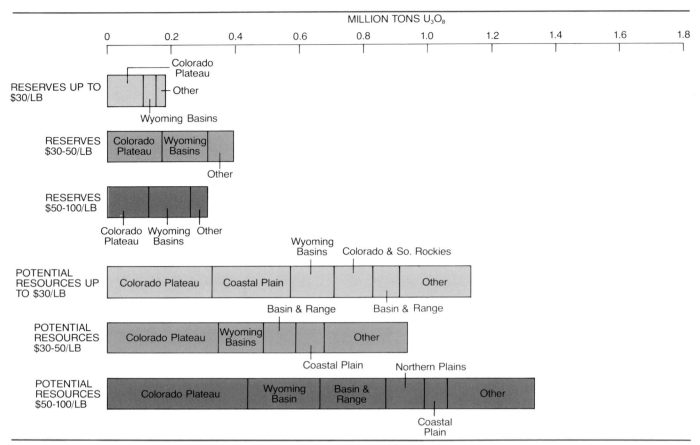

Fig. 4–15 National summary of uranium oxide resources by costs and by DOE resource region as of January 1, 1983, distinguishing reserves from potential resources.
Source: Department of Energy, *Statistical Data of the Uranium Industry*, January 1, 1983.

The Brown's Ferry Nuclear generating plant on the Tennessee River near Athens, Alabama. Its three generating units have a total capacity of 3.3 thousand Megawatts, making it the largest nuclear generating plant in the country. (Courtesy Tennessee Valley Authority.)

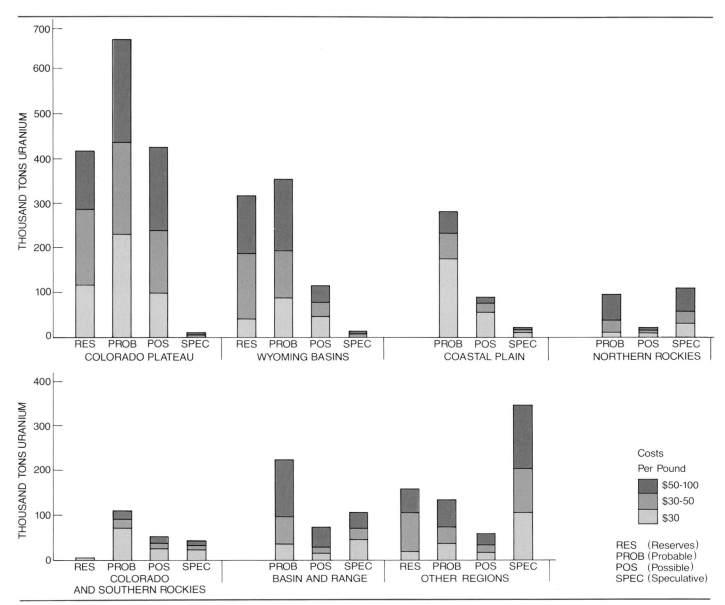

Fig. 4–16 Regional summary of all uranium resources showing costs and degree of certainty as of January 1, 1983.

Source: Department of Energy, *Statistical Data of the Uranium Industry*, January 1, 1983.

percent Uranium-235. At this stage, about 98 percent of the natural uranium is left behind in the form of Uranium-238 and is stored at the gaseous diffusion facilities in stainless steel drums (Woodmanse).

After enrichment, the UF_6 is shipped to plants where it is converted into fuel form for use in the manufacture of fuel elements. The different forms include: uranium oxide (UO_2), metals, alloys, carbides, nitrides, and salt solutions. The most common form used in light water reactors is pelletized ceramic UO_2. The pellets are loaded into steel or zirconium cladding (covered) tubes and fastened together into fuel bundles.

During the operation of a nuclear reactor, fission products accumulate in the fuel elements and reduce reactivity by absorbing neutrons. Because of this, over 25 percent of the reactor fuel core is removed and replaced on an annual basis. These highly radioactive spent fuel rods are submerged in water to allow for cooling and radioactive decay and are generally stored on the reactor site for several months. According to representatives of the Philadelphia Electric Company, 2 to 3 cubic feet of storage space is required for every metric ton of uranium loaded as fuel.

After the cool-down period, the spent fuel may be shipped in specially designed containers to reprocessing plants where uranium and plutonium are separated and recovered for re-use. There are presently no privately

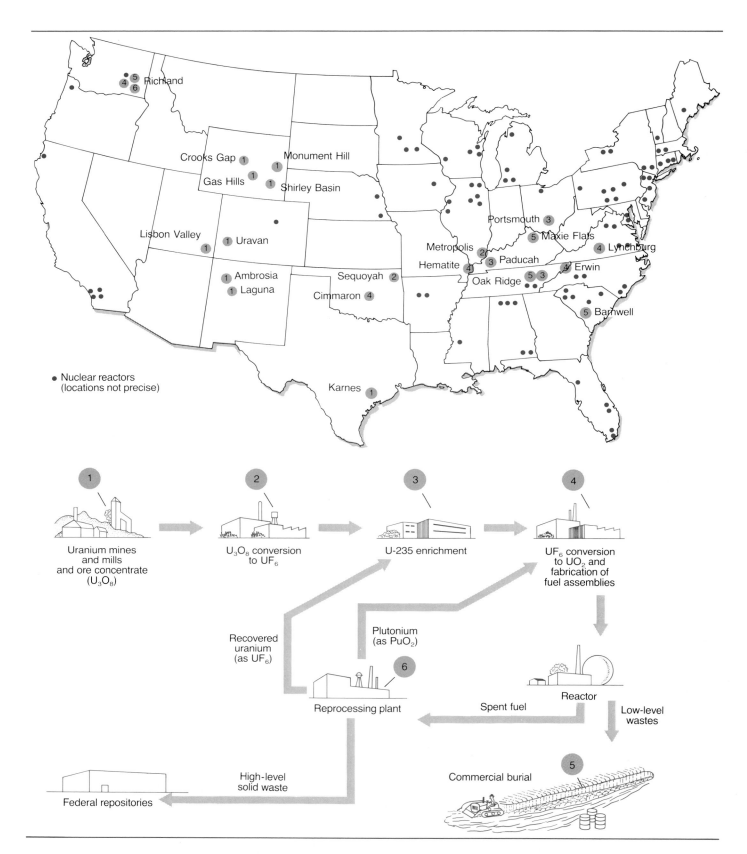

Fig. 4–17 The Nuclear Fuel Cycle showing activities at selected locations. Numbers in map correspond to activities in the cycle.

Sources: Energy Research Group, *Estimates of the Population Served by Nuclear Power Plants in the United States*, February 1984 and Department of Energy, *Statistical Data of the Uranium Industry*, January 1, 1983.

owned reprocessing plants operating in the United States, but some reprocessing is carried out at federal facilities. At one time, a privately owned plant operated for several years at West Valley, New York, but because of numerous containment and handling problems has been shut down. There are plans for additional reprocessing plants to be built in Washington and South Carolina but it is uncertain if and when these plants wil be operable (Fig. 4–17).

Nuclear reactors accumulate radioactive waste and products in liquid, solid, and gaseous forms. These are differentiated as low-level and high-level wastes. The wastes are extracted from the spent fuels, transported, and disposed of by permanent burial or by storage. The low-level wastes include alpha materials contaminated with Plutonium-239, an isotope that does little damage on the outside of the body but is intensively radioactive when deposited on sensitive lung tissue. High-level wastes are accumulated during fuel reprocessing. Presently, there are liquid weapons' waste storage areas in Hanford, Washington, Aiken, South Carolina, and Idaho Falls, Idaho as well as solid waste burial grounds in such places as Oak Ridge, Tennessee, and Maxie Flats, Kentucky (Fig. 4–17).

As of yet, permanent storage methods for high-level radioactive waste have not been developed. The lack of safe methods of storing high-level waste products is a major reason why many people object to nuclear power development, and especially to the development of new-generation breeder reactors which produce highly toxic plutonium waste. The U.S. Department of Energy is considering permanent disposal techniques, such as burial in deep geological formations (salt beds, granites, and basalts), in polar areas under icecaps, or ejection to the sun or outer space (Woodmanse).

Crane lowers spent fuel rods into a pool that will be filled with water for short-term storage at a nuclear reactor. (Courtesy of U.S. Department of Energy.)

Two of the thirty storage tanks for liquid high-level radioactive wastes at the Department of Energy's Savannah River facility at Barnwell, South Carolina. When completed, the double-walled steel tanks will be encased in two to three feet of concrete and covered with soil. Liquid wastes stored at such facilities are produced by uranium enrichment plants, reactors, weapons work and reprocessing spent fuel E.I. Dupont Denemours & Co., (Courtesy of U.S. Department of Energy.)

TYPES OF FISSION REACTORS Most of the nuclear reactors in commercial use in the United States are slow neutron, light water moderated, and use enriched Uranium-235 as fuels.[3] In the nuclear vernacular, these reactors are referred to as "converters." There are two basic types of light water reactors, the boiling water reactors (BWR) and the pressurized water reactors (PWR), whose components are illustrated in Figures 4-18A and 4-18B.

Boiling water reactors were first used commercially in the United States at Commonwealth Edison's Dresden, Illinois plant in 1960. Inside a BWR the water is in contact with the hot fuel rods at a pressure of about 1,000 pounds per square inch. The steam generated rises to the top of the reactor core, and after passing through steam separators and dryers, is piped directly to the steam turbine. Steam bubbles that form around the fuel rods aid in controlling the chain reaction by reducing the moderating effectiveness of water, thus slowing the growth of the chain reaction.[4] As the steam comes into direct contact with the fuel rods in the BWR, there is a possibility that radioactive materials may invade the turbine and leak to the environment. For this reason it is extremely important to seal the turbine.

In the pressurized water reactor (PWR), there are two separate water systems. Water circulates through the reactor core, under high pressure of 2,000 pounds per square inch, and remains in the liquid form up to a temperature of 600 degrees Fahrenheit. Because the steam generator is completely separated from other parts of the system, it is easier to contain the radioactive materials. To compensate for the slowing down of the nuclear reaction in the core as the fuel is expended, boron is added to the cooling water of a PWR at the time the reactor starts up. As the power level decreases, the boron, which has been absorbing neutrons, is gradually removed from the coolant, releasing more neutrons to take part in the reaction (Fowler, 1975).

Figure 4-18C illustrates a relatively new type of reactor, the high-temperature gas-cooled reactor (HTGR). Although the HTGR is classified as a converter, it does produce new fissionable material. In this system, helium at 1400 degrees Fahrenheit is circulated through the reactor core. After passing through a heat exchanger the helium produces steam at approximately 1000 degrees Fahrenheit. In the HTGR, Thorium-232 is added to the reactor to create more fuel. Proponents of the HTGR claim this reactor offers a thermal efficiency of 40 percent which would be equal to the best fossil-fuel plants in operation. In addition, representatives of Philadelphia Electric, which operated a small capacity HTGR for several years, indicate that fuel savings in this type of reactor could amount to about 30 percent because of its ability to produce new fuel from Thorium-232.

THE BREEDER Since the beginning of nuclear reactor development, a great deal of time has been spent considering ways of obtaining energy from the more abundant but nonfissionable Uranium-238. With the scarcity of uranium, attention has begun to focus on the breeder reactor, since it has the capability of converting nonfissionable Uranium-238 into the fissionable isotope, Plutonium-239 (Fig. 4-2C). There are two factors of major importance in a technical evaluation of breeder reactors: the "breeding ratio," or the ratio of Plutonium-239 produced to the nuclear fuel consumed; and the "doubling time," the number of years it takes to double the initial fuel load (Fowler, 1975).

There are two basic types of breeders: thermal breeders that operate with low-energy neutrons and fast breeders that operate with high-energy neutrons. At present, the liquid metal fast breeder reactor (LMFBR) illustrated in Figure 4-18D has generated the most interest among both the proponents and opponents of breeder development. The LMFBR is attractive because its breeding ratio is estimated to be in the range of 1.4 to 1.5, meaning that for every two fissionable nuclei burned as fuel three new ones would be created. In addition, doubling times in the order of 8 to 10 years are expected.

Since the LMFBR can be fueled with the enriched Uranium-238 now stored at enrichment plants, commercial operation could begin without further mining of uranium ore. Lapp 1975 reports that breeder power would allow the United States to tap the 200,000 tons of Uranium-238 presently stored in steel vessels at Oak Ridge, Tennessee. Furthermore, he argues that if the present supply of uranium resources were converted to plutonium, it would yield the energy equivalent of 400 billion tons of coal—the approximate amount the United States has underground in a mineable form. (Lapp, 1975).

A very strong case can be made for the fast breeder on the basis of its efficient use of uranium. Light water reactors (LWR) can use only two percent of uranium mined, whereas the fast breeder cxould use 50 to 70 percent (Fowler, 1975). In addition, the requirements for initial fuel are much lower for the fast breeder. Consider, for example, the fuel needs for a 1000-Megawatt generating plant operating at 80 percent capacity for coal-fired plants, nuclear light water reactors, and fast breeders. The initial fuel requirements for the coal-fired plant would be 200,000 tons; a light water converter would require approximately 200 tons, whereas the LMFBR would need only 1.4 tons (U.S. Atomic Energy Commission, 1973).

It seems clear that if our present assumptions regarding fast breeder reactors were shown to be correct and if the breeder could be developed safely, the United States would have a virtually unlimited supply of energy for several thousands of years.[5]

There are, however, a number of major problems associated with breeders that raise serious and legitimate doubts about a breeder reactor program. One of the prime concerns of opponents of fast breeders is the proliferation of highly toxic PU_{239} which has a half-life (the time period required for one half the atoms to decay into another iso-

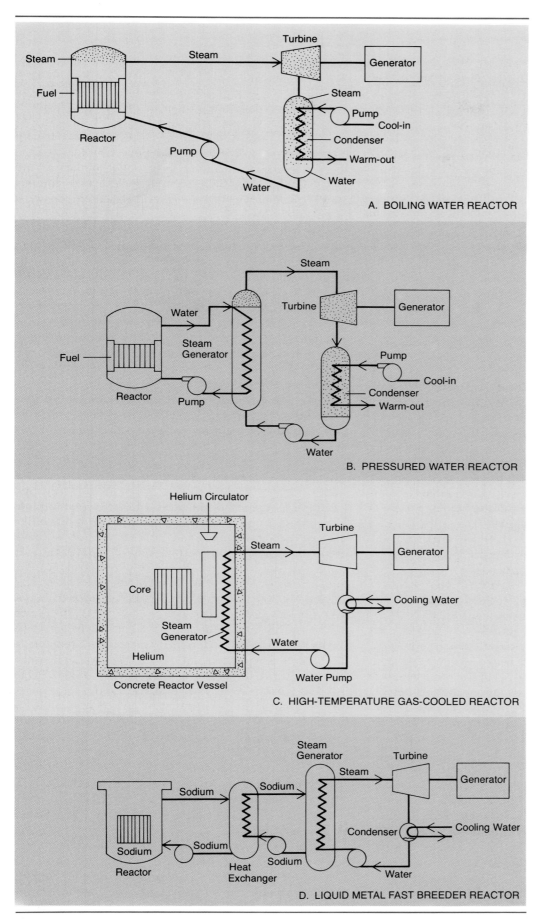

Fig. 4–18 Four types of nuclear reactors for power generation by nuclear fission.

Source: Atomic Energy Commission, *Nuclear Fuel Reserves and Requirements*, 1973.

UPDATE ON BREEDER DESIGNS IN THE UNITED STATES

While breeder reactors play a dominant role in the production of nuclear power in France, there are no large-scale breeders operating in the United States. In fact, the most visible part of the U.S. breeder development program ended when the Clinch River Breeder Reactor Project died in 1983 because of exorbitant costs, uncertainty about safety, and the possible theft of the breeder's plutonium fuel. The possibility of breeders is being kept alive, however, by the Department of Energy, which is funding research on two fronts (See *Science News*, Jan. 26, 1985).

One effort to redesign the breeder is taking place at Argonne National Laboratory near Chicago. The proposed breeder plant is essentially a reworking of a design used 20 years ago at Idaho Falls, Idaho. Its proponents claim the following features and advantages. The liquid sodium, used to transfer heat from the fuel assembly to the working fluid that drives the turbine, will never leave the reactor core, thus reducing the chance of radioactive materials leaking into occupied areas of the plant. Another novelty is the use of uranium alloys as fuel, rather than the usual uranium oxide compounds. Because the alloys are more conductive, they will cool more rapidly in case of an emergency shutdown. Finally, the proposed plant will include its own fuel-reprocessing facility where highly radioactive materials are handled by remote control and where plutonium will not exist in a form that is readily stolen or diverted to military uses

Concurrently, the Department of Energy has sought innovative designs from three established nuclear contractors—General Electric, Westinghouse, and Rockwell International. Since the designs emerging from this competition are likely to emphasize cost-saving as well as safety, the most attractive breeder plant design may combine features of this commercial research with features from the Argonne proposal.

tope) of about 24,000 years. Developing ways to safely handle and store plutonium waste for such a long period is a problem yet to be solved. Furthermore, it has been argued that if the United States and other countries move ahead with fast breeder development, the large stockpiles of plutonium, capable of producing bombs would encourage production of nuclear weapons. Preventing the spread of plutonium throughout the world and protecting it from theft are problems of immense proportion.

CUMULATIVE AND CURRENT PRODUCTION

After a period of rapid production spurred by weapons development and stockpiling from the mid-1950s to the early 1960s the United States' output dropped dramatically until 1966. Between 1966 and 1980 U_3O_8 production rose from 10,100 tons to 23,300 tons. In recent years production has entered a period of decline (Fig. 4–19). In 1982 total production of U_3O_8 was 13,434 tons, down considerably from the record total of 1980. The recent plunge in uranium production is due to the cancellation of numerous nuclear plants and competition from overseas suppliers such as France and the Soviet Union. Through 1982, cumulative production amounted to over 410,000 tons of uranium taken from 230 million tons of uranium ore. The state distribution of cumulative production and the 1982 production by region are shown in Figures 4–20 and 4–21. New Mexico (Colorado Plateau) has dominated past production, accounting for about two-thirds of the total. In 1982, the leading uranium producing states, New Mexico and Wyoming, accounted for 28 and 20 percent respec-

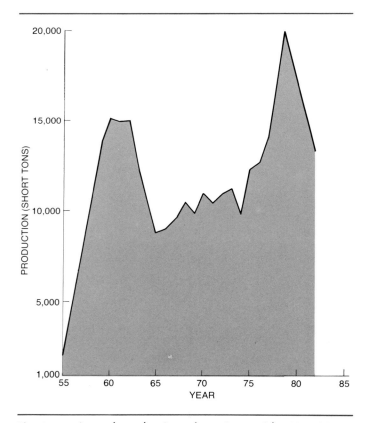

Fig. 4–19 Annual production of uranium oxide, 1955–82.
Source: Department of Energy, *Statistical Data of the Uranium Industry*, January, 1983.

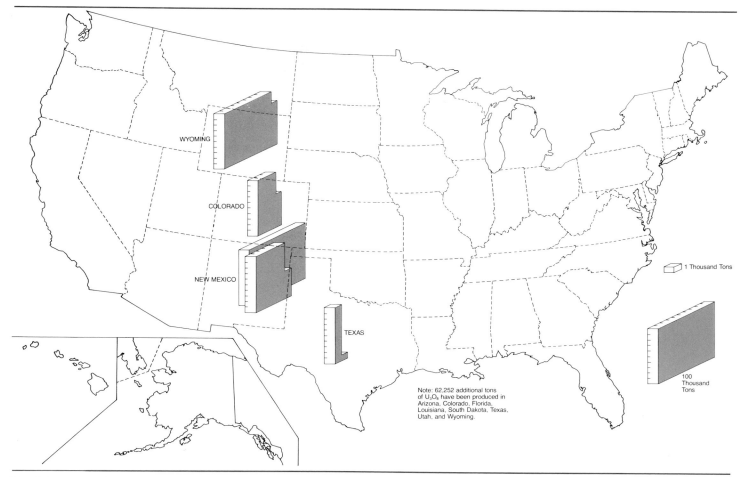

Fig. 4–20 Cumulative production of uranium oxide through 1982 by state.
Source: Department of Energy, *Statistical Data of the Uranium Industry*, January 1983.

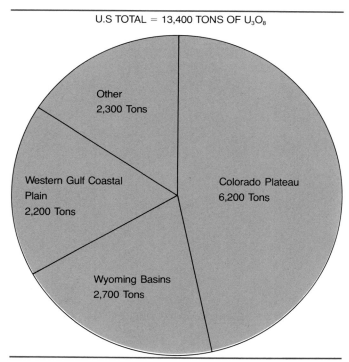

U.S TOTAL = 13,400 TONS OF U₃O₈

Other
2,300 Tons

Western Gulf Coastal Plain
2,200 Tons

Colorado Plateau
6,200 Tons

Wyoming Basins
2,700 Tons

Fig. 4–21 Production of uranium oxide in 1982 showing contributions of leading regions.
Source: Department of Energy, *Statistical Data of the Uranium Industry*, January 1, 1983.

tively. Texas followed in third place with 17 percent of the total production. The remaining 35 percent was produced in Arizona, Colorado, Florida, Idaho, Utah, and Washington. Through 1982, major areas of production were Ambrosia and Laguna in New Mexico and Shirley Basin and Gas Hills in Wyoming (Fig. 4–22).

The Department of Energy reported that uranium ore processed in 1982 was more than double that processed a decade ago. This figure is deceptive, however, because the average grade of ore has decreased, making the increase of uranium concentrate only 33 percent. Between 1966 and 1982 the average grade of processed ore decreased from 0.23 percent to 0.12 percent. The average recovery of the contained U_3O_8 increased from 95 to 96 percent.

STATUS OF U.S. NUCLEAR POWER GENERATION CAPACITY In the period between 1965 and 1984, the number of operable nuclear power plants in the United States grew almost sevenfold, from 12 to 83 plants. During the same period the nation's nuclear power capacity increased from a little more than 1,000 Megawatts to over 66,000 Megawatts. On the average, these plants have operated at about 60 percent of their designed capacity because of frequent breakdowns, lengthy maintenance operations, and federal safety precautions (Fig. 4–23). In addition to the 83 nuclear power plants that were operable

165

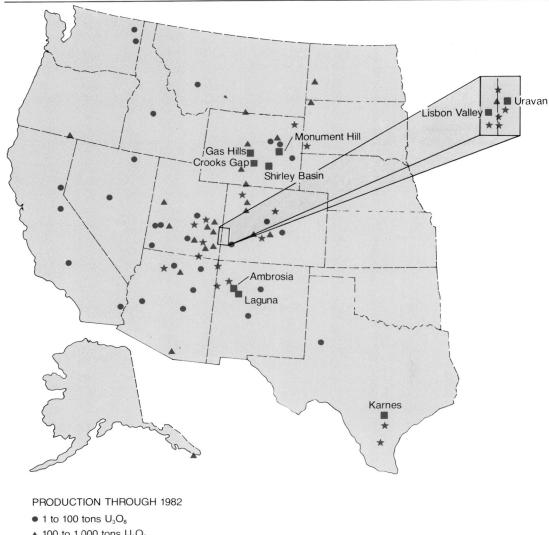

PRODUCTION THROUGH 1982

● 1 to 100 tons U_3O_8

▲ 100 to 1,000 tons U_3O_8

★ 1,000 to 10,000 tons U_3O_8

■ Greater than 10,000 tons U_3O_8

Note: Uranium produced as a by-product from the
processing of phosphates is not included.

Fig. 4–22 Areas of uranium production in the United States.
Source: Department of Energy, *Statistical Data of the Uranium Industry*, January 1, 1983.

as of August 1984, 52 plants with a generating capacity of over 59,000 Megawatts have been granted construction permits. Two additional reactors with a capacity of over 2,000 Megawatts have been ordered. If all the plants under construction and ordered are completed (this would seem unlikely due to the recent increases in cancellations), the nuclear power generation capacity of the United States will amount to over 128,000 Megawatts at the turn of the century. This projected generation capacity has fallen by about 35 percent since 1979 (Table 4–7).

In 1982, nuclear power produced 3 percent of the United States' energy and over 12 percent of its electricity.

Table 4–8 and Figure 4–24 show the proportion of net electric generation produced by nuclear reactors for each state.

Over 80 percent of nuclear power capacity is located east of the Mississippi, and one-half of that capacity is in the populous states of the northeast quadrant (Fig. 4–25). If and when those plants under construction and ordered become operable, the states east of the Mississippi will still house more than three-fourths of the nation's nuclear capacity. Because of the large capacity planned for the southeastern states, however, the northeast quadrant's share will drop to about 40 percent.

PROSPECTS FOR DEVELOPMENT AFTER THREE MILE ISLAND During the last week in March and the first week of April, 1979, the worst commercial nuclear accident in United States' history occurred at the Three Mile Island power plant near Harrisburg, Pennsylvania. A chain of mechanical and human failures triggered by a coolant water valve malfunction led to a series of radioactive steam leaks that spread over an area of up to 20 miles from the plant. There was widespread disagreement among the experts as to the hazard posed by the radiation leak. Many claimed that there was no threat to health, while others warned of the threat of cancer and possible genetic damage. Only time can determine whose claims were accurate.

In the aftermath of the Three Mile Island crisis, nuclear power faces a highly uncertain future. Because of lower forecasted growth in electric needs, construction financing constraints, reversals in the cost advantage of some nuclear units, and a growing antinuclear sentiment, the nation's electric utility industry has substantially reduced its earlier committment to nuclear power. By year-end 1983, the electric utility industry had cancelled 106 nuclear units totalling over 115,000 Megawatts of capacity (Atomic Industrial, Forum, 1984). These cancellations represent over 45 percent of the total commercial nuclear capacity previously ordered. For comparison, only 39 fossil-fuel generating units, totalling about 23,000 Megawatts, have been cancelled since 1972, the year of the first nuclear cancellations (Department of Energy, April 1983).

Despite the woes that have beset the nuclear energy industry in the United States, countries in western Europe

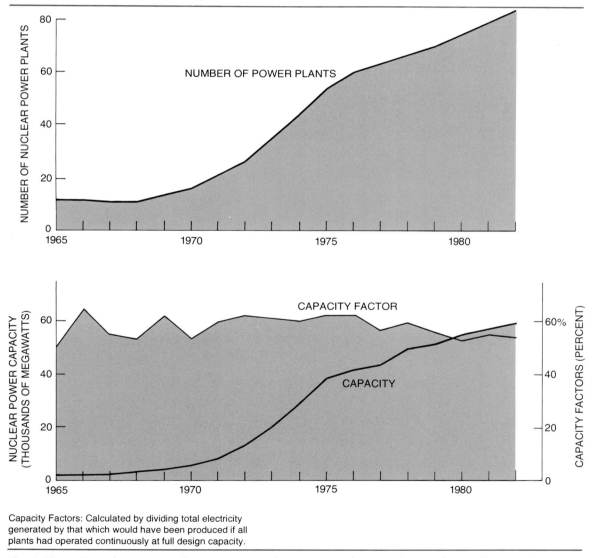

Capacity Factors: Calculated by dividing total electricity generated by that which would have been produced if all plants had operated continuously at full design capacity.

Fig. 4-23 U.S. nuclear power plants 1965–1982 showing growth in capacity and overall capacity factor through time.
Source: Energy Research Group, *Estimates of the Population Served by Nuclear Plants in the United States,* February 4, 1984 and Department of Energy, *Electric Power Annual,* 1982.

NUCLEAR SHARE OF ELECTRICITY GENERATED BY COUNTRY, 1983

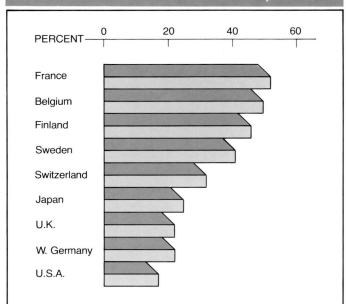

Since 1978, no nuclear power plants have been ordered in the United States because of reduced electricity demand, high construction cost, regulatory uncertainties, and public distrust. However, the international picture is quite different. Twenty-five countries now produce nuclear-generated electricity and eight more have plans to do so by the end of the decade. There are now more than 300 nuclear plants generating electricity worldwide, and almost 200 more are being built. Several European countries produce 25 to 50 percent of their electricity by nuclear generation (see above). Japan expects to double its 1983 nuclear capacity by 1990.

Source: O.E.C.D., 1983

AVERAGE NATURAL BACKGROUND RADIATION BY STATE

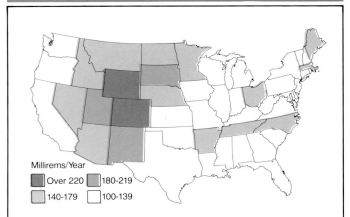

Man is constantly exposed to natural radiation from the sun and outer space, from naturally occurring radioactive materials, and from food and water. Scientists call this "background radiation," and measure it in units called "millirems." The levels of background radiation vary from place to place as exhibited by the above map of the United States. The differences in background radiation from state to state are due to a number of factors including *elevation* (in high elevation there is less overlying air to shield a person from cosmic rays), and *rock and soil type* present. If, for example, a person lives in an area where building materials are composed of rock containing naturally occurring radioactive materials, he or she will be exposed to more radiation than a person living in an area where building materials are less radioactive. Furthermore, a person living in a wooden house will be exposed to about one-half the radiation of that received by an individual residing in a brick house. On the average, a person living in the United States receives about 60 percent of his or her annual radiation dosage from *background* sources.

Map Source: General Electric. *Nuclear Power Quick Reference IV*, San Jose, California: Nuclear Energy Operations, 1984.

Three Mile Island nuclear generating station on the Susquehanna River near Middletown, Pennsylvania, operated by Metropolitan Edison Company. This photo of the station, viewed from the west, was taken before the accident in Spring of 1979 that put Unit Two (right half of photo) out of service. Unit One (located on the left half) was undamaged by the accident. The plant's generating capacity is 800 Megawatts. (Courtesy of Metropolitan Edison Company, Reading, Pennsylvania.)

TABLE 4-7
Number of Nuclear Power Plants in the United States as of August 1984

STATE	NUMBER OF PLANTS BY STATUS				CAPACITY OF PLANTS			TOTAL CAPACITY
	Operable	Being Built	Ordered	TOTAL PLANTS	Operable	Being Built	Ordered	(MW's)
Alabama	5	2	0	7	4,853	1,658	—	6,511
Arizona	0	3	0	3	—	3,810	—	3,810
Arkansas	2	0	0	2	1,762	—	—	1,762
California	4	2	0	6	3,554	2,190	—	5,744
Colorado	1	0	0	1	330	—	—	330
Connecticut	3	1	0	4	2,110	1,156	—	3,266
Florida	5	0	0	5	3,823	—	—	3,823
Georgia	2	2	0	4	1,561	2,320	—	3,881
Illinois	8	6	2	16	6,524	6,492	2,240	15,256
Indiana	0	2	0	2	—	2,260	—	2,260
Iowa	1	0	0	1	538	—	—	538
Kansas	0	1	0	1	—	1,150	—	1,150
Louisiana	0	2	0	2	—	2,038	—	2,038
Maine	1	0	0	1	825	—	—	825
Maryland	2	0	0	2	1,690	—	—	1,690
Massachusetts	2	0	0	2	830	—	—	830
Michigan	4	3	0	7	3,047	2,364	—	5,411
Minnesota	3	0	0	3	1,605	—	—	1,605
Mississippi	1	3	0	4	1,250	3,820	—	5,070
Missouri	0	1	0	1	—	1,120	—	1,120
Nebraska	2	0	0	2	1,256	—	—	1,256
New Hampshire	0	2	0	2	—	2,300	—	2,300
New Jersey	3	1	0	4	2,855	1,067	—	3,922
New York	5	2	0	7	3,749	1,920	—	5,669
North Carolina	4	1	0	5	4,002	900	—	4,902
Ohio	1	2	0	3	906	2,410	—	3,316
Pennsylvania	6	4	0	10	5,757	3,993	—	9,750
South Carolina	5	2	0	7	4,261	2,290	—	6,551
Tennessee	2	4	0	6	2,296	4,820	—	7,116
Texas	0	4	0	4	—	4,720	—	4,720
Vermont	1	0	0	1	514	—	—	514
Virginia	4	0	0	4	3,390	—	—	3,390
Washington	1	3	0	4	850	3,590	—	4,440
Oregon	1	0	0	1	1,130	—	—	1,130
Wisconsin	4	0	0	4	1,579	—	—	1,579
U.S. Totals	81	52	2	138	66,831	59,158	2,240	128,229

Sources: Energy Research Group. *Estimates of the Population Served by Nuclear Power Plants in the United States*, Waltham, Mass, February 1984; and Personal Communications with the Atomic Industrial Forum Washington, July 2, 1984.

are planning to increase their reliance on nuclear power. France, for example, is trying to raise the proportion of nuclear-generated electricity from 30 percent to 70 percent by 1990. When compared to the United States, other nations seem to build nuclear plants faster and more economically. The Organization for Economic Cooperation and development reports that in France nuclear plants cost an average of $680 per Kilowatt of installed capacity compared to $1,434 in the United States.

PROJECTED SUPPLY AND DEMAND FOR URANIUM It is difficult to estimate future demand for uranium resources because of a number of uncertainties associated with the future development of nuclear power. However, Figure 4-26 offers a realistic comparison of uranium supplies with forecasts of domestic uranium requirements prepared by the Department of Energy, Office of Uranium Enrichment and Assessment (Department of Energy, January 1983). These forecasts are based on a 0.20

TABLE 4–8
Proportion of Net Electric Energy Generation produced by Nuclear Reactors in 1982, by State

STATE	NET GENERATION BY ALL FUEL UNITS (MEGAWATT HOURS)	NET GENERATION NUCLEAR UNITS (MEGAWATT HOURS)	PERCENTAGE NUCLEAR GENERATED
Alabama	74,680,141	27,701,065	37.1
Alaska	3,597,379	—	0.0
Arizona	38,719,699	—	0.0
Arkansas	23,478,976	7,482,343	31.9
California	118,320,759	3,734,971	3.2
Colorado	25,577,607	568,851	2.2
Connecticut	24,417,334	13,624,685	55.8
Delaware	7,015,963	—	0.0
District of Columbia	86,826	—	0.0
Florida	92,047,369	19,318,963	21.0
Georgia	59,984,522	6,605,613	11.0
Hawaii	6,366,819	—	0.0
Idaho	11,592,125	—	0.0
Illinois	93,659,643	27,624,984	29.5
Indiana	63,825,809	—	0.0
Iowa	21,867,293	2,269,129	10.4
Kansas	23,220,074	—	0.0
Kentucky	58,111,089	—	0.0
Louisiana	41,534,658	—	0.0
Maine	8,324,010	4,524,226	54.3
Maryland	29,755,160	10,345,444	34.8
Massachusetts	33,986,354	4,173,437	12.3
Michigan	68,703,126	15,002,636	21.8
Minnesota	28,546,903	10,196,944	35.7
Mississippi	16,971,175	—	0.0
Missouri	49,062,897	—	0.0
Montana	14,844,321	—	0.0
Nebraska	18,268,833	8,752,542	47.9
New Hampshire	4,783,684	—	0.0
New Jersey	31,464,547	14,039,107	44.6
New Mexico	23,757,816	—	0.0
New York	101,934,976	14,437,823	14.2
Nevada	15,585,055	—	0.0
North Carolina	74,495,883	9,126,412	12.3
North Dakota	18,313,690	—	0.0
Ohio	105,152,652	3,226,075	3.1
Oklahoma	44,926,475	—	0.0
Oregon	50,770,882	4,792,040	9.4
Pennsylvania	118,973,870	16,472,073	13.8
Rhode Island	441,903	—	0.0
South Carolina	36,777,934	13,156,023	35.8
South Dakota	7,920,104	—	0.0
Tennessee	59,504,897	10,104,304	17.0
Texas	206,392,631	—	0.0
Utah	11,891,628	—	0.0
Vermont	5,052,070	4,174,254	82.6
Virginia	36,599,893	17,420,489	47.6
Washington	96,691,040	3,630,982	3.8
West Virginia	68,284,913	—	0.0
Wisconsin	37,393,198	10,267,832	37.3
Wyoming	27,534,762	—	0.0
U.S. Total	2,241,211,367	282,773,248	12.6

Source: U.S. Department of Energy. *Electric Power Annual, 1982.* Washington, August 1983.

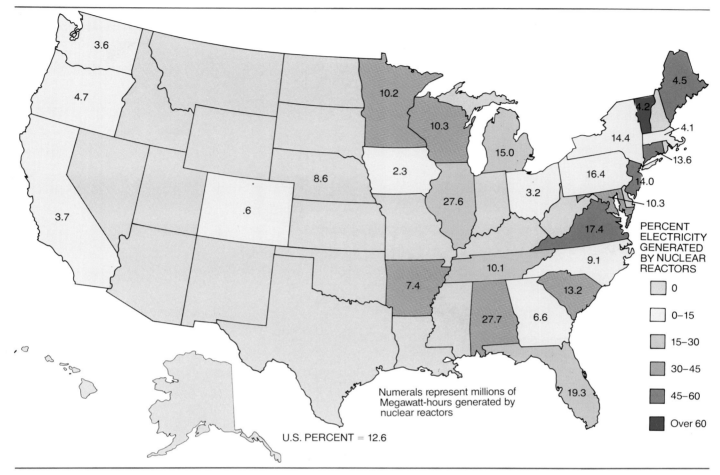

Fig. 4–24   Proportion of net electric energy generation produced by nuclear reactors in 1982 by states.

Source: Department of Energy, *Electric Power Annual*, 1982.

percent enrichment tails assay. They reflect Light Water Reactor use of improved fuel utilization cycles and are based on a nuclear generating capacity of 119,000 Megawatts in 1990 and 133,000 Megawatts in the year 2000.

The bar graph (Fig. 4–26) clearly indicates that the uranium supply is more than enough to meet the near-term demand. There are seven times more uranium reserves than the forecasted demand through the year 2000.[6] Moreover, the cumulative demand can easily be met by reserves in the 30 dollar per pound cost category. The total uranium resource supply is 32 times larger than the projected demand through the remainder of this century.

Low Grade Uranium Ores.   Because fuel costs are such a small part of the total cost of nuclear-generated electricity, it is possible reactors could be fueled with uranium priced at more than 100 dollars per pound. If in the next century there is a need for such resources, very large amounts exist in the Chattanooga shales of Tennessee and adjacent states. Because of its low uranium oxide content, this shale may be most appropriate for supplying breeder reactors

which employ the more abundant isotope, U-238. It may be possible, however, to derive fuel for nonbreeders from such a rock. In one area in eastern Tennessee, the Chattanooga shale contains 60 grams of uranium for every metric ton. Assuming a density of 2.5 metric tons per cubic meter, a vertical column of rock 5 meters tall and one square meter in cross-section would contain 12.5 tons of rock, and consequently 750 grams or 1.65 pounds of uranium (Hubbert, 1971). Since each pound of uranium is only 0.7 percent U_{235}, this 1.65 pounds contains 0.0116 pounds of the fissionable isotope. Using the equivalence, 1 pound U_{235} to 1,360 tons of bituminous coal (Butler, 1967), this column of relatively rich shale 5 feet tall and one square meter in area weighing 12.5 tons holds the energy of 15.77 tons of bituminous coal. Although a ton of the shale, used only for its U-235 content, to fuel nonbreeder reactors, contains slightly more energy than a ton of bituminous coal, there would be a great deal more waste rock from the shales, since only 0.006 percent of it is uranium.

The practicality of such low-grade uranium ore will depend not only upon volumes of rock to be mined, but upon the net energy gain, that is, whether the energy realized in a reactor exceeds energy expended in mining, concentrating, refining, and enriching the uranium compounds. It is interesting that the same Chattanooga shales, being rich in organic matter, are a low-grade source of *kerogen* or shale oil (averaging 10 gallons of oil per ton of shale) which presumably would be recovered if the rock were to be processed for its uranium content.

THE NATURE AND OCCURRENCE OF THORIUM

At present, thorium is an element of limited economic importance. It could, however, become an important nuclear fuel if the nation's uranium resources become more scarce and more costly. As illustrated earlier, Thorium-232 can be transmuted under neutron bom-

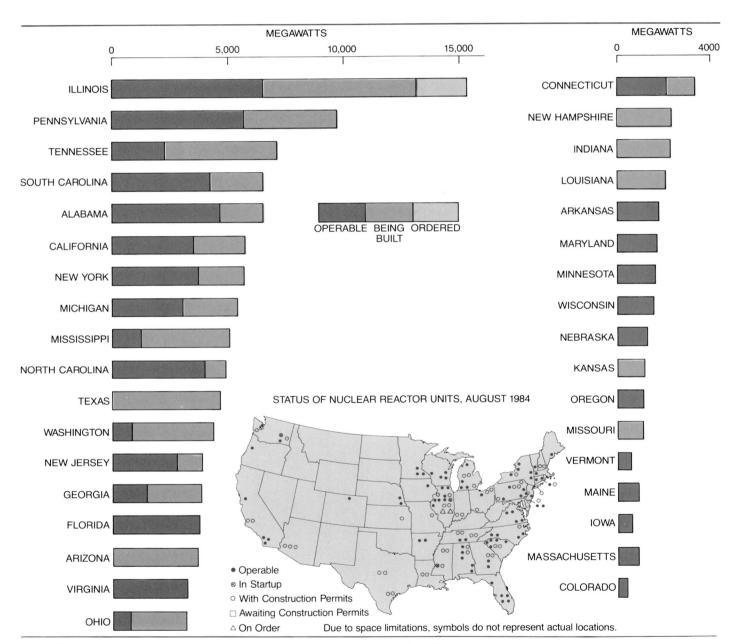

Fig. 4-25 Nuclear reactor sites and state nuclear generating capacity either (1) operable, (2) being built, or (3) ordered as of August 1984.

Source: Energy Research Group, *Estimates of the Population Served by Nuclear Power Plants in the United States*, February 1984; and Personal Communications with the Atomic Industrial Forum, July 1984.

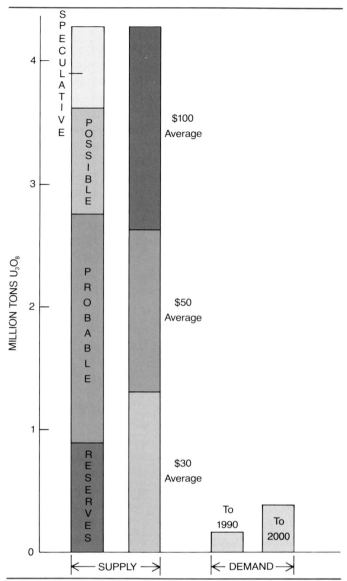

Fig. 4–26 Total uranium resources at costs up to $100 per pound compared with projections of demand to years 1990 and 2000.

Source: Department of Energy. *Statistical Data of the Uranium Industry,* January 1, 1983.

Note: These forecasts of domestic uranium requirements are based on a 0.20 percent enrichment tails assay. They reflect Light Water Reactor use of improved fuel utilization cycles. The forecasts were prepared by the Department of Energy Office of Uranium Enrichment and Assessment as of August 1982 and are based on a nuclear capacity of 119,000 megawatts in 1990 and 133,000 Megawatts in the year 2000.

bardment into Uranium-233, a fissionable isotope in a chain reaction (see Fig. 4–2B). If thorium-fueled reactors, such as the high-temperature gas-cooled reactor (HTGR), the light water breeder reactor (LWBR), and the molten salt breeder reactor (MSBR), become technologically and economically feasible, there will be a greatly increased demand for thorium.

Thorium is a heavy, silver-grey metal widely distributed in nature, and usually found in association with ura-

nium or other rare earth elements. Its geochemical abundance is estimated to range from 6 to 13 parts per million in the earth's crust, which is perhaps three times that of uranium (Staatz *et al.,* 1973). Thorium concentrations of potential economic importance occur in beach sands (placers), in vein deposits in sedimentary rocks, such as thorium-bearing dolomite, and in conglomerates or quartzites enriched in thorium and uranium (Sondermayer, 1975). The chief ore minerals of thorium are monazite, thorite, uranothorite, and bannerite, of which monazite in beach sands is the major source of thorium.

The principal world thorium-bearing deposits are located in the United States (Atlantic Coastal Plain), Canada (Eliot Lake, Ontario), India (Bihar and West Bengal), and Brazil (Atlantic Coast). Excepting the conglomerate deposits of Canada, these deposits are *fluviatile,* or those produced by river waters, and *beach placers,* or those produced by ocean waters. Additional thorium deposits are found throughout the world in a variety of geological settings in Australia, Africa, Greenland, and Asia.

Reserves and Resources

Total world resources of thorium amount to almost 3.9 million metric tons (4.3 million short tons) with 1.2 metric tons of that in Reserves (reasonably assured resources) (Fig. 4–27). The Middle East, North African region, Western Europe, and South Asia house almost 80 percent of the Reserves. The African Reserves are found primarily in Egypt, those of Western Europe are found in Denmark and Norway, and the Asian Reserves are housed principally in India. The 2.6 million metric tons of estimated additional resources are found primarily in Latin America (Brazil) and North Africa (Egypt) with these two regions accounting for almost 60 percent of the total.

Due to the limited demand, thorium resources of the United States are not well known. Presently identified thorium resource areas of the United States are shown in Figure 4–28 and resource amounts are summarized in

THORIUM	
Thorium is a grey radioactive metal which occurs in nature as Thorium-232. Other isotopes include Thorium-233 and Thorium-234.	
Symbol	Th
Atomic number	90
Atomic weight	232.04
Melting point	1700°C
Boiling point	4500°C
Density	11.725 g/cm³

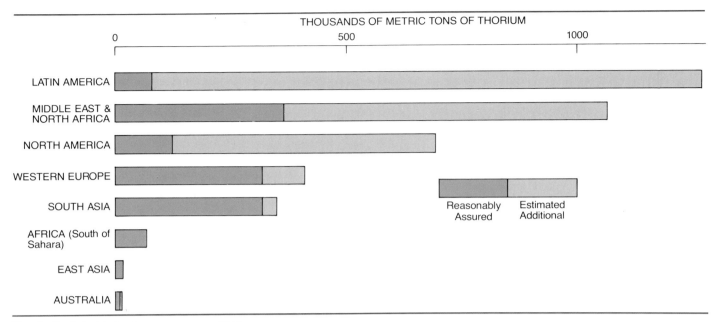

Fig. 4–27 World thorium resources, 1980.

Source: World Energy Conference, 1980.

Notes: 1. The resource amounts are based on recovery costs up to $75/kg (approximately $34/lb).
2. To convert to short tons multiply resource amounts by 1.1.
3. Reasonably assured resources are roughly equivalent to *reserves.*

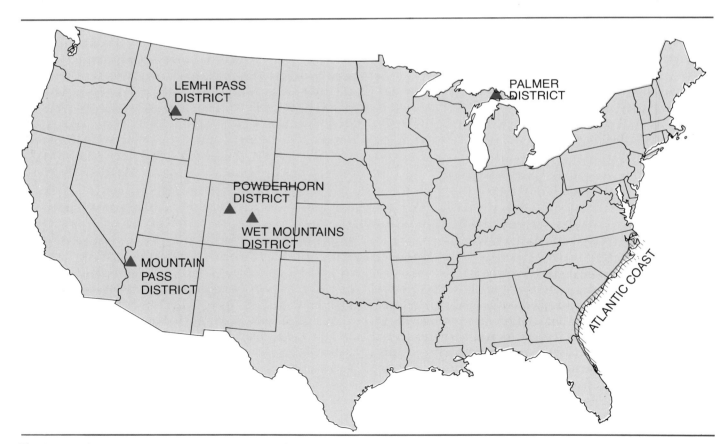

Fig. 4–28 Thorium resource areas in the United States.

Source: Sondermayer, 1975.

TABLE 4-9
Identified Thorium Resources of the United States as of 1975
(in thousands of short tons ThO_2 estimated recoverable)

AREA[1]	TYPE OF DEPOSIT	AS BY-PRODUCT	PRIMARILY FOR ThO_2 Ore Grade	
			Over 0.1%	Under 0.1%
Atlantic Coast	Beach Placer	16	0	0
North and South Carolina	Fluviatile Placer	0	0	56
Idaho and Montana	Fluviatile Placer	2	0	38
Lemhi Pass District, Idaho and Montana	Veins	0	100	0
Wet Mountains	Veins	0	4.5	0
Powderhorn	Veins	0	1.5	0
Mountain Pass, California	Veins	0	0.5	0
Mountain Pass, California	Carbonatite	28	0	0
Palmer, Michigan	Conglomerate	0	0	46
Bald Mt., Wyoming[2]	Conglomerate	0	0	2
U.S. Total		46	106.5	142

[1]See accompanying map.

[2]Not shown on accompanying map.

Source: Sondermayer, Roman V. "Thorium." *In Minerals Facts and Problems. Bureau of Mines Bulletin, 667,* Washington: 1975.

Table 4-9. At present, the only mineable reserves are localized beach placers along the Atlantic Coast, where monazite containing about 4 percent thorium oxide (ThO_2) is produced as a by-product of titanium mining (Staatz *et al.,* 1973). These monazite sands in Green Cove Spring, Florida, Hilton Head Island, South Carolina, and Folkston, Georgia are estimated to contain 16,000 tons of ThO_2. In addition, there are approximately 28,000 tons ThO_2 that are recoverable as a by-product of rare earth mining in carbonatite deposits at Mountain Pass, California. Larger amounts of ThO_2, over 100,000 tons, are located in vein deposits at several locations in the western United States. The principal vein deposits in the Lemhi District of Idaho and Montana, contain ore of more than 0.1 percent ThO_2 which is considered recoverable for its thorium content alone. Lower grade ThO_2 ores (less than 0.1 percent ThO_2) amounting to over 140,000 tons are considered recoverable from fluviatile placers in North Carolina, South Carolina, Idaho, and Montana.

Considering the present low demand, it would appear that Atlantic Coast placer deposits will remain the major source of thorium in the United States for some time. It is also likely that new placer deposits, located below sea level and off present Atlantic Coast beaches, will be discovered in the future. One estimate of monazite content in undiscovered offshore deposits on the United States continental shelf is about 4 million tons (McKelvey, 1968). If these monazite sands contain 4 percent ThO_2 as do other known deposits of this type, undiscovered continental shelf deposits could amount to 160,000 tons of ThO_2.

United States Supply and Demand

Total United States thorium consumption in all forms in 1980 was estimated at 33 short tons. Most of this was used in nonenergy industries, the major nonenergy uses being mantles for incandescent lamps, magnesium thorium alloys, refractories, and thoriated tungsten welding rods. Future thorium demand will depend on the demand for thorium-fueled reactors in the electrical generation industry. Presently, there is only one thorium-fueled nuclear reactor in the United States. The Fort St. Vrain thorium-fueled, HTGR reactor in Colorado is rated at 330 Megawatts and is producing power for the Public Service Company of Colorado. Until 1982, the Department of Energy's experimental Light Water Breeder (LWBR) reactor was operating at Shippingport, Pennsylvania. The reactor, which used the thorium–uranium-233 fuel system, was retired in 1982 after 25 years of operation.

Another type of thorium-fueled reactor considered for commercial application is the molten salt breeder (MSBR). However, it is not, at present, being looked upon favorably by the nuclear power industry because when compared to the liquid metal fast breeder reactor (LMFBR), it has a relatively low breeding ratio (1.05) and a long doubling life (20 years).

It would appear that thorium-fueled reactors will play a minimal role in the production of electrical energy in the near future. United States' thorium resources should be more than adequate to meet the demands of both nonenergy and energy uses well into the next century.

FUSION

Thermonuclear fusion is viewed by a large segment of the scientific community as a very attractive, long-term solution to the world's energy supply problem. The fusion reaction was first demonstrated about 45 years ago. To date, the only manmade, self-sustaining fusion reaction has been the explosion of the hydrogen bomb. Fusion may use fuels which are essentially inexhaustible and this process appears to be environmentally safe. A great deal of research is needed, however, before even a working model can be built. Electrical power from fusion may not be commercially available until well after the year 2000.

The fusion process, like that of fission, releases energy by the conversion of heavy nuclei to nuclei of intermediate weight (see Fig. 4–2). The two fundamental differences between fusion and fission are the nature of the products of the reaction and the techniques used to make the two processes occur. For the purpose of this discussion, emphasis will be placed on the basic fusion processes and their fuel requirements.

Three reactions involving isotopes of hydrogen appear to offer the greatest potential for the creation of energy to be used in a thermonuclear fusion power plant (Fig. 4–29). In equations in Figures 4–29A and 4–29B, two deuterium nuclei (Hydrogen-2) combine to form either tritium (Hydrogen-3) or helium (Helium-3) and release 4.0 and 3.2 millions of electron volts (MeV) of energy respectively (Fowler, 1975). The equation described in Figure 4–29C uses deuterium and tritium to produce ordinary helium (Helium-4). In this deuterium–tritium (D–T) reaction, 17.6 MeV of energy are released, a considerably larger amount

than is released in either of the two deuterium–deuterium (D–D) reactions.

The three fusion reactions described in Figure 4–29 appear rather simple and direct. Several decades of research, however, have raised a number of major problems that must be solved before useful energy can be obtained from the fusion process. The central problem is how to bring the hydrogen nuclei close enough for fusion to take place. To produce this reaction, the two nuclei, which have the same electrical charge, must approach each other with speeds sufficient to overcome the repulsive force between them (Hulme, 1969). The only practical way to achieve this is to heat the matter to incredibly high temperatures. The required temperatures are in the order of 40 million degrees Celsius for the D–T reaction and 400 million degrees Celsius for the D–D reaction (Turk 1974). Clearly, containment of these reactions at such temperatures is a problem of immense proportion. Nevertheless, present fusion research indicates that the reactions may be contained through the use of a very strong magnetic field. If and when the scientific feasibility of fusion is demonstrated a large number of engineering problems will inhibit its commercial applications.

Based on our present knowledge, the most promising way of achieving fusion energy will be by means of the D–T reaction. When compared with the D–D reaction, it releases significantly more energy and can be achieved at much lower temperatures and faster rates. For the D–T reaction, the deuterium supply is virtually unlimited: one cubic kilometer of seawater contains an amount of deuterium equivalent to the energy potential of 300 billion tons of coal or 1,500 billion barrels of crude oil. The total

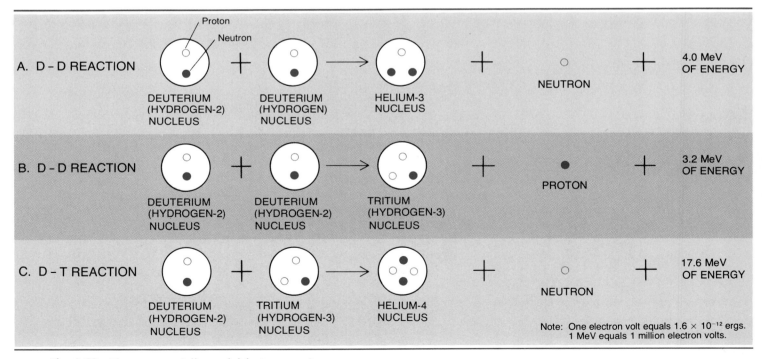

Fig. 4–29 Three potentially useful fusion reactions.

ADVANTAGES OF FUSION

1. If the deuterium–deuterium reaction is perfected, the fuel source will be limitless and cheap.
2. Reactor meltdown, with release of radioactive materials, cannot occur.
3. Radioactive wastes produced would be smaller in amount and also less hazardous than those produced by fission reactors.
4. Fuel materials will not be suitable for weapons use.
5. Because of their safety, fusion plants could be located close to settlements, so their waste heat could be used for space heating.

DISADVANTAGES OF FUSION

1. The greatest hazard would be the release of radioactive tritium as gas or in tritiated water. It has a short half-life (12 years) but as a gas is extremely difficult to contain.
2. Plants will hae to be very large and expensive.
3. After 10 years of operation, the plant structure itself may be weakened and radioactive and may have to be replaced.
4. Staggering technical problems stand in the way of a working model, let alone an electrical generating plant of commercial scale.

volume of the oceans is about 1.5 billion cubic kilometers (Hubbert, 1971). On the other hand, tritium exists in only small amounts in nature, and tritium fuel requirements for fusion power must be produced by Lithium-6 by means of a lithium blanket surrounding the reactor vessel. Only 7.4 percent of natural lithium is Lithium-6, limiting still further the potential for producing tritium. It is possible, however, to use the neutrons produced by the D–T reaction to produce even more tritium than is used in the reaction. Ideally we can get both energy and replacement of the expensive and scarce portion of the fuel requirements. Nonetheless, the future of a commercial fusion power industry depends to a great extent on the nation's ability to find, recover, and process lithium resources.

FUSION ENERGY DEVELOPMENT IN THE UNITED STATES Fusion research efforts in the United States are focused on two types of confinement; magnetic and inertial. Magnetic confinement research dates back to the 1950s while serious inertial confinement research started in the 1970s.

The magnetic confinement approach is illustrated in Figure 4–30. In this technique the plasma in which the fusion reaction takes place is confined within a doughnut-shaped magnetic field. The energy released by the fusion reaction is transmitted through heat exchangers to a steam generator. After the steam generator stage, the process is the same as either a fission or a fossil-fuel electrical generation plant. Magnetic confinement fusion reactors are referred to in the industry as "tokamaks" a term which originated in the Soviet fusion program. The largest experimental Tokamak Fusion Reactor in operation today is the "Doublet III" machine built by General Atomic in San Diego. The first fusion reactor designed to provide a

The experimental Tokomak fusion reactor at Princeton University, N.J. (Courtesy of U.S. Department of Energy.)

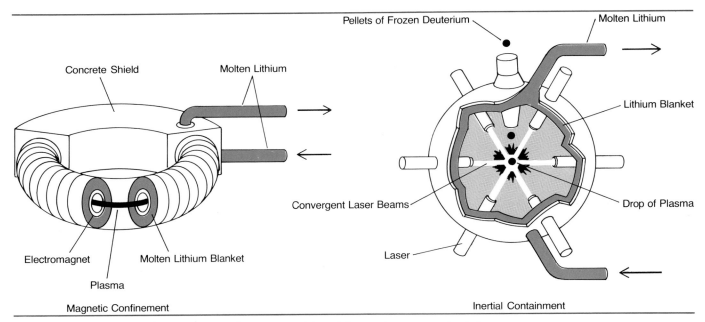

Fig. 4–30 Comparison of magnetic confinement and inertial containment methods for initiating a nuclear fusion reaction.
Source: Adapted from *Living in the Environment: Concepts, Problems, and Alternatives* by G. Tyler Miller, Jr. Copyright 1975 by Wadsworth Publishing Company, Inc. Reprinted by permission of the publisher.

positive net energy output, the Tokamak Fusion Test Reactor, is near operational at Princeton, New Jersey.

In the other experimental fusion approach, inertial confinement, the fuel is contained in a solid pellet rather than in plasma. The pellets are injected with large amounts of energy by means of lasers producing fusion reactions. This inertial confinement concept is being developed at the Lawrence Livermore Laboratory in California and Los Alamos and the Sandia laborotories in New Mexico. An inertial confinement device has been operational at the Kurchatov Institute in the Soviet Union since 1979 (Pryde, 1983).

Although fusion energy research has made great strides since the 1950s, it will be decades before it can become a commercial alternative to other types of electrical generation facilities. At the present, the Department of Energy estimates that a commercial fusion reactor will not be available until after 2020.

THE NATURE AND OCCURRENCE OF LITHIUM

Lithium is a very light, soft, and ductile metal (see Glossary), found principally in granitic pegmatites, subsurface brines, and salt water. In granitic pegmatites, lithium minerals are usually found in association with sodic and potassic feldspars, quartz, and muscovite (Singleton *et al.* 1975). Spodumene is the most plentiful lithium-bearing mineral in pegmatites and typically makes up 20 to 25 percent of the rock mined. The spodumene belt near

Kings Mountain, North Carolina is the richest known lithium deposit in the world. In addition to the North Carolina deposits, major occurrences of lithium-bearing spodumene are found in South Dakota (Black Hills), Canada (Bernie Lake, Manitoba), Rhodesia (Bikita Tin Fields), and the Soviet Union. Other lithium-bearing minerals found in granitic pegmatites are lepidolite, petalite, amblygonite, and eucryptite. All successfully mined lithium-enriched pegmatites contain at least 1.0 percent, but normally no more than 2.0 percent Li_2O (Singleton *et al.* 1975).

Subsurface brines contain more of the world's known lithium resources than do pegmatites, but lithium concentrations in these brines are low, ranging from 0.02 to 0.2

LITHIUM	
Lithium is the lightest of all metals. Bombardment of lithium with neutrons in a nuclear reactor yields a heavy isotope of hydrogen, tritium, a fuel for use in fusion reactions.	
Symbol	Li
Atomic number	3
Atomic weight	6.94
Density	0.534 g/cm³
Melting point	180.5°C
Boiling point	1347°C

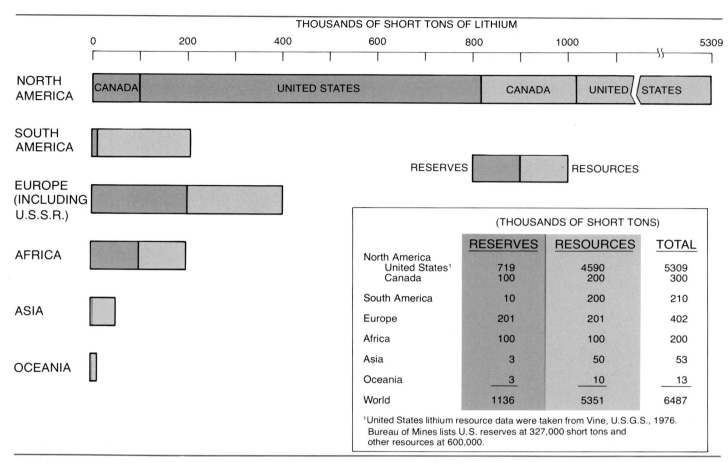

THOUSANDS OF SHORT TONS OF LITHIUM

	(THOUSANDS OF SHORT TONS)		
	RESERVES	RESOURCES	TOTAL
North America			
United States[1]	719	4590	5309
Canada	100	200	300
South America	10	200	210
Europe	201	201	402
Africa	100	100	200
Asia	3	50	53
Oceania	3	10	13
World	1136	5351	6487

[1]United States lithium resource data were taken from Vine, U.S.G.S., 1976. Bureau of Mines lists U.S. reserves at 327,000 short tons and other resources at 600,000.

Fig. 4-31 World identified lithium resources.
Source: Vine, 1976.

percent. Lithium-rich brine deposits occur in Clayton Valley, Nevada and Salar de Atacama, Chile. In addition, potentially large deposits of lithium occur worldwide in geothermal and oil well brines. The geothermal brines in the Imperial Valley in California contain an estimated million tons of lithium. Major engineering problems, however, inhibit the realization of their potential.

The ocean waters of the world are estimated to contain approximately 300 million tons of lithium, but because of the low concentration, about 0.00002 percent, this source is not considered exploitable in the near future. In the United States, the Great Salt Lake in Utah constitutes a significant resource of this type.

World Lithium Resources

Figure 4-31 summarizes Identified world lithium resources. Identified world lithium resources amount to almost 6.5 million tons, with Reserves accounting for 18 percent of the total and other resources for 82 percent. The United States houses the vast majority of lithium resources, with 63 percent of the Reserves and 86 percent of the other resources. The United States' share of Identi-

fied resources, however, may be smaller than indicated because of the inclusion of more recent data for the United States than for other world regions. Canada has about 9 percent of the Reserves and 4 percent of the other resources. Outside of North America, most of the remaining lithium resources are found in Eastern Europe (including the Soviet Union) and Africa (Rhodesia and Zaire).

United States Lithium Resources

The U.S. Geological Survey has classified the nation's lithium resources into three categories: (1) *Identified resources,* which are profitable to mine at present prices and technology; (2) *Identified Low-Grade resources,* which are not economic at today's relatively low prices and the present level of technology but may become so with increased prices and improved technology; and (3) *Hypothetical* and *Speculative* resources which have yet to be discovered and developed (Vine, 1976).

Lithium resources of the United States are found in five principal locations: (1) Kings Mountain, North Carolina; (2) Clayton Valley, Nevada; (3) Searles Lake, California; (4) Imperial Valley, California; and (5) the Black Hills of South Dakota. Additional resources of significant quantity are known to exist in oilfield brines and borate mine waste at various other locations (Fig. 4–32). As of early 1976, the total United States' lithium resources were estimated at 7.9 million short tons. According to the Geological Survey's classification scheme, Reserves account for 9 percent of the total; Low-Grade resources for 58 percent; and Hypothetical and Speculative resources for 33 percent.

Lithium Reserves are estimated at 719,000 short tons of lithium (see Fig. 4–31) the vast majority of which (85 percent), are located in the spodumene belt of North Carolina. The remainder is found in subsurface brines of Clayton Valley, Nevada (13 percent), and potash-bearing brines at Searles Lake, California (2 percent). Identified Low-Grade resources are estimated to be approximately 4.6 million tons, 90 percent of which are in the spodumene belt of North Carolina, Clayton Valley, Nevada, the Imperial Valley of California, and in oilfield brines. Hypothetical and Speculative resources, estimated by the Geological Survey at 2.6 million tons of lithium, are thought to occur mostly in association with clays and brines at various locations, with about 20 percent of the estimated amount found in pegmatites.

Recoverability of Lithium Resources

Although the total of lithium resources appears to be more than adequate to meet the demands of alternative energy applications, the figures are, in fact, misleading. They fail to forecast amounts which will actually be available to the marketplace (Vine, 1976). In reality, a large part of the lithium resources are inaccessible as a result of economic, technical, and political problems as well as environmental restrictions. In order to allow a reasonable evaluation of the future supply and demand for lithium, the Geological Survey has estimated the percent of lithium recoverable and the probable yield by the year 2000 from various resource locales (Table 4–10).

From the three categories of resources, it is estimated that 1.3 million tons of lithium will be recovered by 2000. Reserves are projected to yield 276,000 tons, or 23 percent, of the lithium. Low-grade resources and Hypothetical and Speculative resources are expected to yield 30 and 47 percent respectively. In summary, more than three-fourths of the lithium recovered by the year 2000 is expected to be extracted from subeconomic and undiscovered resources. If this forecast is correct, its realization will require a great

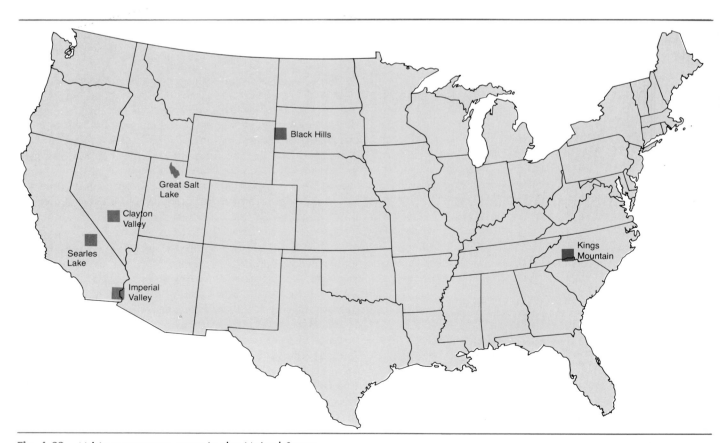

Fig. 4–32 Lithium resource areas in the United States.
Source: Vine, 1976.

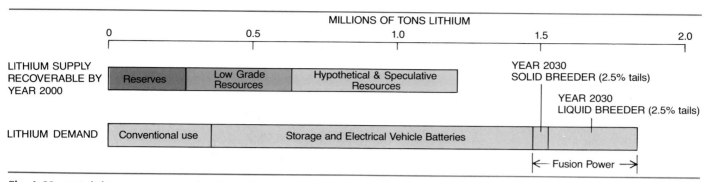

Fig. 4-33 U.S. lithium resources of various degrees of certainty compared with cumulative requirements for conventional uses, batteries, and fusion reactors.

Source: Vine, 1976.

Notes: 1. In the year 2000 lithium requirements for energy storage and electric vehicles would be about equally divided.
2. Fusion power demand for lithium assumes a national electric capacity of 2,010 Gigawatts in the year 2030.

TABLE 4-10
United States' Lithium Resources, Classified by Geologic Certainty (in millions of shorts tons of contained lithium)

SOURCES	IN-PLACE	ESTIMATED PERCENT RECOVERABLE BY 2000	PROBABLE YIELD
Reserves (Economic)			
Kings Mt., N. C.	.612	35	.222
Clayton Valley, Nev.	.096	50	.048
Seales Lake, Calif.	.011	50	.006
Total	.719		.276
Low-Grade Resources (Subeconomic)			
Kings Mtn., N. C.	1.008	25	.240
Clayton Valley, Nev.	.750	10	.075
Searles Lake, Calif.	.032	10	.003
Imperial Valley Calif.	1.200	1	.012
Oilfield Brines	1.200	1	.012
½ Great Salt Lake	.316	5	.012
Borate Mine Waste	.072	7	.005
Black Hills, S. D.	.012	30	.004
Total	4.590		.363
Hypothetical and Speculative Resources			
Clays	.960	25	.240
Brines	1.200	20	.240
Pegmatites	.420	20	.084
Total	2.580		.564

Source: Vine, James D., "The Lithium Resource Enigma." *Lithium Resources and Requirements by the Year 2000. U.S.G.S. Professional Paper 1005.* Washington: Government Printing Office, 1976.

deal of advanced planning on the part of the lithium industry, and a healthy economic climate which provides incentives for exploration and development of undiscovered lithium deposits.

Lithium Uses and Projected Requirements

Presently, lithium-bearing minerals are used in the manufacture of a number of products including glass for special uses, glass ceramics, porcelain enamels, and refractories. In addition, they are used in the production of lithium metals and chemicals. Domestic production of lithium minerals and chemicals is carried on primarily by two companies, Foote Mineral Company and Lithium Corporation of America. At present, the United States produces and consumes more than one-half of the world's lithium supply. In 1980 United States' production was withheld to avoid disclosing company proprietary data (Searls, 1980). However, in that same year lithium production in the rest of the world totaled 2,250 short tons. World consumption was estimated at 7,400 tons. Based on conventional uses only, the United States' demand for lithium has been projected to increase at an average annual rate of almost 9 percent each year to the year 2000 (Vine, 1976). If this is the case, by the year 2000 conventional uses will consume over 360,000 short tons of lithium, one-third of the country's recoverable resources (Fig. 4-33). Anticipated lithium requirements for energy-related uses are difficult to predict because of many technological uncertainties. If, however, the lithium anode battery is developed for commercial use in electric vehicles and off-peak storage (see Glossary), lithium supplies could be seriously threatened by the year 2000. Furthermore, the commercial development of nuclear fusion in the early decades of the twenty-first century could further complicate the lithium supply and demand situation.

LITHIUM REQUIREMENTS FOR LITHIUM ANODE BATTERIES For the past several years, Argonne National Laboratory has been working on the development of lithium-aluminum/iron-sulfide batteries for use as energy storage devices on electric vehicles (Chilenskas *et al.*, 1976). Argonne anticipated that by the year

2000, as much as three percent (3×10^8 Kilowatt hours) of the total United States' energy consumption may be supplied by such batteries. At the same time, as many as 18 million electric vehicles may be powered by the lithium-aluminum/iron-sulfide battery. Based on the Argonne projections, which assume a 10-year battery life, no recycling of the battery itself, and no recycling of the lithium, the maximum cumulative lithium requirement will be over 1.1 million tons up to the year 2000 (Fig. 4–33). This amount can be reduced to .97 million tons with 90 percent recycling of the lithium contained in the spent batteries. Anticipated demand for these same uses, through 1990, is expected to be approximately 156,000 tons.

A comparison of cumulative demand for lithium use in batteries only, and the cumulative amounts thought to be recoverable by the year 2000 reveals that the supply will be 184,000 tons greater than the demand (Fig. 4–33). When the requirements for conventional uses are also considered, the demand will exceed the supply by more than 170,000 tons, assuming no recycling, and by 32,000 tons assuming 90 percent recycling.

It is clear that the anticipated demand for lithium use in batteries by the end of the century could create a major supply problem. If lithium anode batteries are to be used to the extent projected by Argonne, and if they are to make a significant impact upon petroleum conservation, the lithium industry will have to work at accelerated rates to identify new resources and expand their production facilities to meet the demand.

LITHIUM REQUIREMENTS FOR FUSION POWER

The development schedule for fusion power calls for a test reactor to operate in the 1980 and 1981 time period; an experimental power reactor in 1986 to 1988; a small fusion electric plant in 1990 to 1991; and a demonstration power plant by the end of the century. This development plan is, of course, tentative and depends on satisfactory solutions to a number of technical problems. Without adequate funding on the part of both private and public sectors, the research needed for demonstrating the feasibility of commercial fusion power in this century will be impossible. For the sake of providing an assessment of expected lithium requirements for fusion power, however, let us assume that adequate funding will be found for research and development.

Fusion's demand for lithium is highly dependent on the reactor design. For example, in a liquid breeder, where lithium (metallic or salt) is used as both a breeder and coolant, the amount of lithium required, or *inventory,* is very high. On the other hand, in solid breeder reactor designs, the initial lithium fuel needs are less than half

those for the liquid breeder, but they must be highly enriched in the Lithium-6 isotope. Furthermore, because of a high burn-up rate in the solid breeder, periodic refueling is necessary and will affect the annual net lithium requirements for fusion power. The Energy Research Development Administration (ERDA) has developed two scenarios which extrapolate cumulative needs to the year 2030. The first scenario is designated the *Base Case* and assumes a continuation of present electric energy growth trends, with the effects of conservation slowing down the growth rate after 1990. It calls for a 2010-Gigawatt (Billion watts) electric capacity by the year 2030 and a 270-Gigawatt fusion capacity. The second scenario, defined as the *Massive Shift Case,* assumes a significant increase in the use of electricity and a decrease in the use of conventional fuels such as oil and gas. The Massive Shift Case also assumes a national electric capacity of 3000 Gigawatts in the year 2030 and a fusion capacity of 614 Gigawatts. In both cases, lithium demand was calculated for two tail fractions, 2.5 and 0.3 percent, for the solid breeder and one tail fraction, 2.5 percent, for the liquid breeder.[7]

Figure 4–33 summarizes the projected lithium requirements for the Base Case. For this projection the lithium requirements are: 1) 373,000 short tons using liquid breeders and assuming 2.5 percent Lithium-6 remaining in the tails; 2) 27,000 short tons using solid breeders and assuming 2.5 percent tails; and 3) 14,000 short tons using solid breeders and assuming 0.3 percent tails.

The results of these extrapolations clearly demonstrate the much larger fuel demands of the liquid breeder concept. Not only does the solid breeder require much smaller initial inventories of fuel but it also allows most of the lithium to be re-used after the Lithium-6 isotope is removed. For example, over 94 percent of the calculated lithium demand may be returned to the market place for use in other industries, assuming a tails fraction of 2.5 percent.

In conclusion, the use of liquid breeders for fusion power could place an intolerable strain on our presently defined supply of lithium resources. Although solid breeders would require far less lithium, the technological uncertainties associated with this concept make it too early to tell which of the two reactor concepts (if either) might be the best choice for commercial fusion power production. In the near term, lithium requirements for use in electric vehicles and off-peak storage batteries are a greater threat to the nation's lithium supply than are the demands from fusion power. Nevertheless, the availability of lithium is fundamental to the use of fusion as an energy supply option, and a strategy needs to be developed to assess the future supply and demand for lithium fully.

NOTES

1. The half-lives of U-235 and U-238 are 7.13×10^8 and 4.51×10^9 years respectively. Because of these slow rates of radioactive decay, natural uranium is only mildly radioactive.

2. The bulk of the Reserves in the up to 30 dollars per pound class range from 0.08 to 0.11 percent U_3O_8, (average 0.10 percent), lie less than 700 feet below the surface, and are found exclusively in

sandstone. In many instances low-grade materials are included in the low-cost Reserve amounts because they are co-located with adjacent high-grade ores and it is economic to mine them together.

3. The exception is a 330-Megawatt high-temperature gas-cooled reactor in operation at Fort St. Vrain, Colorado.

4. As the chain reaction grows, the fuel rods get hotter and more bubbles are formed. The increase in space occupied by the steam bubbles reduces the moderating effectiveness of the water, reduces the number of slow neutrons, and thus slows the growth of the chain reaction.

5. Assuming that it would take 3.5 million tons of uranium to fuel 800 converter reactors for a forty-year life, Lapp estimates that the same amount of fuel would supply 800 breeder reactors for 3700 years.

6. Each additional 1000-Megawatt nuclear generating plant would add 5,300 tons of U_3O_8 to the total requirements over thirty years of operation.

7. The tails fraction represents the percentage of Lithium-6 remaining in the depleted lithium after fuel processing.

References

Atomic Industrial Forum. "Historical Profile of U.S. Nuclear Power Development." *Background Info,* Bethesda, MD: Atomic Industrial Forum, January 1, 1984.

Bogart, S. Locke. "Fusion Power and the Potential Lithium Requirement." In *Lithium Resources and Requirements by the Year 2000.* U.S.G.S. *Professional Paper 1005,* 1976.

Chilenskas, A.A. et al. "Lithium Requirements for High-Energy Lithium-Aluminum/Iron Sulfide Batteries for Load-Leveling and Electric Vehicle Applications," *Lithium Resources and Requirements by the Year 2000. U.S. Geological Surgey Professional Paper 1005.* Washington, DC: Government Printing Office, 1976.

Energy Research Group. *Estimates of the Population Served by Nuclear Power Plants in the United States,* Waltham, MA: Energy Research Group, February, 1984.

Finch, Warren I. et al. "Uranium." *In United States Mineral Resources, U.S.G.S. Professional Paper 820,* 1973.

Fowler, John M. *Energy and the Environment.* New York: McGraw-Hill, 1975.

Hubbert, King. "The Energy Resources of the Earth." *Scientific American, 224* (3), September, 1971.

Hulme, H.R. *Nuclear Fusion.* London: Wykeham Publications, 1969.

Lapp, Ralph E. "We may find ourselves short of Uranium." *Fortune* October, 1975.

McKelvey, V.E. "Mineral Potential of the Submerged Part of the U.S." *Ocean Industry,* 3 (9), 1968.

Nuclear Energy Policy Study Group. *Nuclear Power Issues and Choices.* Cambridge, MA: Ballinger, 1977.

OECD Nuclear Energy Agency and The International Atomic Energy Agency. *Uranium, Resources, Production and Demand.* Paris: OECD, February, 1982.

Pryde, Philip. *Nonconventional Energy Resources.* New York, NY: John Wiley Interscience Series, 1983.

Searls, James P. "Lithium." In *Minerals Yearbook, Metals and*

Minerals. (Vol. I) Washington, DC: Government Printing Office, 1980.

Singleton, Richard H. and Hiram Wood. "Lithium." *Mineral Facts and Problems. Bureau of Mine's Bulletin* 667 Washington, DC: Government Printing Office, 1975.

Sondermayer, Roman V. "Thorium." In *Mineral Facts and Problems, Bureau of Mine's Bulletin 667* Washington, DC: Government Printing Office, 1975.

Staatz, Mortimer, et al. "Thorium." *United States Mineral Resources,* U.S.G.S. *Professional Paper 820,* 1973.

Turk, Amos. *Environmental Science.* Philadelphia, PA: W.B. Saunders, 1974.

U.S. Atomic Energy Commission. *Nuclear Fuel Reserves and Requirements,* WASH-1234. Washington, DC: Government Printing Office, 1973.

U.S. Department of Energy. *Electric Power Annual, 1982.* Washington, DC: Government Printing Office, August, 1983.

———. *Inventory of Power Plants in the United States, 1982.* Washington, DC: Government Printing Office, June 1983.

———. *Nuclear Power Cancellations: Causes, Costs, and Consequences.* Washington, DC: Government Printing Office, April, 1983.

———. *Statistical Data of the Uranium Industry.* Washington, DC: Government Printing Office, January, 1983.

U.S. Energy Research and Development Administration. *National Uranium Resource Evaluation Preliminary Report.* Grand Junction, CO: Energy Research and Development Administration, 1976.

Vine, James D. "The Lithium Resources Enigma." In *Lithium Resources and Requirements by the Year 2000.* U.S.G.S. *Professional Paper, 1005,* 1976.

Woodmanse, Walter C. "Uranium" *Mineral Facts and Problems, 1975, Bureau of Mine's Bulletin 667.* Washington, DC: Government Printing Office, 1975.

5 Geothermal Heat

◄Workers surrounded by steam from a geothermal well in The Geysers area north of San Francisco. (Courtesy of the American Petroleum Institute and Aminoil USA Inc.)

THE NATURE OF THE RESOURCE

Geothermal heat is one of three types of *primary* energy that flow continuously to the surface of the earth; the other two are solar radiation and tidal energy. Despite its constant flow from the earth's interior, geothermal heat is not defined here as a renewable resource but rather is placed under the broad heading of *nonrenewable* energy sources.

The sources of earth heat are the molten core of the earth and the heat generated by radioactive decay of elements in the crust. Both sources are diminishing very slowly, and the amounts of heat within the earth are virtually unlimited. This immensity and virtual permanence of the earth's heat supply are not so relevant, however, as the size and the renewability of *occurrences that offer accessible high temperatures* and applications such as space heating or electrical generation. Such occurrences are limited in size, and are located only at certain sites. More importantly, these occurrences are not created rapidly by nature. As a result, an intensive program of exploitation could deplete these usable resources, even as the vast reservoir of heat within the earth continues practically undiminished. If depletion is defined as drawing down local temperatures until they are no longer usable, then the resource is not renewed until natural processes restore the original temperatures. For very large occurrences, renewal of usable temperatures in a few decades is a reasonable expectation; for small occurrences, however, renewal is more doubtful. For resources of the geopressured type (see later section on *exploitable occurrences*) there is no prospect for natural renewal.

After a brief discussion of world geothermal patterns and development, geothermal energy in the United States is reviewed according to the following plan. First, the whole country and its various physiographic regions are viewed from the perspective of geothermal potential. Next, the major types of usable, or exploitable, occurrences are explained. After this, the locations and the potency of these occurrences are mapped, with *accessible heat in-place* being summed up by state. Then, the amounts of *energy recoverable* from the various types of occurrence are mapped by state and expressed as electrical generating potential or as

185

fuel saved through the use of geothermal energy for space heating. Finally, recent and current developments in utilization of U.S. geothermal resources are shown.

WORLD PATTERNS

At any location, heat flow from the interior to the surface of the earth can be deduced from the temperature gradient, or rate of change of temperature with depth, and the conductivity of the rock or sediments in which temperatures are measured.

Heat flow measurements from over 5,000 sites on continents and ocean floors around the world show a pattern that is best explained by the suggestion that large plates of the earth's crust—each plate often embracing both continent and sea floor—drift slowly across the globe in response to internal forces. Where they collide, they give rise to compressional forces that build mountains; where they separate, they cause tensional forces, rift valleys, and volcanic trends; and where they grind together, they cause earthquakes.

Areas of tension, that is, zones of spreading plates, allow upwelling of molten rock through the crust, causing such volcanic mountain trends as the mid-Atlantic ridge. Not surprisingly, heat flow is high along such trends in the Indian and Atlantic Oceans. It is especially high in the Pacific west of South America, and where the Pacific ridge system intersects North America along the Gulf of California (Fig. 5-1).

Areas of compression, where one crustal plate is forced under an opposing plate, lead to belts of both low and high heat flow. The plate that is forced down (subducted) introduces as a cool mass of rock that weakens the heat flow, but at the same time, some of the plate melts and rises buoyantly to cause volcanoes and high heat flow. Parallel elongate zones of high and low heat flow, suggesting this process, are seen in the vicinity of the Japanese island chain, in which the islands themselves were formed by such upwelling and volcanism.

The association of high heat flow and volcanism confirms a common-sense interpretation of geothermal patterns. Where hot igneous rock is brought from the depths to near surface level by volcanic activity, heat flow anomalies result. Volcanic activity coincides with plate edges where there is communication with the interior. The opposite characteristics, low heat flow and lack of recent volcanism, occur in stable continental areas well removed from the edges of presently active plates. Shield areas of North America, South America, and Africa exemplify this best (see Fig. 5-1).

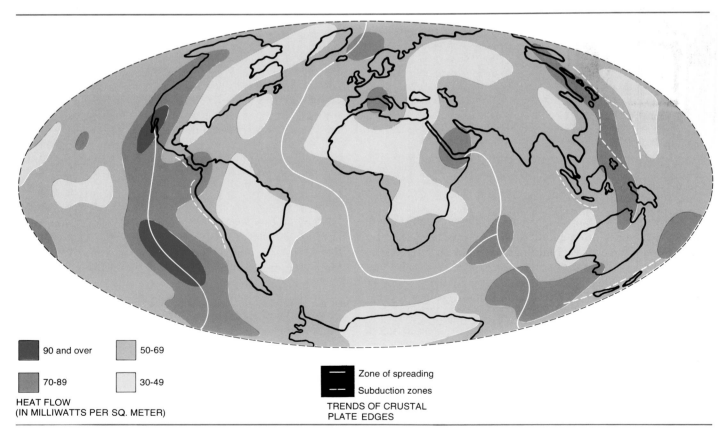

■ 90 and over	■ 50-69
■ 70-89	□ 30-49

HEAT FLOW
(IN MILLIWATTS PER SQ. METER)

—— Zone of spreading
-- -- Subduction zones

TRENDS OF CRUSTAL
PLATE EDGES

Fig. 5-1 World heat flow areas.
Source: Redrawn from Pollack and Chapman, 1977.

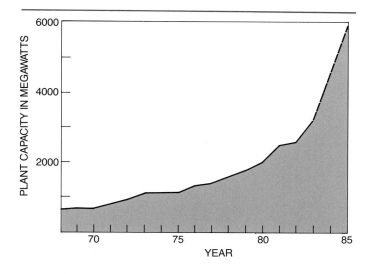

| COUNTRY | NO. UNITS | GENERATING CAPACITY (MW) | |
		As of June 1983	Expected 1985
UNITED STATES	24	1283.7	2122.3
PHILIPPINES	14	593.5	1718.5
ITALY	41	457.1	502.1
JAPAN	8	227.5	282.5
MEXICO	10	205.0	700.0
NEW ZEALAND	14	202.6	202.6
EL SALVADOR	3	95.0	95.0
ICELAND	5	41.0	41.0
INDONESIA	3	32.25	32.25
KENYA	2	30.0	45.0
SOVIET UNION	1	11.0	21.0
CHINA	10	8.136	11.386
PORTUGAL (AZORES)	1	3.0	3.0
TURKEY	1	0.5	40.5
NICARAGUA	0	0	35.0
FRANCE (GUADELOUPE)	0	0	6.0
TOTALS:	137	3190.286	5858.136

Current and Projected Geothermal Generating Capacity

Source: Years 1968 to 1977, United Nations, 1978.
Years 1979 to 1985, de Pippo, 1983.

Fig. 5-2 World geothermal electrical generating capacity, 1969 to 1985.

In a number of countries, such as Iceland, New Zealand, and the Soviet Union, geothermal heat has been used for many years for heating buildings. Generation of electrical power from geothermal heat was begun in Italy in 1904 at the Larderello field. World growth of generating capacity was slow until the 1970s (see Fig. 5-2), when developments at the Geysers in California and at three sites in the Philippines began. Now the growth is more rapid, with the United States, the Philippines, Italy, Japan, Mexico, and New Zealand being the leading nations. Rapid growth in generating capacity in the Cerro Prieto area, near the California border, will give Mexico a generating capacity that surpasses Italy's in the year 1985. The world total of geothermal electrical generating capacity, as of June, 1983, was 3,190 Megawatts—roughly equivalent to the generating capacity of three very large coal-burning or nuclear power plants. The capacity in

1985 is roughly twice that of 1983, with the equivalent of two large nuclear plants operating in the United States, mainly in the Geysers area just north of San Francisco.

UNITED STATES REGIONS: THE BACKGROUND HEAT

Within the United States, substantial information on temperature gradients is available in data obtained from oil and gas drill holes and from test holes drilled expressly for geothermal measurement. Analysis of this information points to three broad types of regions in the country and allows for the estimate of the total amounts of heat stored in those regions. The following materials on background heat are an interpretation of an article dealing with this aspect of the resource in the U.S. Geological Survey Circular which is currently the best source of geothermal information (Diment, *et al*, 1975).

Character of Heat Flow Regions

In the United States, as elsewhere, heat flows to the earth's surface from two different sources: one is in the lower crust, or in the *mantle*, the semi-liquid layer that underlies the crust; the other source is radioactive decay of elements in the crust itself.

The first of these two sources is more important in determining the character of a region. While in some areas the base of the crust may be deep, in areas of recent or current volcanic activity molten rock at the base of the crust is much closer to the surface. In such an area, high temperatures are encountered at relatively shallow depths and are, therefore, much more accessible.

The second component, heat due to radioactive decay, varies throughout a region according to the character of the crustal rocks. It is measured in heat generation units (HGU) which can range from 1 to 20, but most often are in the vicinity of 4 or 5. One example of extremely high radioactive heat generation is the igneous rocks of the White Mountain Series in New Hampshire, whose HGU values are over 20. The Conway granite in that area is well known as a potential low-grade ore of uranium.

Figure 5-3 shows three types of regions classified according to the deep crustal component of heat flow: (1) Eastern or Normal type; (2) Basin and Range, or Hot type; and (3) Sierra Nevada, or Cold type. The Eastern type typifies the stable continental interior with no recent volcanic activity, which, as noted in Figure 5-1 occurs near the middle, not the edges, of crustal plates. The Basin and Range type characterizes much of the western part of the country, where volcanic activity is much more recent. The Basin and Range map pattern is modified by two elongate areas of Eastern or Normal type: one coincides with the Wyoming Basins and Colorado Plateau; while that along the Pacific Northwest and the

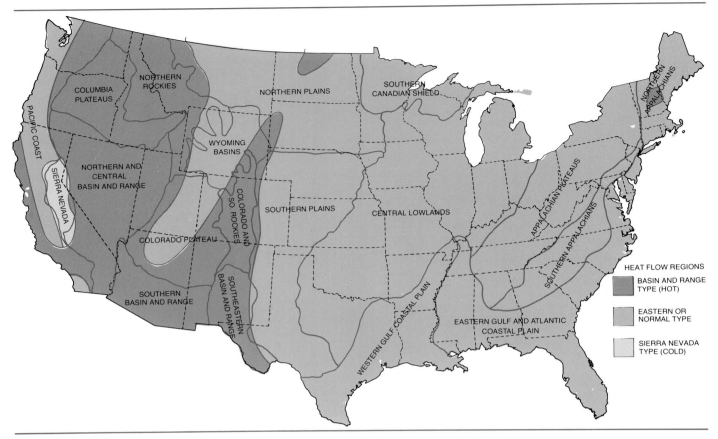

Fig. 5-3 Physiographic provinces and three types of heat flow regions.
Source: U.S.G.S. *Circular 726*, 1975.

Great Valley of California may be due to an ancient subduction zone like the one recognized near Japan (Fig. 5-1). Associated with the same trend is the single region of abnormally low heat flow, the Sierra Nevadas. Despite the volcanic activity and heat flows along its western edge, the mountain range appears to be the top of a Cold block subducted into the mantle, causing molten rocks to be more remote.

Evidently, the most important characteristic that distinguishes among the three types of regions is the proximity to the surface of hot molten rocks. In the Normal or Cold type they are deep at the base at the crust, and in the Basin and Range (Hot) type are closer to the surface because of recent volcanism. The various *temperature gradients* that accompany the different types are all-important to the possibilities of using earth heat for they determine the drilling depth required to reach useful temperatures. Of course, local hot spots are in reality the sites where drilling would be undertaken. Nevertheless, it is helpful to examine the depths dependent on the general character of a region.

Figure 5-4 is a rearrangement of conventional temperature gradient plots, and is designed specifically to show how depths of useful temperatures vary from one type of region to another. Considered in the plots are the effects of varying radioactive heat generation, so that for any chosen HGU value, 0 through 20, the depths necessary to reach temperatures of 90 or 150 degrees Celsius may be estimated.

Assuming, for simplicity, that 5 is the HGU value for local rocks, part A of the figure shows that in the Sierras a depth of about 7 kilometers or 23,000 feet would be necessary to reach 90 degree temperatures. In this region, 150 degrees is out of reach since drilling beyond 30,000 feet is impossible with current technology. In the Eastern or Normal type of region (part B) the lower temperature would be reached at 12,000 feet and the higher temperature at 25,000 feet. In the Basin and Range type of region (Fig. 5-4, parts C and D), the higher temperature can be expected at depths less than 17,000 feet, and in the more favorable subtype (D), at less than 14,000 feet. One revealing feature of the charts is the differing slopes or "droop" of the reference lines on parts A through D. In the Cold type of region, a change in the degree of radioactive heat generation is of vital importance. It is much less important in a region favored by high rate of flow from molten rock at depth: that flow becomes the dominant factor as shown by the slight change in depth for changing HGU values. Also noteworthy is the vertical spacing between the two reference lines on the series of plots. The progressively closer spacing on plots A through D reveals steeper temperature gradients in the more favored regions, that is, a shorter vertical (depth) difference between 90 and 150 degree temperatures.

Heat In-place In the Country As A Whole

Using information that reveals at what depth temperatures occur (similar to that shown in Fig. 5–4) and data on thermal conductivity of rocks, it is possible to estimate the *heat content* per unit area of a region to a chosen depth. Then the content for a geologic province can be obtained by multiplying by the area of the province. Only the temperatures exceeding mean annual temperatures at the earth's surface (assumed to be 15 degrees Celsius) are used in this calculation in order to distinguish earth heat from heat due to solar radiation.

Table 5–1 shows the 17 geologic provinces that make up the conterminous United States. These provinces are divided into three region types for the sake of the calculations. Alaska and Hawaii are treated separately because geologically they are so different from the continental United States. The table shows estimates of stored heat from surface to 3 kilometers depth, and from 3 to 10 kilometers. The total for each province, and for the country as a whole, therefore, represents stored heat to a depth of 10 kilometers (6.25 miles, or 33,000 feet). This is the practical limit to drilling in the foreseeable future.

The relative contributions of various regions and the provinces they comprise is shown graphically in Figure 5–5. From this figure it is clear that the Eastern type region contributes greatly to the total of available geothermal heat because of its large area. Provinces of the Basin and Range type, despite their shallow high temperatures, contribute less to the national total because of smaller areas. As noted there, the best estimate of the United States' total stored heat is about 800×10^{22} calories, not the 680.2×10^{22} calories, which is the sum of province estimates. This revised total is used later in a summation of heat in-place.

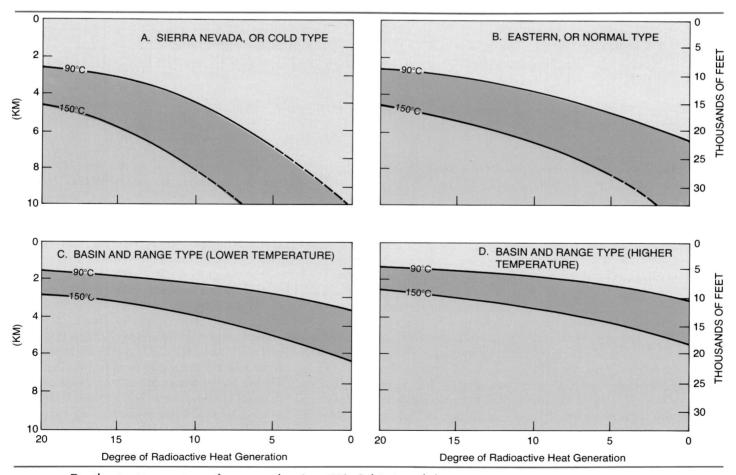

Fig. 5–4 Depths to temperatures for space heating (90° Celsius) and for power generation (150° Celsius) in four types of regions.
Source: Data from U.S.G.S. *Circular 726*, 1975.

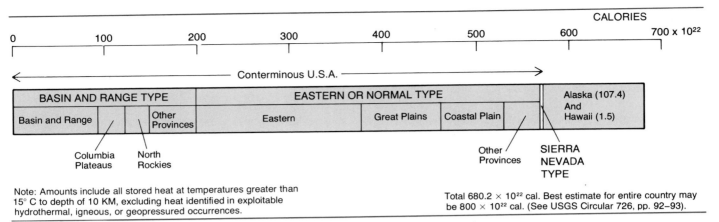

CALORIES

| 0 | 100 | 200 | 300 | 400 | 500 | 600 | 700 x 10²² |

← ——————————— Conterminous U.S.A. ——————————— →

| BASIN AND RANGE TYPE | EASTERN OR NORMAL TYPE | Alaska (107.4) And Hawaii (1.5) |

| Basin and Range | | | Other Provinces | Eastern | Great Plains | Coastal Plain | | |

Columbia Plateaus   North Rockies

Other Provinces   SIERRA NEVADA TYPE

Note: Amounts include all stored heat at temperatures greater than 15° C to depth of 10 KM, excluding heat identified in exploitable hydrothermal, igneous, or geopressured occurrences.

Total 680.2 × 10²² cal. Best estimate for entire country may be 800 × 10²² cal. (See USGS Circular 726, pp. 92–93).

Fig. 5–5 Estimated heat content in crust to depth of 10 km, showing physiographic regions.
Source: Data from U.S.G.S. *Circular 726*, 1975.

TABLE 5–1
Stored Heat (Resource Base) in 19 Provinces, Grouped into 3 Types of Regions

REGION TYPE	PROVINCE	STORED HEAT (CALORIES × 10²²)		
		0–3 Km.	3–10 Km.	Total
Sierra Nevada	Sierra Nevada	0.31	2.80	3.11
Eastern or Normal	Eastern	17.20	160.40	177.60
	Coastal Plain	6.70	62.40	69.10
	Great Plains	8.30	77.60	85.90
	Wyoming Basins	0.625	5.84	6.47
	Peninsular Range	0.046	0.43	0.49
	Pacific Northwest	0.468	4.37	4.84
	Klamath Mts.	0.254	2.38	2.63
	Great Valley	0.295	2.76	3.06
	Colorado Plateau	1.990	18.60	20.60
Basin and Range	Basin and Range	8.72	83.70	92.40
	Northern Rockies	2.49	23.90	26.40
	Central Rockies	1.26	12.10	13.40
	Southern Rockies	1.13	10.90	12.00
	Columbia Plateaus	2.70	25.90	28.60
	Cascade Range	1.07	10.30	11.40
	San Andreas Fault Zone	1.26	12.10	13.40
Total Conterminous United States		54.80	516.90	571.30
Alaska		10.20	97.20	107.40
Hawaii		0.14	1.37	1.50
U.S. Total		65.14	615.47	680.2[1]

[1] This total was revised to 800 × 10²² calories in the source document U.S.G.S. *Circular 726*, p. 92–93.

EXPLOITABLE OCCURRENCES

The foregoing massive heat in-place is interesting, but not very useful. It is true that the warmth found in shallow wells can be used by *heat pumps* as a source of energy for space heating. However, if earth heat is to be used *directly* for heating buildings, the temperatures must be around 90 degrees Celsius—temperatures found at depths around 10,000 feet in the most favorable of heat flow regions. Drilling to such depths for modest return of heat is not practical. But, of course, there are a number of occurrences in which high temperatures in the earth are more accessible.

There are three broad types of such occurrence: hydrothermal, igneous (volcanic), and geopressured. The most attractive of the hydrothermal type, and all of the igneous are found in the Basin and Range type of province, associated either directly or indirectly with volcanism that has left molten rock relatively close to the surface. The third type, involving geopressured reservoirs, is quite different and not dependent on volcanism.

Hydrothermal Systems

Natural hot springs occur where waters from the surface flow down through fractures or aquifers into deeply buried rock where they are heated. These heated waters are then forced to the surface through buoyancy, and underground pressure. Although hot springs occur throughout the country, those in which the waters encounter very high temperatures are located in those areas in the West where molten or very hot igneous rocks are unusually near the surface.

Figure 5–6 shows a hypothetical circumstance in which the prime source of heat is the magma chamber, a body of molten rock embedded near the surface as a remnant of volcanic activity. Surface waters descend down an aquifer to the level of very high temperatures, and a fault allows water to rise to the surface. Not all hydrothermal systems depend on molten rock near the surface. Apparently, in some systems, water is carried by faults to great depths where, in a region of the Basin and Range type, sufficiently high temperatures are encountered simply by virtue of the steep temperature gradient (Renner *et al.*, 1975).

In general, hydrothermal systems are extremely attractive for geothermal heat applications because the circulating water serves to bring the high temperatures within reach, at or near the surface. The magma or other "source rocks" may be so deeply buried that drilling to reach them is far more expensive than drilling to the hot water or steam. Most important, the circulating waters provide a system by which heat is drawn continuously from a large volume of hot rock. The most promising of these natural hydrothermal systems are those that can be readily *developed* by drilling a number

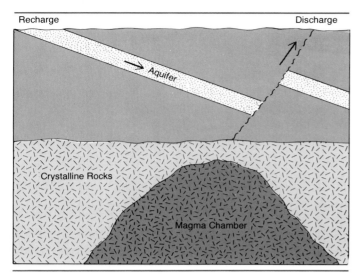

Fig. 5–6 Elements of a hydrothermal convection system.

At The Geysers geothermal field, north of San Francisco, a development well encounters a steam-producing zone. (Courtesy of American Petroleum Institute and Union Oil Company of California.)

of wells to increase the flow of steam or hot water to the surface.

The most important variables in hydrothermal systems are the fluid temperatures and whether the fluid circulating to the surface is hot water or steam. *High-temperature* systems, with temperatures over 150 degrees Celsius, can be used easily for electrical power generation. Those of *intermediate temperatures*, between 90 and 150 degrees Celsius, can be used directly for heating buildings (i.e., space heating) and can also be used for power generation in *binary* power plants where a low boiling-point fluid actually drives the turbines. *Low-temperature* waters, at 90 degrees Celsius and below, can be used for a variety of agricultural tasks.

Hydrothermal systems yielding dry steam are very rare: Only three have been identified in the United States. They are much more desirable than those yielding hot water because steam can be fed directly into turbines, whereas hot water usually must be routed through heat exchangers in order to boil water for steam. Furthermore, hot geothermal waters often are laden with dissolved minerals that cause corrosion of equipment, as well as pollution.

Igneous or Volcanic Potential Systems

Although often referred to as "igneous systems," these occurrences are better understood as *potential* systems because they lack the natural plumbing by which to extract heat from hot rock and carry it to the surface.

Figure 5–7 shows a hypothetical occurrence like that in Figure 5–6, but without permeable rocks overlying the magma. Since there are neither aquifers nor fractures to convey fluids, the embedded magma mass is without communication to the surface. Such occurrences are often very large and hold a great store of heat undiminished *because* of the lack of communication.

These occurrences of hot dry rock are less attractive than

hydrothermal systems, and therefore, they have not yet been used commercially. Their exploitation will demand some method of artificially inducing circulation through the rock to create a heat-extraction system comparable to that in hydrothermal systems. Some very promising experiments have been conducted by the Department of Energy (see later section), so possibly these occurrences *can* be utilized. Because the technology is so young, it is impossible to make estimates of what portion of the in-place heat can be recovered by induced circulation.

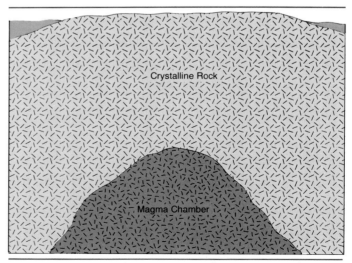

Fig. 5–7 Elements of an igneous (volcanic) potential system.

Geopressured Reservoirs of the Gulf Coast

Whereas the foregoing two types of exploitable occurrence depend on the location of hot or molten igneous rocks unusually near the surface, the third type of occurrence is in a sedimentary rather than igneous environment. Furthermore, it depends upon deep rather than shallow burial of the critical rocks.

Figure 5-8 represents a cross-section of a thick sequence of young sedimentary rocks in the Gulf coast region. At depths ranging from 6,000 feet (1,800 meters) to 15,000 feet (4,600 meters) abnormally high subsurface pressures have been encountered in sand bodies penetrated by holes drilled for oil and gas exploration. The sands were initially of interest because they were a problem for petroleum engineers who had to contain the pressures when drilling through the sands. They have since been recognized as a source of usable energy (Dickinson, 1953).

The pressures labelled "abnormal" in the figure are greater than "hydrostatic." This means that they exceed the pressure exerted by a column of water reaching from the surface to the depth of the sand body. The inset in Figure 5-8 shows one interpretation. Pressure in the thick and continuous main sand series is *not* abnormal because formation waters have been able to drain out and allow the sandstones to be compacted by the weight of overlying sediments. In sand bodies isolated by faults, or by enclosing shales, formation waters are trapped so that compaction is prevented. Pressures build as more sediments are deposited at the surface and the weight of overburden increases.

In addition to pressures that will raise water to the surface in artesian fashion and offer usable mechanical energy, these waters are hot and also contain dissolved methane (natural gas). Their temperatures, usually in excess of 150 degrees Celsius, are *considerably higher than expected for the depth*, according to the Eastern or Normal type of region in which the Gulf Coast province occurs (Table 5-1). In that type of province, a depth of 5 kilometers (16,000 feet) is consistent with 150 degree Celsius temperatures only if the radioactive heat flow is extremely high, that is, between 15 and 20 heat generation units (HGU; Fig. 5-4B). With no evidence of highly radioactive rock, the abnormally high temperatures at depth are apparently due to the low thermal conductivity of overlying porous sediments, which constitute an insulating blanket.

Development of this geopressured type of energy will be

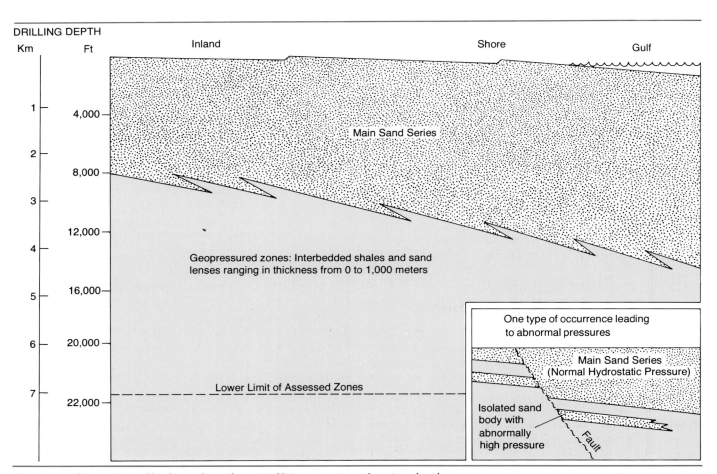

Fig. 5-8 Schematic profile through onshore Gulf Coast region, showing depths to geopressured zones.
Source: Derived from information in Dickinson, 1953, and U.S.G.S. *Circular 726*, 1975.

accomplished by drilling through the isolated sand bodies, or "reservoirs" and allowing the hot methane-laden salt water to rush into the drillhole and up to the surface. The methane will be separated at the surface and used in the same way as any natural gas. The waters are hot enough to raise steam for electrical generation or to be used in industrial processes.

ACCESSIBLE HEAT IN-PLACE IN IDENTIFIED HYDROTHERMAL, VOLCANIC, AND GEOPRESSURED OCCURRENCES

Just as in the previous estimates of background heat for provinces and regions, the amount of heat in-place is only the portion at greater than 15 degrees Celsius, (the temperature that represents average conditions at the surface). Furthermore, it is only that heat which is *accessible* in identified occurrences (Fig. 5–9). For hydrothermal systems this estimate considers heat in rocks down to a depth of 3 kilometers (9,843 feet). For volcanic type the depth limit is 10 kilometers (33,000 feet). For geopressured reservoirs the limit is 6.86 kilometers (22,500 feet). *Accessible heat in-place*, defined this way, is comparable to tonnage of coal that is *mineable* by underground methods (see Chapter 1). The geothermal heat actually *recoverable* at well-head is markedly less than the amounts accessible. Not shown on Figure 5–9 is the fact that the amount of heat *applied* to some task is less than the amount recovered at well-head because of energy lost in conversion or exchange of heat. The same is true of the heat value in coal or oil: Much of it is lost, whether it is burned in a residential furnace or a power plant. For geothermal heat, however, the loss tends to be larger because the relatively low temperatures make for low conversion efficiencies.

In the materials that follow, amounts of heat in-place and recoverable are tabulated in units of 10^{18} *calories*—the heat equivalent of 690 million barrels of crude oil or 154 million short tons of bituminous coal.

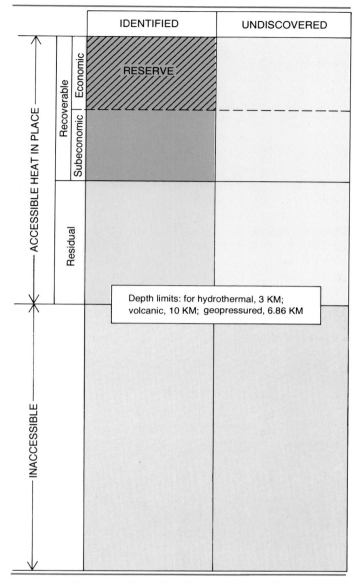

Fig. 5–9 Conceptual setting for amounts of geothermal energy in-place and recoverable. The proportion of heat in-place that is recoverable varies with the type of occurrence.

Hydrothermal Systems of the High-temperature Type

Figure 5–10 shows identified systems in western states and in Alaska with average reservoir temperatures over 150 degrees Celsius. The remainder of the country is not shown because neither high- nor intermediate-temperature systems occur there.

Among the systems mapped, there is some variety in reservoir temperatures, some being barely over 150 degrees Celsius, while many exceed 200 degrees. One reservoir in the Salton Sea area of southern California has temperatures of 340 degrees Celsius (Renner *et al.*, 1975). Systems with higher temperatures are the more desirable, both because they tend to increase total heat content and because they

lead to higher efficiencies in application of the energy. Generally, heat in-place is based upon reservoir temperatures, specific heat of the rocks, approximate area of the subsurface reservoir, and a lower depth limit of 3 kilometers, which is the current limit of geothermal drilling.

A good proportion of the systems mapped are fairly large, having stored heat greater than 1×10^{18} calories. There are fewer systems with stored heat greater than 10×10^{18} calories. The largest of them is Yellowstone, with a heat content of 296×10^{18} calories, while Long Valley and Coso Hot Springs in California each contain about 20×10^{18} calories.

Systems with temperatures over 150 degrees Celsius can be used to raise steam for electrical power generation, unlike those of intermediate temperature. Among these high-temperature systems are those dominated by steam, rather

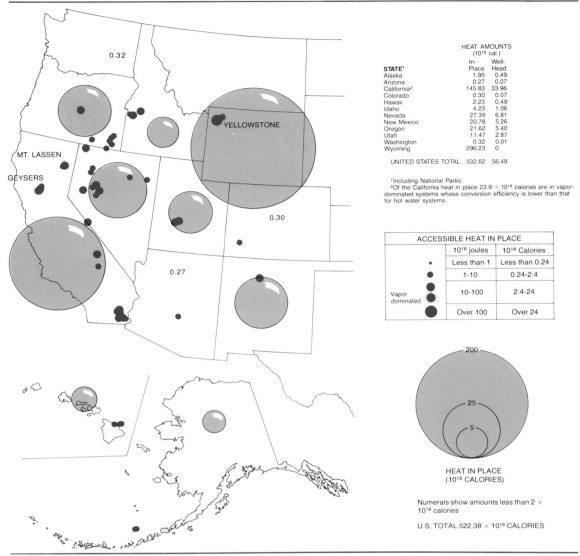

HEAT AMOUNTS
(10^18 cal.)

STATE[1]	In-Place	Well-Head
Alaska	1.95	0.49
Arizona	0.27	0.07
California[2]	145.83	33.96
Colorado	0.30	0.07
Hawaii	2.23	0.49
Idaho	4.23	1.06
Nevada	27.39	6.81
New Mexico	20.78	5.26
Oregon	21.62	5.40
Utah	11.47	2.87
Washington	0.32	0.01
Wyoming	296.23	0
UNITED STATES TOTAL	532.62	56.49

[1]Including National Parks.
[2]Of the California heat in place 23.9 × 10^18 calories are in vapor-dominated systems whose conversion efficiency is lower than that for hot water systems.

ACCESSIBLE HEAT IN PLACE

	10^18 joules	10^18 Calories
·	Less than 1	Less than 0.24
●	1-10	0.24-2.4
●	10-100	2.4-24
Vapor dominated ✹	10-100	2.4-24
●	Over 100	Over 24

HEAT IN PLACE
(10^18 CALORIES)

Numerals show amounts less than 2 × 10^18 calories

U.S. TOTAL 522.38 × 10^18 CALORIES

Fig. 5–10 Identified high-temperature hydrothermal systems, and accessible heat in-place (by state) to depth of 3 km.
Source: U.S.G.S. *Circular 726,* 1975.

than hot water, making them more favorable for power generation. The three identified systems of this type, the Geysers, Mt. Lassen, and the mud volcano system at Yellowstone, are labelled on Figure 5–10. Two of the three are quite large, Geysers containing 24, and Mt. Lassen containing 10 of the 10^18 calorie units, while the steam-dominated portion of Yellowstone accounts for only 1.1 units (U.S.G.S. *Circular* 790, 1979). The heat contents of the Mt. Lassen and the Yellowstone systems are included in the state summary of heat in-place, but are not in later assessments of *recoverable* energy because they are in national parks where no exploitation is allowed.

When heat content is summarized by state (Fig. 5–11) it is clear that Wyoming and California dominate. However, none of Wyoming's 296 units of heat will be available for extraction (because of national parks) whereas California's total of 146 includes only 10 units untouchable at Mt.

Lassen. Although Oregon, New Mexico, Utah, and Nevada all have very substantial amounts of heat in-place, the amounts are tiny in comparison to those of California and Wyoming.

Hydrothermal Systems of the Intermediate-temperature Type

Figure 5–11 shows systems with intermediate reservoir temperatures—between 90 and 150 degrees Celsius. These are more plentiful than the high-temperature type—both in the contiguous states and in Hawaii and Alaska, with a total of 164, as opposed to 56 of the high-temperature type. None of the intermediate-temperature systems mapped has a reservoir temperature under 100 degrees Celsius. Many have temperatures of 145 and 150 degrees, and may in fact be usable for electric power generation. For the most part, however, these systems are suited to tasks such as space heating.

Large systems are less prevalent than small ones. One of the largest is Klamath Falls, Oregon, which has 7×10^{18} calories in-place and is being studied as a source of energy for space heating in Portland. The Bruneau-Grandview system near Boise, Idaho, is a giant among hydrothermal systems, covering 1,483 square kilometers and holding an estimated 107×10^{18} calories as accessible heat in-place, or the equivalent of more than 74 billion barrels of crude oil.

The distribution of heat in-place according to state is quite different from that for the high-temperature resource—although both are restricted to western states. For intermediate temperatures the dominant state is Idaho, with 122.22×10^{18} calories of accessible heat in-place. This is the energy equivalent of 18.65 billion tons of coal—roughly the amount of bituminous coal accessible to mining in the state of Pennsylvania. Of the remaining states, Oregon has the largest amount of energy in-place, though the magnitude of its 13×10^{18} calories is not obvious on Figure 5–11. Apparently, the states of Idaho, Nevada, Oregon, and California have substantial amounts of accessible heat in-place in *both* high- and intermediate-temperature occurrences (Figs. 5–10, 5–11).

Low-temperature Geothermal Waters

Very significant amounts of heat exist in fluids that are neither high- nor intermediate-temperature, but are at temperatures lower than 90 degrees Celsius. Where such

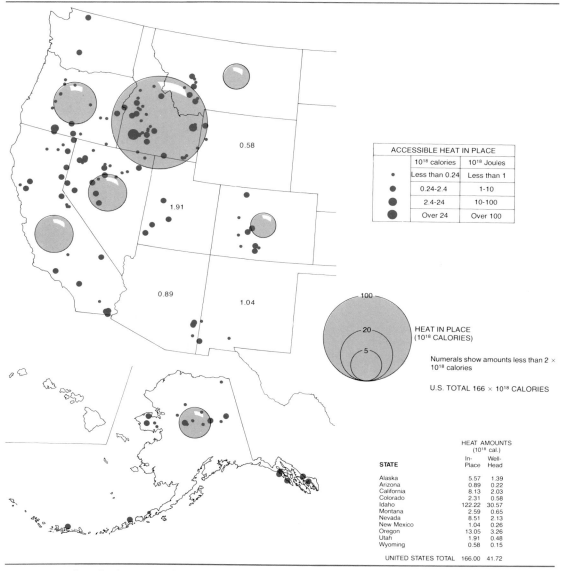

ACCESSIBLE HEAT IN PLACE		
	10^{18} calories	10^{18} Joules
·	Less than 0.24	Less than 1
●	0.24-2.4	1-10
●	2.4-24	10-100
●	Over 24	Over 100

HEAT IN PLACE
(10^{18} CALORIES)

Numerals show amounts less than 2 × 10^{18} calories

U.S. TOTAL 166 × 10^{18} CALORIES

STATE	HEAT AMOUNTS (10^{18} cal.) In-Place	Well-Head
Alaska	5.57	1.39
Arizona	0.89	0.22
California	8.13	2.03
Colorado	2.31	0.58
Idaho	122.22	30.57
Montana	2.59	0.65
Nevada	8.51	2.13
New Mexico	1.04	0.26
Oregon	13.05	3.26
Utah	1.91	0.48
Wyoming	0.58	0.15
UNITED STATES TOTAL	166.00	41.72

Fig. 5–11 Identified intermediate-temperature hydrothermal systems, and accessible heat in-place (by state) to depth of 3 km.
Source: U.S.G.S. *Circular 790*, 1979.

waters emerge in hot springs, they have been used for bathing. They have many other applications, though. Waters with temperatures near 90 degrees can be used for heating buildings, and those at 60 degrees or below can be used to warm the soil in greenhouses, to promote the growth of fish in hatcheries, or to stimulate the production of crude oil.

These waters occur in two different kinds of environments. In some areas they come to the surface as springs, that is, as low-temperature hydrothermal occurrences. In other areas the waters are buried in aquifers that have no connection with the surface and must be reached by drilling. Until recently, neither kind of occurrence had been assessed quantitatively, but a U.S. Geological Survey study (*Circular 892*) published in 1983 did estimate the amounts of energy in low-temperature waters, as a sequel to the work that evaluated high- and intermediate-temperature geothermal resources. The study dealt with the United States in three parts, Western, Central, and Eastern. Here, we have brought those parts together and summarized the results.

First, the limits of the resource should be defined. To qualify as useful, fluids at the surface must have a temperature that is a least 10 degrees Celsius higher than the average air temperature at the location. For fluids at depth, the difference between fluid temperature and average air temperature must increase by 25 degrees for every kilometer of depth. That means if air temperature is 12 degrees, the minimum temperature for a hot spring is 22 degrees, while for water at a depth of 2 kilometers the minimum temperature is 72 degrees. This requirement expresses the idea that if drilling must be done, then the temperatures of the deep water must be high enough to justify the effort, as well as high enough to remain hot after being lifted to the surface. In some parts of Alaska, where average air temperature is 0 degrees, the 25 degrees per km formula will include a reservoir with temperature of 89 degrees at a depth as great as 3.2 km . In Florida and Hawaii, where air temperature is 23 degrees, any site that meets that 25 degree per km gradient requirement will have temperatures of *90 degrees* at a depth of only 2.2 km . Any attractive reservoirs below that depth would be excluded from this energy assessment because the plus-90 temperatures would throw them into the intermediate-temperature category. Therefore, the "floor" limiting what reservoirs are included in this assessment is deeper in cooler climates than in warmer ones. In this respect, the study differs from the one that dealt with high- and intermediate-temperature resources, because that study applied a floor of 3 km uniformly across the nation.

As in the foregoing study, the primary estimate is for amounts of *accessible heat in-place* in reservoirs that are judged to be sufficiently porous and permeable to produce the water. Next, an estimate was made of the *Resource* defined as the amount of energy *recoverable* over a period of 30 years. Finally, the *beneficial heat* is estimated, that is, the amounts of energy that would actually be delivered to a task, such as space heating, when the temperature of the water falls to the temperature of the environment or the material being heated.

Figure 5–12 combines the areas of occurrence with estimates of accessible heat in-place. In the western third of the country, there are hundreds of hydrothermal convection systems with low-temperature waters in relatively small volumes. (These, incidentally, are in addition to all the hydrothermal systems holding high- and intermediate-temperature waters.) The locations of all these sites are generalized on the map as patches or swarms. The sphere symbols show the amounts of heat in-place in these hydrothermal systems are not very large when compared to the heat in some of the Central states. Comparison with the heat in high- and intermediate-temperature systems in the western states will show, however, that the magnitudes are similar.

The massive amounts of low-temperature heat are in the Central part of the country, in very extensive reservoirs in undisturbed sedimentary rocks underlying the Great Plains. In Montana, there are two aquifers: limestones and dolomites of the Madison formation (Paleozoic age) hold about 87 percent of the accessible heat in-place for that state, while the balance is mostly in sandstones of the Dakota formation (Mesozoic age). In the other Central states that show significant amounts of heat, the reservoirs are, again, Paleozoic carbonate rocks and Mesozoic sandstones. The information on all these buried reservoirs comes from the hundreds of oil and gas wells drilled through the Mesozoic rocks and into the underlying Paleozoic. Hot springs do play some part in this central part of the country—in Wyoming, Arkansas, and Colorado, especially—but the amounts of heat in-place are very small compared with those in the buried aquifers.

In the Eastern United States, thermal gradients are low, and only a few scattered hot springs provide areas of interest. There are, though, some deeply buried sedimentary rocks that may hold waters of useful temperatures because they are overlain by rocks of low conductivity whose insulating effect may keep those waters hot. On Figure 5–12, the area centered in western Pennsylvania is shown as possibly attractive for this reason, as is a comparable area involving Texas, Louisiana, and Mississippi in the Central region.

The very large contribution of deep aquifers to the amounts of heat in-place is undeniable, however it diminishes sharply when amounts of energy *recoverable* are considered (see Table 5–2). The development plans assumed by the U.S. Geological Survey entail very low recovery rates from these extensive aquifers, so that, while they account for 99 percent of the low-temperature heat in place for the nation, they account for only 65 percent of the recoverable resource.

When the nation's total low-temperature energy estimate is compared to estimates for the high- and intermediate-temperature waters (Table 5–2) the recovery factor is again crucial. And the matter of identified versus undiscovered occurrences can alter the balance.

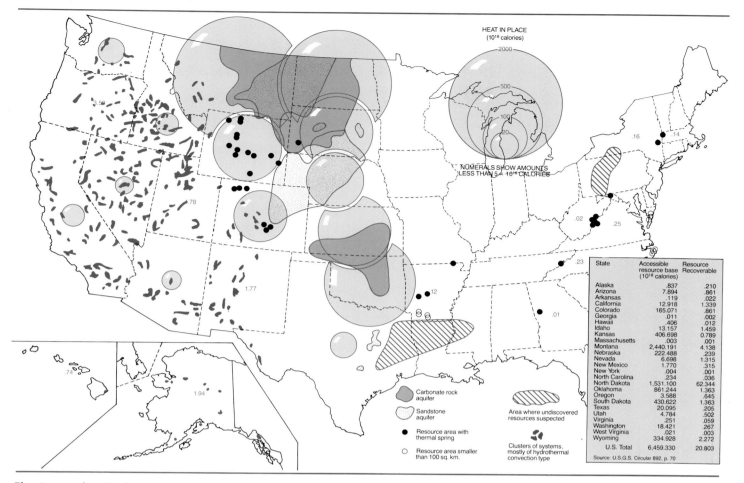

HEAT IN PLACE
(10^18 calories)

2000
500
100
20

NUMERALS SHOW AMOUNTS
LESS THAN 5 × 10^18 CALORIES

State	Accessible resource base (10^18 calories)	Resource Recoverable
Alaska	.837	.210
Arizona	7.894	.861
Arkansas	.119	.022
California	12.918	1.339
Colorado	165.071	.861
Georgia	.011	.002
Hawaii	.406	.012
Idaho	13.157	1.459
Kansas	406.698	0.789
Massachusetts	.003	.001
Montana	2,440.191	4.138
Nebraska	222.488	.239
Nevada	6.698	1.315
New Mexico	1.770	.315
New York	.004	.001
North Carolina	.234	.036
North Dakota	1,531.100	62.344
Oklahoma	861.244	1.363
Oregon	3.588	.645
South Dakota	430.622	1.363
Texas	20.095	.205
Utah	4.784	.502
Virginia	.251	.059
Washington	18.421	.267
West Virginia	.021	.003
Wyoming	334.928	2.272
U.S. Total	6,459.330	20.803

Source: U.S.G.S. Circular 892, p. 70

Legend:
- Carbonate rock aquifer
- Sandstone aquifer
- ● Resource area with thermal spring
- ○ Resource area smaller than 100 sq. km.
- Area where undiscovered resources suspected
- Clusters of systems, mostly of hydrothermal convection type

Fig. 5-12 Identified low-temperature occurrences, hydrothermal and deep aquifer, showing locations and accessible heat in-place (by state) to depths of about 3 km.
Source: U.S.G.S. *Circular 790,* 1979.

In identified occurrences, those with low-temperature waters account for 90 percent of all the accessible heat in-place, but only 18 percent of recoverable energy because so much of the low-temperature energy is in deep aquifers with the low recovery rates. If undiscovered occurrences are included, the proportions in low-temperature waters drop to 81 percent of heat in-place and 6 percent of energy recoverable. This drop occurs because very large amounts of energy are postulated in undiscovered occurrences of hydrothermal systems holding high- and intermediate-temperature waters in systems that will permit relatively high recovery rates.

Igneous or Volcanic Potential Systems

The potential of igneous or volcanic occurrences depends on molten rock (magma) masses in the upper crust. Some of these pockets of molten rock contribute to hydrothermal systems assessed in the previous section; others have no connection to the surface of the earth, as suggested by Figure 5–7. Since the heat content of such a molten mass is so large, it is virtually unaffected by heat drawn off through natural hydrothermal systems.

Only molten rock of a certain character is likely to form storage chambers which are in upper levels of the crust and therefore accessible to drilling and exploitation. Magmas of *basic* composition, that is, those which form basalts, andesites and comparable rocks, have low viscosity and tend to flow directly from the lower crust and exit through pipes and fissures. On the other hand, magmas of more *acidic* composition such as those that form granites, are more viscous and tend to create storage chambers within 10 kilometers of the surface. It is from these chambers that eruptions occur. Coincidentally, the acidic magmas are known to cause explosive volcanic activity, whereas the low-viscosity basic magmas are associated with quiet eruptions and widespread flows such as those in the Columbia plateaus of the state of Washington.

The volcanic hot spots mapped and assessed here are

TABLE 5-2

National Summary of Energy in Identified and Undiscovered Geothermal Waters of High-, Intermediate-, and Low-Temperatures, Excluding those in National Parks and in Geopressured Reservoirs

ENERGY SOURCE	ACCESSIBLE RESOURCE BASE (10^{18} cal.)		RECOVERABLE ENERGY (10^{18} cal.)	
IDENTIFIED				
High				
Vapor		24		2
Hot Water		203		50
Intermediate		167		42
Low				
Hydrothermal		48		7.4
Deep Aquifer		6,459		13.4
Total Identified		7,201		115
UNDISCOVERED				
High	669 to 1172		167 to 294	
		1,914		478
Intermediate	741 to 1244		184 to 298	
Low		1,722		16
Total Undiscovered		3,636		494
Grand Total		10,095		609

Sources: USGS *Circular 790, Assessment of Geothermal Resources of the United States*, 1979, and USGS *Circular 892, Assessment of Low Temperature Geothermal Resources of the United States*, 1983. Values corrected from joules to calories.

those deduced from surface evidence of acidic eruptions, which provide a hint that a chamber exists below. Critical to estimates of present temperature and heat content is a knowledge of the age of the eruption. The molten rock in the chamber is assumed to have been cooling at a given rate from a temperature of approximately 850 degrees Celsius at the time of eruption until the present. The size of the chamber is estimated on the basis of surface indications of its area and on the assumption that its thickness ranges from 2.5 to 10 kilometers, which is the depth limit for the volumes that contribute to heat content calculations. The heat contents reported are "that part of the heat content . . . that still remains in the ground, both within and around the original magma chamber" (Smith and Shaw 1975). Of the total heat content in all occurrences, about half exists in molten or partially molten magma while half is in solid or "dry" hot rock surrounding the magma chamber. An exception to this occurs in Alaska, where practically all the heat in-place is contained in molten or partially molten masses.

Figure 5-13 shows the locations of identified volcanic "systems" in the conterminous United States, Alaska, and Hawaii. Because of the volcanic character of the Aleutian chain, there are 88 identified occurrences in Alaska. Hawaii,

with large oceanic volcanoes, is credited with five despite the low-silica character of the magmas. In western states there are 61 identified, with nineteen in California and 20 in Oregon. A substantial number of these do support hydrothermal systems.

Occurrences which have been studied completely enough to determine an estimate of heat in-place are classified according to size in Figure 5-13. Because of the great magnitude of heat content in these magma masses, the size classification is quite different from that used on maps of hydrothermal systems. The largest are in Yellowstone and the neighboring area in Idaho, where one Idaho occurrence holds estimated heat of greater than $4,000 \times 10^{18}$ calories, that is, *more than 30 times the size* of the gigantic Bruneau-Grandview hydrothermal system.

The summary of heat content by state makes it clear that by far Wyoming and Idaho have the highest totals on the basis of volcanic occurrences of known size. There are a number of identified occurrences of unknown size, however, whose heat content could change the relative resource standing of some states. Alaska, in particular, has a large number of such occurrences in the Aleutian chain.

Geopressured Reservoirs of the Gulf Coast

There are a number of sedimentary basins in the United States where usable geopressured reservoirs *may* exist. The thick sedimentary rocks do occur in those areas: It is the formation pressures and the ability to produce hot water that are not well-known (Fig. 5-14). Part of the Gulf Coast area, however, is more thoroughly understood, because the U.S. Geological Survey has studied hundreds of oil and gas well records in the study area shown.

Sandstone bodies holding methane-rich hot water occur in a thick zone whose top ranges from about 1.8 kilometers (6,000 feet) below the surface at the inland edge of the study area in Texas, to as deep as 4.6 kilometers (15,000 feet) off the coast near New Orleans. Sands between these depths and a lower limit of 6.86 kilometers (22,500 feet) have been studied in an area covering 310,000 square kilometers and extending offshore into federal waters where it is limited by the 600 feet depth line marking the outer edge of the continental shelf.

Mechanical energy associated with the flow of fluids driven to the surface by high pressures constitutes less than 1 percent of the total accessible fluid resource, so it was ignored in this study. The energy in hot water and dissolved methane was estimated for the study area by using more than 3,500 oil and gas well records from which data on sand thicknesses, sand porosity, fluid pressures, and fluid salinity were extracted. To organize the intricately interbedded sand and shale in which the geopressured sands occur, the volume of sediments studied was hypothetically sliced into 14 horizontal intervals, each 457 meters (1,500 feet) thick. Average values of rock and fluid characteristics were assigned to each interval. On the basis of a number of simplifying assumptions

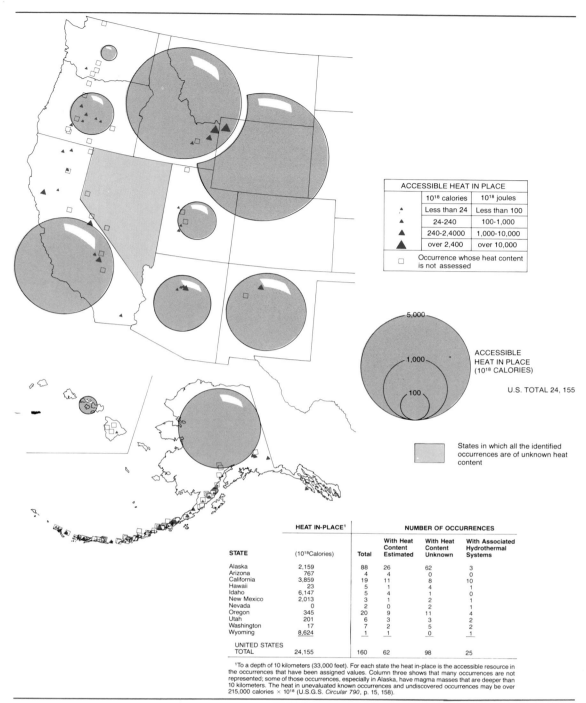

ACCESSIBLE HEAT IN PLACE

	10^{18} calories	10^{18} joules
▴	Less than 24	Less than 100
▲	24-240	100-1,000
▲	240-2,4000	1,000-10,000
▲	over 2,400	over 10,000
☐	Occurrence whose heat content is not assessed	

ACCESSIBLE
HEAT IN PLACE
(10^{18} CALORIES)

U.S. TOTAL 24, 155

States in which all the identified
occurrences are of unknown heat
content

STATE	HEAT IN-PLACE[1] (10^{18}Calories)	NUMBER OF OCCURRENCES			
		Total	With Heat Content Estimated	With Heat Content Unknown	With Associated Hydrothermal Systems
Alaska	2,159	88	26	62	3
Arizona	767	4	4	0	0
California	3,859	19	11	8	10
Hawaii	23	5	1	4	1
Idaho	6,147	5	4	1	0
New Mexico	2,013	3	1	2	1
Nevada	0	2	0	2	1
Oregon	345	20	9	11	4
Utah	201	6	3	3	2
Washington	17	7	2	5	2
Wyoming	8,624	1	1	0	1
UNITED STATES TOTAL	24,155	160	62	98	25

[1]To a depth of 10 kilometers (33,000 feet). For each state the heat in-place is the accessible resource in the occurrences that have been assigned values. Column three shows that many occurrences are not represented; some of those occurrences, especially in Alaska, have magma masses that are deeper than 10 kilometers. The heat in unevaluated known occurrences and undiscovered occurrences may be over 215,000 calories × 10^{18} (U.S.G.S. *Circular 790*, p. 15, 158).

Fig. 5-13 Identified volcanic occurrences, and accessible heat in-place (by state) to depth of 10 km.
Source: U.S.G.S. *Circular 790*, 1979.

and calculations, the volumes of hot water, volumes of dissolved methane, and the associated thermal and methane energy in-place were estimated for each depth interval, then summed for Texas, Louisiana, and federal areas on the continental shelf (U.S.G.S. *Circular 790*, 1979).

Figure 5-15 shows the state totals of accessible energy in-place, with symbols scaled the same as those showing accessible heat in-place in igneous (volcanic) occurrences. The amounts are very large: The smallest total, for the state of Louisiana, is about the same size as the total volcanic heat in-place in Idaho (Fig. 5-13); the largest, for Texas, is over twice the magnitude of the volcanic heat in-place in the state

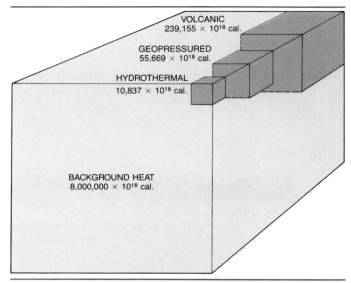

Fig. 5–16 National summary of accessible heat in-place, both identified and undiscovered, shown as part of nation's background heat to depth of 10 km.
Source: U.S.G.S. *Circular 790,* 1979.

For hydrothermal systems, *undiscovered* refers to concealed natural systems of circulation that are postulated on the basis of surface indications. For volcanic occurrences, undiscovered amounts include identified magma bodies for which no estimate has been made, and others not yet identified. For geopressured reservoirs, there are possibilities beyond the Gulf Coast area in other sedimentary basins (see Fig. 5–14). Amounts of energy thought to be in those "other

TABLE 5–3
National Summary of Energy in All Types of Geothermal Occurrence, Showing Accessible Energy In-Place and Estimated Recoverable

TYPE OF OCCURRENCE	ENERGY AMOUNTS (10^{18} CALORIES)	
	Accessible Resource Base	Estimated Recoverable Energy
Background Heat (to 10 km.)	8,000,000	not applicable
Hydrothermal Waters of high-, intermediate- and low-temperature (includes deep aquifers for low-temperature)		
Identified	7,201	115
Undiscovered	3,636	494
Total	10,837	609
Volcanic		
Identified	24,155	no estimates
Undiscovered	215,000	no estimates
Total	239,155	
Geopressured Reservoirs		
Gulf Coast	40,669	102 to 1,060
Other Areas	11,000	28 to 287
Total	55,669	
Total Identified	72,025	217 to 1,175
Total Undiscovered	229,636	522 to 781

Sources. USGS Circular 790, *Assessment of Geothermal Resources of the United States,* 1979, and Circular 892, *Assessment of Low Temperature Geothermal Resources of the United States,* 1983.

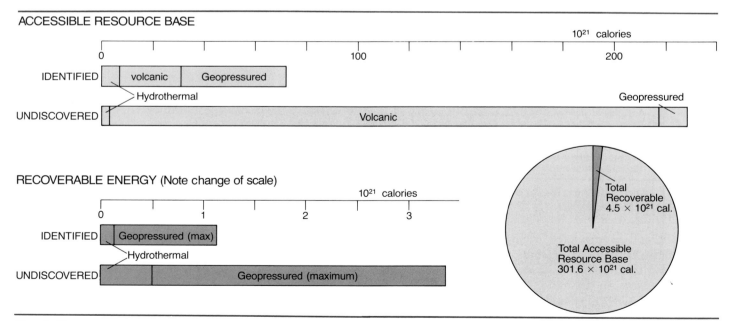

Fig. 5–17 National summary of accessible heat in-place and recoverable energy, showing identified and undiscovered components.
Source: U.S.G.S. *Circular 790,* 1979.
Note: The discrepancy between accessible and recoverable energy is exaggerated by the very large identified and undiscovered amounts of accessible heat in volcanic occurrences for which *no estimate* of recoverable energy has been made.

sedimentary basins" of the U.S. Geological Survey, are included here as undiscovered.

Energy *recoverable* is a small fraction of the resource base, largely because volcanic occurrences are excluded until more is known about the technology of inducing the circulation of water through hot dry rock. Energy recoverable from hydrothermal and geopressured occurrences is shown (with change of scale) in the lower part of Figure 5–17. The geopressured reservoirs seem to offer much more energy than the hydrothermal systems, but these figures assume the most ambitious development plan for Gulf Coast reservoirs. Table 5–3 shows the *range* of recoverable energy estimates, with the more modest estimate being roughly one-tenth of the more ambitious.

The relationship between estimated resource base and the recoverable energy (shown in Fig. 5–17) is affected by the recovery rate: For volcanic systems it is unknown; for hydrothermal systems it is around 25 percent; and for geopressured reservoirs it is around 2.5 percent.

The total amounts of energy recoverable are quite large, when compared to other kinds of energy resources. The higher of the two possible totals for energy recoverable from identified occurrences, that is, $1,175 \times 10^{18}$ calories, is around 4,600 Quads of energy, and is of the same magnitude as the energy in all coals recoverable from the Demonstrated Reserve Base (see Chapter 1). Such a comparison between geothermal energy recoverable and amounts of a fossil fuel recoverable must be qualified by two observations. First, at this time it is not clear whether any of the hydrothermal or geopressured energy recoverable should be called *reserves*, with the exception of the hottest and most desirable occurrences that are being exploited currently. Coal from the Demonstrated Reserve Base, however, *is* economically recoverable. The second factor to consider is the *efficiency* of using geothermal fluids, whose temperatures are far lower than those attained by burning fossil fuels. Lower temperatures imply lower efficiency, so it is misleading to simply translate geothermal well-head heat into BTUs and compare that figure with the energy content of coal, crude oil, or natural gas. A more sound approach is adopted here: that is, to estimate *how much of the traditional fuels would be displaced* if geothermal heat were used for certain tasks, such as heating buildings or generating electrical power.

Space Heating with Intermediate- and Low-temperature Geothermal Waters

If all the *identified* occurrences of intermediate-temperature hydrothermal systems and all the low-temperature waters in hydrothermal and deep aquifer occurrences were applied to the heating of buildings a great deal of fuel oil would be displaced. The amount, for illustrative purposes only, can be estimated by ignoring economic factors and simply working with the total *beneficial heat* which the U.S. Geological Survey has

calculated on the basis of how the water temperature compares with the temperature of the environment being heated (see notes Fig. 5–18). The amounts of fuel oil that would be burned to supply an equivalent amount of heat are estimated, for Figure 5–18, by assuming a furnace efficiency of 85 percent. The resulting fuel oil amounts, expressed as 10^{12} gallons (trillions of gallons) are concentrated in western states, as would be expected. The largest amount is in Idaho, because of the massive hydrothermal systems of intermediate temperature. Montana and North Dakota show very substantial amounts due to low-temperature waters—largely in deep aquifers.

Electrical Energy from High-temperature Hydrothermal Systems and from Geopressured Reservoirs

These two types of occurrences are not the only ones with potential for electrical power generation. Volcanic (hot dry rock) occurrences will be used this way when the technology has been refined and the economic climate is favorable. Hydrothermal waters of intermediate-temperature can also be used for power generation, but only if *binary* power plants are employed. Binary plants are those using geothermal water to boil a special working fluid, such as freon, whose boiling point is low. If those two possibilities are ignored, then the geothermal electric generating potential can be estimated on the basis of high-temperature hydrothermal systems and geopressured reservoirs.

In Figure 5–19, part A shows the hydrothermal contribution would total over 23 thousand Megawatts of generating capacity, running steadily for 30 years. As the illustration implies, this figure is equivalent to 23 large nuclear or coal-fired power plants running at 100 percent capacity for that period. California shows a total of almost 14,000 Megawatts, part of which (1,340 Megawatts) was already developed as of 1984, mostly at the Geysers area north of San Francisco. The amounts of fuel saved (see tabulation) are calculated directly from average fuel needs of coal-burning and nuclear power plants. The total coal saved would be over 3 billion tons, which is roughly three times the annual coal production of the nation.

The contribution from geopressured reservoirs depends upon the development plan assumed. If the most ambitious plan were implemented, the electrical output would be as tabulated and mapped in part B of Figure 5–19, assuming that only thermal energy, not methane, were applied to the task. On this basis, the geopressured contribution would be roughly 10 times the hydrothermal. However, if the least ambitious development plan were adopted, the two contributions would be roughly equal.

SUMMARY OF FUELS SAVED For the sake of comparison with other resource amounts, the potential fuel savings from space heating and electrical applications are

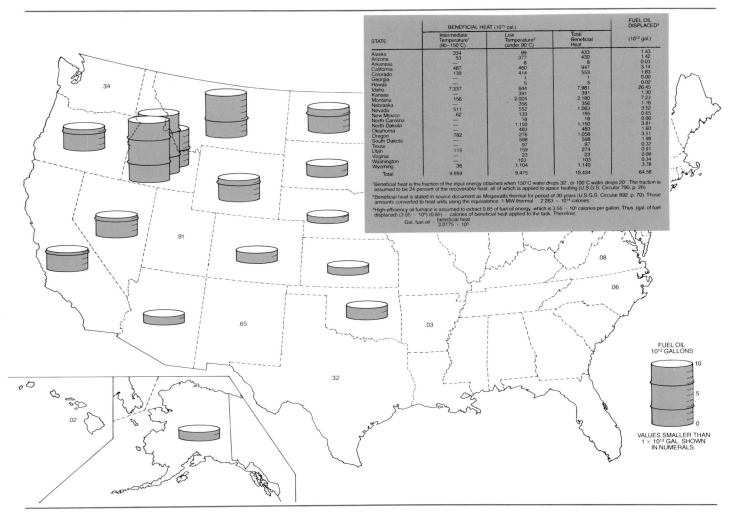

STATE	BENEFICIAL HEAT (10¹⁵ cal.) Intermediate Temperature[1] (90–150°C)	Low Temperature[2] (under 90°C)	Total Beneficial Heat	FUEL OIL DISPLACED[3] (10¹² gal.)
Alaska	334	99	433	1.43
Arizona	53	377	430	1.42
Arkansas	—	8	8	0.03
California	487	460	947	3.14
Colorado	139	414	553	1.83
Georgia	—	1	1	0.00
Hawaii	—	5	5	0.02
Idaho	7,337	644	7,981	26.45
Kansas	—	391	391	1.30
Montana	156	2,024	2,180	7.23
Nebraska	—	356	356	1.18
Nevada	511	552	1,063	3.52
New Mexico	62	133	195	0.65
North Carolina	—	18	18	0.06
North Dakota	—	1,150	1,150	3.81
Oklahoma	—	483	483	1.60
Oregon	782	276	1,058	3.51
South Dakota	—	598	598	1.98
Texas	—	97	97	0.32
Utah	115	159	274	0.91
Virginia	—	23	23	0.08
Washington	—	103	103	0.34
Wyoming	36	1,104	1,140	3.78
Total	9,959	9,475	19,434	64.58

[1]Beneficial heat is the fraction of the input energy obtained when 150°C water drops 32° or 100°C water drops 20°. The fraction is assumed to be 24 percent of the recoverable heat, all of which is applied to space heating (U.S.G.S. Circular 790, p. 26)

[2]Beneficial heat is stated in source document as Megawatts thermal for period of 30 years (U.S.G.S. Circular 892, p. 70). Those amounts converted to heat units using the equivalence: 1 MW thermal = 2.263 × 10¹⁴ calories.

[3]High-efficiency oil furnace is assumed to extract 0.85 of fuel oil energy, which is 3.55 × 10⁵ calories per gallon. Thus, (gal. of fuel displaced) (3.55 × 10⁵) (0.85) = calories of beneficial heat applied to the task. Therefore,

$$\text{Gal. fuel oil} = \frac{\text{beneficial heat}}{3.0175 \times 10^5}$$

FUEL OIL 10¹² GALLONS

VALUES SMALLER THAN 1 × 10¹² GAL. SHOWN IN NUMERALS.

Fig. 5–18 Fuel oil saved if all identified intermediate- and low-temperature waters were applied to space heating.

Source: Derived from data on beneficial heat in U.S.G.S. *Circular 790*, 1979 (for intermediate-temperature) and *Circular 892*, 1983 (for low-temperature).

One of the wellheads at The Geysers, California. Pipes carry steam from wells to a number of separate power plants that together have the capacity to generate over 1,000 Megawatts of electricity as of 1984. (Courtesy of Pacific Gas and Electric Company.)

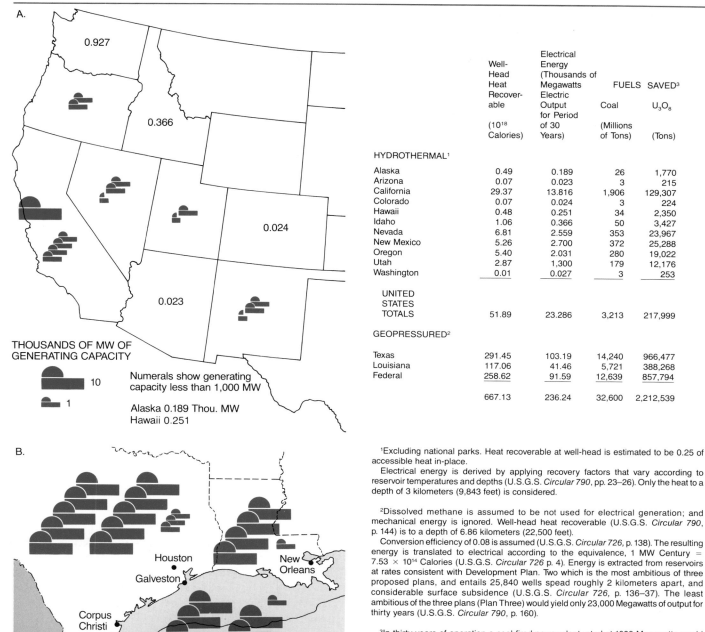

	Well-Head Heat Recoverable (10¹⁸ Calories)	Electrical Energy (Thousands of Megawatts Electric Output for Period of 30 Years)	FUELS SAVED[3] Coal (Millions of Tons)	U₃O₈ (Tons)
HYDROTHERMAL[1]				
Alaska	0.49	0.189	26	1,770
Arizona	0.07	0.023	3	215
California	29.37	13.816	1,906	129,307
Colorado	0.07	0.024	3	224
Hawaii	0.48	0.251	34	2,350
Idaho	1.06	0.366	50	3,427
Nevada	6.81	2.559	353	23,967
New Mexico	5.26	2.700	372	25,288
Oregon	5.40	2.031	280	19,022
Utah	2.87	1,300	179	12,176
Washington	0.01	0.027	3	253
UNITED STATES TOTALS	51.89	23.286	3,213	217,999
GEOPRESSURED[2]				
Texas	291.45	103.19	14,240	966,477
Louisiana	117.06	41.46	5,721	388,268
Federal	258.62	91.59	12,639	857,794
	667.13	236.24	32,600	2,212,539

THOUSANDS OF MW OF GENERATING CAPACITY

10

1

Numerals show generating capacity less than 1,000 MW

Alaska 0.189 Thou. MW
Hawaii 0.251

[1]Excluding national parks. Heat recoverable at well-head is estimated to be 0.25 of accessible heat in-place.

Electrical energy is derived by applying recovery factors that vary according to reservoir temperatures and depths (U.S.G.S. *Circular 790*, pp. 23–26). Only the heat to a depth of 3 kilometers (9,843 feet) is considered.

[2]Dissolved methane is assumed to be not used for electrical generation; and mechanical energy is ignored. Well-head heat recoverable (U.S.G.S. *Circular 790*, p. 144) is to a depth of 6.86 kilometers (22,500 feet).

Conversion efficiency of 0.08 is assumed (U.S.G.S. *Circular 726*, p. 138). The resulting energy is translated to electrical according to the equivalence, 1 MW Century = 7.53 × 10¹⁴ Calories (U.S.G.S. *Circular 726* p. 4). Energy is extracted from reservoirs at rates consistent with Development Plan. Two which is the most ambitious of three proposed plans, and entails 25,840 wells spead roughly 2 kilometers apart, and considerable surface subsidence (U.S.G.S. *Circular 726*, p. 136–37). The least ambitious of the three plans (Plan Three) would yield only 23,000 Megawatts of output for thirty years (U.S.G.S. *Circular 790*, p. 160).

[3]In thirty years of operation a coal-fired power plant rated at 1000 Megawatts would burn roughly 90 million tons of bituminous coal (Nuclear Energy Policy Study Group, p. 78) and a light water reactor of the same rating would need roughly 6090 tons of unenriched U₃O₈ (Sondermayer, p. 1118). If the plants are assumed to run at 65 percent of capacity, then these fuel requirements are for only 650 Megawatts of output. The fuels needed for each 1,000 Megawatts of output would be roughly 138 million tons of coal, and 9,366 tons of U₃O₈; these are the figures used to convert geothermal electrical output into coal and uranium oxide saved.

Fig. 5–19 Potential generating capacity for 30 years from energy in identified occurrences. A. High temperature hydrothermal systems, B. Gulf Coast geopressured reservoirs.
Source: U.S.G.S. *Circular 790*, 1979.

brought together in Table 5-4, which includes the use of methane from geopressured reservoirs. The grand total of fuel energy saved, roughly 1,600 Quads, is the amount of all primary energy that the nation uses in about 21 years, at 75 Quads per year.

DEVELOPMENT ACTIVITIES

One indication of recent development work is the number of holes drilled to explore for and to better define geothermal prospects. In Figure 5–20 all deep holes drilled

ELECTRICAL POWER FROM GEOTHERMAL HEAT

As in fuel-fired generating plants, the generator is driven by a turbine that is spun by steam (or some other vapor) *expanded* through the turbine, that is, rushing through the turbine toward lower pressures on the downstream side. Just how this is accomplished depends largely upon the kind of hot fluids carrying geothermal heat out of the earth.

1. In some hydrothermal systems there is reduced pressure at depth that allows a zone of steam to form above hot water in a reservoir. If this dry steam is produced through pipes it can be fed directly into a turbine. In the simplest turbine arrangement, the spent steam is vented to the atmosphere. It is usual, though, to create a more effective drop in pressure by condensing the steam and disposing of the resulting heat by use of cooling towers, ponds, or river water.

2. Much more common in hydrothermal systems is hot water under high pressure in the earth. If the water is allowed to flow to the surface, part of it (10 to 15 percent) will immediately flash to steam as it encounters surface pressures. If this steam is not corrosive or heavily laden with polluting gases, it may be fed directly into a turbine. If the steam is undesirable, it—and the accompanying very hot water—can be used in a heat exchanger to boil clean water for steam.

3. An alternative that has a number of advantages is to *pump* the very hot water to the surface, keeping it under pressure and preventing the formation of steam, and thus reducing pollution and scaling problems. This hot water under pressure is circulated through a heat exchanger where it boils clean water for steam.

If the fluid temperatures are not high enough to boil water effectively, a special *working fluid* such as isobutane or freon is used to spin the turbine; It is "boiled" at a temperature around 90 degrees Fahrenheit, expanded through the turbine, condensed on the far side, and returned to the heat exchanger to be vaporized again. This two-fluid, or *binary*, plant opens the door to much wider use of hydrothermal fluids of intermediate-temperature.

4. Hot igneous rocks that have no associated hydrothermal system can be fractured so an artificial circulation of water is provided (see section on experiments at Los Alamos). The resulting hot water will be pumped to the surface, used to boil some working fluids, then injected back into the reservoir, travelling in a closed loop under pressure.

5. Geopressured reservoirs yield hot water with significant amounts of dissolved methane. At the Brazoria test well in Texas, some methane will be separated from the water, then the hot water will be used in a binary plant to vaporize a working fluid for one generator turbine. Low-grade methane will be burned in a gas turbine, whose waste heat will be applied to superheat the working fluid vapor, thus increasing the efficiency of that turbine. Prior to the methane separation, the high-pressure fluid from the well-head will drive a compressor that restores the clean methane to a pressure suitable for sale to a pipeline company.

in the period from 1978 through 1982 are mapped. They are called deep *exploratory* wells, but, in fact, many were drilled in areas of known prospects, such as the Geysers and Imperial Valley in California, and might be better labeled *development* wells. The footage in Maryland was drilled by the Department of Energy to evaluate low-temperature deep aquifers; the holes in Texas and Louisiana were drilled by the federal government and by private investors for geopressured reservoirs. All the others were probably directed toward hydrothermal targets.

Various geothermal activities are shown in Figure 5-21, which includes, for reference, the estimated heat flow rates in different parts of the western states. One group of symbols on that map identifies projects supported by the Department of Energy and aimed at promotion of the direct use of geothermal heat for space heating and agricultural applications. The locations lettered A through E represent only *some* of the places where geothermal heat has been used for many years, and in many cases, without federal government support. Also shown in Figure 5-21 are the locations of geothermal-electric projects, which are discussed below.

Rapid growth in geothermal electric power capacity in the United States began in the 1970s with development of the steam-dominated high-temperature hydrothermal occurrences in the Geysers area of northern California (see graph in Fig. 5-22). The map and table in that illustration show that this area, where development began in 1960, is expected to dominate through the year 1992.

The other areas developed in 1984 are also noteworthy. At Roosevelt Hot Springs the 20 Megawatt plant that began operation in 1984 is utilizing a reservoir (at 2,000 feet. depth) which in the general area has an electric potential of 200 to 300 Megawatts for a period of 35 years. In the Imperial Valley, the small power plant at Heber is of the *binary* type,

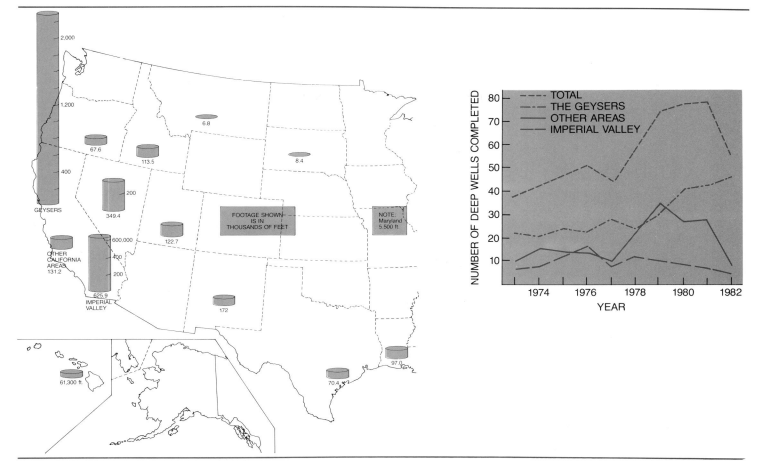

Fig. 5-20 Exploratory wells drilled 1978 to 1982, showing total footage by state. Source: DOE, *Geothermal Progress Monitor*, Nov. 1983.

testing the possibility of using intermediate-temperature waters for power generation. One plant that has been operating for a few years in the Imperial Valley is the 10 Megawatt pilot plant operated by Southern California Edison. It uses hot brines that turn into steam when their pressure is reduced (see photo).

The 10 Megawatt geothermal-electric pilot plant near Brawley in the Imperial Valley of Southern California. Hot water produced by Union Oil Company of California from wells three to six thousand feet deep is purchased by Southern California Edison Company for the power plant. (Courtesy of American Petroleum Institute and Union Oil Company of California.)

TABLE 5-4
National Summary of Fuel Energy Saved through Space-heating and Electrical Applications of *Identified* Geothermal Energy, Assuming All Recoverable Amounts Applied to Tasks

GEOTHERMAL OCCURRENCE	SPACE HEATING		ELECTRICAL GENERATION	
	Oil or gas Displaced	Quads	Coal Displaced	Quads
Low- and intermediate-temperature waters	64.58×10^{12} gallons fuel oil	91.1		
High-temperature waters			$3,213 \times 10^6$ tons	83.5
Geopressured Reservoirs				
Methane for heating	617×10^{12} cubic feet gas	638.6		
Thermal for Electrical			$32,600 \times 10^6$ tons	847.6
Totals		729.7		930.1
Grand total	1,659.8 Quads			

Note: Geopressured energy amounts assume the most ambitious development plan. On the basis of a more modest plan the amounts would be roughly one-tenth of those noted here. Energy amounts are expressed as Quads, i.e., 10^{15} BTUs, as well as in fuel units.

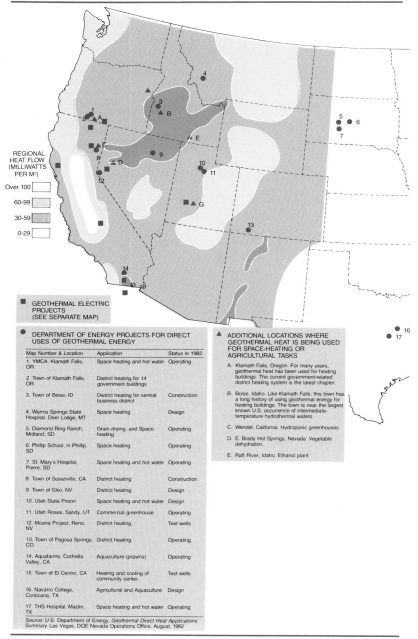

REGIONAL HEAT FLOW (MILLIWATTS PER M²)

Over 100

60-99

30-59

0-29

■ GEOTHERMAL ELECTRIC PROJECTS (SEE SEPARATE MAP)

● DEPARTMENT OF ENERGY PROJECTS FOR DIRECT USES OF GEOTHERMAL ENERGY

Map Number & Location	Application	Status in 1982
1. YMCA. Klamath Falls, OR	Space heating and hot water	Operating
2. Town of Klamath Falls, OR	District heating for 14 government buildings	
3. Town of Boise, ID	District heating for central business district	Construction
4. Warms Springs State Hospital, Deer Lodge, MT	Space heating	Design
5. Diamond Ring Ranch, Midland, SD	Grain drying, and Space heating	Operating
6. Phillip School, in Philip, SD	Space heating	Operating
7. St. Mary's Hospital, Pierre, SD	Space heating and hot water	Operating
8. Town of Susanville, CA	District heating	Construction
9. Town of Elko, NV	District heating	Design
10. Utah State Prison	Space heating and hot water	Design
11. Utah Roses, Sandy, UT	Commercial greenhouse	Operating
12. Moana Project, Reno, NV	District heating	Test wells
13. Town of Pagosa Springs, CO	District heating	Operating
14. Aquafarms. Cochella Valley, CA	Aquaculture (prawns)	Operating
15. Town of El Centro, CA	Heating and cooling of community center	Test wells
16. Navarro College, Corsicana, TX	Agricultural and Aquaculture	Design
17. THS Hospital, Maclin, TX	Space heating and hot water	Operating

Source: U.S. Department of Energy, *Geothermal Direct Heat Applications Summary* Las Vegas, DOE Nevada Operations Office, August, 1982

▲ ADDITIONAL LOCATIONS WHERE GEOTHERMAL HEAT IS BEING USED FOR SPACE-HEATING OR AGRICULTURAL TASKS

A. Klamath Falls, Oregon. For many years, geothermal heat has been used for heating buildings. The current government-related district heating system is the latest chapter.

B. Boise, Idaho. Like Klamath Falls, this town has a long history of using geothermal energy for heating buildings. The town is near the largest known U.S. occurrence of intermediate-temperature hydrothermal waters.

C. Wendel, California. Hydroponic greenhouses

D. E. Brady Hot Springs, Nevada. Vegetable dehydration.

E. Raft River, Idaho. Ethanol plant.

Fig. 5-21 Activities using intermediate- and low-temperature waters. Also shown are heat flow rates taken from U.S.G.S. *Circular 790,* 1979.

Among developments now under construction is a hybrid power plant at Susanville, California that will combine geothermal energy with heat from burning wood chips. At Lakeview, Oregon, a binary plant is being built.

Developments anticipated between 1984 and 1992 will add generating capacity roughly equal to the 1984 capacity, making a total of around 3,000 Megawatts—the equivalent of three very large nuclear or coal-fired power plants. This developed capacity will be about one-eighth of the potential in identified hydrothermal systems with temperatures over

150 degrees Celsius. Further exploration may increase that potential by four or five times (Nathenson, 1983) so there is still room for development of hydrothermal resources. Additonal, and quite separate, potential exists in geopressured reservoirs and in hot dry rock, both of which are undergoing some research.

Activity In Geopressured Reservoir Development

Since 1978, the U.S. Government (through the Department of Energy and its predecessor, the Energy Research and Development Administration) has been conducting tests to learn more about the character of geopressured sand bodies and the flow of water from them. The rate of flow, and whether it can be sustained over time, are crucial to the use of these reservoirs for electrical generation and other applications.

The Department of Energy has been involved with 18 different test holes, 5 of them drilled expressly for geopressured reservoirs, and 13 of them originally drilled for oil and gas, then acquired by the government when abandoned by their operators. Because of either mechanical failure or drilling problems, the Department of Energy has succeeded in carrying out testing in only 4 of the 5 new holes, and only 7 of the 13 abandoned oil and gas wells. The locations of these 11 holes are shown on Figure 5–23, which also maps the "fairways," or areas of most attractive reservoir rocks.

As of the autumn of 1984, only two test wells were active. The well near Houston, Texas (Fig. 5–23) often referred to as the Brazoria well, is the first in the United States to be drilled for geothermal testing. Recently, it has produced gas-laden

Nero Geothermal well on the island of Hawaii. (Courtesy of U.S. Department of Energy.)

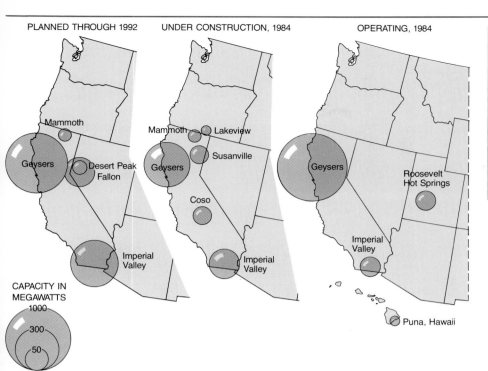

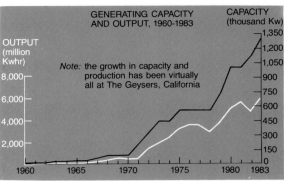

Source: DOE, *Annual Energy Review, 1983*, April 1984

AREA AND LOCATION	STATUS, NUMBER OF PLANTS AND CAPACITY IN MEGAWATTS						
	Operating		Under Construction		Planned Through 1992		
	(No.)	MW	(No.)	MW	(No.)	MW	TOTAL CAPACITY
Milford (Roosevelt Hot Springs) UT	(1)	20	—		—		20
Puna, Hawaii	(1)	3	—		—		3
Geysers, CA	(19)	1,310	(6)	385	(15)	850	2,545
Imperial Valley, CA							
Brawley	(1)	10	—		(2)	88	98
East Mesa	(1)	10	—		—		10
Salton Sea	(1)	10	—		—		10
Geothermal	—		(2)	14	(4)	187	201
Niland	—		(2)	49	(2)	74	123
Heber	—		(1)	45	(1)	49	94
Mammoth, CA	—		(1)	7	(1)	7	14
Susanville (Honey Lake), CA	—		(1)	19	—		19
Coso, CA	—		(1)	20	—		20
Lakeview, OR	—		(1)	3	—		3
Fallon, NV	—		—		(1)	75	75
Desert Peak, NV	—		—		(1)	9	9
Imp. Valley Total	(3)	30	(5)	108	(9)	398	536
U.S. Total	(24)	1,363	(15)	542	(27)	1,339	3,244

Source: DOE, *Geothermal Progress Monitor*, November, 1983.

Fig. 5–22 Geothermal electric generating capacity, current and anticipated.
Source: Derivedf from data in DOE, *Geothermal Progress* monitor, Nov. 1983. Overview, 1960–1983, taken from
DOE, *Annual Energy Review, 1983*, April 1984.

salt water at rates near 40,000 barrels per day. However, the accumulation of lime scaling in the tubing forced a shutdown in late 1984. Future plans include rejuvenating the well and installing an experimental power plant of 1 Megawatt capacity. This novel power plant will make use of all three of the energy forms that reside in the reservoir water: the mechanical force in the pressurized water, the heat, and the methane dissolved in the water.

The second active test well is near Grand Chenier about 30 miles south of Lake Charles, Louisiana. Production testing, with flows in the vicinity of 25,000 barrels per day, is continuing, but there are no plans for installing a power plant.

Another well has also been acquired by the government, an abandoned gas well near Lafayette, Louisiana, that Superior Oil Company turned over to the Department of Energy. It will be the third active test well and is especially

interesting because it penetrates a reservoir sand at a depth of 20,000 feet where waters are at a temperature of 175 degrees Celsius—about 25 degrees higher than waters in the other two test holes where the reservoirs are only 10 to 15,000 feet deep. (This account of current activity is based on personal communications during 1984 with James Bresee of the Department of Energy, Office of Geothermal and Hydropower Technologies.)

Heat Extraction from Hot Dry Rock

These volcanic occurrences are being tested by the Department of Energy at Fenton Hill, New Mexico, 30 miles west of Los Alamos (see Fig. 5–24). Hot rocks at depth are fractured by massive hydraulic pressure applied through perforations in the longer pipe shown in the diagram. Then,

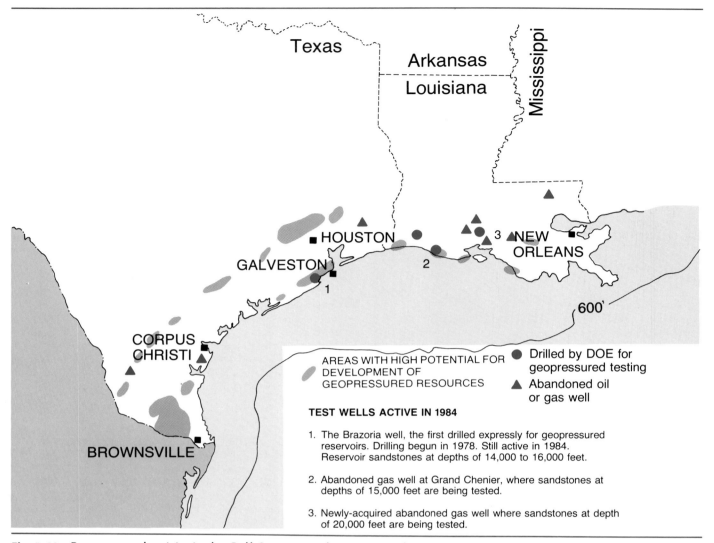

Texas

Arkansas

Louisiana

Mississippi

■ HOUSTON

GALVESTON

3 ▲ NEW ORLEANS

2

1

600'

CORPUS CHRISTI

BROWNSVILLE

AREAS WITH HIGH POTENTIAL FOR DEVELOPMENT OF GEOPRESSURED RESOURCES

● Drilled by DOE for geopressured testing

▲ Abandoned oil or gas well

TEST WELLS ACTIVE IN 1984

1. The Brazoria well, the first drilled expressly for geopressured reservoirs. Drilling begun in 1978. Still active in 1984. Reservoir sandstones at depths of 14,000 to 16,000 feet.

2. Abandoned gas well at Grand Chenier, where sandstones at depths of 15,000 feet are being tested.

3. Newly-acquired abandoned gas well where sandstones at depth of 20,000 feet are being tested.

Fig. 5–23 Prospects and activity in the Gulf Coast area of geopressured reservoirs.
Sources: Areas of high potential, from U.S.G.S. *Circular 790*, 19XX; test hole locations from Swanson *et al.*, 1983; latest activity and plans from Bresee, 1984.

cold water from the surface is injected through the longer pipe to percolate through the network of fractures where steam is produced to be drawn off through the shorter pipe. The pancake-shaped reservoir zone, 3,000 feet in diameter, would be formed if fractures continue to grow in response to stresses caused by the rapid cooling of hot rock.

The initial project, completed in 1978, used a pair of holes drilled to about 8,000 feet, and succeeded in recovering heat at a rate of 5 Megawatts. This was used to run a small electrical generator—producing, therefore, the first electrical power from this type of geothermal heat source.

Now, with contributions of funds and technical personnel from West Germany and Japan, the Department of Energy is proceeding with a larger scale operation, near the first project, but at a depth of 11,000 feet (the holes were drilled to 14,000 feet, but temperatures there were too high for the down-hole monitoring equipment). These holes were completed in 1981, but the fracture system has not joined the two holes as expected—though it has established a large reservoir with fluid capacity of nearly 6 million gallons.

To make connection with the reservoir, and thereby complete the *loop*, a hole will be started from a point *within* the second hole, and drilled at an angle until it meets the reservoir. This work, begun in October of 1984, is expected to end in the fall of 1985, at which time a 1-year period of testing will be undertaken to learn what rate of steam withdrawal can be sustained and how the rock temperature falls during an extended period of exploitation.

The Department of Energy expects that heat can be withdrawn at a rate of near 50 Megawatts. With production test data in hand, they will then see whether some utility company is interested in using the steam for commercial generation of power.[1]

NOTE

1. Information on the second project was obtained from Jim Reynolds, U.S. Department of Energy, Office of Geothermal and Hydropower Technologies, September 27, 1984.

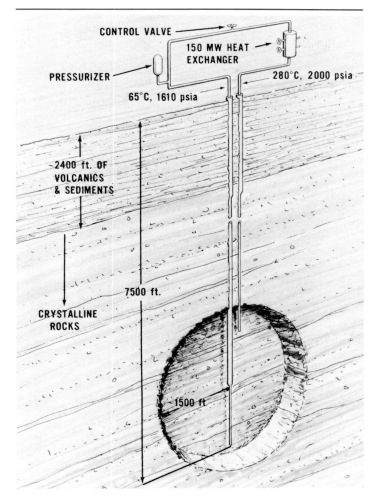

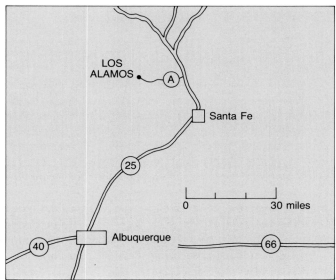

Fig. 5–24 The Fenton Hill site, near Los Alamos, New Mexico, where the U.S. Department of Energy is successfully recovering heat from hot igneous rock by inducing water circulation.
Source: Schematic from U.S. Department of Energy photo services.

References

Bresee, James. U.S. Department of Energy, Geothermal and Hydropower Technologies Office, Washington, DC. Personal communication, Sept. 1984.

De Pippo, Ronald. "Worldwide Geothermal Power Development: An Overview and Update," *Proceedings, Seventh Annual Geothermal Conference and Workshop,* Palo Alto, CA: Electric Power Research Institute, Sept. 1983.

Dickinson, George. Geological Aspects of Abnormal Reservoir Pressures in Gulf Coast Louisiana." In *Bulletin of American Association of Petroleum Geologists,* 37:2, 1953, 410–432.

Diment, *et al.* "Temperatures and Heat Contents Based on Conductive Transport of Heat." *U.S. Geological Survey Circular 726.* Alexandria, VA, 1975.

Nathenson, Manuel. "High-Temperature Geothermal Resources in Hydrothermal Convection Systems in the United States," *Proceedings, Seventh Annual Geothermal Conference and Workshop,* Palo Alto, CA: Electric Power Research Institute, Sept. 1983.

Pollack, Henry N., and D.S. Chapman. "The Flow of Heat from the Earth's Interior." *Scientific American,* 237:2, 1977, 60–76.

Renner, J. L., D. E. White, and D. L. Williams. "Hydrothermal Convection Systems." In White, D. E., and D. L. Williams. *U.S.G.S. Circular 726,* 1975.

Smith, R. L., and H. R. Shaw. "Igneous-Related Geothermal Systems." In White, D. E., and D. L. Williams. *U.S.G.S. Circular 726,* 1975.

Swanson, R. K., J. S. Osoba, and W. J. Bernard. "Geopressure Revisited, 1983." In *Proceedings, Seventh Annual Geothermal Conference and Workshop.* Palo Alto, CA: Electric Power Research Institute, Sept. 1983.

United Nations. *U.N. Statistical Yearbook,* various years.

U.S. Department of Energy. *Annual Energy Review,* 1983, DOE/EIA-0384(83) Washington, DC; Government Printing Office, April 1984.

U.S. Department of Energy. *Geothermal Direct Heat Applications Program Summary.* From program review meeting in Las Vegas, Nevada. Las Vegas; Nevada Operations Office, DOE, Aug. 1982.

U.S. Department of Energy. *Geothermal Progress Monitor No. 8, Progress Report.* Washington, DC; Government Printing Office, 1983.

U.S. Geological Survey. *Assessment of Low-Temperature Geothermal Resources of the United States—1982. U.S.G.S. Circular 892.* Alexandria, VA, 1983.

U.S. Geological Survey, Circular 790. *Assessment of Geothermal Resources of the United States—1978.* Washington, DC; Government Printing Office, 1979.

U.S. Geological Survey, Circular 726. *Assessment of Geothermal Resources of the United States—1975,* Washington, DC; Government Printing Office, 1975.

Part Two

RENEWABLE RESOURCES

This part of the atlas explores a variety of energy sources that either flow continuously (for example, wind) or are being slowly replenished by natural processes and thus may be gathered for thousands of years through judicious management (for example, energy crops).

The ultimate dependence of most of these upon solar energy will be explained in Chapter 6, which deals with solar radiation. Chapters 7 and 8 focus on wind and hydropower and are followed by Chapter 9 on ocean resources, which brings together tidal power, wave and current energy, and ocean thermal gradients. Finally, Chapter 10 presents information on biomass.

Of all these sources, only hydropower is represented by firm numbers to express total potential and the portion now developed. For other energy sources, estimates of total potential are difficult and can be made only if certain economic assumptions are adopted.

Energy sources such as wind, hydropower, and ocean thermal gradients are regarded as producers of electricity and are assessed, therefore, in units of generating capacity and electrical output. Solar radiation and biomass, though, have varied applications, so their potential contributions can be expressed as quadrillion BTUs of energy or, in some cases, as amounts of mineral fuels that would be displaced by using the renewable source.

6 Solar Radiation

- The Resource and Its Distribution
- Hot Water Heating
- Space Heating and Cooling
- Electrical Generation
- The Future Role of Solar Energy

The term, *solar energy*, is sometimes used broadly to encompass a number of energy flows that are derived from solar radiation. Here, we distinguish the raw resource, *solar radiation*, from the various physical and biological effects which can be harnessed or gathered for energy purposes (see Table 6–1A.).

Radiation can be used for either its thermal or its photovoltaic effects. With low-temperature devices it can heat buildings. In addition, it can generate electricity either by high-temperature thermal devices or by photovoltaic installations. The three major physical effects that depend on the sun—wind, hydro energy, and ocean thermal gradients—all can be used for generating electricity. The biological phenomena based on photosynthetic conversion of radiation by green plants vary greatly and can be used to produce a rich variety of fuels. Although some of those fuels could be used to generate electricity, they are usually regarded as substitutes for coal, oil, and natural gas, in nonelectric applications.

Part B of the table identifies the three major applications that are recognized by the Department of Energy: (1) thermal applications, such as space heating; (2) electrical generation; and (3) production of alternate fuels. In the right-hand column the resources drawn upon for those applications are arranged to coincide with the order of the chapters in Part Two.

This chapter is concerned only with solar radiation used directly for its thermal and photovoltaic effects. It is arranged according to applications, as in Table 6–1B. After a discussion of radiation, its measurement, and its distribution, hot water heating is reviewed. This is followed by a more lengthy section dealing with solar space heating and the climatic and economic factors that determine its contribution to heating needs and its economic feasibility in different parts of the United States. Then electrical power generation by thermal and photovoltaic means is discussed. Finally, some estimates of the future contributions of these solar technologies are presented.

The major purpose of this chapter is to promote an understanding of the resource, its distribution, and the geographic factors that condition its use. Secondarily, the status of the different applications is clarified, and some

◄Part of an experimental solar collector that concentrates the sun's energy to obtain very high temperatures for power generation or industrial purposes. In this design, the receiver moves during the day to stay at the focal point of the collector. (Courtesy of U.S. Department of Energy.)

TABLE 6–1
Solar and Solar Related Phenomena, and Their Applications

A. ARRANGED ACCORDING TO ORIGINS

RADIATION		Photo-Voltaic	SOLAR-DEPENDENT PHYSICAL EFFECTS	SOLAR-DEPENDENT BIOLOGICAL EFFECTS
Thermal				
Low-Temperature	High-Temperature	Electrical generation	Wind electric	Photosynthesis, supporting all plant and animal life
Hot water and space heating	Industrial Electrical generation		Waves Ocean currents	
Agricultural and industrial			Hydroelectric energy	Biomass: organic wastes and energy crops yielding alternate fuels
			Ocean thermal electric	

B. ARRANGED ACCORDING TO APPLICATIONS

APPLICATION OR PRODUCT	RESOURCE USED	CHAPTER
Hot water and space heating (and cooling)	Radiation	6
Agricultural and industrial process heat	Radiation	6
Electrical generation	Radiation Thermal Photovoltaic	6
	Wind	7
	Hydropower	8
	Ocean energy	9
Alternate fuels: solid, liquid, and gaseous	Biomass: wastes and crops	10

important projects are illustrated. The sections covering different applications vary in length because there is less information available on some locational aspects, whereas for others (such as active space heating) some very useful studies exist whose principles are applied in this chapter.

THE RESOURCE AND ITS DISTRIBUTION

Incoming solar radiation, or *insolation*, is measured by instruments designed to sense either *global radiation*, the total energy that would be intercepted by a flat collector, or *direct-normal* radiation, which is the rays coming directly from the sun that can be focused by mirrors or lenses in special collectors (see photos of instruments).

Insolation is always expressed as energy amounts per square unit of surface. Tabulated or mapped data for a location can be used, therefore, to determine how large a collector should be to gather enough energy for a given task—provided the efficiency of the device is taken into account. Various energy units are used by different authorities. The unit common among meterologists is the Langley, which is one calorie (small calorie) per square centimeter. This is readily converted into BTUs per square foot of area. In this edition of the atlas, most of the energy amounts are in

megajoules per square meter of area, that is, MJ/m^2, which can be converted by using these relationships:

1 Megajoule = 1 million joules $\quad$ $1 MJ/m^2 = 24$ Langleys
$\quad$ 1 Joule = 0.24 calories $\quad$ $1 MJ/m^2 = 88.11$ BTU/ft^2

Radiation Data in the United States

There are few meteorological stations with long records of high-quality solar radiation data. Therefore, researchers preparing the latest insolation data for the United States (see Department of Energy/Solar Energy Research Institute, (DOE/SERI), *Solar Radiation Energy Resource Atlas*, 1981) made use of radiation records from only 26 stations across the nation, recorded over a 23-year period ending in 1975.

For direct-normal (focusable-beam) radiation, the only possible procedure was to refine and correct the actual measurements from those 26 stations, then use them as the basis for contour-style maps like those presented in following pages. For global radiation, however, a procedure could be followed that took advantage of the numerous weather stations which record conditions such as cloud cover and precipitation that affect the amount of global radiation received. First, at each of the 26 stations recording both

This radiometer instrument is an *Epply 8-48 Pyranometer* that measures *global radiation*, i.e., the total of direct-beam (focussable) radiation and diffuse sky radiation. (Courtesy of Solar Energy Research Institute.)

global radiation *and* the other meteorological variables, the relationships between global radiation and the other variables were studied until formulas were obtained that would predict global radiation satisfactorily from the other variables. Then these formulas were applied to information from 222 stations that have good meteorological records, *but no data on radiation.* This was done for each of the 222 stations by choosing the nearest and most appropriate of the 26 control stations, then applying the relationships derived from that station's records. In this way, estimates of global radiation were obtained for a relatively dense network of stations. These data then served as the basis for contour-style maps of expected radiation amounts.

Selection and Presentation of the Radiation Data

The originators of this latest data have manipulated it and mapped it in a number of different ways to anticipate the needs of the homeowner, architect, and engineer.

An important distinction is that between *global*, or total, radiation and the *direct* component. For direct radiation, the amounts made available are relatively easy to interpret, because they are taken from an instrument that presents a surface always perpendicular to the sun (hence, "direct-normal") and because the designer knows his collector will track the sun just as the measuring instrument did.

For global radiation, the instruments measure energy falling onto a *horizontal* surface. This is the information mapped in most meteorology and climatology books, incidentally. Until now the designer of a solar heating system

was obliged to convert these values into the energy amounts expected on a flat collector tilted from the horizontal for a specific application. Now, the SERI data for global radiation is already converted into estimates for each of three different tilt angles. Furthermore, to consider the fact that some collectors cannot use small amounts of energy, SERI presents both global and direct-normal radiation as amounts that exceed certain threshold values.

In the following pages, this atlas explains the geometry and logic of the common collector tilt angles (Fig. 6-1). Then we present maps of *selected data* for the conterminous states, using only the months that are likely to most useful (Figs. 6-2, 6-3, 6-4, 6-5). Radiation for stations in Alaska and Hawaii is presented in Table 6-2, along with a location map. Complete information for each radiation category, based on data for every month of the year as well as annual totals, is available in the source document.

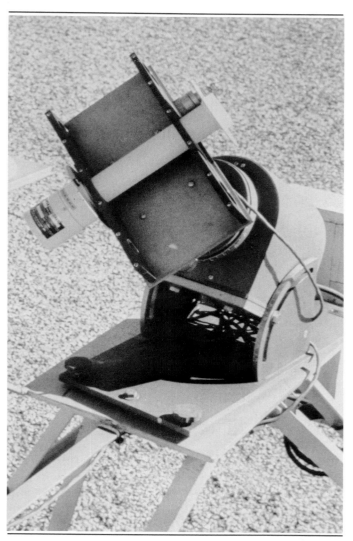

This is an *Epply Normal Incident Pyrheliometer.* Its narrow gathering tube is on a clock-driven mount, so it points directly at the sun and records only direct-beam (focussable) radiation. (Courtesy of Solar Energy Research Institute.)

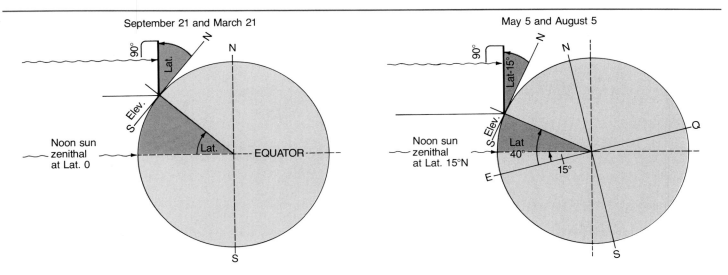

September 21 and March 21

Noon sun zenithal at Lat. 0

On Equinox days the noon sun is zenithal (perpendicular to a horizontal surface) at the Equator. At any other location on those days a collector should be raised by an angle equal to the location's latitude. This suggests a rule that applies to any day of the year: the collector should be raised by an angle equal to the arc between the location and the latitude where noon sun is zenithal.

May 5 and August 5

Noon sun zenithal at Lat. 15°N

On these days, chosen to represent summer months, the zenithal sun has shifted into the Northern Hemisphere to a point near Lat. 15 degrees north. A location at 40 degrees north is, therefore, only 40-15 = 25 degrees removed from the zenithal sun; so the collector should be raised by 25 degrees. As a general rule for summer collection of energy in the mid-latitudes, the collector is raised by an angle equal to the Latitude minus 15 degrees.

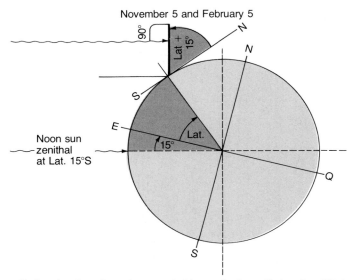

November 5 and February 5

Noon sun zenithal at Lat. 15°S

On these two days, chosen to represent winter months, the zenithal sun has shifted to near 15 degrees south. Any location in the Northern Hemisphere is removed from the zenithal sun by an arc equal to the location's latitude *plus* 15 degrees. In winter the collector is usually raised toward the south by that angle.

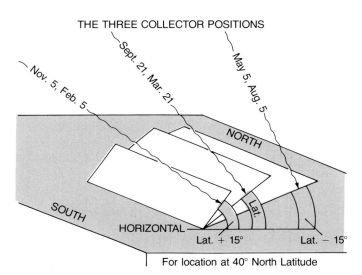

THE THREE COLLECTOR POSITIONS

For location at 40° North Latitude

The three collector positions are illustrated here for a location at 40 degrees north. The most appropriate tilt will depend on how the energy is to be applied. For heating domestic hot water, needed throughout the year, the collector is usually raised to the intermediate position (tilt equals latitude). For winter heating, the choice will be Lat. + 15 degrees; and for energy applied to air conditioning the choice would be Lat. − 15 degrees. As well, the collector can be adjusted periodically during the year to follow the changing sun elevations. If that is done, it should be designed to accommodate the extremes of summer and winter: on June 21, for instance, the ideal tilt would be Lat. − 23½ degrees; and on Dec. 21 it would be Lat. + 23½ degrees.

Fig. 6-1 Tilting a solar collector to increase the energy received. The amount of energy gathered on each square foot of collector surface will be increased if the collector intercepts the sun's rays at an angle that is close to perpendicular. The preferred collector tilt varies for different days of the year.

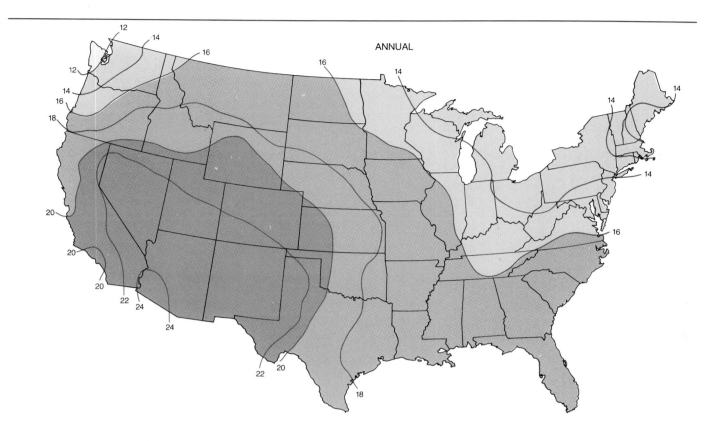

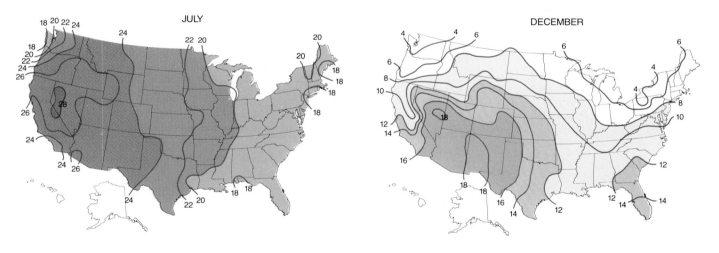

GLOBAL RADIATION

This is an estimate of all solar energy that would be intercepted by a flate collector, such as those used to warm water or air for space heating, or to hold photovoltaic arrays.

On a horizontal surface, global radiation is the combination of *direct solar beam* radiation and *diffuse sky radiation,* both of which have an impact on a flat-plate collector. Instruments measure the global radiation received on a horizontal surface; but that amount can be adjusted to make estimates of energy received on a collector tilted to meet the sun. Such an estimate considers two factors: the benefit of meeting the sun at a more advantageous angle, and also the slight additional energy that is reflected to the collector from the ground. Estimates for a tilted collector, therefore, include direct, diffuse, and reflected radiation. Amounts greater than those mapped could be obtained if a reflector were added to the collector.

For year-round applications, such as domestic water heating, this atlas maps daily totals of the energy expected on a collector whose tilt equals the latitude of the location. Daily radiation is shown on the basis of annual data, and then for the months of July and December which represent, roughly, the periods of greatest and least radiation that must be considered when sizing a collector for year-round use.

Fig. 6–2 Global radiation. Average daily total received on collector with tilt equal to latitude. Units are MJ/m².

Source: DOE/SERI, *Solar Radiation Energy Resource Atlas,* 1981.

Global Radiation For Summer. The energy amounts shown are those expected on a flat collector, tilted up toward the South by an angle equal to the location's latitude minus 15°. The energy may be used in either thermal or photovoltaic applications.

The benefit of tilting the collector to the summertime angle is not dramatic, and it varies from month to month. Comparison of the July daily values on this page with those mapped with the assumption of Tilt=Latitude shows values that are roughly 2 units higher in most parts of the map, that is, the values are 8 to 10 percent greater with the specific seasonal tilt than with the year-round tilt.

Through this period of generally high sun combined with long days in the higher latitudes the pattern of radiation values is *not* dominated by differences in latitude between northern and southern parts of the nation, but shows, instead, the differences in cloudiness across the map. High values are not simply in the South, but in the Southwest. Lows are in the coastal Northwest and the Great Lakes areas. The patterns for July and August demonstrate this most clearly with contour lines whose major orientation is North-South. In July, the area of great radiation is a North-South belt that coincides roughly with Great Plains and Mountain states, while the low areas are in the Northwest and on the East coast.

Fig. 6–3 Global radiation for summer months. Average daily total received on collector with tilt = Lat. −15°. Units are MJ/m².

Source: DOE/SERI, *Solar Radiation Energy Resource Atlas*, 1981.

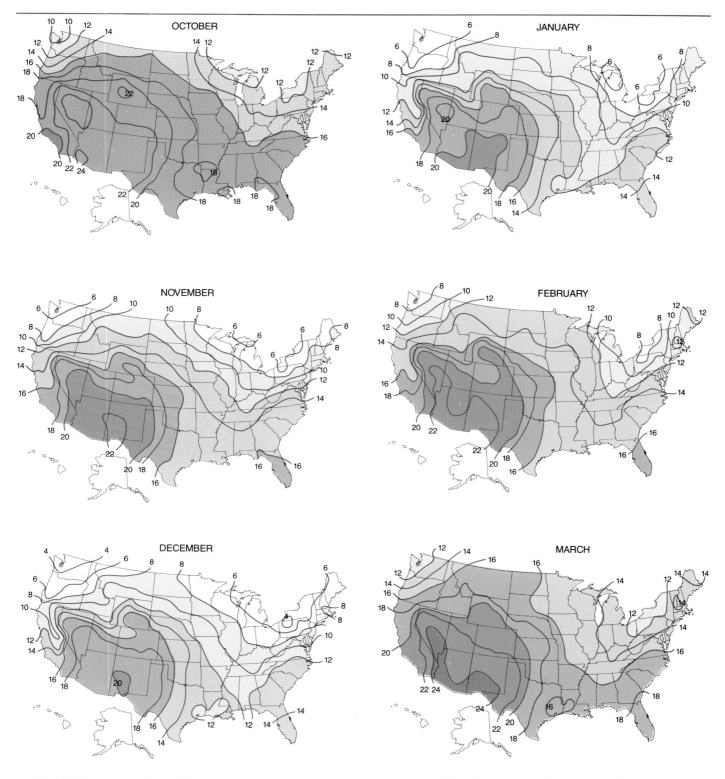

Global Radiation For Winter. The energy amounts shown are those expected on a flat collector tilted up toward the South by an angle equal to the location's latitude plus 15°. The energy may be further increased by using reflectors.

The benefit of tilting the collector to the seasonal angle, as opposed to using the year-round Tilt=Latitude, is more pronounced than in the summer months. In December, for instance, the values mapped here are roughly 2 units greater than those with the assumption of Tilt=Lat. In the northern part of the country that difference amounts to a 20 percent increase.

Through this winter period, the pattern of radiation is strongly affected by differences in latitude, values in Montana being less than half those in New Mexico during the month of December. The difference between clear-sky areas of the Northern Great Plains and the more cloudy Northwest and Great Lakes areas is evident, but less important in these months than in the months of summer.

Fig. 6–4 Global radiation for winter months. Average daily total received on collector with tilt = Lat. +15°. Units are MJ/m².

Source: DOE/SERI, *Solar Radiation Energy Resource Atlas,* 1981.

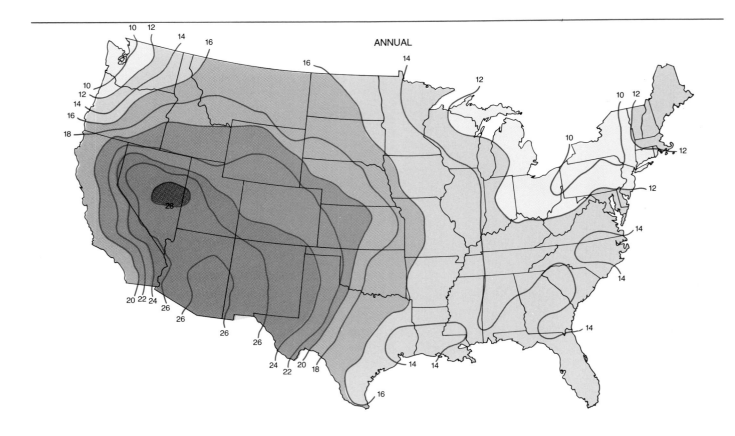

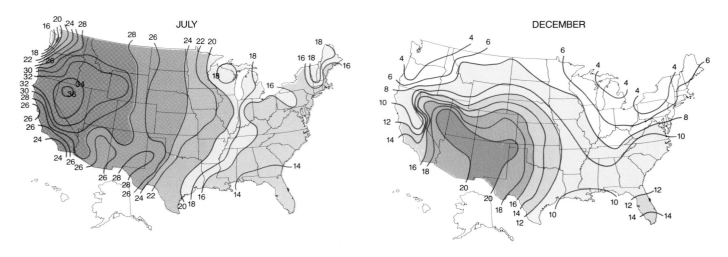

DIRECT-NORMAL RADIATION

This energy is received as parallel rays directly from the sun and can be focussed by lenses or mirrors to attain high temperatures that may be applied to a boiler or to an engine of the Stirling type. The mapped estimates assume the collector, like the measuring instrument, will track the sun continuously through the day.

Because the application of this energy is likely to be for electrical power generation on a *year-round* basis, this atlas shows estimated daily totals of radiation on three maps. On the basis of *Annual* records, the large map provides amounts expected on an average day—so the annual total can be estimated by multiplying by 365. The map based on July records represents the average daily total that can be expected during the period of greatest energy supply. The map based on December records show the average daily total that can be expected during the time of least energy supply.

Fig. 6–5 Direct-normal radiation. Average daily total received on collector that tracks the sun for both elevation and direction. Units are MJ/m^2.

Source: DOE/SERI, *Solar Radiation Energy Resource Atlas*, 1981.

TABLE 6–2
Global and Direct-Normal Radiation for Stations in Alaska and Hawaii. Categories correspond to those on foregoing maps. Units are MJ/m².

	ALASKA						HAWAII		
	Nome	**McGrath**	**Homer**	**Big Delta**	**Yakutat**	**Annette**	**Lihue**	**Honolulu**	**Hilo**
Global Radiation Tilt=Latitude									
Annual	9.7	9.3	10.8	10.6	8.1	9.8	17.9	19.2	16.2
July	14.7	14.0	16.8	16.8	12.1	14.9	19.7	21.1	17.4
Dec.	0.04	0.3	1.4	0.1	1.0	2.5	14.4	15.6	13.4
Global Radiation Tilt=Lat−15°									
April	17.2	16.5	17.0	17.6	13.6	14.8	18.7	20.5	16.3
May	19.0	17.5	18.6	20.1	14.8	17.1	20.5	21.9	17.6
June	19.7	17.2	19.3	19.7	14.5	16.1	21.1	22.5	18.8
July	16.1	15.4	18.4	18.5	13.2	16.3	20.9	22.5	18.4
Aug.	12.3	12.3	14.8	15.9	11.3	14.1	20.6	22.3	18.1
Sept.	10.2	10.1	11.3	12.0	8.5	11.1	20.1	20.9	17.7
Global Radiation Tilt=Lat.+15°									
Oct.	5.9	5.8	8.1	6.4	5.6	6.1	18.3	19.4	16.7
Nov.	1.3	2.5	4.3	2.5	2.9	4.2	15.9	17.7	14.6
Dec.	0.0	0.3	1.4	0.1	1.0	2.5	15.2	16.6	14.1
Jan.	0.5	1.3	2.9	1.1	2.2	3.9	15.5	16.6	15.5
Feb.	5.4	5.9	7.0	5.8	4.8	6.4	16.7	18.0	15.6
Mar.	11.1	12.0	12.4	13.0	9.2	9.9	17.0	18.8	15.5
Direct-Normal Radiation									
Annual	8.8	7.9	9.3	10.0	5.9	7.5	13.8	16.0	11.2
July	13.3	11.6	15.2	16.6	8.2	11.9	15.3	18.2	11.7
Dec.	0.0	0.1	1.0	0.0	0.6	1.8	12.8	14.7	11.2

Source: DOE/SERI, *Solar Radiation Energy Resource Atlas*, 1981.

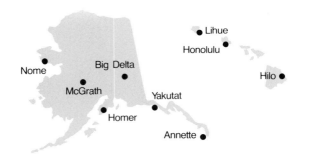

HOT WATER HEATING

Hot water heating and space heating are *low-temperature* applications of solar radiation. As such, they are very logical ways of using the sun's energy because solar radiation is converted to low-temperature heat very efficiently. A solar hot water heater can absorb as heat 50 to 70 percent of the radiation falling on it. In contrast, photoelectric cells convert only about 15 percent of the radiant energy to electrical energy; and photosynthesis is only 1 per-

cent efficient. Substituting solar radiation for fuels in low-temperature applications is wise because it matches the energy source to the task. Fuel oil and natural gas are capable of producing very high temperatures and are wasted on applications such as hot water or space heating.

Heating of domestic (or commercial) hot water is a solar application even more logical than space heating, since the need (or the "load") for domestic hot water is constant, whereas a need for heating exists only during the winter months. This means that for domestic hot water an installation properly sized will be fully used throughout the year and will lead more quickly to fuel savings that will repay the initial investment.

The collector and other elements of an active hot water heating system are sketched in Figure 6–6. Optimum collector size, and hence the cost of the system, depends on a number of factors, including: (1) the demand and the implied size of the storage tank; (2) the radiation expected at the location in question; and (3) the price of fuel otherwise burned to heat water.

A rough guide to the sizing, without regard for optimizing the system in the face of specific fuel costs, is as follows.

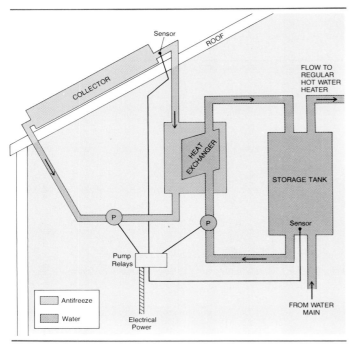

Fig. 6–6 Elements of an active domestic solar hot water system added to a conventional hot water heater.

Assuming there is a backup system (using a fuel), the size of tank need not consider storage for cloudy days, and can be arrived at simply by per capita use. On the basis of 20 gallons per person per day, a family of four could use a system of 80 to 100 gallons capacity. While this factor of need is essentially constant throughout the country, the radiation varies greatly and dictates that a larger collector is needed where fewer BTUs are received per square foot of surface area. Figure 6–7, based on zones implicit in a map of mean daily radiation from annual records suggests how collector size must vary in order to provide enough energy for a system in various parts of the country. The *standard* collector assumed for the left-hand set of multipliers is an active flat-plate collector, with double glazing, and flat black paint. The *high efficiency* collector, that requires smaller collector area for every gallon stored, would track the sun and focus it on to tubes painted with selective black coating to reduce radiation of heat back into the atmosphere.

A much more precise estimate of the system size that is best for a given location, taking into account the anticipated fuel savings, can be obtained by the use of simulations such as SOLCOST, a computer-aided analysis originally offered by the federal government, but now available through consultants who are listed with the Department of Energy's

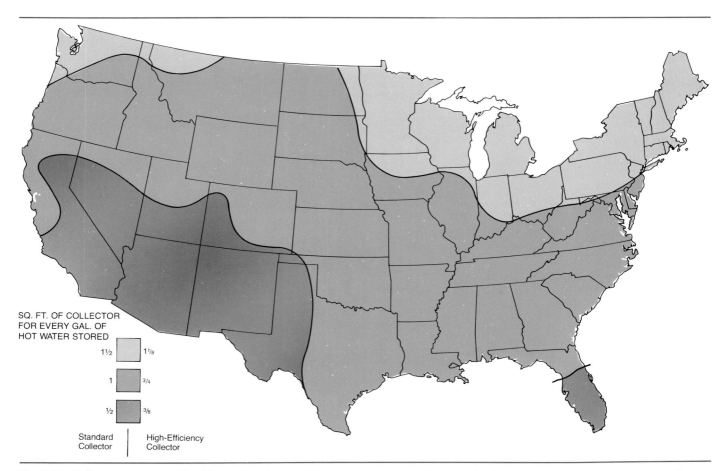

Fig. 6–7 Collector areas required for active domestic hot water systems using standard and high-efficiency collectors.

Source: Ridenour, 1976; personal communication, Ridenour, 1984.

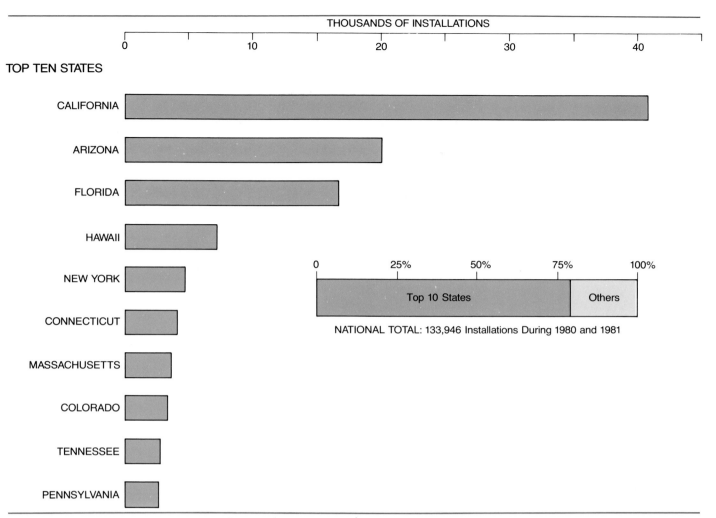

THOUSANDS OF INSTALLATIONS

0   10   20   30   40

TOP TEN STATES

CALIFORNIA

ARIZONA

FLORIDA

HAWAII

NEW YORK

CONNECTICUT

MASSACHUSETTS

COLORADO

TENNESSEE

PENNSYLVANIA

0   25%   50%   75%   100%

Top 10 States | Others

NATIONAL TOTAL: 133,946 Installations During 1980 and 1981

Fig. 6-8 Solar domestic hot water installations during 1980 and 1981, top ten states.
Source: DOE/EIA, *Active Solar Installations Surveys*, 1980 and 1981.
Note: These numbers exclude installations that combine domestic hot water with space heating.

Conservation and Renewable Energy Inquiry and Referral Service.[1]

Figure 6-8 shows that the distribution of recent water-heating installations is very uneven—79 percent of them concentrated in the 10 states illustrated, and 30 percent of them in California. California's large population is partly responsible for that dominance, but its generous tax incentive (discussed later) encourages these and other solar installations.

One installation made *without* benefit of tax incentives is the one at the White House (see photo).

SPACE HEATING AND COOLING

Application of solar energy to space heating and cooling is more interesting and complex than its use for hot water heating: The need for either heating or cooling varies throughout the country, and is seasonal. Also, the periods of greater need and greater supply coincide for cooling, but not for heating. Finally, more energy is used for heating and cooling than for water heating, so there are greater opportunities for saving fuel.

The White House solar hot water system. Installed in April, 1979, on the West Wing of the White House, this system provides hot water mostly for the kitchen of the Staff Mess. The solar panels preheat water before it enters the regular hots water heating tanks and are expected to save about $1,000 dollars in fuel costs every year. Collector area: 611 square feet; Collector tilt: 33 degrees, i.e. Lat. −5°; Heating Fluid: 50 gallons ethylene glycol solution; System Output: 75–130 thousand BTU's per year.

225

A commercial application of solar hot water heating. The 144 panels are on the roof of the Red Star Industrial Service Laundry in Fresno, California. (Courtesy of U.S. Department of Energy.)

The Need for Heating and Cooling

The distribution of *need* is fundamental to any study of heating and cooling, regardless of the energy source, and is expressed in maps of *degree days. Heating degree-days* (Fig. 6-9, Part A) are the total for a whole year, of the daily needs, each of which is derived this way: (65 degrees Fahrenheit) minus (daily mean temperature). For example, a day with a mean temperature of 45 degrees Fahrenheit would contribute 20 degree days to the annual total. In this way the annual or seasonal need for heating is captured in a single number that can be used by fuel suppliers to anticipate customers' needs and by meteorologists to express the severity of a month or a winter. The pattern, considerably simplified in Figure 6-9 Part A, is dominated by greatest values in northern areas and by an intrusion southward of greater values in highland areas, especially the Rocky Mountains and the Sierra Nevada.

Cooling degree-days are derived by a similar procedure which uses the "base" of 65 degrees Fahrenheit, and sums the daily differences between that and the daily mean, that is: (daily mean temperature) minus (65 degrees Fahrenheit). When these values are mapped (Fig. 6-9, Part B) it is not surprising that greater values occur in the South. The pattern is modified, however, by greater values extending northward in the middle of the country where the lack of highland or ocean moderation causes higher summer temperatures.

The Needs and the Solar Supply

How the supply of energy from the sun matches and satisfies the needs for heating and cooling is one of the most important geographic phenomena of our time. Responding

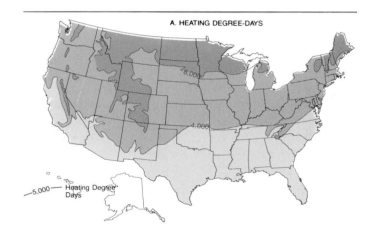

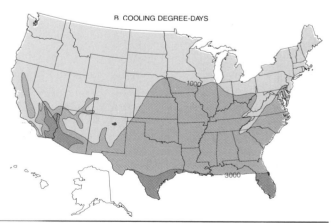

Fig. 6-9 Needs for heating and cooling—Expressed as degree-days based on 65°F (19°C).

Source: U.S. Department of Commerce, *Climates of the United States*, 1973.

226

to the research opportunity offered by the federal government in the past few years, a number of engineers have tackled the problem. Here is the result of one such study by the TRW Systems Application Center in their "Phase Zero" report on impacts of solar heating and cooling (TRW Systems Group, 1974).

HEATING DEMAND AND SUPPLY In winter, it is obvious that areas of greatest heating need are at the same time deficient in solar radiation. Actual conditions within the country, however, are more complex than might be expected, because of certain combinations of demand and supply. Figure 6–10, Part A shows two degree-day values, 2,500 and 5,000 degree days, that divide the country into three belts with regard to heating need. At the same time, two radiation values, 250 Langley and 350 Langley, slice the country into three belts with regard to supply of solar energy *during the heating season*. The combination of the two sets of three belts leads to 9 theoretical region types.

Type 1, in the extreme southeastern and southwestern parts of the country, is where the heating need is least and the supply of solar energy is greatest. At the other extreme is Type 9 in the North, where heating need is greatest and solar supply is least. Type 7 is nonexistent, because it combines very little heating need with very low radiation values. Type 8 is unique to the Northwest where ocean moderation causes a mild winter of intermediate heating need, and at the same time cloudy skies in the winter reduce radiation to the lowest category. Type 6 is unique to the Northern Plains where heating need is great and the solar supply is greater than the latitude would suggest. Both these areas will appear in a later section in the more thorough analysis of heating supply and demand.

COOLING DEMAND AND SUPPLY Using an approach similar to that for heating demand and supply, three zones of cooling demand taken from a cooling degree-days map may be combined with three zones of greater to lesser solar radiation received during the cooling season. Again the theoretically possible combinations are reflected in 9 types in Figure 6–10, Part B.

In summer the cooling need is well-matched to the supply. Logically, the areas of high temperatures tend to be areas of great radiation (Type I) and areas of lower temperatures tend to be areas of lower radiation (Type 9). Clear skies in the West allow the condition of greatest radiation to sweep into northern areas, while in the East the high radiation values are restricted to southern latitudes. Once again, not all theoretical types actually exist, Types 7 and 8 being the unlikely combinations of relatively great cooling demand and lowest radiation.

Although the use of heat for cooling is not new (for example, in the 1940s refrigerators were gas-fired) the equipment for solar-fueled absorption-type air conditioning is nevertheless unsatisfactory. Solar-driven air conditioning requires high-temperature operation which renders collectors inefficient. Concentrating collectors are suited to the

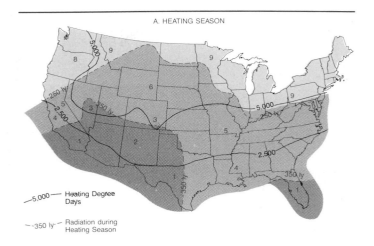

A. HEATING SEASON

—5,000—— Heating Degree Days

— -350 ly- — Radiation during Heating Season

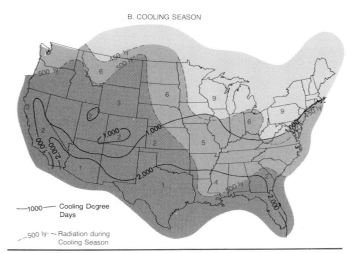

B. COOLING SEASON

—1000—— Cooling Degree Days

— 500 ly- — Radiation during Cooling Season

Fig. 6–10 Heating and cooling needs and solar energy supply during heating and cooling seasons.
Source: TRW Systems Group, 1974.

task, and have been used, but the technology is regarded as immature, and for that reason no thorough geographic analysis of cooling feasibility in various parts of the country is available. A new kind of solar-assisted air conditioner, using *dessicants* may be more practical than the high-temperature type (Lindsley, 1984).

Solar Space Heating and the Effects of Climatic Variation

Solar energy may be applied to heating buildings in a number of different ways. *Passive* solar heating refers to the use of architectural design that properly orients the building and makes deliberate use of south-facing surfaces either to admit or to absorb radiation. The structure itself is used as the solar collector and storage device and can attain a large part of its heating needs this way. *Active* solar heating systems entail a separate collector, a storage device, and

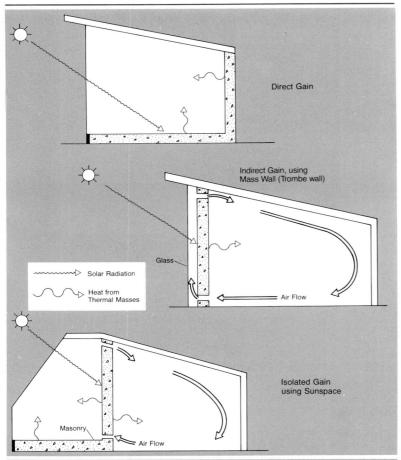

Direct Gain

Indirect Gain, using
Mass Wall (Trombe wall)

Glass

⟶ Solar Radiation

⟶ Heat from
Thermal Masses

Air Flow

Isolated Gain
using Sunspace

Masonry

Air Flow

Direct Gain. Radiation penetrates south-facing windows and is absorbed as heat by masonry floor and wall. These thermal masses store heat then release it overnight.

Indirect Gain. A massive south wall is covered with glass, creating an air space through which air circulates. Cool air from the building enters near the floor, is warmed, and flows out at the top. Vents are closed overnight to prevent circulation of air. The glass reduces radiation loss from the wall overnight when it is giving up its stored heat to the adjacent room.

Isolated Gain. A "sunspace" room that may be used as a greenhouse which collects heat and circulates warm air during the day. Excess heat may be stored in a rock bed (not shown) and drawn upon at night. Glass areas may be covered at night to reduce heat loss. If the sunspace is located downsloped from the building, heat will flow very effectively upward into the building, while cool air drains down, in a convective circulation called *thermosiphon.*

Fig. 6-11 Three types of passive design for space heating.

controls linked to pumps or fans that draw heat from storage when it is available. Systems that are dominantly passive, but make use of fans to move air are often called "hybrids." In severe climates, both passive and active systems are backed up by a fuel-burning heater of some kind.

Passive Solar Heating

Passive designed buildings can achieve great fuel savings with relatively small investment and little or no complicated apparatus. It can be argued, in fact, that *every* building should be oriented and built with solar radiation and other climatic factors as paramount considerations in design.

Buildings can intercept and store solar energy in three basic ways: direct gain, indirect gain, or isolated gain (Fig. 6-11). A building may employ more than one of the

three strategies. In *direct gain*, the sun enters through a south-facing window and strikes "thermal masses" deliberately placed to absorb radiation. These usually are a heavy concrete floor, and sometimes masonry walls, fireplaces, or water stored in drums. Thermal masses store the heat, then release it at night. *Indirect gain* is typified by a glazed masonry wall. *Isolated gain* uses a south room as a collector and a heavy masonry wall to absorb heat and protect the dwelling space from wide swings of temperature.

The home illustrated here (Fig. 6-12), like most solar homes, conserves heat by thorough insulation of walls and roof and by reducing window area on the north wall. Partly because the heating need is reduced, a very substantial portion of it can be met by the sun alone. Table 6-3 shows that a number of passive homes, in a variety of climates, have obtained 65 to 95 percent of their heating needs without burning any fuel.

In the mid-1980s, there has been a strong movement away from active systems toward passive solar design because of its simplicity and reliability. While fans may be used to move heat to and from storage, thus making the system *hybrid* by some definitions, most new solar buildings gather the sun's energy in the building's structure, not on collectors attached to the building.

Despite that trend, this atlas explores the geographic patterns of solar heating possibilities by study of *active systems*, because the costs of adding-on such a system can be readily specified and balanced against energy savings in various parts of the nation. Therefore it is possible to design a system that is optimal in net savings, while seeing how climatic factors (heating needs and solar energy supply) dictate that the installations be of different sizes in different parts of the country.

Active Systems, Their Components, and the Method of Anticipating Performance

The schematic, Figure 6-13, shows how an active solar system works. The collector feeds heat through an exchanger into a storage tank from which heat is drawn for domestic hot water and for the major task of heating the building. After a sunny period hot water will be available in storage, and the system will draw upon it for heating. Overnight, or during a cloudy period, stored heat may be exhausted, requiring the system to draw heat from the auxiliary furnace. Active systems may, in fact, be used in many localities in combination with *heat pumps* to aid those heat-transfer devices when outdoor temperatures are too low for their efficient operation (Gilmore, 1978). The following analyses, however, assume active solar heating is backed up by a fuel-burning heater. Collectors use either hot air or hot water as the medium for moving and storing heat. While the following materials are derived from studies that specify hot water, systems using hot air are not very different in their economic feasibility (Duffie, Beckman, and Dekker, 1976).

Predicting the performance of a solar space-heating

Designed by architect Nels Larson of Bryn Mawr, Pa. and built by Glenn Garis, Inc., of Souderton, this residence for the Fedde family uses *direct gain*. The two-story glass sunspace on the south side of the building is accessible to residents but is separated from the living spaces by a 14-inch concrete wall that soaks up sun and then radiates heat to every major room.

In winter, hot air is drawn down from the attic level to a space under the first floor. There, its heat is stored in water containers and allowed to rise through the floor.

In summer, outside air enters the sunspace and is exhausted by an attic fan, which also draws fresh air into the house through windows.

Over a period of 3 years the home has been monitored and has consumed an average of $40. of electricity per YEAR for space heating. Roughly 97 percent of the building's heating needs are met by the sun.

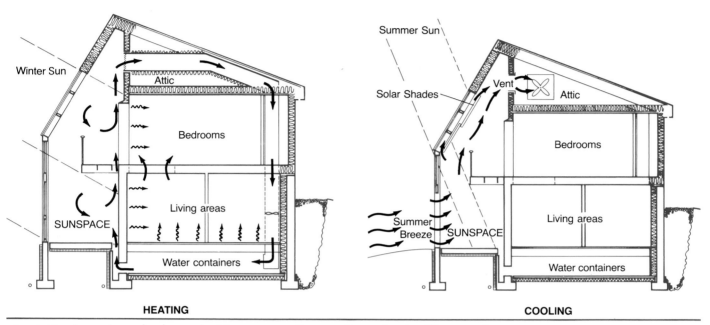

Fig. 6–12 A passive solar home in Montgomery County, Pennsylvania, (Photo and sketches courtesy of Nels Larson.)

TABLE 6–3
Selected Buildings of Passive Solar Design, and the Proportion of Heating Needs Gained from the Sun

BUILDING	HEATING DEGREE-DAYS	SOLAR PROPORTION OF HEATING ENERGY NEEDS (Percent)	HEATING STRATEGY
Sea Ranch Sundown, 100 miles north of San Francisco, California	2,969	95	Direct gain
Kelbaugh residence, Princeton, New Jersey	5,100	75	Indirect gain (mass wall)
Whitcomb residence Los Alamos, New Mexico	6,500	30	Indirect gain (water mass wall)
The Rural Center, Northern California	7,520	100[1]	Indirect gain (roof pond)
Erwin residence Nacogdoches, Texas	1,500	75–90	Isolated gain (sunspace)
P. Davis residence, Albuquerque, New Mexico	4,348	75	Isolated gain (thermosiphon)
Crowther residence Denver, Colorado	5,500	65	Indirect gain

[1] Tolerating 55° F indoors
Source: Department of Housing and Urban Development, Nov. 1978.

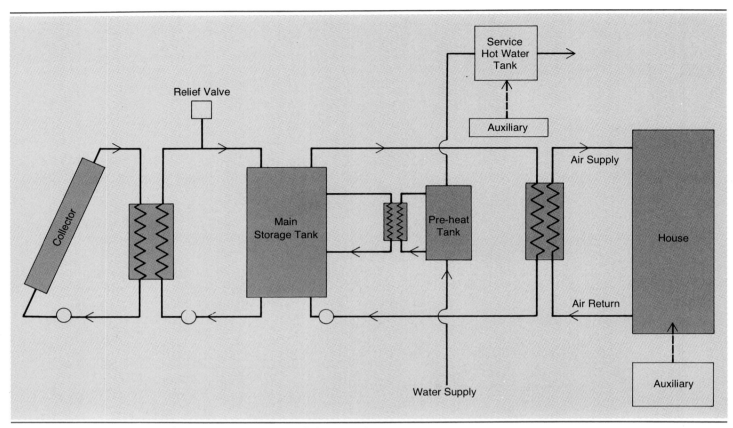

Fig. 6–13 An active system for space heating and domestic hot water.

system in a variety of climates is complicated, because both heat demand and radiation per square unit of collector vary throughout the country. An important goal in such a study is an optimal design—one in which the collector and associated storage are large enough to meet the winter's demands, but not so large as to make the installation uneconomic. It would be possible to build a collector and storage large enough to warm the building on the coldest day of the year. The system would be unnecessarily large, however, for heating on most winter days. Of course, rising fuel prices will encourage the development of larger solar installations which allow greater fuel savings, and which would then offset the cost of installation and maintenance.

Evaluating the performance of existing active space-heating systems can provide some guidance for the design of new systems. This is only a starting point, however, and will not yield principles that have application in any area of the country. What is needed is a procedure for *simulating* the performance of a solar heating system as it responds to the temperatures and the radiation that characterize a number of sites across the country. The following are the main factors in such an analysis.

DEMAND ESTIMATES A building's demand for heat can be expressed by a single figure called a *heat loss factor*. This figure is arrived at by conventional heating and ventilation formulas which consider areas of floors, walls, and windows, along with their materials and the degree of insulation. Heat loss for a building of a certain size may be expressed as *heat needed per degree-day*. Thus, a given mean temperature for a day implies a certain heat demanded by the building. The question of what climatic data best *represents* a locality for the sake of such simulation work is vital, and a number of different studies have used different approaches to this problem.

SUPPLY ESTIMATES Solar radiation records provide the amount of solar energy per square unit of tilted collector. With knowledge of the collector's tendency to reflect some radiation and to radiate heat to the environment, an estimate can be made of how much heat will be gathered and stored for every square unit of collector. Various collector sizes may be tried in a simulation to see how the heat gained matches the heat needs for the building in question.

SOME IMPORTANT MEASURES RELATED TO SYSTEM PERFORMANCE As the hypothetical heating system performs in response to climate data that drive the simulation, a number of quantities can be noted.

1. *Solar Fraction or Solar Dependency.* This is the proportion of the building's heating load supplied by the sun. Usually it is reported as an average for the heating season, though it

will be a greater proportion in the mild portions of the season and a smaller proportion in severely cold weeks. Solar fraction will rise as collector area increases, but, as suggested, it is not necessarily economical to strive for very high solar fraction, since the increased cost of a larger solar installation may not be offset by fuel savings.

2. *Fuel Savings.* For every heat unit supplied by the solar system, some fuel is not burned. The total value of such fuels is the fuel savings. These may be considered *gross* fuel savings or *gross savings* to avoid confusion with an economic criterion, *savings*, which is used elsewhere.

3. *Critical Fuel Cost.* Peculiar to a location and its climate, this is the price level to which competing fuels must rise before a solar installation of specified cost is likely to be economically attractive.

4. *Optimum Collector Area.* As progressively larger collectors and associated storage are used, either in practice or in simulation, the (gross) fuel savings increase along with the cost of the system. The collector size which appears to maximize the *net savings* is the optimum.

Designs for Various Climates

There have not been many comprehensive studies of how the many variables in hypothetical solar heating systems interact in various parts of the country. This lack exists largely because of the massive calculation task required to follow a system through a heating season, while experimenting with alternative designs.

Two earlier studies made an impact (Löf and Tybout, 1970, 1974), but the definitive analysis was made by Professors Duffie, Beckman, and associates at the Solar Energy Laboratory, University of Wisconsin, Madison. While processing results from a large number of performance simulations based on *half-hourly* radiation and temperature data, they discovered an empirical relationship that makes possible the estimation of system performance from *monthly data* (Klein, Beckman, and Duffie, 1975). A full explanation of their procedures will not be attempted here. Their paper, however, clarifies how to modify local radiation data to consider the tilted collector surface, and it shows how that information along with collector characteristics and heating load for a particular building—all derived from *monthly data*—may be entered into a simple graph, now widely known as the *F-Chart*, in order to learn the *solar fraction* for a collector of a given size. The solar fraction implies a certain amount of fuel saved and that can be balanced against installation and other costs. A series of ever-larger systems (collectors) can then be tested in the simulation in order to arrive at the optimum size.

The F-Chart shortcut has made possible the hypothetical testing of various systems at a large number of stations—leading, therefore, to a national picture of variations due to climatic differences. Most studies of solar system feasibility made since the F-Chart appeared in 1976 have made use of

it, though often coupling it with an economic analysis different from that used by Duffie and associates.

THE ECONOMIC CRITERION   To determine economically optimal system size Duffie and associates used the single value, *average annual savings*, which is the total fuel savings in a year minus the annual cost of the system. Fuel savings is simply the amount of fuel which would have otherwise been used multiplied by its price (assumed to escalate over time). The annual cost of the system is calculated to be the total installation cost multiplied by a single factor called the *annual payment rate*. A payment rate of 10 percent, which is the one used in generating the results mapped here, could represent a loan of 5.57 percent for 15 years, at 7.75 percent for 20 years, or at 10.31 percent for 30 years.

ANNUAL SAVINGS AND FUEL PRICES   Figure 6-14 shows (for Madison, Wisconsin) how annual savings

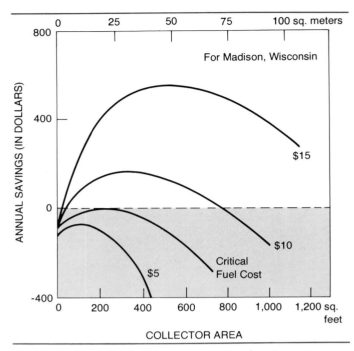

Fig. 6-14 Annual savings in an active space heating system are maximized at different collector sizes for four different fuel costs, installation costs held constant.
Source: Duffie, Beckman, and Dekker, 1976.

increase with larger collector area—up to a point—and how the savings are greater and smaller according to the price of fuel saved.

In any study of a system at a particular site, local fuel prices are used. For the sake of the national study, however, *fuel prices were held constant* at future prices for all locations. The price used was $10 per million BTUs. Considering the conversion losses for gas and oil furnaces, this price for BTUs delivered in the heating system corresponds to 3.4 cents per Kilowatt-hour for electricity, 98 cents per gallon for fuel oil, and $7.25 per thousand cubic feet for natural gas.

An important principle is evident in Figure 6-14: For higher fuel prices, the optimal collector size is larger. This means that any system built to optimal size at today's fuel prices will be undersized as fuel prices rise. Some studies have built in an annual escalation of fuel prices, but the approach in this study was to choose a single value that represented average price of fuel energy delivered over the period under consideration.

The critical fuel cost is labeled in Figure 6-14. This is the fuel price (price and cost are used interchangeably here) that allows a system just to break even. At fuel prices lower than critical, even the optimized system yields annual savings that are negative.

GEOGRAPHIC VARIATIONS IN DESIGN AND PERFORMANCE FACTORS

Collector area. Figure 6-14 indicates that at the $10 price for fuel, savings are maximized with a collector of 30 square

meters (323 square feet) in Madison, Wisconsin. Similar analysis at 86 other stations led to optimal collector areas throughout the United States (Fig. 6-15). (The square symbols on that map are *not* made with areas proportional to collector areas.) Largest collectors (over 400 square feet in area) are appropriate in the Northern Plains; moderately large (300 to 400 square feet) are indicated in northern and northeastern states; and small systems are indicated in the extreme South.

Solar Fraction. At each location, the series of trials with ever-larger collectors led to an increase in solar fraction, as shown for Madison in Figure 6-16. When the savings equation indicates optimal collector size, the corresponding solar fraction is identified; at Madison, for instance, it is 50 percent. At the other locations it is the value mapped in Figure 6-17.

Plains and Mountain states and extreme southern areas show fractions of greater than 60 and 70 percent, while systems in the Pacific Northwest and Middle Atlantic states carry only 20 to 40 percent of the season's heating load.

Annual Savings. The value of fuels saved, diminished by the annual cost of the installation, yields annual savings— the quantity maximized in successive trials of system size, and the major criterion used here to show relative economic feasibility. For optimal systems based on assumed system cost and assumed fuel cost, annual savings are the amounts mapped in Figure 6-18.

The savings pattern shows a very clear maximum in the Northern Plains and Rocky Mountain states where a simplified approach to heating need and solar supply (Fig. 6-10, Part A) revealed a region combining substantial heating need and radiation amounts exceptionally high for the latitude (Type 6). In addition, savings would be relatively high in Maine. *The substantial heating need* is a key factor, because part of the installation costs are for the "fixed" items which do not produce solar BTUs. In an installation with a small collector (for a small heating need), these fixed costs are large in relation to the amounts of fuel saved.

Savings in Figure 6-18 are calculated on the basis of collectors costing $110 per square meter, and competing fuels delivering energy at $10 per million BTUs. Figure 6-19 shows how savings would rise if fuel costs increased to $15 per million BTUs, and also shows how savings would be "shifted down" if collectors were to cost $150 per square meter rather than $110.

Factors Promoting the Feasibility of Solar Heating Systems

Three separate factors would tend to make active solar heating systems competitive: falling costs of installation combined with rising fuel prices, tax breaks that would reduce the costs of installation, and low-cost loans for the solar investment.

COLLECTOR AREA
IN SQ. FEET

◻ UNDER 100

◻ 100–199

◻ 200–299

◻ 300–399

◻ 400–450

Fig. 6–15 Optimal flat-plate collector areas vary with climatic factors. Areas shown are valid only if certain costs are assumed.
Source: Duffie, Beckman, and Dekker, 1976.

INSTALLATION COSTS AND OTHER ASSUMPTIONS IN THE UNIVERSITY OF WISCONSIN STUDY

The following list presents the specific costs assumed in the study at the Solar Energy Laboratory at the University of Wisconsin (Duffie, Beckman, and Dekker, 1976).

Fixed Costs   These are costs of the parts of the installation that are independent of collector size, and are inevitable in every system. They include piping, pumps, controls and valves, exchangers, and the labor to install them. The amount used was $1,000 (which by 1978 standards was low).

Variable Costs   Costs that vary with the size of a system were as follows:

1. **Collector.** The type of collector used was a flat plate, with a single glass cover and a selective coating to reduce long-wave radiation. The price used was $110 per square meter ($10.22 square foot) which was lower than the $15 to $20 per square foot being quoted in the summer of 1978, and anticipated prices falling as a result of mass production.

2. **Storage cost** was estimated at $15 for every square meter of collector ($1.39 per square foot of collector area)—roughly the cost being quoted in 1978.

Other Assumptions of the Study   The hypothetical building to be heated had a floor area of 150 square meters (1,615 square feet) and a wall area of 120 square meters (292 square feet). It was insulated to meet ASHRAE standards, which are different in different parts of the nation; therefore no single heat-loss factor can be stated for the building. Domestic hot water heating load was assumed to be 30 liters (79 gallons) per day which had to be from an inlet temperature of 11 degrees Celsius (53 degrees Fahrenheit). Operation and maintenance costs were ignored, and the study assumed no increase in property tax due to the installation, no income tax credit for interest paid on the loan, and no government investment tax credit or other incentive to the buyer.

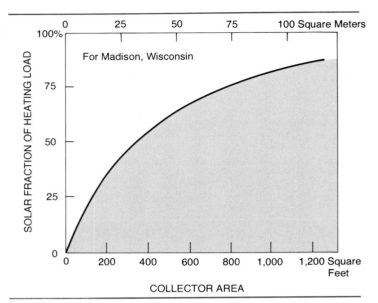

Fig. 6-16 Solar fraction of heating load varies with collector area, for Madison, Wisconsin.
Source: Duffie, Beckman, and Dekker, 1976.

FALLING INSTALLATION COSTS

It is often assumed that mass production of solar collectors will markedly reduce their cost. Recent experience calls into question that prospect. The rise in fuel prices, also assumed in the conventional wisdom of late 1970s, has *certainly* not occurred without interruption. The world oil glut that kept fuel prices low through the period 1981–1984 may well be temporary and would surely end with the first crisis that closed the Straits of Hormuz. Nevertheless, worldwide efforts at conservation have contributed to the reduced demand that caused the glut. Further conservation measures and use of alternative energy sources will tend to keep oil demand low and perpetuate the lowered fuel prices.

TAX BREAKS FOR THE PURCHASER

There are both federal and state-level tax incentives that, since the National Energy Bill in 1978, have made solar installations much more attractive economically. In fact, some analyses show that without these incentives solar heating could not have competed with even the most expensive alternative,

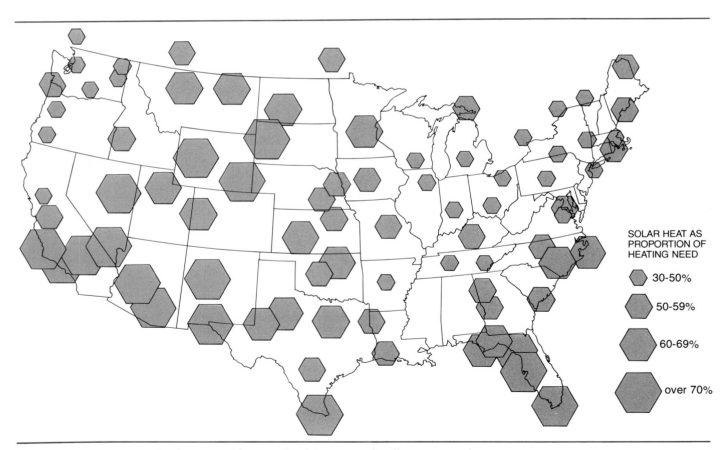

Fig. 6-17 Variation in solar fraction of heating load for optimal collector size. These fractions are valid only if certain costs are assumed.
Source: Duffie, Beckman, and Dekker, 1976.

Fig. 6–18 Variation in annual savings for optimal collector size. Actual savings depend on installation cost and fuel cost.
Source: Duffie, Beckman, and Dekker, 1976.

electrical resistance heating, in some areas of the nation (see Bezdek, 1978; also Bezdek, Hirshberg, and Babcock, 1979).

The federal government has allowed individuals and businessmen to make use of the Investment Tax Credit

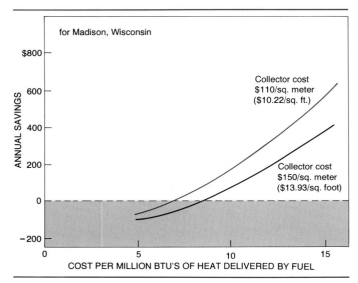

Fig. 6–19 Annual savings vary with fuel costs and collector costs.
Source: Duffie, Beckman, and Dekker, 1976.

whereby a given proportion of an installation cost can be subtracted directly from the income tax liability. This device is potent, especially since a homeowner may subtract 40 percent of the first 10,000 dollars of installation costs, while a business may subtract up to 15 percent, depending on the useful life of the property. The solar industry rightly regards this federal tax credit as a vital incentive (see Johnson, 1984) and views with alarm the fact that these terms applied only through the tax year 1984. It is not yet certain whether there will be a renewal of the credit for the tax year 1985 and beyond.

A number of states also provide substantial incentives—related to state income taxes. Figure 6–20 shows the 29 states offering tax credits as of April 1984. The terms and the limits of these incentives vary greatly from state to state: California, is exceptional, with a 50 percent tax credit for residential installations. Most states offer 20 or 25 percent and are more generous for residential than for commercial projects (terms for each of the 29 states can be found in the Malloy article in *Solar Age*, June 1984.)

LOW-COST LOANS The interest paid on a loan for a solar installation on either a new or an old building is important to the total installation cost. An analysis for the Joint Economic Committee of the U.S. Congress studied the effects of both interest rates and future fuel costs on the feasibility of solar heating (Schulze, Shaul, Balcomb, *et al.*,

A commercial application of active solar hot water for space heating, at the East Norristown branch of the First Pennsylvania Bank near Philadelphia. (Courtesy of U.S. Department of Energy.)

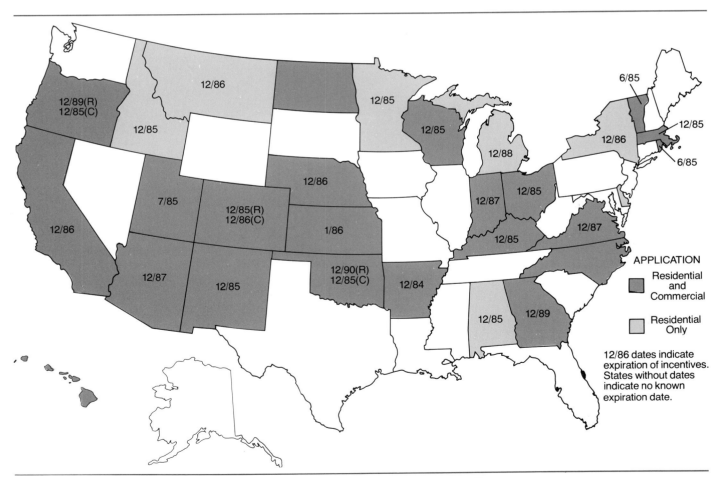

Fig. 6–20 States offering income tax incentives for solar water and space heating installations, as of April 1984.
Source: Malloy, 1984.

1977). Differing interest rates were found to be of crucial importance to either slowing or promoting the spread of solar heating in various parts of the country. It was apparent that either rising fuel prices or lowered interest rates would promote solar installations. However, since higher fuel prices have greater impact on low-income families, the study recommends government intervention to provide low-interest loans.

Economic Feasibility in Various Areas

It was pointed out earlier that fuel costs must be at a certain level or else even an optimized system will not yield positive savings. That level—the *critical fuel cost*—varies across the country with changes in temperatures and solar radiation, and can be calculated for any location by using the relevant climatic data and assumptions about the fixed and variable costs of an installation. Because those costs change from year to year, a map of critical fuel costs is soon obsolete. The patterns on the map, however, since they depend on climatic factors, are a useful summary of the *relative* suitability of various parts of the nation.

Figure 6-21 shows those patterns, generated from installation costs for the year 1978. The positions of these boundaries are arbitrary; nevertheless the areas of relative suitability are valid. Greatest feasibility is in those clear-sky areas where demand is high enough. Lowest feasibility is generally where the radiation is inadequate or the demand is too low to justify investment in an active space-heating system. Miami and Brownsville, for instance, need so little heat that fuel displaced by solar would not be great enough. Coastal Washington and Oregon are deficient in radiation, especially during the heating season. The belt of low feasibility that joins Louisiana to Michigan is due, in its southern part, to low demand and, in its northern part, to reduced insolation during the heating season, as can be seen in Figure 6-4.

For current feasibility, it is necessary to calculate critical fuel costs for a specific location, and then compare with local cost of the fuel being considered as backup. Another option is to undertake an optimization analysis, using F-chart logic. This can be accompanied by a projection of fuel savings expected and the number of years to pay back. As of 1984, this sort of analysis was no longer available through the U.S. Department of Energy, but could be obtained from consultants listed with the *Conservation and Renewable Energy Inquiry and Referral Service.*[1]

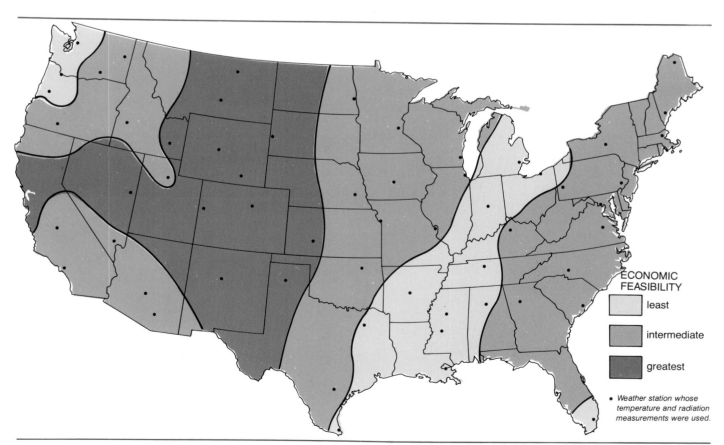

ECONOMIC FEASIBILITY

least

intermediate

greatest

• Weather station whose temperature and radiation measurements were used.

Fig. 6-21 Relative feasibility of active space heating, on the basis of critical fuel costs. Actual fuel costs assumed uniform across country.
Source: Calculations based on Duffie, Beckman, and Brandemuehl, 1978.

Distribution of Active Installations

The Department of Energy has conducted two surveys of firms that install solar heating apparatus. A 1981 survey gathered data on installations pre-1980 and during 1980. The second survey, conducted in 1982, gathered data on installations made during 1981.

When the numbers of installations for all three periods and for all applications are taken together (Fig. 6–22) a number of vital facts emerge. First, about the same numbers of systems were installed in each of the time periods. Second, the majority of installations by far, 64 per cent, were for heating hot water alone, while only 14 percent were for space heating. Third, the top ten states (see Fig. 6–22) accounted for 71 percent of all the installations.

The numbers per state show California, Florida, and Arizona as the leaders. These numbers are influenced by state populations, by the proportion of the population living in single-family dwellings (where most installations tend to be made), and by policies that encourage individuals to invest. The map in Figure 6–22 accounts for the population factor by converting the numbers into installations per thousand persons. On this basis, Arizona and Hawaii are most prone to undertake active solar installations, while Florida and Wyoming fall into the second category. California is not exceptional on this measure. States with fewest per capita are clustered in eastern, and especially southeastern, states.

Installations that provide space heating are only 14 percent of the total for the years 1980 and 1981, according to this Department of Energy survey. Because the survey was discontinued, no comparable data for later years is available, but information from a survey of *collector shipments* (Department of Energy/Energy Information Administration, June 1984) suggests that in the years 1982 and 1983, roughly 12 percent (by area) of all collectors shipped were destined for space-heating applications.

The distribution of active installations for space heating in 1980 and 1981 is not logical (Fig. 6–23). Michigan is the leading state, while none of its more populous neighbors

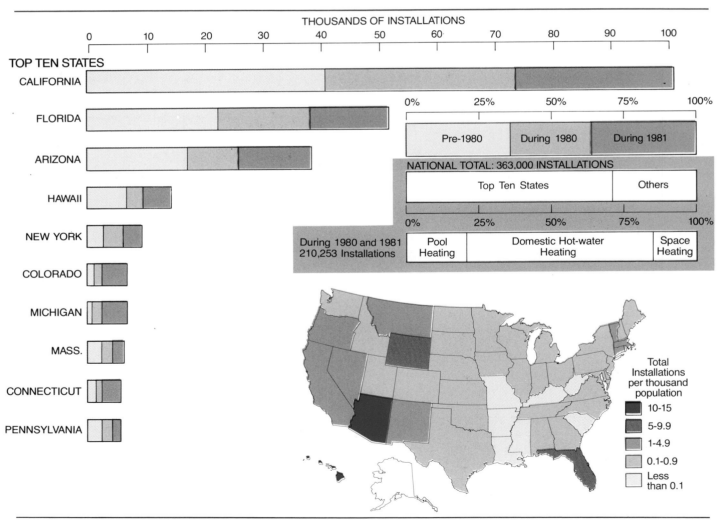

Fig. 6–22   Active solar installations for all applications—pre-1980, during 1980, and during 1981.

Source: DOE/EIA, *Active Solar Installations Surveys,* 1980 and 1981.

Suntree Apartment Complex, in Davis, California, showing how a solar housing complex is oriented toward the sun. Each of the 95 units makes use of solar-heated water for bathing and laundry and for space heating.

The City of Davis is a community ahead of its time that promotes energy conservation and the use of solar energy through various programs including building codes that specify orientation, insulation, window areas, heat loss and gain, and shading of south-facing windows. (Courtesy of City of Davis, California.)

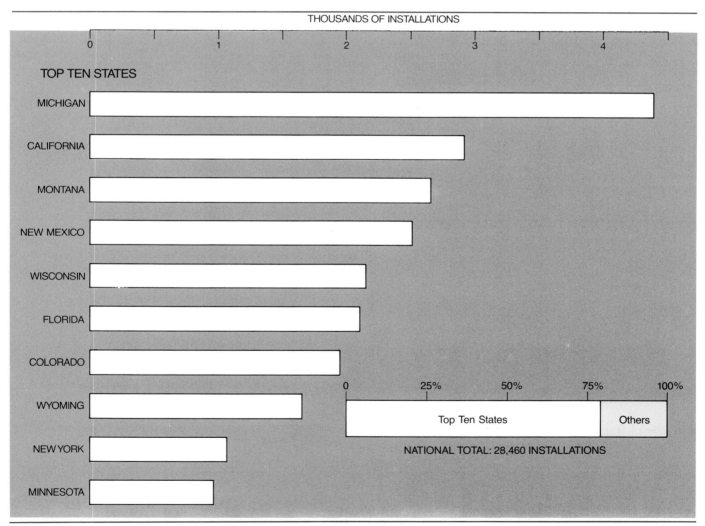

Fig. 6–23 Active solar installations for space heating and combined water and space heating during 1980 and 1981, top ten states.

Source: DOE/EIA, *Active Solar Installations Surveys*, 1980 and 1981.

comes close; a 30 percent state tax credit, with a generous 5,000 dollar maximum credit may be responsible (Malloy, 1984). Montana, however, figures in the top ten states mapped, despite a small population and a very stingy state tax policy.

ELECTRICAL GENERATION

Generation of electrical power can be accomplished by two quite different manifestations of solar energy. One is the ability of the sun to heat the surface of a collecting device. Schemes that use this are *thermal* systems. The other, and quite different trait of solar energy (and other radiant energy) is its ability to dislodge electrons from certain prepared materials. This is the *photovoltaic* effect.

Thermal Systems

There is surprising variety among thermal systems for generating electrical power, but most of them entail high temperatures and require, therefore, the *concentration* of sun's rays onto the collecting surface by either mirrors or lenses. All such applications can utilize only the *direct,* focusable, portion of insolation, so the relevant maps for raw resource are those in Figure 6-5, Average daily total of direct-normal radiation. Those amounts assume the collector, like the measuring instrument, is tracking the sun and intercepting the sun's rays as directly as possible. Energy amounts on the map are expressed as Megajoules per square meter of collecting surface, which can be converted into Kilowatt hours per square meter by multiplying by 2.228×10^{-1}. The resulting value will be Kilowatt hours *thermal*. If all the energy losses incurred when gathering and transporting the heat, then converting it into electrical energy are taken into account, then an estimate of collecting area required to do a certain generating job can be made.

The maps in Figure 6-5 show greatest amounts of direct-normal radiation in the clear-sky desert areas of Arizona and Nevada. Southern California enjoys only slightly less of the resource, and it is there that some important developments have taken place.

THE POWER TOWER In the United States, the solar thermal-electric device to progress most rapidly is the type referred to as *central-receiver* because the sun's rays are gathered by a large expanse of mirrors and focused on a central tower where they heat fluid to high temperatures for steam generation. The concept was tested at SANDIA Labs, in Albuquerque, New Mexico, and now has been developed in a pilot plant near Barstow, California (see Fig. 6-24). The plant, with generating capacity of 10 Megawatts, is a cooperative effort between the Department of Energy and a utility, Southern California Edison—along with the Los Angeles Department of Water and Power and the California Energy Commission.

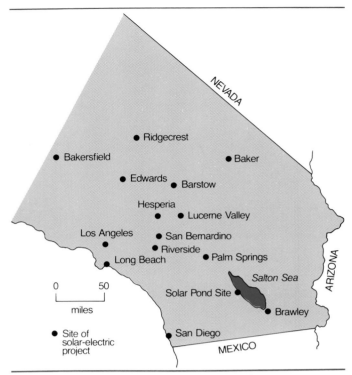

Fig. 6-24 Locations of solar-electric projects in southern California.

At this plant, called "Solar One," steam is made in the receiver unit and used directly to drive the turbine to generate power. Any excess steam (during shutdown of the generator) can be routed to an exchanger where it heats oil to be stored in an insulated tank. Later, that oil can be pumped through another exchanger where it boils water for steam (see Fig. 6-25 and photo).

The first generation of electrical power for the commercial grid was in April 1982. The end of test and evaluation phase for the plant was July 1984, and the start of continuous power production was in August 1984.

Plans are underway to follow this 10 Megawatt pilot plant with a 100 Megawatt commercial demonstration plant at Lucerne Valley, 40 miles south of Barstow, California. The utility expects this project to be on-line late in 1988 (Southern California Edison press release, Feb. 16, 1984).

THE PARABOLIC TROUGH This is an example of a *distributed receiver* system in which the conversion of radiation to high temperatures is accomplished, not at a central location, but on extended surfaces that are distributed over a large area in some way.

Troughs with piping laid along their axes were used in the 1920s to boil water that drove steam engines. The same idea can be used to drive a generator's turbine: On a small scale, power is generated like this to run irrigation pumps at Coolidge, Arizona, for example.

An application at the *utility scale* has now been undertaken at Daggett, California (near Barstow) by

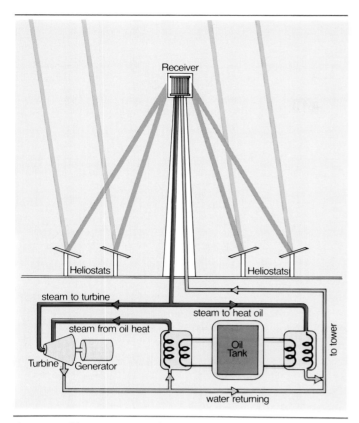

Fig. 6–25 The operation of *Solar One*, a central receiver solar-thermal power plant.
Source: Southern California Edison Company, 1984.

Southern California Edison. The plant started operation in 1984 with a generating capacity of 13 MW and will be expanded eventually to a capacity of 43 MW. The pipes that run along the axes of troughs carry a heat-absorbing oil that reaches high temperatures and is used to boil the working fluid that drives generator turbines (see photo).

Solar One, near Barstow, 75 miles northeast of Los Angeles, is the first commercial plant to generate electricity by solar thermal methods. The array of mirrors (heliostats) focusses the sun onto the boiler atop the central tower. At its daily peak power the plant can generate at 10 Megawatts. (Courtesy of Southern California Edison Company.)

Parabolic troughs at the Daggett, California 13 MW generating station. A special oil in the piping is heated to high temperatures and used to boil the fluid that drives turbines. (Courtesy of Southern California Edison Company.)

Parabolic dish solar-electric generator installed at Rancho Mirage (near Palm Springs, California). At the focal point of the 36-foot diameter mirror is the self-contained generator unit. (Courtesy of Southern California Edison Company.)

THE PARABOLIC DISH AND STIRLING ENGINE
This is a significant departure from other solar-thermal devices because it *does not use a turbine*. Radiation falling onto the 36-foot diameter dish is focused on to the receiver unit (see photo) where its heat is applied to hydrogen gas which expands to drive the *Stirling* engine coupled to a generator. The engine runs without combustion, using and re-using the hydrogen as a working gas. Since the generator is driven directly, without the boil-and-condense cycle typical of most solar-thermal devices, the overall efficiency can be as high as 27 percent, whereas a turbine-driven generator converts only around 15 percent of solar energy into electrical.

If a large group of these dishes were arranged to feed power into a common line, then the array could be an example of a *distributed receiver* system. As of now, a single unit, sponsored by the Department of Energy and Southern California Edison, is providing 25 Kilowatts of electrical power to residents of Rancho Mirage, near Palm Springs, California. Earlier, an experimental unit was installed at Edwards Air Force Base. Another, operated by a different utility, functions at Shenandoah, Georgia.

THE SOLAR POND IDEA This scheme makes use of intermediate-temperature water held in very large ponds and warmed by the sun without benefit of mirrors or other concentrating devices. It is akin, therefore, to a flat-plate

collector, because it makes use of direct *and* diffuse solar radiation.

Water at the bottom of the pond is heated by radiation striking the bottom surface. In any normal body of water, this would lead to rising of the bottom water and convective mixing, so the whole mass of water would be warmed. Ponds are used this way to provide 500,000 gallons of domestic hot water per day at Fort Benning, Georgia.

For power generation, higher temperatures are needed. For this purpose, saline water is essential—dictating that solar-electric ponds be located near natural sources of salt. If the bottom layers of water are rich in salt, they will be more dense and *will not* rise buoyantly, even though strongly heated. In this way, brine at temperatures up to 200 degrees Fahrenheit can be produced—hot enough to boil a special working fluid, such as ammonia, which drives the turbine. An upper layer of fresh water remains cool enough to serve as cooling water to condense the working fluid on the downstream side of the turbine (Fig. 6-26). One significant feature of this method of generating power from the sun is the great thermal mass in the hot brine: It is enough to run the power plant through the night. Solar ponds used for generating electricity are, in principle, very much like the devices that exploit natural thermal gradients in tropical oceans (see OTEC under Ocean Energy).

Such generating units have been operating successfully near the Dead Sea in Israel since 1975. The largest Israeli unit, with capacity of 5 Megawatts, started in 1983 just as its contractor signed an agreement with Southern California Edison to build a very large unit on the west shore of the Salton Sea in southern California (see Fig. 6-24). This plant will have a generating capacity of 48 Megawatts.

Solar pond generating station in Israel. Nets cover the pond surface to reduce wind-mixing. Large pipes draw off hot bottom water that is used to boil the working fluid that drives turbines.

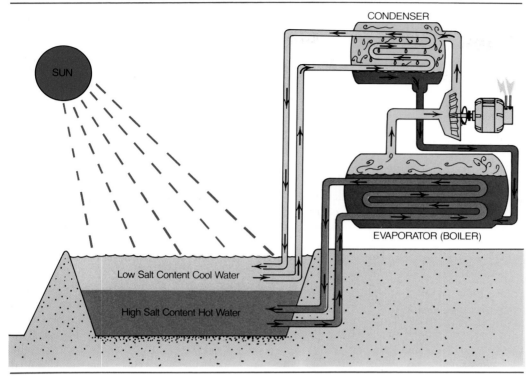

CONDENSER

EVAPORATOR (BOILER)

Low Salt Content Cool Water

High Salt Content Hot Water

SUN

Fig. 6-26 The operation of a solar pond for generating electrical power.
Source: Southern California Edison Company, 1984.

Photovoltaic Power Generation

In photovoltaic systems, solar radiation is converted directly into electricity. Although there are several ways of doing this, solar cells, or photovoltaic cells, appear to be the most promising (see diagram).

These cells can make use of both *direct and diffuse* radiation; therefore, the relevant resource maps are those for *global radiation*, as in Figure 6-2. That map assumes collector tilt = Latitude, which is appropriate for year-round gathering of energy for electrical power. The energy amounts on those maps, expressed as Megajoules per square meter per day, can be converted to Kilowatt hours per day, then diminished to around 12 percent of that value to ascertain how much electrical output can be obtained for every square meter that is covered with cells and exposed to the sun.

That 12 percent efficiency is one of the traits researchers are striving to improve, for if it were possible to double the efficiency, only half the number of cells would have to be used. That saving would be vital, because the cells are still very expensive.

When photovoltaic cells were first used beyond the laboratory (in the *Vanguard I* space satellite in 1958) the cost was $600 per watt of power delivered under peak conditions. In 1973, it was $100 per watt; in 1977, $22 per watt; in 1984, around $7. A further decline to around $1 per peak watt will be necessary before photovoltaic power plants are competitive with fuel-burning power plants.

Photovoltaic power does compete with conventional power supplied by a utility when the application is in some remote area, and especially if the application demands an electrical device that is independent of power lines. A good example of a remote, but prosaic, installation is the federally sponsored 3.5 Kilowatt system at the village of Schuchuli, Arizona, on the Papago Indian reservation. Running transmission lines to the village would have cost hundreds of thousands of dollars, so the alternative for power would have been a generator run on diesel fuel or gasoline. More exotic examples are emergency telephones in desert areas, marine navigation markers or foghorns on buoys. These would have to depend upon batteries frequently checked and replaced, so a photovoltaic power source would be economical in the long run, though the initial cost is high.

RESIDENTIAL APPLICATIONS The roof area of most houses is large enough to accommodate a photovoltaic array that will power the home. The U.S. Department of Energy is now testing photovoltaic arrays for domestic use at three Residential Experiment Stations—in Las Cruces, New Mexico, Cape Canaveral, Florida, and Concord, Massachusetts. Different installations being tested in Cape Canaveral range in peak generating capacity from 2.9 to 4.6 Kilowatts. Residential use of photovoltaic power has progressed beyond such experimental stations into homes which are lived in. Many of these homes were built as federally funded development projects, but some, such as one in Carlisle, Massachusetts (see photo), was built and sold without

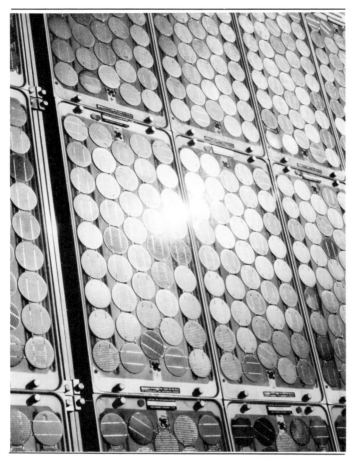

An array of photovoltaic cells. Each is 2 to 3 inches in diameter and has a peak output of roughly 0.25 watts. (Courtesy of U.S. Department of Energy.)

OPERATION OF A SILICON PHOTOVOLTAIC CELL

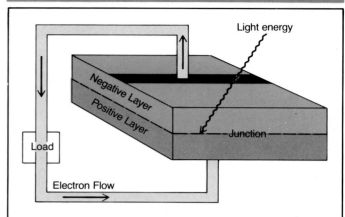

Unlike *photoconductive* cells (electric eyes) which require batteries for their operation, photovoltaic cells actually generate a flow of electricity.

Purified silicon is treated to form two layers which are placed in contact with one another. The negative layer is formed from silicon that has been deliberately contaminated (doped) with an element such as phosphorus. The adjacent positive layer is made from silicon doped with an element such as boron. Because the *electron energy levels* are different in these two doped strata, electrons will not flow across the junction between them. But, when light energy strikes the junction, it dislodges electrons which will flow from the negative to the positive layer if electrical contacts have been provided.

Since light must travel through to the junction, the silicon layers must be very thin wafers, produced either by sawing a mass of silicon or by growing very thin cyrstals or ribbons. Advances in the production of such ribbons or wafers are most important to making photovoltaic cells more economical.

The country's first all-solar residence in Carlisle, Massachusetts, generates its own electricity from a roof-top array of photovoltaic panels.

The solar electric system produces 7.5 peak kilowatts of electricity under bright sun resulting in an annual surplus of electricity for the home which the owner sells to local utility.

In addition to the solar electric system, the home, designed by Solar Design Associates of Lincoln, Massachusetts, incorporates solar domestic water heating and state-of-the-art passive solar and energy conservation features. (Courtesy of Solar Design Associates.)

government aid. With the price of cells near $7 per watt, an array large enough to meet most of the home's electrical needs will cost over 20,000 dollars, so the photovoltaic addition is not economically attractive by itself.

UTILITY APPLICATION A number of intermediate-sized installations, with capacities ranging from 30 to 225 Kilowatts, have been built at colleges, national park visitor centers, and air fields—usually with some funding from the U.S. Department of Energy.

Ventures of larger scale, and especially those undertaken without government subsidy, are unusual. A striking example is the 1 Megawatt (1,000 Kilowatts) project at Hesperia, southern California (see photo). This was built by ARCO SOLAR, under an agreement with the purchaser of the power, Southern California Edison. For the sake of comparison, the photo caption suggests how much fenced area would be required if such an installation were to have generating capacity of 1,000 Megawatts. In fact, utility photovoltaic installations may continue to be relatively small, and placed near the demand, or else used to supplement fuel-burning plants.

Photovoltaic power generation is ideal for placement at

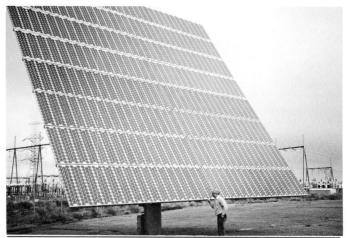

Photovoltiac power installation near Hesperia, in Southern California, built and operated by ARCO Solar. It began operation in January of 1983. There are 108 pedestal-mounted panels, which track the sun, and together have a peak generating capacity of 1 Megawatt (1,000 Kilowatts).

The fenced area is about 20 acres, or 300 yards on a side. If it were simply *scaled-up* to match the 1,000 Megawatt capacity of a large fuel-burning plant, the land area required would be, theoretically, 31.25 square miles, i.e., an area 5.6 miles square. (Courtesy of Southern California Edison Company.)

One of the units at the Hesperia site. Every two minutes, roughly, the panel adjusts its position according to a programmed sequence determined by the height and compass direction of the sun on each day of the year. (Photo by Brian H. Cuff.)

or near the site of power consumption. The systems are nonpolluting, and they can be scaled to various sizes for different consumers. Another attractive aspect may develop as *concentrating* photovoltaic arrays are further refined. These use lenses to focus the sun on to cells, thus making better use of their potential. To avoid overheating the cells, there must be cooling system which could provide *hot water* to the building being served with electricity.

Despite this appeal of small installations at the location where power is needed, the greatest development work (as measured by the generating capacity of photovoltaic modules manufactured and shipped) in 1983 was central power stations (Table 6–4A). A single commercial utility-scale project can strongly bias the figures in any one year,

however, because it will consume an extremely large number of cells: Nearly one million were used in the ARCO SOLAR project at Hesperia, California. During the year 1982, the balance was different, because 47 percent of all cells were for stand-alone systems (water pumping, for instance) while only 44 percent were for central power stations. When the photovoltaic units shipped are divided according to *market sector*, rather than type of end-use (Table 6–4B) the year-to-year change in emphasis is evident. Still, commercial and industrial sectors are dominant in most years. The totals of generating capacity manufactured and shipped show very strong growth through the 3-year period, rising from 2,000 Kilowatts in 1981 to 12,000 Kilowatts (12 Megawatts) in 1983.

SUMMARY OF CALIFORNIA PROJECTS In California alone, there are numerous solar-electric projects underway, though the total generating capacity is not very large, Table 6–5 shows that, as of December 31, 1983, there were 28 projects. Of these, 24 involve arrangements with a contractor who will supply power to the utility (such as the photovoltaic plant built and operated by ARCO SOLAR for Southern California Edison), while 4 projects are owned by utilities themselves. A total of 180 Megawatts of generating capacity is anticipated when these projects all are completed. At the end of 1983, 13 Megawatts of generating capacity was in operation from solar-thermal and photo-

This photovoltaic generating system, rated at 225 Kilowatts in peak sun, was built in 1982, near Sky Harbor International Airport, Phoenix, Arizona, to supply power for Arizona Public Service Company. Each of the sun-tracking arrays of photovoltaic cells is covered by fresnel lenses to *concentrate* the sun's energy 35 times. To obtain the same electrical power with normal photovoltaic cell arrays, the installation would have to be 35 times as large. In this installation, excess heat is dissipated from the back of each cell through an aluminum plate cooled by simple air convection. (Courtesy of U.S. Department of Energy. Additional information supplied by Mr. Joe McGuirk, project manager.)

TABLE 6–4
Photovoltaic Modules Manufactured in the United States,
1981–1983, Showing End-Uses and Market Sectors

A. END-USES, 1982 and 1983
PROPORTION IN VARIOUS END-USES (PERCENT)

YEAR	Central Power	Stand-Alone	Specialty	Utility-Connected Residential	Utility-Connected Intermediate
1982	44.1	47.3	1.3	0.7	6.6
1983	69.7	22.1	4.8	3.3	0.1

B. MARKET SECTORS, 1981, 1982, 1983
GENERATING CAPACITY IN VARIOUS MARKET SECTORS (KILOWATTS PEAK POWER)

YEAR	NUMBER OF MANUFAC-TURERS	Resi-dential	Commer-cial	Indus-trial	Agri-cultural	Other[1]	Total
1981	16	311.2	656.9	1,514.3	76.5	247.2	2,806.1
1982	19	826.8	3,482.7	1,648.8	219.2	719.4	6,897.0
1983	18	207.7	10,555.3	593.5	109.9	1,153.9	12,620.3

[1] Used mostly by government

Source: (A) Taken from graphics in DOE/EIA, *Solar Collector Manufacturing Activity, 1982* and *1983*; (B) DOE/EIA, *Solar Collector Manufacturing Activity, 1983.* June, 1984.

TABLE 6–5
Solar Thermal and Photovoltaic Electric Utility Projects
Planned or Built in California as of December 31, 1983

TYPE OF OWNERSHIP	NUMBER OF PROJECTS	CAPACITY IN MEGAWATTS
Contractor Signed Contracts	21	164
Letters of Intent	3	6
Utility-owned	4	10
Total	28	180
In Operation		13

Source: Papay, Lawrence T. "Status and Outlook for Solar and Wind Systems," March 1984.

voltaic utility projects in California. Of this, 11 Megawatts belonged to Southern California Edison Company.

THE FUTURE ROLE OF SOLAR ENERGY

It is possible to estimate the contribution of solar energy on the basis of three factors: (1) the projected total demand in various sectors of the economy; (2) the expected rise in fuel prices; and (3) the government incentives which will make solar energy competitive. Since fuel prices and the suitability of regions to solar applications vary throughout the country, solar energy will play different roles in different parts of the country. To be consistent, one projection is used here as the major source of information.

The projection reported here was made in 1977 by the METREK division of the Mitre Corporation under contract to the Energy Research and Development Administration (ERDA), now the U.S. Department of Energy (METREK, 1978). Two scenarios are employed throughout the analysis: one based on policies specified in the National Energy Plan as proposed in April 1977; the other based on recent trends in energy prices, policies and technology. The National Energy Plan scenario presented here consistently leads to greater solar contributions than the recent trends scenario because of various government incentives. Both projections are inaccurate to the extent that Presidential and Congressional initiatives and new technical information have altered policy and have made inappropriate some assumptions in the mathematical model used to generate the two cardinal outputs: the extent of market penetration (proportion of need served by solar) and the amounts of solar energy expected to be delivered.

Assumptions of the National Energy Plan proposal are as follows.

- Tax credits for solar investments in residential, commercial, and industrial applications
- Federal demonstration projects in residential and commercial sectors
- Gas and oil prohibited in new utility electric installations
- Oil and gas burning restricted in commercial buildings
- Escalation of fuel prices
- Taxes on oil and gas used by industrial sector

The assumptions that *have not materialized* in recent

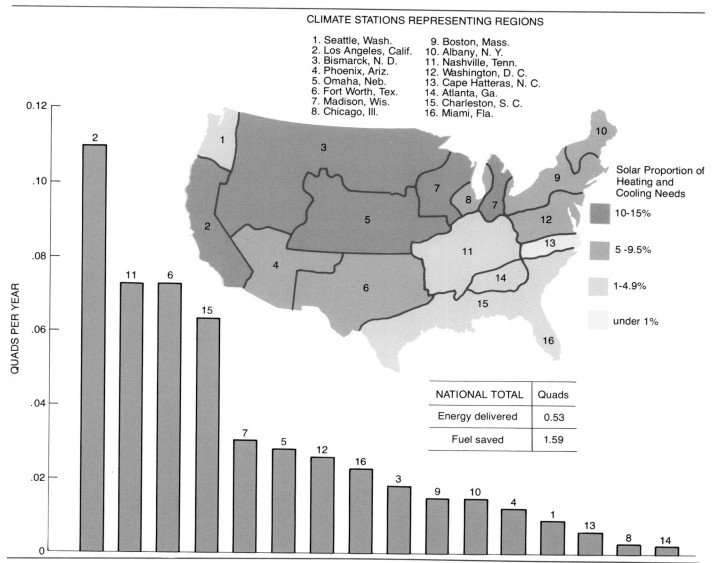

CLIMATE STATIONS REPRESENTING REGIONS

1. Seattle, Wash.
2. Los Angeles, Calif.
3. Bismarck, N. D.
4. Phoenix, Ariz.
5. Omaha, Neb.
6. Fort Worth, Tex.
7. Madison, Wis.
8. Chicago, Ill.

9. Boston, Mass.
10. Albany, N. Y.
11. Nashville, Tenn.
12. Washington, D. C.
13. Cape Hatteras, N. C.
14. Atlanta, Ga.
15. Charleston, S. C.
16. Miami, Fla.

Solar Proportion of
Heating and
Cooling Needs

10-15%

5 -9.5%

1-4.9%

under 1%

NATIONAL TOTAL	Quads
Energy delivered	0.53
Fuel saved	1.59

Fig. 6–27 Solar energy for buildings in year 2000. Estimates of percent market penetration and Quads of energy delivered through waterd and space heating, and space cooling. Solar competing with electricity only.
Source: METREK, 1978.

years are those associated with fuels. Prices have not, in fact, continued to escalate without interruption, and the federal government has not imposed taxes to discourage use of oil and gas in commercial buildings. Nevertheless, the patterns illustrated here are useful because they reveal how the solar alternatives would blossom in various regions, if a single-minded promotional policy *were* adopted in Washington.

The Future of Hot Water Heating, Space Heating and Space Cooling

In these applications, referred to by METREK as the "buildings sector," solar energy is allowed to compete only with electrical energy (the role of heat pumps is not specified). In the regional and national totals shown below for the year 2000, residential installations are expected to be about three times the number of commercial installations; retrofit are roughly 1.5 times the number of new installations; and about 75 percent of all installations are expected to be for hot water heating.

Figure 6–27 assumes 16 regions, each of which has some climatic "uniformity" and is represented by only one climatic station. The map shows projected *market penetration* percentages, the highest of which is 13.2 percent for Region 5, represented by Omaha, Nebraska, and the lowest is 0.5 percent for Region 13 represented by Cape Hatteras. These proportions when applied to demand estimates lead to the *projected energy delivered*, which is greatest in Region 2 where a high penetration rate combines with a large

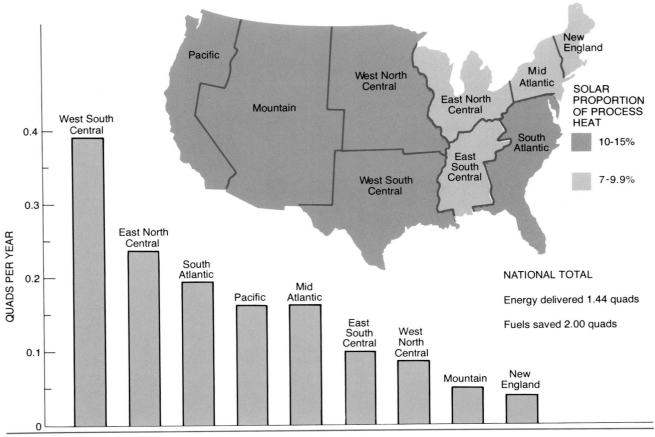

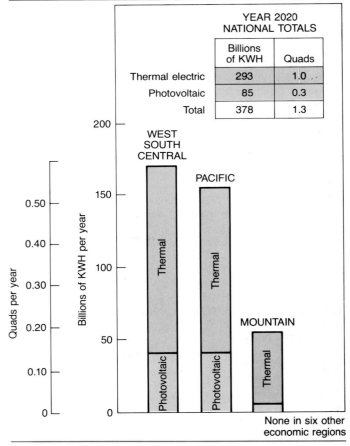

Fig. 6–28 Solar energy for process heat in year 2000. Estimates of percent market penetration and Quads of energy delivered.
Source: METREK, 1978.

population and, consequently, a large demand. The regional energy amounts and the national total of 0.53 Quads are all estimates of *annual* amounts delivered.

The Future of Agricultural and Industrial Process Heat

Air or water heated by the sun can be used for a variety of agricultural and industrial applications. Hot air is needed to dry tobacco, fruit crops, and lumber. Hot water is needed for drying textiles, for laundries, and for food processing operations. Altogether, heat supplied for such processes is referred to as agricultural and industrial process heat. As in domestic applications, the extent to which solar energy will displace fuels depends upon fuel prices.

Fuel, oil, natural gas, and coal are the alternatives considered in this analysis. Low-temperature solar applications are expected to displace fuels early in the scenario, but high-temperature applications—using concentrating collectors—constitute the majority of solar installations by the year 2000.

Figure 6–28 shows the market penetration of solar energy for process heat to be generally much higher than for the buildings sector. In addition, its distribution is more uniform throughout the nine economic regions: The highest penetration is 14.5 percent in the Pacific region, while the lowest, 7.2 percent, is in the East North Central. In estimated amounts of solar energy delivered, the West South Central

Fig. 6–29 Solar electric in year 2020, estimated annual output from thermal and photovoltaic installations. Only three economic regions are expected to produce.
Source: METREK, 1978.

region clearly leads, by virtue of a 12.2 percent penetration and a substantial total demand expected. The expected national total of energy delivered annually, 1.44 Quads, is almost three times that anticipated in the buildings sector.

The Future of Electrical Generation By Utilities

The four solar-related contributors to future electrical generation are radiation (thermal and photovoltaic), wind power, ocean thermal gradients, and biomass. Hydroelectric power is excluded simply because it is rather thoroughly developed. Projected electrical contributions from the four new sources and their combined market penetration are shown in the final chapter; here, only the projected thermal and photovoltaic output are considered.

Electrical generation from radiation is considered in the METREK projection only in the economic regions which embrace substantial areas of desert, West South Central, Pacific, and Mountain (Fig. 6-29). Expected output

is roughly the same in the first two regions, and less than half that amount in the Mountain region where total demand is smaller. In all three cases, solar-thermal generation is expected to overshadow photovoltaic for utility application and will be used largely as fuel savers, that is, not as independent installations but auxiliary to fuel-burning power plants. The (annual) national total electrical output from thermal and photovoltaic together *in the year 2020* is thought to be 378 billion Kilowatt-hours, or 1.3 Quads. This is roughly the same as solar energy to be delivered as process heat in the year 2000 (Fig. 6-28), but the fuel saved will be greater because electrical generation rejects more fuel energy in conversion than does the direct burning of fuels for process heat.

NOTE

1. This service was formerly the National Solar Heating and Cooling Information Center, Telephone 1-800-523-2929; in Pennsylvania, 1-800-462-4983; in Alaska and Hawaii, 1-800-523-4700.

References

Bezdek, Roger H. *An Analysis of the Current Economic Feasibility of Solar Water and Space Heating,* Washington, DC: U.S. DOE, Division of Solar Application, Jan. 1978.

Bezdek, Roger H., Alan S. Hirshberg, and William H. Babcock. "Economic Feasibility of Solar Water and Space Heating." *Science,* 203, March 1979, 1214-1220.

Duffie, J. A., W. A. Beckman, and M. J. Brandemuehl "A Parametric Study of Critical Fuel Costs for Solar Heating in North America." Paper obtained from authors, Solar Energy Lab. University of Wisconsin, Madison, 1978.

Duffie, J. A., W. A. Beckman, and J. R. Dekker "Solar Heating in the United States." Paper presented at winter Annual Meeting Amer. Soc. Mechanical Engineers, New York City, Dec. 5, 1976.

Gilmore, C. P. "Solar Assisted Heat Pumps." *Popular Science,* May, 1978, 86-90.

Johnson, Wayne C. "The 40% Solution: Habit-forming but Still Vital to Solar," *Solar Engineering and Contracting,* March-April, 1984.

Klein, S. A., W. A. Beckman, and J. A. Duffie. "A Design Procedure for Solar Heating Systems," Paper presented at meeting of International Solar Energy Society, July 1975.

Lindsley, E. F. "Solar Air Conditioners," *Popular Science,* July 1984, 64.

Löf, G. O. G., and R. A. Tybout. "Solar Energy Heating." *Natural Resources Journal,* 10, 1970.

———. "The Design and Cost of Optimized Systems for Residential Heating and Cooling by Solar Energy." *Solar Energy,* 16:9, 1974.

Malloy, Molly. "The Solar Age of 1984 State Tax Credit Survey," *Solar Age,* June 1984, 24-26.

Metrek. *Solar Energy: A Comparative Analysis to the Year 2000.* Prepared for ERDA under contract No. E-(4818)-2322, MITRE Technical Report MTR-7579. The Mitre Corporation, Metrek Division, McLean, Va., 1978.

Papay, Lawrence T. "Status and Outlook for Solar and Wind

Systems." Paper presented at Electric Power Research Institute Conference on Solar and Wind Power, San Diego, Cal., March 14-16, 1984.

Pryde, Philip R. *Nonconventional Energy Resources,* New York, Wiley-Interscience Series, 1983.

Ridenour, Steve. "Solar Water Heaters." In Eccli, Eugene, ed. *Low-Cost Energy-Efficient Shelter.* Emmaus, Pa.: Rodale Press, 1976.

Schulze, William D., B. Shaul, J. D. Balcomb, *et al.* *The Economics of Solar Home Heating.* Report prepared for the Joint Economic Committee of the U.S. Congress, Jan. 1977.

Southern California Edison Company. Press release, Feb. 16, 1984.

TRW Systems Group. *Solar Heating and Cooling of Buildings (Phase Zero).* Vol. 1. Executive summary, prepared for NSF-RANN, Washington, DC, May 31, 1974.

U.S. Department of Commerce. *Climates of the United States.* Washington, DC: Government Printing Office, 1973.

U.S. Department of Energy/Energy Information Administration. *1980 Active Solar Installations Survey.* Washington, DC: Government Printing Office, Oct. 1982.

U.S. Department of Energy/Energy Information Administration. *1981 Active Solar Installations Survey.* Washington, DC: Government Printing Office, Dec. 1982.

U.S. Department of Energy/Energy Information Administration. *Solar Collector Manufacturing Activity, 1983.* Washington, DC: Government Printing Office, June 1984.

U.S. Department of Energy/Solar Energy Research Institute. *Solar Radiation Energy Resource Atlas.* Washington, DC: Government Printing Office, Oct. 1981.

U.S. Department of Energy/Energy Information Administration. *Solar Collector Manufacturing Activity, 1982.* Washington, DC: Government Printing Office, May 1983.

U.S. Department of Housing and Urban Development. *Regional Guidelines for Building Passive Energy Conserving Homes.* HUD-PDR-355. Washington: Government Printing Office, Nov. 1978.

7 Wind Power

Windmills first appeared around 200 B.C. in Persia, and possibly as long ago as 1700 B.C. in Babylon and China. The use of windmills persisted in Europe during the Middle Ages, and by the fourteenth century the Dutch had used them extensively for pumping water during the reclamation of Rhine delta polders. In the sixteenth century windmills drove sawmills and paper mills in Europe, and by the middle of the nineteenth century there were an estimated 9,000 being used for various purposes in the Netherlands. With the introduction of the steam engine, reliance on wind diminished. In the Netherlands, for example, the number of operating windmills dropped from 9,000 to 2,500 by the year 1900, and to less than 1,000 by 1960 (Eldridge, 1975, Reed, Maydew, and Blackwell, 1974).

In the United States, large numbers of small windmills have been used for pumping water, for powering sawmills, and for generating electricity in later decades. In addition to numerous small electric generators at rural sites, one giant windmill of 175-foot blade diameter, the Smith-Putnam machine near Rutland, Vermont, fed power into the network of the Central Vermont Public Service Company during the period 1941 to 1945. An estimate of wind's contribution to total United States' energy needs (Fig. 7–1) shows the sudden decline around 1860 when steam engines became widely used. Another drop occurred around 1940 as the Rural Electrification Administration introduced to many areas cheap electrical power generated by fossil-fueled and hydroelectric plants.

Ironically, wind power was being abandoned at the beginning of the post-World War II period, which saw an immense increase in absolute and per capita use of energy in various forms—especially electrical. Now, with recurrent fuel shortages and rising electrical rates, wind power is being rediscovered. Already there is a resurgence in the use of small generators on farms and residences, and a growing industry providing wind power to utility companies.

THE CHARACTER OF WIND

Wind offers ready-made mechanical energy because nature has converted the sun's energy into moving masses of air. As an energy resource, wind has many advantages: it is

◄Wind farms invade cattle country in the Tehachapi Mountains of Southern California. These fourbladed horizontal-axis machines are each rated at 50 KW. At full output, each could meet the needs of 15 to 20 homes. (Courtesy of Southern California Edison.)

251

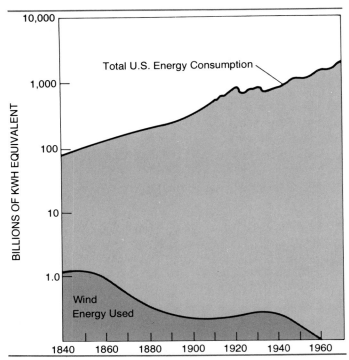

Fig. 7–1 Contributions of wind to U.S. energy needs, 1840–1970.
Source: Eldridge, 1975.

Power Available Calculation

Since wind energy is the kinetic energy of moving air, it is proportional to air density, wind speed, the volume of air moving, and hence proportional to the cross-sectional area of flow. Thus, Power Available, P_A (per unit of time, t) may be derived as follows (Reed, 1975, p. 6):

$$P_A = \frac{\text{kinetic energy}}{t} = \frac{m\,V^2}{2t} = 1/2\,\rho AV^3$$

in which
$\rho =$ air density
$m =$ mass of air
$R =$ flow rate
$V =$ wind speed
$A =$ cross-sectional area perpendicular to flow direction

renewable, nonpolluting, and in some areas quite reliable. Unfortunately, it has daily and seasonal periods of lesser intensity; and, like the sun's radiation, it is rather diffuse, so that large installations are needed to gather enough energy to generate massive amounts of electrical power.

Winds flow in response to differences in atmospheric pressure, and those differences are due to unequal heating of the earth by the sun. Most relevant to the strength of wind is the rate of change of pressure, or the *pressure gradient* as revealed by the spacing of *isobars,* the lines that join points assumed to have equal pressure. In winter, because of strong temperature contrast the isobars are more closely spaced, indicating steeper gradients (Fig. 7–2). Winter winds, therefore, tend to be stronger than summer winds.

Most desirable, of course, are winds that are relatively strong, and at the same time, steady—as indicated by meteorological records. Just how such records of wind speeds can be used to estimate the potential of an area is discussed under the *Power Available* section following. The matter of how much of that energy can be extracted from the wind to turn the shaft of a turbine is dealt with later under the *Power Extractable* section. Following that, we present the highlights of a remarkable study that set out to estimate the maximum wind-electric potential in the United States (land areas only) and also to project how long it might take to saturate all suitable areas with wind machines if this were judged to be a workable strategy. Finally, we review the latest developments in the federal government's program to develop large wind machines and the developments in the private sector, where progress has been rapid.

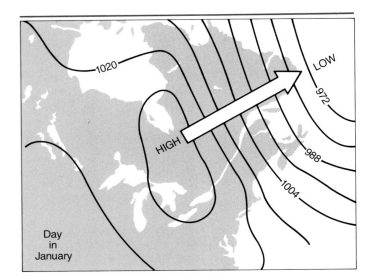

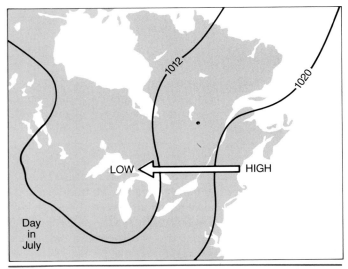

Fig. 7–2 Generalized pattern of atmospheric pressure at surface in northeastern North America in January and July. Steeper *gradient* (arrow) in January is due to more rapid change of pressure across the map, as revealed by closely spaced isobars. Pressure values are in millibars (mb).

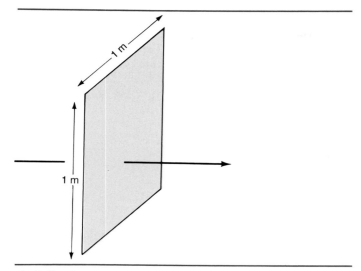

Fig. 7-3 One square meter perpendicular to the wind flow—as assumed when expressing Power Available as Watts per square meter.

It is useful to reduce the expression to one that applies to a cross-sectional area of one unit, such as a square meter. For sea level density of air, the expression becomes:

$P_A = 0.05472 \ V^3$, if V is measured in miles per hour (mph)

or $\quad 0.08355 \ V^3$, if V is measured in knots

or $\quad 0.6125 \ V^3$, if V is measured in meters per second

It is significant that power is proportional to the *cube* of wind velocity. In all cases the Power Available is expressed in Watts of power for every square meter of area perpendicular to the wind flow (Fig. 7-3). Analogous formulas could be derived to express power in other units such as Watts per square foot.

Meteorological Data

Because Power Available is proportional to the cube of wind speed (Fig. 7-4), detailed wind data and careful analysis are necessary for a fair assessment of power. The average or mean wind speed at a site throughout a given period is not adequate if the total power in the wind is to be calculated. An average speed (V) of 15 mph used in the above formula would yield power of 185 Watts per square meter; and if the wind in fact blew steadily at that speed the calculated power would be representative. But if winds blow at 10 mph half the time, and 20 mph half the time, the sum of the power calculated from the two speeds would be 245 Watts per square meter (Reed, 1975, p. 2).

It is apparent that Power Available must be calculated from data that reveals the *frequency of winds of various speeds.* Fortunately, wind data gathered by observers in the network of U.S. Weather Bureau stations (now National Oceanographic and Aeronautical Administration or

N.O.A.A.) is organized by wind speed categories. In addition, upper-air balloon observations can also be assigned to speed categories.

In response to federal research contract opportunities, three separate national assessments of wind power in the United States have been conducted in the past few years—by SANDIA Labs, by Lockheed-California, and by General Electric. Most recently, an overview of the three assessments was prepared by Batelle Pacific Northwest Laboratories. As will be shown, the wind power estimated for different regions of the country varies considerably from study to study because of different data employed and different treatments of the data.

THE USE OF SURFACE WIND DATA Central to any national assessment of wind power is wind speed data collected at hundreds of stations throughout the country. The nature of that data and the general procedure for deducing available wind power are outlined here.

The pioneering study employing large numbers of station records was conducted by Jack Reed of SANDIA

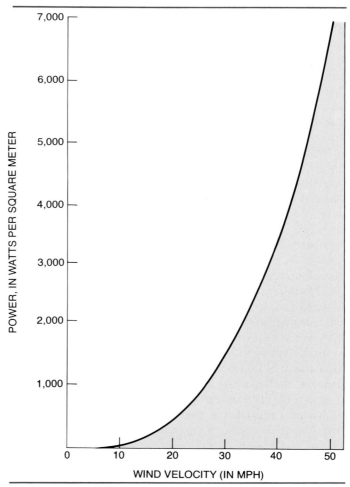

Fig. 7-4 Theoretical power available at various wind velocities. Data derived from the Power Available formula (see text).

TABLE 7-1
Frequencies of Winds in Ten Velocity Categories at Cheyenne Airport (Percent values on monthly basis[1])

PERIOD	\multicolumn{10}{c	}{VELOCITY RANGES (KNOTS)}	POWER
	1-3	4-6	7-10	11-16	17-21	22-27	28-33	34-40	41-47	48-55	(Watts/m²)
Jan.	5.5	10.7	24.7	27.2	15.8	10.7	3.7	1.1	0.1	0.0	433.1
Feb.	4.7	10.7	24.5	26.5	16.3	11.4	4.5	0.9	0.1	0.0	453.5
Mar.	5.1	11.9	25.6	27.4	14.5	9.5	4.0	1.3	0.2	0.0	434.6
Apr.	5.3	12.0	25.1	28.3	15.1	8.1	3.2	1.2	0.2	0.0	399.8
May	6.0	13.7	30.8	31.3	11.6	4.5	1.3	0.2	0.0	0.0	242.6
June	6.2	16.0	35.3	30.5	8.4	2.4	0.5	0.1	0.0	0.0	176.2
July	8.5	19.4	38.4	26.3	5.7	1.2	0.1	0.0	0.0	0.0	125.6
Aug.	8.2	18.8	38.0	26.6	6.2	1.3	0.2	0.0	0.0	0.0	132.4
Sept.	7.2	17.8	35.2	28.7	7.9	2.1	0.3	0.0	0.0	0.0	157.1
Oct.	6.0	16.0	35.6	27.5	8.9	3.7	1.3	0.4	0.0	0.0	228.7
Nov.	5.7	11.7	26.8	26.8	15.2	3.1	3.1	1.2	0.0	0.0	402.6
Dec.	5.3	11.1	24.0	26.6	15.6	11.3	4.2	1.0	0.1	0.1	463.9
ANNUAL	6.1	14.1	30.4	27.8	11.8	6.2	2.2	0.6	0.0	0.0	302.7

[1]Not shown is that on annual basis winds are calm 0.7 percent of the time.
Source: Reed, 1975, p. 152.

Laboratories (Reed, 1975). In preparation for that analysis, wind speed data for 758 stations were assembled from records at the National Climatic Center, Asheville, N. C.[1] Table 7-1 shows the organization of data by wind speed categories, the tabulated numbers in the ten columns expressing the percentage of time (during the month) that speeds were in the range shown by the column heading. Monthly power available was calculated according to the formula above, using the *median value* in each range as the velocity factor, and the frequency value to weight the power derived when summing the contributions from each speed category (procedure outlined in Table 7-2). Annual power is not the total of all months, but is computed by following the steps in Table 7-2, using the annual percent frequencies (mean of the twelve monthly frequencies) for each speed category. Similarly, seasonal Power Available can be deduced by using an average frequency for a group of months: for instance, for the winter period, November through February, the frequency of winds in the 1 to 3 knot range would be 5.3 percent.

A similar procedure was used in the Lockheed and General Electric studies but with variations as to which station records were used and whether median speed or some other value was employed to represent a speed category.

UPPER AIR DATA In order to approximate the wind potential in highland areas where weather station data is extremely sparse, the velocities of upper-air winds as revealed by balloon soundings is useful. Speeds are mapped in a publication used by all three of the above-mentioned studies for this purpose (Crutcher, 1959).

OFFSHORE WIND INFORMATION Coastal stations provide some indication of wind conditions in the near offshore areas. A most useful supplement is actual wind measurements made at sea, and generalized into values applied to "Marsden Squares" (rectangular areas one degree of latitude by one degree of longitude). Of the three studies, only the one conducted by General Electric made use of such data.

POWER AVAILABLE

The General Electric study includes a very thorough examination of Power Extractable based on its maps of Power Available, (shown here in Figs. 7-5 and 7-6). To avoid confusion, maps of Power Available from the other three

TABLE 7-2
Calculation of Power Available in the Month of January at Cheyenne Airport

SPEED CATEGORY (Knots)	MEDIAN OF CATEGORY (V)	P_A, i.e., $0.08355 \times V^3$	PERCENT FREQUENCY FOR CATEGORY	P_A MULTIPLIED BY FREQUENCY (Watts/sq. meter)
1-3	2.0	0.67	5.5	.037
4-6	5.0	10.44	10.7	1.117
7-10	8.5	51.31	24.7	12.674
11-16	13.5	205.56	27.2	55.912
17-21	19.0	573.07	15.8	90.545
22-27	24.5	1,228.70	10.7	131.471
28-33	30.5	2,370.53	3.7	87.710
34-40	37.0	4,232.06	1.1	46.550
41-47	44.0	7,117.12	0.1	7.117
48-55	51.5	11,412.17	0.0	0.0
POWER AVAILABLE FOR MONTH				433.135[1]

[1]See Table 7-1.

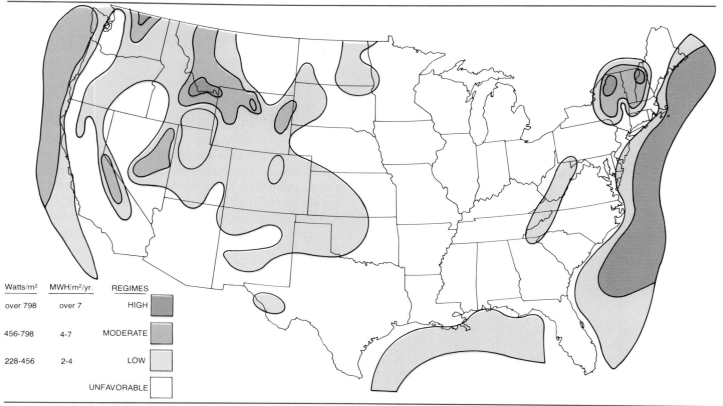

Watts/m²	MWH/m²/yr.	REGIMES	
over 798	over 7	HIGH	
456-798	4-7	MODERATE	
228-456	2-4	LOW	
		UNFAVORABLE	

Fig. 7–5 Favorable wind regimes *at surface level,* onshore and offshore, showing Power Available as Watts per square meter and as Megawatt hours per square meter per year.
Source: General Electric, *Final Report,* 1977.

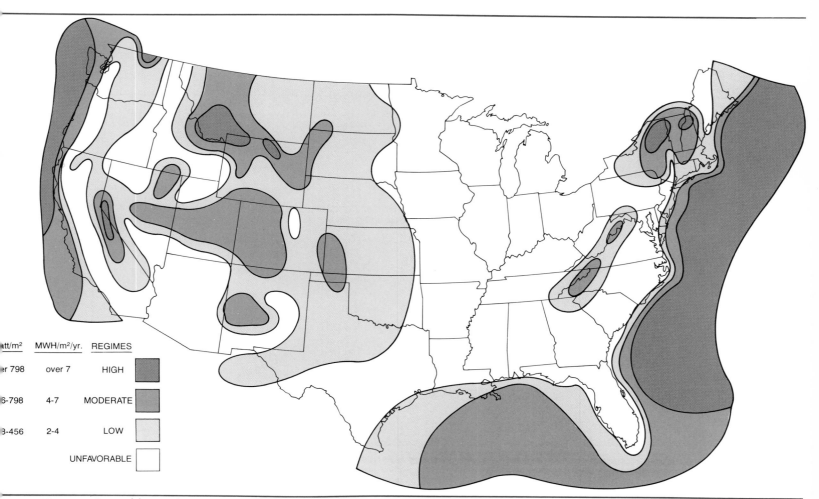

att/m²	MWH/m²/yr.	REGIMES	
r 798	over 7	HIGH	
6-798	4-7	MODERATE	
8-456	2-4	LOW	
		UNFAVORABLE	

Fig. 7–6 Favorable wind regimes *at 50 meter height,* onshore and offshore, showing Power Available as Watts per square meter and as Megawatt hours per square meter per year.
Source: General Electric, *Final Report,* 1977.

Mountains. Their higher values for the coastal region may be due to the better data employed, and the higher mountain values due to the assumption that mountain winds are equivalent to those in the "free circulation," that is, the upper-air winds.

[1] SANDIA values extrapolated from 10 meter height to 50 meters, using law of 1/7 power.

[1] High and low values are extreme because Lockheed maps show no isolines (isodynes) but only spot locations.

Source: Elliot, 1977.

the Watts per square meter value multiplied by 8,760 hours in a year. A quick conversion from one measure to another uses one Megawatt hour per square meter per year equals 114 Watts per square meter. The odd values in Watts per square meter for each region are due to the fact that G.E.'s

DETAILED LOCAL INFORMATION ON WIND POWER

Anyone planning to install a wind machine must have the best possible information on wind power available at the site. While this atlas uses maps of national scale, there is one recent study offering comprehensive wind data at state and regional levels.

The study was conducted for the U.S. Department of Energy by Pacific Northwest Laboratory, in Richland, Washington. There are 12 volumes, one for each of twelve regions that cover the nation—including a volume for Alaska, one for Hawaii, and one for Puerto Rico and the Virgin Islands.

For each state, the study combines meteorological data with landform and elevation data to come up with a map showing the wind power expected in various parts of the state, the state being divided according to its land form and elevation areas. In addition, a map shows the certainty (reliability) of the wind resource data in different parts of the state.

The report by Dennis L. Elliot et al. is *Wind Energy Resource Atlas*, Volumes 1 through 12, Richland, Washington, Pacific Northwest Laboratory. It is available from National Technical Information Service, Alexandria, Virginia. Telephone: 703-487-4650. The NTIS numbers are PNL-3195-WERA-1 through WERA-12. The volumes, and the states included in them, are as follows.

VOLUME	REGION	STATES INCLUDED
1	Northwest	Washington, Oregon, Idaho, Montana, Wyoming
2	North Central	North Dakota, South Dakota, Nebraska, Minnesota, Iowa
3	Great Lakes	Wisconsin, Michigan, Illinois, Indiana, Ohio
4	Northeast	Maine, Vermont, New Hampshire, New York, Connecticut, Rhode Island, Massachusetts, Pennsylvania, New Jersey
5	East Central	Maryland, Delaware, Virginia, West Virginia, North Carolina, Kentucky, Tennessee
6	Southeast	Mississippi, Alabama, Georgia, South Carolina, Florida
7	South Central	Kansas, Missouri, Arkansas, Oklahoma, Texas, Louisiana
8	Southern Rocky Mountain	Utah, Colorado, Arizona, New Mexico
9	Southwest	California, Nevada
10	Alaska	
11	Hawaii and Pacific Islands	
12	Puerto Rico and Virgin Islands	

POWER EXTRACTABLE

While the foregoing analysis provides a useful translation of wind velocities into wind Power Available, the results indicate only areas relatively more attractive or less attractive. What can be accomplished with the power, or conversely how many windmills would be required in certain areas to do a given job, are questions that remain to be answered.

Two approaches to the question of Power Extractable seem to be necessary in light of present energy conversion technology: Here they are called the *idealized* and the *electrical* approaches.

Idealized Power Extractable

This approach, used here for the sake of understanding fundamentals, assumes that virtually all the Power Available, *is in fact available* for extraction. Because Power Available, as derived and mapped in the earlier discussion, is the sum of contributions from winds of various speeds, it is only truly available if a wind energy conversion device will operate at low, medium, high, and very high speeds. This is not consistent with generation of alternating electrical current (AC), but is compatible with generation of direct current (DC) power, and with the execution of mechanical tasks such as pumping water uphill, or producting hot water through turbulence. Assuming such tasks are relevant, the mapped Power Available at a selected site must be subjected to the following considerations: (1) the efficiencies of energy conversion, and (2) the size of energy conversion devices.

EFFICIENCIES OF ENERGY CONVERSION The primary consideration here is how much of the wind's mechanical or kinetic energy can be translated into rotary power on the shaft of a windmill or turbine. Not all the power in the windstream can be extracted, for if it were, the wind would stop flowing at the extraction device. The problem of how much power can be extracted without interrupting the flow of air is susceptible to theoretical analysis. This analysis yields a maximum efficiency of 0.59, often called the *power coefficient* or the Betz coefficient (Betz, 1927). As suggested in Figure 7-10 the maximum efficiency for various types of machines is achieved at various combinations of blade speed and wind speed. On that graph, the most relevant entries are for the "high-speed two-blade type" (horizontal axis machine) and the Darrieus rotor, which is the vertical axis or

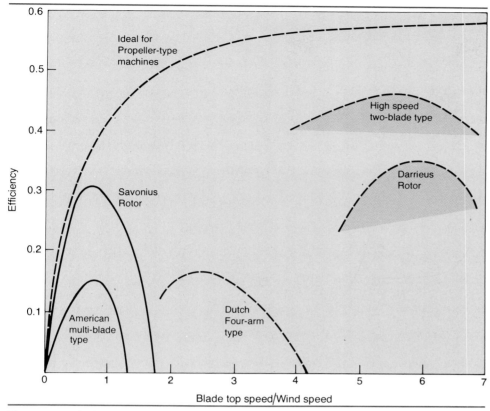

Fig. 7–10 Efficiency of conversion of wind power into rotary power on the shaft of a turbine. For each type of machine efficiency is greatest at a certain ratio of blade tip speed to wind speed.
Source: Eldridge, 1975.

eggbeater-style machine. Since maximum efficiencies are around 0.35 and 0.45 respectively, the use of 0.40 conversion efficiency is realistic. (Conversion from mechanical to electrical power entails another loss which may be expressed as an efficiency of 0.85 to 0.90, but will be ignored in this idealized analysis).

SIZE OF THE CONVERSION DEVICE Since Power Available has been expressed in *Watts per square unit of swept area*, the Power Extractable is proportional to the area swept by the "blades" of the wind machine, and a machine with a blade radius twice that of a smaller one will intercept *four times* the power. There is another effect of machine size. If the blade radius is great, the device must be so tall that part of the swept area is at a height greater than that at which wind velocity was measured and will encounter winds faster than those measured. If performance of a large machine is estimated by reference to wind measured at or near the ground, this factor for "height effects" should be considered.[3]

It follows that if the goal is converting mapped Power Available to Idealized Power Extractable, the following formula may be applied (Reed, Maydew, and Blackwell, 1974, p. 31).

$$P_E = (0.40) \times (\pi R^2) \times (R/10)^{3/7} \times (P_A)$$

Power
Extract-
able =   Efficiency   Area   Height   Power
          of 40%   Swept   Effects   Available
                                    Factor

The expression may be reduced to:

$P_E = 0.469 \ R^{17/7} \ (P_A)$ so that only the blade radius of the machine and the Power Available need be known.

The simpler expression.

$$P_E = (0.40) \ (\pi R^2) \ (P_A)$$

will be used here because consideration of height effects is inappropriate when Power Available is gleaned from a generalized map, and particularly because the concept will be illustrated with power deduced from the map showing power at 50 meters, the hub height for large machines.

Figure 7–11 plots the results of a series of calculations according to that formula, that is, simply multiplying power per square meter by the machine's swept area, and reducing

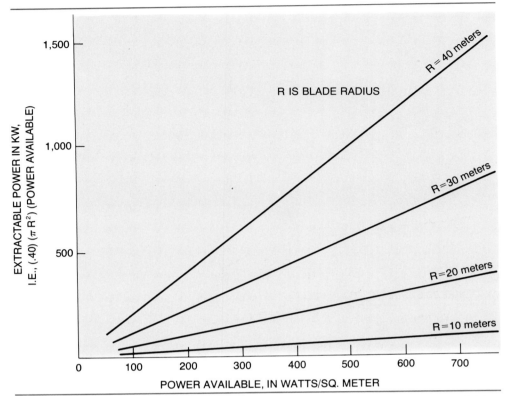

Fig. 7–11 Idealized Power Extractable at hypothetical sites where Power Available ranges from 100 to 700 watts per square meter. Four different blade sizes are considered. Data derived from the idealized Power Extractable formula.

the amount by the factor of 40 percent to consider extraction efficiency. For Power Available of 500 Watts per square meter, which prevails as an annual average in a Moderate regime such as in Western Colorado (Fig. 7–6) it is apparent that a giant machine with a blade radius of 40 meters could capture approximately 1,000 Kilowatts of power. At this point, the analysis might conclude that, based on the capacity of one such machine, it would take 1,000 of them to equal the generating capacity of a 1,000-Megawatt coal-fired plant, but, for reasons pointed out below, this idealized analysis is not relevant to electrical power generation.

Electrical Power Extractable

WIND CHARACTERISTICS AT THE SITE The main factor that undermines the idealized method of calculating Power Extractable is the fact that wind machines, especially those now built in the United States to produce alternating current, do not make use of all the Power Available in the winds because they do not respond to winds of all speeds. First, a large machine will not start when winds are under 5 to 8 mph, and any Power Available calculated on the basis of light winds will be false for this reason.

More important, given the impact of the V^3 factor in deriving Power Available, are the higher winds. These may account for a very large part of the calculated Power Available, but their power will not be used by horizontal axis machines

which feather their blades (adjust blade pitch) in order to maintain a constant rpm on the shaft.

These considerations demand that the winds of a particular site, or a wind regime, be described by the very information that is used (then lost) in calculating total Power Available at a site, namely, the frequency of winds of certain speeds. Table 7–1, along with Figures 7–12 and 7–13, show how that information may be used to better grasp the potential for electrical generation at a particular site.

Annual percent frequencies in the form shown in Table 7–1 can be manipulated to make a plot such as in Figure 7–12, which shows proportions of the year during which winds *exceeding* certain speeds may be expected to occur. For any chosen series of wind speeds, then, these frequencies may be converted to hours per year (as in Table 7–4), and the speeds converted to their implied power available per square meter. The derived plot of *power duration* (Fig. 7–13A) can then be constructed using the hours at certain Power Available values. This is extremely useful information, for the *expected hours* at various values of Power Available are crucial to wind machine output and design. One further graph (Fig. 7–13B) is derived from Figure 7–13A by noting the hours at various power (wind speed) values, expressing the product as Kilowatt hours, and plotting the results opposite wind speeds. This plot of *power density* reveals in which wind speeds the greatest amount of power may be expected. *The speeds that do contain the most power are*

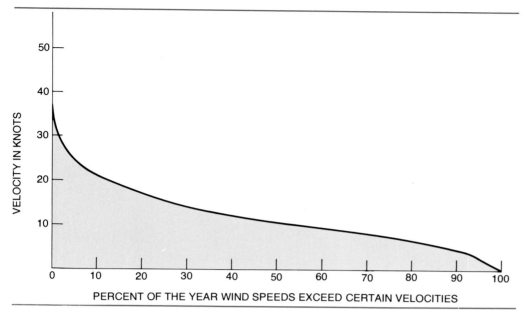

Fig. 7–12 Surface winds at Cheyenne Airport, Wyoming, represented by a cumulative percent frequency plot.
Source: Reed, 1975.

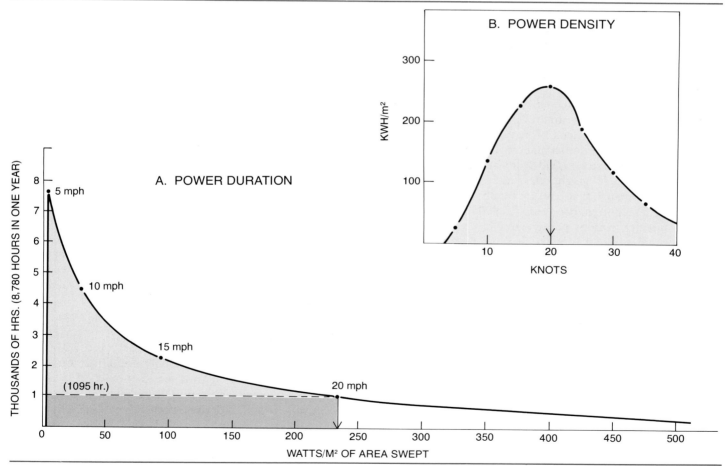

Fig. 7–13 Wind characteristics at Cheyenne Airport, Wyoming, as shown by *power duration* and *power density* curves.
Source: from Reed, 1975.

TABLE 7-4
Frequency of Wind Speeds and the Power Associated with each.
Surface data from Cheyenne, Wyoming

	SPEEDS (VELOCITIES) IN KNOTS							
	5	10	15	20	25	30	35	40
Percent frequency, taken from Fig. 7-12	86	52	25.2	12.5	4.5	1.7	0.6	0.2
Hours per year on the basis of percent frequency	7,534	4,555	2,234	1,095	394	149	53	18
Velocity cubed	125	1,000	3,375	8,000	15,625	27,000	42,875	64,000
Power at each speed, in Kilowatts	3.6	29.24	98.69	233.9	456.9	789.5	1,253.7	1,871.4
Thousands of kilo-watt hours per year (KW × 8760 × % Frequency)	27.1	133.2	220.5	256.1	180.0	117.6	66.4	33.6

those of sufficiently high velocity and sufficiently high frequency to garner the highest product: They are neither the highest velocities, because those are rare, nor the most frequent, because those winds are light. In the Cheyenne, Wyoming case illustrated in Figure 7-13B it is winds of 20 mph that contain most power in the average year.

Hypothetically, a wind turbine might be designed for peak output in winds of 20 mph velocity. If it were, then Figure 7-13A would reveal the number of hours during the year the winds may be expected at that velocity, and therefore, the number of hours the turbine would be operating at peak output. The shaded rectangle represents the product of peak power (234 Watts per square meter) and the hours at that power (1,095). The stippled area represents hours of operation at speeds of greater than 5 mph but less than 20 mph. If blades feather so that winds of greater velocity than 20 mph are treated as if they were 20 mph, there are no more hours of operation to account for. Since power, or energy, is now expressed in Kilowatt hours per square meter per year, the expected output from machines with larger and smaller rotors might be deduced.

THE GENERAL ELECTRIC APPROACH TO POWER EXTRACTABLE FOR UTILITIES

Although the wind characteristics as described by graphics such as Figures 7-12 and 7-13 are fundamental to the design of wind turbines, there are various approaches to a practical and economical turbine design. Lockheed's analysis of costs for utility applications led them to recommend rotor diameters of nearly 350 feet (107 meters) at most locations, with various transmissions and generator capacities to suit variations in wind speed (Lockheed-California Co., 1976). The General Electric approach employs generators of the same capacity in various locations, and varies rotor diameters in order to keep shaft rpm constant, and to gather enough energy to power the standard dynamo in three different wind regimes.

In order to assess the potential of the three regimes, High,

Moderate, and Low, the G.E. study chose for each regime one or more stations whose wind characteristics were thought to be representative of that regime. Economical wind turbines were then designed by procedures that take into account the need for efficient blade tip speed during operation in various wind speeds and seek to find the minimum investment per Kilowatt of output.

Table 7-5 shows the features of turbine units designed for three median wind speeds frequently occurring in each of the three regimes. In each of the three cases, the "nameplate" capacity of the turbines chosen for electrical utility application is 1,500 Kilowatts (that is, 1.5 Megawatts). Rotor speeds diminish from High to Low regimes, and the rotor diameter needed to achieve the desired rpm is smallest in the High regimes, largest in the Low regime (Fig. 7-14). The *rated wind velocity*, that is, the velocity at which peak power of 1,500 Kilowatts is realized, is always higher than the *design velocity* which represents either mean or median wind speed for a site (both terms are used in the G.E. report). The *capacity factor* is extremely important because, on the basis of velocity duration curves (Fig. 7-12), it expresses the proportion of the time during which winds blow at the rated velocity and at speeds less than that. Differing velocity duration curves characteristic of higher and lower mean wind sites lead to highest capacity factor (0.50) at the 18-mph site, and lowest (0.40) at the 12-mph site.

In estimating the output of the 1,500-Kilowatt units located in each of the three regimes, the study recognized that the wind speeds designed for, namely 18, 15, and 12 mph do not occur with the same frequency at all sites in the regimes. A more conservative capacity factor for each regime was arrived at, therefore, that better represents velocity duration in the regime—in every case smaller than the factor applied to the three design cases.[4]

The revised capacity factors (Table 7-6) are critical to the remainder of the analysis, for they determine the estimated annual output of the standard 1,500-Kilowatt machines in each of the three regimes. Any analysis of

TABLE 7–5
**Characteristics of 1,500-Kilowatt Units Designed for Three Median
Wind Speeds**

| | MEDIAN (DESIGN) SPEEDS | | |
	18 mph	15 mph	12 mph
Rotor diameter	190 ft[1]	219 ft[1]	278 ft[1]
Rotor speed	40 rpm	31.5 rpm	20.6 rpm
Rated wind velocity	22.7 mph	20.3 mph	16.8 mph
Design annual output	6.62 KWh $\times$ 10^6	5.85 KWh $\times$ 10^6	5.29 KWh $\times$ 10^6
Capacity factor, *i.e.* proportion of time the winds flow at rated velocity	0.50	0.45	0.40

[1]Optimum rotor diameters to maintain steady rotor rpm in different winds are not known precisely. Elsewhere, these diameters, 218 feet, 278 feet, and 331 feet are used for High, Medium, and Low wind regimes when defined packing density.

(1) how many machines are needed to accomplish a given task, or (2) the maximum electrical energy extractable in an area depends on those estimates.

THE QUESTION OF PACKING DENSITY Equally important is the matter of how closely the wind ma-chines may be planted in an area without deleterious effects on the wind field. Lacking experience with this, industry researchers estimate that a spacing of 10 to 15 rotor diameters may be necessary. Lockheed assumed spacing of 10 times the diameter while General Electric chose the more conservative spacing of 15 times the diameter, which results

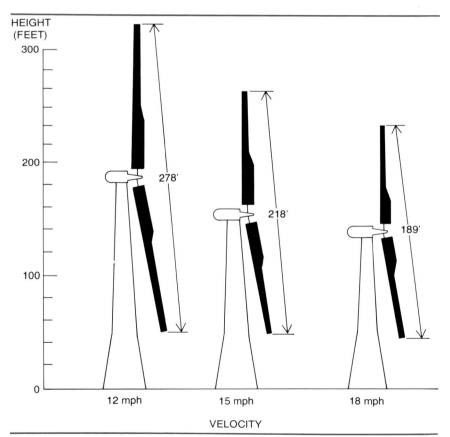

Fig. 7–14 Three rotor diameters that are consistent with the three mean velocities representative of Low, Moderate, and High regimes in the General Electric analysis.
Source: General Electric, *Final Report*, 1977.

TABLE 7–6
Operating Characteristics of 1500-Kilowatt Units in Three Wind Regimes

	REGIME		
	High	Moderate	Low
Annual energy output (KWh × 10⁶)	5.01	4.33	2.92
Regime capacity factor considering variety of winds	0.38	0.33	0.22

in fewer machines planted per unit area. Considering the three rotor sizes in the three regimes, this rule of thumb leads to the following densities: High regime, 3.01 per square mile; Moderate, 1.851 per square mile; and Low, 1.306 machines per square mile.[5]

INSTALLATION REQUIRED TO MATCH CAPACITY OF A 1,000-MEGAWATT ELECTRIC PLANT

An intriguing question is the nature and size of a wind-electric installation which generates as much power as the competition, that is, a very large fuel-burning steam-electric plant. The comparison is not entirely appropriate, since wind power installations would not be employed in such a concentrated or independent fashion. Nevertheless, the wind installation(s) may be construed as scattered through an area that would otherwise be served by the steam-electric plant.

Assuming the conventional power plant operates in the average year at 50 percent capacity, the actual capacity to be matched by wind installation is not 1,000 Megawatts, but 500 Megawatts. In the most favorable, that is, the High wind regime, the average capacity factor of 38 percent (Table 7–6) suggests that for an effective capacity of 500 Megawatts, the installed capacity must be about 1,300 Megawatts. The space requirements for such an installed capacity can be found in a G.E. plot which bypasses the number of machines and directly shows the square miles needed (Fig. 7–15). If the conservative spacing of 15 rotor diameters is assumed, the need is for 300 square miles; but if spacing of 10 rotor diameters is assumed the space required drops to about 140 square miles. Thus, the best of the wind cases demands far more land area than the solar-thermal electric plant of the power-tower or central receiver type, which requires only 18 square miles for a comparable output (see Chapter 6 on solar energy). This suggests that research should be directed toward devices or structures that focus and concentrate the wind flow as well as dynamos that make use of the power residing in the higher winds at a given site, rather than spilling the winds that exceed the design speed.

It must be recognized that, as with solar-electric power installations, the area is committed only once, and thereafter produces clean energy without fuel. The coal or nuclear plant, in contrast, requires continual mining which may disturb as much as 15 to 20 square miles during the life of the plant.

Net energy balance for wind-electric systems according to both Lockheed and G.E. studies is not immediately attractive when compared to fuel-burning plants. General

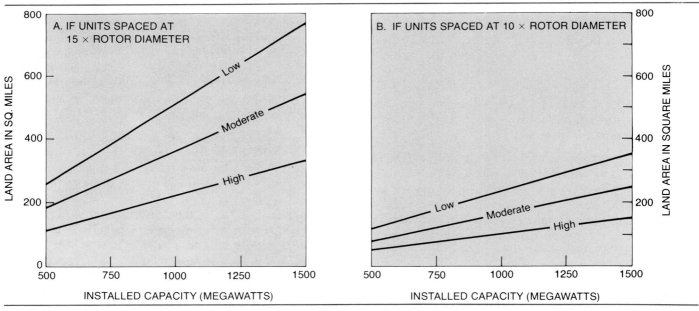

Fig. 7–15 Land areas required for generating capacities comparable to those of large conventional power plants. Plots show areas in three wind regimes, using two different assumptions about dynamo spacing.
Source: General Electric, *Final Report*, 1977.

Electric estimates the *energy payback period*, that is, the time for energy generated to compensate for the energy expended in materials for the installation, is 0.29 to 0.89 years for wind machines in the three regimes, whereas for a 700-Megawatt coal plant the period is 0.076 years, and for a 1,150-Megawatt nuclear plant it is 0.109 yrs. To put it more favorably, the wind machines pay back the energy cost of their materials within 3½ to 11 months, depending on regime, and from then on consume virtually no energy. Fuel-burning plants pay back their material energy costs much sooner, but continue to consume energy. Obviously this "net energy" analysis is incomplete and should be improved by a *life cycle* approach that considers all energy expended and produced during the installation stage and the expected life of the installations. The results of such analysis should supercede any analysis of dollar economics in deciding whether electric power from the wind is desirable.

MAXIMUM POTENTIAL OF THE THREE REGIMES

A corollary to the question of how many machines are needed to accomplish a given generating task is the question of how great is the total potential for electrical generation. Some stimulating estimates of the power output from large numbers of machines in offshore areas and in the Great Plains have been made in the past (Heronemus, 1972), but no thorough study of the actual number of machines that *could* be planted in various parts of the country had been made until the Lockheed and General Electric studies referred to here. The following account is entirely based on the General Electric study which did not attempt to estimate installations in offshore areas, but focused only on land areas in the contiguous states (General Electric, *Final Report*, 1977).

With a view to electrical utility application of wind power, the vital assumptions have already been stated.

1. The annual output of the selected 1,500-Kilowatt machines in each of the three regimes is as noted in Table 7–6.
2. The number of machines that can be planted per square mile in each of the three regimes, considering the three rotor diameters, is noted above in the section on *Packing Density*.

It is necessary, therefore, only to estimate the *land areas available* in each of the three regimes, to arrive at the maximum number of machines possible in each regime, and hence in the 48 states as a whole. This maximum number of machines and the generating output that follows are considered the *saturation case*. The procedure used to arrive at available land areas is as follows:

1. In each regime, land area was diminished by any institutional lands, military bases, national parks, national monuments, and Indian reservations.
2. In each regime all areas were assigned to one of a number of land use categories, such as Irrigated Land, Swamp, or Urban Areas, by reference to land

TABLE 7-7 Land Use Factors used by General Electric to Determine Land Available for Wind Machines	
LAND USE CATEGORY	**PROPORTION OF AREA CONSIDERED AVAILABLE FOR WIND MACHINES**
Alpine meadows, mountain peaks	1.0
Swamp	1.0
Marshland	1.0
Moist tundra and muskeg	1.0
Forest and woodland, mostly ungrazed	0.9
Sub-humid grassland	0.9
Desert shrubland, mostly ungrazed	0.9
Desert shrubland, grazed	0.8
Open woodland, grazed	0.8
Forest and woodland, grazed	0.8
Forest and woodland with some cropland	0.8
Cropland with pasture, woodland and forest	0.5
Cropland with grazing land	0.2
Mostly cropland	0.2
Urban areas	0.01
Irrigated land	0.0

use maps in the *National Atlas of the United States of America* (U.S.G.S., 1970, p. 158).

3. For each land use type a factor was assigned to denote the proportion of land which could be used for wind machines (see Table 7–7).
4. By reference to Landform classes in the *National Atlas* (U.S.G.S., 1970, p. 61) slope factors were applied to reduce the available land to that with suitably gentle slopes.
5. The resulting "net" land areas were then filled according to the packing density appropriate to the regime, and the resulting number of 1,500-Kilowatt units subjected to the regime's capacity factor to yield total Kilowatts, which multiplied by 8,760 hours yields Kilowatt hours per year.

Results of the land use analysis are summarized in Table 7-8, which shows for each regime the total land area, amounts not available for land use and slope reasons, and the net available land in square miles and as a proportion of the total. Altogether 18 percent of the land was lost to institutional uses, with the least impact of this factor occurring in the High regime which is largely mountainous. Similarly, the overall loss to competing land use is 26 percent, with the greatest impact in the Low regime, which is, coincidentally, of lower elevation and more heavily farmed. The factor of slope deleted 42 percent of the land overall, and 70 percent in the High wind regime. The last column in Table 7-8 shows that of all the remaining land area 74 percent is in the Low regime, 25 percent in the Moderate, and only *1 percent in the High wind regime.*

TABLE 7-8
Net Land Areas Available for Wind Power Systems (in thousands of square miles)

WIND REGIME	TOTAL LAND AREA	INSTITUTIONAL LANDS	NOT AVAILABLE (OTHER LAND USE)	NOT AVAILABLE (POOR TOPOGRAPHY)	NET AVAILABLE LAND	FRACTION OF REGIME AREA (Percent)	DISTRIBUTION OF AVAILABLE AREA (Percent)
Low	1,104	201.5	325.5	420.4	156.6	14.2	74
Moderate	415	85.6	79.7	195.8	53.9	12.0	25
High	55	2.2	10.8	39.0	3.0	5.5	1
Total	1,574	289.3	416.0	654.7	213.5	13.5 (mean)	100

Source: General Electric. *Wind Energy Mission Analysis, Final Report,* February 18, 1977.

Apparently the question of whether sloping land can be employed is critical to a fuller exploitation of areas where wind potential is greatest.

Table 7-9 summarizes the results of hypothetical planting of 1,500-Kilowatt capacity wind machines in the available land in the three regimes, noting the number of units at saturation and their output in Watts and in Watt-hours per year. The relative capacity (or output) of the three regimes is roughly as suggested by relative net land areas in Table 7-8. The more dense packing of machines in the High regime, however, allows 3 percent of the total capacity to be installed there despite its having only 1 percent of the available land area. This more effective use of available land in the higher wind regimes is indicated by the *Efficiency Factor* column in Table 7-9, which shows 14.90 millions of Kilowatt hours per square mile in the High regime, versus only 3.78 per square mile in the Low. The overwhelming majority of the generating capacity resides in the Low wind regime, with about half that amount in the Moderate. The total capacity of 470.3 Gigawatts, that is, 470,300 Megawatts, is only for land areas, excluding therefore, the considerable potential in Pacific, Atlantic and Gulf offshore areas.

IMPACT OF DIFFERENT ASSUMPTIONS In such an assessment of national potential output at satura-

tion, certain assumptions have a great impact on the resulting numbers. It is clear, for instance, what a drastic effect the slope requirements have upon the numbers of machines in the High regime, much of which is mountainous.

The matter of packing density, which follows from the estimated spacing required between machines is critical. Figure 7-15 shows how the factor can affect land areas required for certain generating capacities, and, logically, the effect on total capacity at saturation is profound. Figure 7-16 shows how spacing of 10 times rotor diameter could boost the national total at saturation from 1,069.5 to 2,460 billion of Kilowatt hours per year. Installing machines of 3,000 instead of 1,500-Kilowatt rating has a much smaller impact.

Not demonstrated here are the effects of choosing *land use factors* other than those noted in Table 7-7.

Implementation Rates and the Impact on National Utility Needs

Figure 7-17 shows, for the electric utility sector only, three possible futures for the use of wind power, in which the numbers of 1,500-Kilowatt units installed increases at rates consistent with Rapid, Medium, or Slow implementation cases.[6] The land area of all favorable

TABLE 7-9
Total Wind Energy in each Regime and Available Wind Energy at Saturation (Electric utilities only)

WIND REGIME	TOTAL LAND AREA IN REGIME (Sq. Miles × 1,000)	TOTAL ENERGY IN REGIME (Billions KWh/Yr.)	AVAILABLE AREA FOR WECS (Sq. Miles × 1,000)	ENERGY OUTPUT (Billions KWh/Yr.)	REQUIRED NUMBER OF 1,500 KW UNITS	INSTALLED ELECTRICAL CAPACITY (Gigawatts)	EFFICIENCY FACTOR (Millions KWh/mi²)	PROPORTION OF TOTAL CAPACITY
Low	1,104	4,168	156.6	591.8	204,800	307.2	3.78	0.65
Moderate	415	3,331	53.9	433.0	99,800	149.7	8.03	0.32
High	55	827	3.0	44.7	8,960	13.4	14.90	0.03
Total	1,574	8,326	213.5	1,069.5	313,500	470.3	5.01	

Source: General Electric. *Wind Energy Mission Analysis, Final Report,* February 18, 1977.

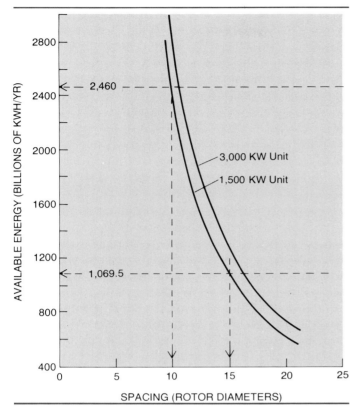

Fig. 7-16 Estimated wind-electric energy that could be produced at saturation will be greater if larger machines and closer spacing are assumed.
Source: General Electric, *Final Report*, 1977.

regimes is theoretically saturated by the year 2015, 2025, and 2035 according to the Rapid, Medium, and Slow cases respectively.

The estimated generating capacity implied by the numbers of machines, and how that capacity relates to national needs are shown in Figure 7-18. Wind generating capacity *at saturation*, 470 Gigawatts (taken from Table 7-9), would be just under one-half of the country's estimated needs in the year 1990; but a substantial number of wind machines cannot be installed that soon. By the year 2000 at Rapid implementation approximately 250 Gigawatts of wind generating capacity might be installed, by which time the country's electrical needs, increasing at between 5 and 6 percent, would have reached 1,650 Gigawatts. At most, therefore, wind power could contribute 15 percent of the projected 1,650-Gigawatt need. If the nation's total generating need were to remain constant at the 1980 level of just over 600 Gigawatts, the 470-Gigawatt capacity of wind machines could account for over two-thirds, assuming saturation.

More relevant than installed generating capacity is the *expected output* from electrical plants, and in this regard, wind power from the machines assumed in the foregoing analysis is less productive than most other generating installations. Fuel-burning plants put out roughly 45 percent of their rated capacity, nuclear plants about 40 percent, and hydropower plants about 50 percent; but the capacity factor for the foregoing wind machines is only 22 to 38 percent in Low to High wind regimes respectively (Table 7-6). The

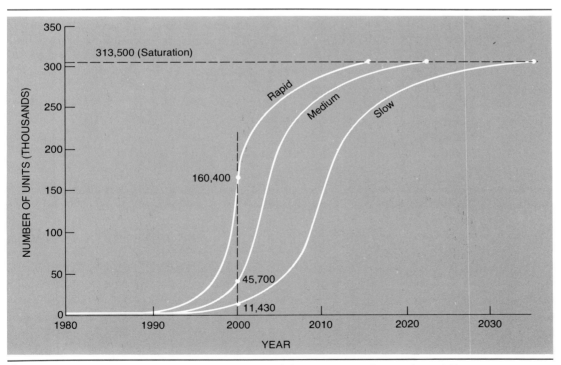

Fig. 7-17 Growing numbers of 1,500-Kilowatt wind dynamos according to three different implementation rates. Saturation of available land area is reached around the year 2015 in the case of rapid implementation.
Source: General Electric, *Final Report*, 1977.

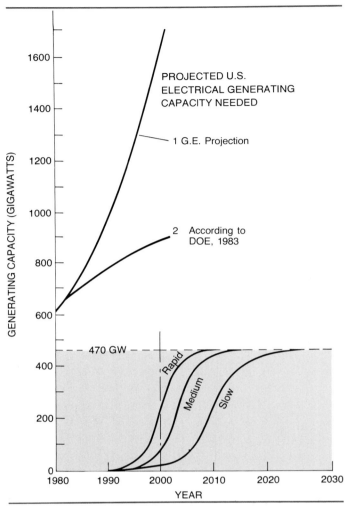

Fig. 7–18 Maximum impact of wind-electric generating capacity compared with projected need for generating capacity in the United States.
Sources: General Electric, *Final Report*, 1977. Revised projection of national need from DOE, *Annual Energy Outlook, 1983*, 1984.

Economic Feasibility and Numbers of Machines Produced

General Electric uses the concept of *break-even costs*, that is, the dollar cost per installed Kilowatt of wind-generating capacity that would make wind power economical for a utility company, assuming 1975 conditions of fuel costs and electric rates as represented by four regions of the country surveyed (General Electric, *Final Report*, 1977, pp. 5–16).

Included in any consideration of the installed costs is the matter of how per unit costs may be expected to drop with mass production, the rate of drop in per unit cost being expressed in "learning curves" (Fig. 7–19). The 0.90 learning curve applies if doubling of cumulative amounts produced leads to per unit costs that are 90 percent of the predoubling cost. Similarly, if doubling of cumulative production reduces per unit costs to only 85 percent of their former level the failing costs are described by the 0.85 learning curve.

In Figure 7–20, falling costs according to both 0.90 and 0.85 learning curves determine the *number of units required* in order to reach break-even cost. Because of the greater capacity factors in higher wind regimes, break-even costs can be higher (see table in Fig. 7–20) and the required numbers of units for break-even correspondingly lower. The logarithmic vertical scale reveals the numbers of units required, and the dates noted at the head of each bar show the years in which those numbers of units might be installed according to the Rapid implementation case only. Thus, in the High wind regime the units required would be reached by the year 1988 or 1993, depending on the learning curve that is chosen; in the Moderate regime break-even would be achieved in 1991 or in 2000; and in the Low regime breakeven is reached in 1998 only by virtue of the more favorable learning curve. At the 0.90 learning curve, per unit costs do not fall rapidly enough, so the required number of units far exceeds saturation for the Low regime, and breakeven is not achieved there with the assumptions of Figure 7–20.

Various Other Applications

While the foregoing analysis has focused on utility application, and has therefore depended in its geographical aspect upon the rotor requirements for the 1,500-Kilowatt dynamo chosen by G.E., there are other applications of wind-generated electricity that have in common an independence from utility networks, specifically in residential, industrial, agricultural, and remote community uses.

General Electric has chosen wind machines suitable for these applications for each of the three wind regimes and has specified rotor diameters—all of which is considered not essential here since the geographic dimension of extractable wind power is clearly expressed in the analysis of utility power. A comprehensive summary of various applications showing numbers of units and the resulting installed

expected annual output from all wind machines in the United States *at saturation* is 1,069.5 billion Kilowatt-hours (Table 7–9) while the national need in the year 1980 is 2,769 billion Kilowatt hours (U.S. Department of the Interior, 1975). Apparently, wind machines could provide roughly 38 percent of the needed 1980 output if all favorable land regimes were saturated according to the General Electric assumptions. An even larger contribution of wind power can be anticipated if any of the following are realized:

1. **More machines are packed per square mile by virtue of new designs.**

2. **The offshore potential is used, especially in New England.**

3. **The needed output is reduced through conservation, and by using electricity only for those tasks that cannot be accomplished by any other means.**

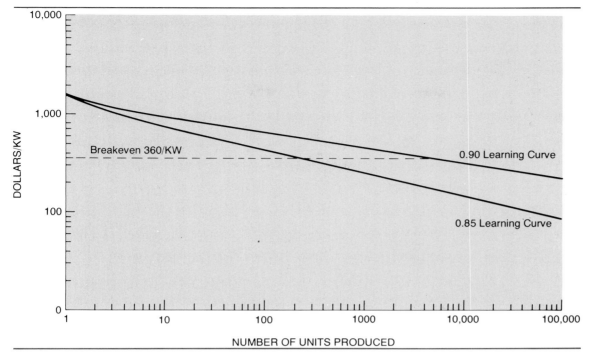

Fig. 7-19 Relationship between costs of a single 1,500-Kilowatt wind turbine and the number of turbines produced, according to two learning curves.
Source: General Electric, *Final Report*, 1977.

capacity and annual output values for the whole country is provided in Table 7-10. The two parts of the table are *for saturation*, in the first case, and *for the year 2000* at Rapid implementation rates in the second case.

On the basis of market potential it was determined that the tabulated numbers of units would be installed. In both the "saturation" and the "year 2000" cases, the numbers of residential machines are about three times the number of those run by utilities. However, the annual energy output from utility machines is about five times that of all the residential machines. The other applications are relatively small, agriculture being conspicuously larger than the single industry chosen (paper mills) or the remote community application.

RECENT DEVELOPMENTS

Progress in the development of wind machines is taking place on two fronts—the federal program and the private sector.

The Federal Program

The U.S. Department of Energy now continues the Federal Wind Energy Program begun under the Energy Research and Development Administration in 1975. The program's objective is to accelerate development and utilization of both small and large wind systems. Applica-

TABLE 7-10
(Part A)   Potential at Saturation for Various Applications

APPLICATION	NUMBER OF UNITS	SIZE OF UNITS (KW)	INSTALLED CAPACITY (Gigawatts)	ANNUAL ENERGY OUTPUT (Billions KWh)
Electric utility	313,503	1,500	470.3	1,069.5
Residential	9,300,000	10	93.0	207.0
Paper industry	1,000	1,500	1.9	2.9
Agriculture	780,000	35	27.0	43.0
Remote communities	1,000	1,500	1.9	2.9
Total	10,395,503		594.1	1,325.3

Part B)   Potential in Year 2000 for Various Applications—Rapid Implementation

APPLICATION	NUMBER OF UNITS	SIZE OF UNITS (KW)	INSTALLED CAPACITY (Gigawatts)	ANNUAL ENERGY OUTPUT (Billions KWh)
Electric utility	160,000	1,500	240.6	581.3
Residential	4,600,000	10	46.0	103.0
Paper industry	1,000	1,500	1.9	2.9
Agriculture	360,000	35	13.0	21.0
Remote communities	1,000	1,500	.8	1.45
Total	5,121,900		302.3	709.65

Source: General Electric, *Wind Energy Mission Analysis, Final Report*, February 18, 1977.

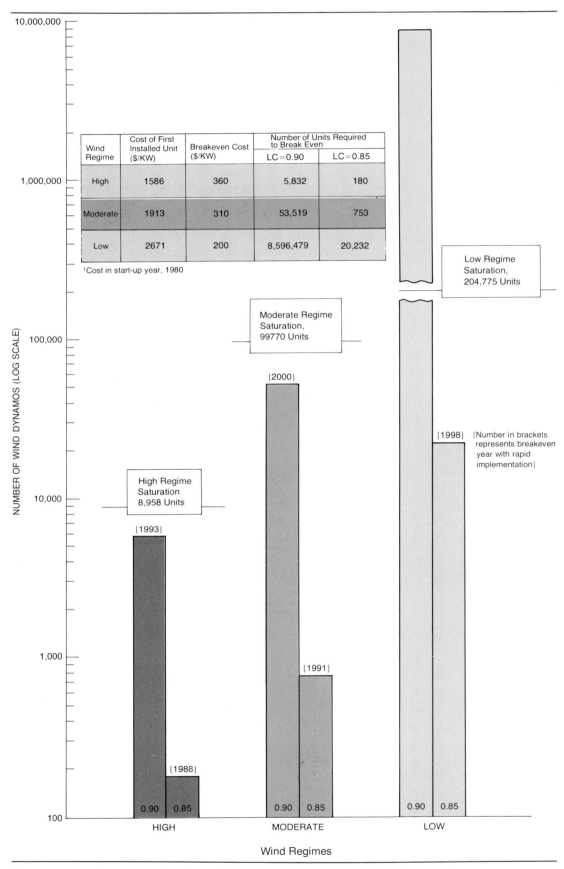

Wind Regime	Cost of First Installed Unit ($/KW)	Breakeven Cost ($/KW)	Number of Units Required to Break Even	
			LC=0.90	LC=0.85
High	1586	360	5,832	180
Moderate	1913	310	53,519	753
Low	2671	200	8,596,479	20,232

¹Cost in start-up year, 1980

Low Regime Saturation, 204,775 Units

Moderate Regime Saturation, 99770 Units

[Number in brackets represents breakeven year with rapid implementation]

High Regime Saturation 8,958 Units

NUMBER OF WIND DYNAMOS (LOG SCALE)

Wind Regimes

HIGH   MODERATE   LOW

Fig. 7–20 Numbers of wind turbines needed to reach break-even costs. In each wind regime, the left bar assumes 0.90 learning curve, and the right bar assumes the more rapid cost reductions of a 0.85 learning curve. Dates of break-even are entered in parentheses.

Source: Compiled from data in General Electric, *Final Report*, 1977.

MOD-Zero machine rated at 100 Kilowatts was built at the Plumbrook site near Sandusky Ohio in 1976. Three similar machines, termed MOD-Zero-A and rated at 200 Kilowatts, were built at Clayton, New Mexico, Culebra Island, Puerto Rico, and Block Island, Rhode Island, in 1978 and 1979 (see Fig. 7–21). The third phase, MOD-One, entailed a single very large machine with a blade diameter of 200 feet and a rating of 2,000 Kilowatts (2 Megawatts). At the time of its completion near Boone, North Carolina, in 1979, this was the largest ever built, exceeding by 25 feet the diameter of the

The first of the utility-scale wind turbines sponsored by the federal government was this machine, installed in 1977 near Sandusky, Ohio. With a blade diameter of 125 feet, this turbine is rated at 100 Kilowatts in winds of 18 miles per hour. (Courtesy of U.S. Department of Energy.)

tions being studied involve *small systems* for farm and rural home uses, *100 KW-scale systems* for irrigation and industry uses, *Megawatt-scale* systems for utility use, and *Multi-unit systems*, which are arrays of Megawatt-scale systems used for a large proportion of a utility's generating capacity.

Small systems are being studied at the Department of Energy test center at Rocky Flats, just west of Denver, and numerous studies have been sponsored on novel conversion devices, system technology, and wind characteristics (see Department of Energy, Wind Energy Systems Program, 1980).

The work on horizontal-axis large and Megawatt-scale turbines has produced some exciting results. First, the

This is one of three MOD-ZERO-A wind machines in the DOE-NASA program. Blade diameter is 125 feet; the rating is 200 KW. This one is on Block Island, R.I. (Courtesy of U.S. Department of Energy.)

NOTE: The WTS4 at Medicine Bow, Wyoming, is owned by Bureau of Reclamation.

HORIZONTAL AXIS
- MOD 0, 100 KW
- MOD 0–A, 200 KW
- MOD 1, 2,000 KW
- MOD 2, 2,500 KW
- WTS 4, 4,000 KW

VERTICAL AXIS
- 100 KW

Fig. 7–21 Wind machines owned by the Department of Energy. Compiled from various sources.

Smith-Putnam machine that generated power near Rutland, Vermont for a few years in the 1940s.

The fourth phase saw the construction of three even larger horizontal axis machines, with blades of *300 feet* diameter. These MOD-Two turbines, rated at 2.5 Megawatts are in a cluster at Goodnoe Hills in southernmost state of Washington near John Day dam on the Columbia River. They are connected to the Bonneville Power Administration Network. One other MOD-Two machine is located at Medicine Bow, Wyoming, and a third is at a site in Solano County, near San Francisco, where is it being tested by Pacific Gas and Electric Company.

After about 7 months of operation, the first of the Goodnoe Hills machines suffered an *overspeed* event due to hydraulic valve failure. Next, the machine developed a crack in the propellor rotor. These failures led to modifications of all five MOD-Two machines, keeping them idle for many months.

These setbacks, and a number of other factors have retarded the government's program which was to have leapt ahead to the giant MOD-Five. (Three and Four are not mentioned.) These machines, with rotor diameters of 400 feet and ratings of around 7 Megawatts were to be built with guidelines derived from the operation of the MOD-Two's. Meanwhile, the new administration in Washington reduced the wind energy budget from $ 60 million dolllars in 1982 to less than $ 20 million in 1983. This transformed the MOD-Five development effort from a government demonstration project to a *cost-sharing arrangement* among turbine manufacturers, interested utilities, and the Electric Power Research Institute, which is funded by the utilities. On this basis, the effort depends on the turbine manufacturers' conviction that utilities will eventually want to buy such machines. This interest, however, has faded. General Electric, the contractor who was to have worked on the first MOD-Five (to be located in Hawaii) pulled out of the project, citing its concern with falling world oil prices and forecasts for reduced growth in electrical needs (Moore, 1984). Another contractor, Hamilton Standard, who built the 4 Megawatt machine operated by the Bureau of Reclamation at Medicine Bow, Wyoming, has stopped looking for buyers for that machine. Meanwhile, Westing-

THE BATTLE OF THE WIND MACHINES

An amazing variety of wind machines have been proposed and, to varying degrees, tested. There is the Savonius, which resembles a collage of oil drums that have been sliced vertically. There is a machine that links airfoils into an endless belt, resembling a giant Venetian blind, which rolls vertically in response to light winds. There is the cylindrical tower with strategically placed windows which encourage an updraft that is harnessed by a turbine in the center of the tower's base. There is the Flettner rotor, a tall cylinder that rotates slowly, creating (like a spinning golf ball) a low pressure area where the spin direction is the same as the wind direction, and a higher pressure where the spin and wind directions are opposed. This causes a force that can be translated into motion of a ship or a vehicle that runs on an endless track and generates power by the rotation of its wheels. Then, there are devices that focus the wind, or at least create enhanced pressure, by using a shroud around a conventional bladed rotor.

But the two main contenders, in the United States and on the world scene, appear to be the conventional horizontal-axis machine and the vertical-axis Darrieus rotor that resembles an inverted eggbeater.

Horizontal-axis machines have been the favorite, so far, of the DOE—NASA program to develop large machines for utility application (see photo). Most of the machines installed in wind farms across the country also are that general type. However, their sizes vary greatly, from 15 to 20 Kilowatts to 300 and 400 Kilowatts, with most machines falling in the 25 to 75 Kilowatt range.

The very large horizontal-axis machines being tested by DOE-NASA now are rated at 2,500 Kilowatts (2.5 Megawatts). Although they are intended to show the way toward using such machines for commercial electrical power, these big wind turbines have been troubled by frequent breakdowns. These breakdowns indict the complexity of the large "windmill" that requires microprocessors and hydraulic controls to alter propeller pitch, further controls to turn the big machine into the wind, and a system to shut the machine down safely in case of very high winds. The failure of one component can render a very large investment useless. The recent troubles with these large machines suggests that their future is in doubt (Moore, 1984).

Meanwhile, the unconventional Darrieus, or vertical-axis wind turbine is gaining some favor as it is tested by the federal government, and by utilities (see photo). It is now being used in privately funded wind farms in different parts of the country.

In the United States the Darrieus turbine was studied and refined at SANDIA Labs serving the Department of Energy in Albuquerque, while in Canada, the Canada Research Council tested and refined the machine. Now, large numbers are installed in wind farms in California's Altamont Pass, Salinas Valley, and the San Gorgonio Mountains, and soon will be used in farms underway at Cape Blanco, Oregon, Kahuku, Hawaii, Ellenville, New York, and in the Tehachapi Mountains of Southern California. In all four of these farms nearing completion, the turbines are quite large: 300 Kilowatts at Ellenville and Cape Blanco, 500 Kilowatts at the Tehachapi location, and up to 750 Kilowatts at Kahuku, Hawaii.

The vertical-axis machine has the following advantages:

- It does not need to swing to meet the wind, because it utilizes winds from any direction.
- It has no tower to interfere with the wind flow.
- The generator is placed at the ground, therefore the machine is stable, and furthermore has no need for expensive lightweight materials in the generator.
- The Installation uses much less material because there is no supporting structure.

Held against the Darrieus is its slightly lower efficiency of energy attraction, and the fact that it must be started electrically if for any reason it stops.

---

house, who built the MOD-Zero machines, is independently developing an *intermediate-size* machine with capacity of 500 Kilowatts that it plans to sell to wind farm developers.

The Private Sector and Wind Farms

While the federal program to develop large turbines for utility use faltered, some utilities have been experimenting on their own. Southern California Edison, for example, is independently testing a large 1.3 Megawatt machine, and a number of vertical-axis machines (see photo).

Beyond those experiments, something profoundly different has occurred—a development that relieves the utility of the need to purchase any wind machines—large or small. Instead, they purchase electric power from developers who, encouraged by lenient federal and state tax provisions (see box) have launched into the wind-electric business. These developers have, so far, favored mostly small- to intermediate-

Artist's rendering of one of the very large horizontal-axis wind machines visualized in the DOE-NASA program. This design, conceived by Boeing Engineering and Construction, was to have a blade diameter of 300 feet and stand on a tower 200 feet tall. In a wind of 14 miles per hour it conceivably would generate electricity at a rate of 2.5 Megawatts per hour. (Courtesy of U.S. Department of Energy.)

This vertical-axis—or Darrieus-type—wind machine was built by the Department of Energy's SANDIA laboratories in Albuquerque, New Mexico. It is seven stories tall, and rated at 30 Kilowatts in winds of 22 miles per hour. The blades are airfoils that move *into the wind*. A very large version of this kind of rotor, built by the Canadian firm, DAF-Indal, and rated at 500 Kilowatts, has been operated by Southern California Edison at its test site in San Gorgonio Pass in Southern California since 1982.

size machines of 25 to 75 Kilowatts, and have planted them in groups known as wind farms.

Wind farms, first seen in 1980, have expanded at a remarkable rate—especially in California, where favorable winds combine with an attractive investment climate and an affluent population with dollars in search of tax shelter. In some wind farms, the wind machine manufacturer is the developer (see Marier, 1983). Other farms have developed by the formation of a limited partnership to raise capital, after which, a long-term contract is signed with a utility company, and the wind farm is built. In California the utilities have been dependent largely upon fuel oil for their power plants. Committed by federal law to buy wind power at a price that reflects their *avoided costs* (costs of fuel not burned) the utilities have offered a price per Kilowatt hour that is more attractive than those in some other states.

The California Energy Commission has promoted the enterprise by studying the wind resource, defining favored areas, and helping businesses put deals together (California Energy Commission, 1984).

California, in fact, accounts for 76 percent of all wind farm projects either in-place or announced as of March, 1984 (Table 7-11 and Fig. 7-22). On the basis of actual the General Electric analysis.

California's two major areas are Altamont Pass, near San Francisco, where Pacific Gas and Electric buys the power, and the Tehachapi Mountains, near Los Angeles, where the utility involved is Southern California Edison (see photos). These two utilities are responsible for virtually all the wind power purchased in the state, the exception being the 1.5

This Bendix-Schachle turbine generator, rated at 1.3 Megawatts in a wind of 30 miles per hour, is a very large horizontal-axis machine comparable in size to the Mode-One machine in the DOE-NASA program. This one, owned by Southern California Edison Company, is at its test site near Palm Springs in Southern California. To follow the wind, the whole tower rotates on the circular track. For scale, notice the light standard in foreground and the fuel barrels beyond the wind machine. (Courtesy of Southern California Edison Company.)

TAX CREDITS AND THE FUTURE OF WIND POWER

Federal and state tax incentives have played a vital role in the rapid development of wind machines for utility application, just as they have in promoting the use of solar hot water and space heating. The impact of these incentives has been not only to promote wind-electrical in general, but to steer the activity to those states whose policies are more attractive.

The federal government allows an enterprise producing electric power for sale to subtract 15 percent of installation costs from taxes owing. This deduction is *in addition to* the standard 10 percent investment tax credit for a business investing in equipment of various kinds. Furthermore, rapid depreciation of capital costs are allowed under the Energy Recovery Tax Act. That policy was in force through the tax year 1985 but may or may not be extended.

California is among 16 states that include wind machines in their renewable energy incentives (see Malloy, 1984). This state has a particularly attractive policy that allows a wind power producer to subtract 25 percent of installation costs from state taxes (in addition to the federal credit). California's policy will expire at the end of the 1986 tax year. Its future depends in part on the extension of federal tax credits past the end of 1985. California's incentive may, in any case, take a different form—one possibility being a credit that depends upon Kilowatt hours of power produced (Batham, 1984).

These incentives have encouraged the formation of *limited partnerships* (like those formed for real estate investing) which invest in wind-electric installations and sell their power under contract to a utility company. Commenting on this phenomenon, John Cummings, director of the Electric Power Research Institute's Renewable Resource Systems Department says: "The tax incentives are providing a window, a period of time when a lot of small turbines are being deployed. The bulk of the limited partnership tax advantages accrue in the first three years. The important question is how many of the developers will be around after the tax advantages expire" (Moore, 1984).

generating capacity installed, California accounts for 96 *percent* of the national total (Fig. 7–23). Some of the wind farms tabulated and mapped are more like wind gardens with as few as two machines. But some hold over 800 machines, and, when adjoining lease areas are considered together, there are thousands of wind machines in one area. This is the realization of the scenario suggested earlier in this chapter—planting wind machines in favored areas until those areas are saturated. One big difference is that the machines, so far, are not of the 1.5 Megawatt size assumed in

TABLE 7–11
Installed and Planned Capacity at Wind Farm Sites as of March 1984

STATE AND LOCATION	INSTALLED CAPACITY (Megawatts)[1]	POTENTIAL OF PROJECTS UNDERWAY (Megawatts)[2]	NUMBER OF PROJECTS
CALIFORNIA	282.86	1,745.62	78
Northern Part			
Pacific Gas and Electric Company			
Altamont Pass	177.67	439.80	27
Pacheco Pass	6.19	111.20	3
Salinas Valley	1.38	11.1	3
Bodega	None	3.0	1
Solano Co.	None	126.79	3
Southern Part			
1. *Southern California Edison Company*			
Tehachapi Mts.	84.29	677.49	28
San Gorgonio Pass	11.85	371.26	16
2. *San Diego Gas and Electric*			
San Diego and Boulevard	1.48	5.78	6
DELAWARE			
Sussex Co.	none	10.0	1
KANSAS			
Garden City	no data	no data	1
MASSACHUSETTS	0.3	4.62	
Nantucket	0.3	0.3	1
Princeton	0.0	0.32	1
Windsor	0.0	4.0	1
MONTANA	0.225	0.225	
Livingston	0.1	0.125	3
Ulm	0.125	0.125	1
NEW HAMPSHIRE			
Crotched Mt.	0.6	0.6	1
NEW YORK	0.0	23.7	
Ellenville	0.0	21.3	1
Riverhead	0.0	2.4	1
OREGON	1.25	26.25	
Cape Blanco	0.0	25.0	1
Coos Bay	1.25	1.25	1
TEXAS	0.250	0.250	
Freeport	no data	no data	1
Lubbock	0.125	0.125	1
Pampa	0.125	0.125	1
VERMONT			
Manchester	0.80	0.80	1
WASHINGTON			
Goodnoe Hills[3]	7.5	7.5	1
HAWAII			
Six Sites	1.65	84.75	6
U.S. Totals	295.21	1,894.31	102

[1] This is (Number of machines) $\times$ (Rated capacity of machines)

[2] For some projects, potential was obtained from (total number of machines planned) $\times$ (rated capacity of machines). If that information was not available, the stated site capacity was used, although it may *overstate* the actual plans for development.

[3] This is part of the Department of Energy program to develop large wind turbines. Three MOD-Two machines, rated at 2.5 Megawatts each, are connected to the Bonneville Power Administrations network. The project is included here because it involves a *group* of wind machines.

Source: from data provided by American Wind Energy Association, 1984.

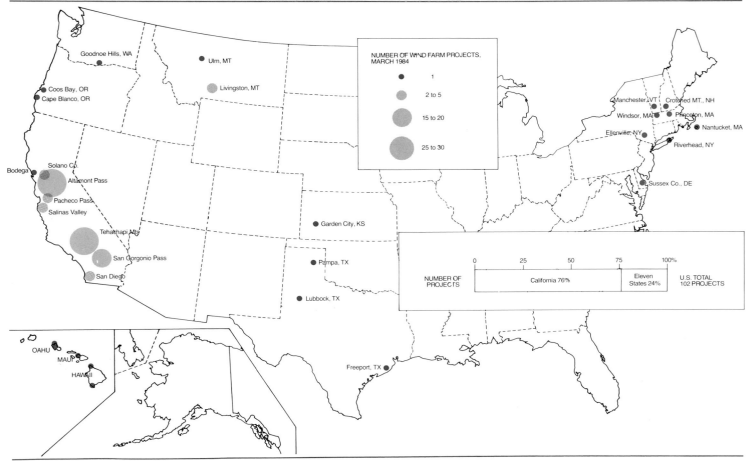

Fig. 7-22 Wind farm projects in the United States as of March 1984.
Source: From data provided by American Wind Energy Association, 1984.

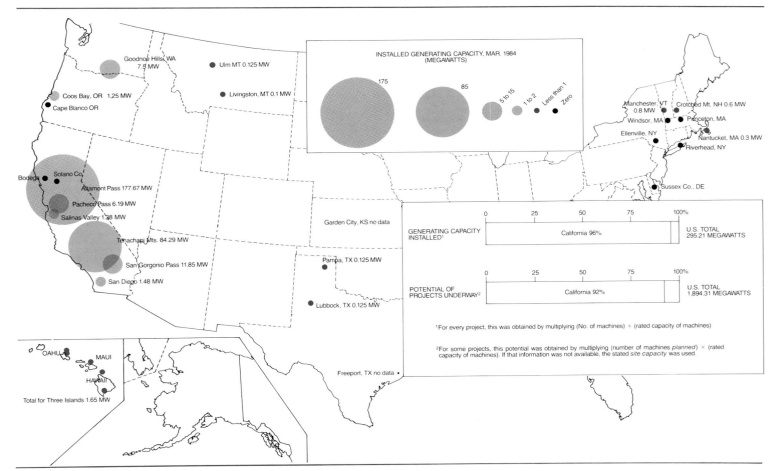

Fig. 7-23 Generating capacity installed in U.S. wind farms as of March 1984.
Source: From data provided by American Wind Energy Association, 1984.

Windfarms in the Altamont Pass area, southeast of San Francisco. In the foreground are five wind machines of the Darrieus, or vertical-axis, type. In the background, groups of more conventional horizontal-axis machines cover the hills. (Courtesy of Sandia National Laboratories.)

Megawatt of capacity near San Diego, where San Diego Gas and Electric is the utility.

As of March, 1984, the northern utility had contracts for 65 percent of the state's capacity, but when the planned potential is installed as suggested in Table 7–11, Southern California Edison will control 60 percent of the generating capacity.

Apparently, the total wind resource of California is large enough to afford much more opportunity for development. There are extensive areas of good and excellent resources along the high elevations of the Sierra Nevada and the Coastal Ranges, though many of the Sierra sites are unattractive because they are so remote and rugged (Fig.

7–24). In areas where the resource has been thoroughly studied, a total capacity of 6,680 Megawatts could be tapped. If areas with poorly defined resources are included, the estimate rises to 13,000 Megawatts (California Energy Commission, 1984). In light of these numbers, the capacity reached when today's plans are realized, that is, 1,746 Megawatts (Table 7–11) is a substantial proportion of California's total potential. The development is proceeding quite rapidly.

There is a striking similarity between wind and geothermal resources in California. First, the estimated 13,000 Megawatts of generating potential is remarkably similar to the 16,000 Megawatts of estimated generating

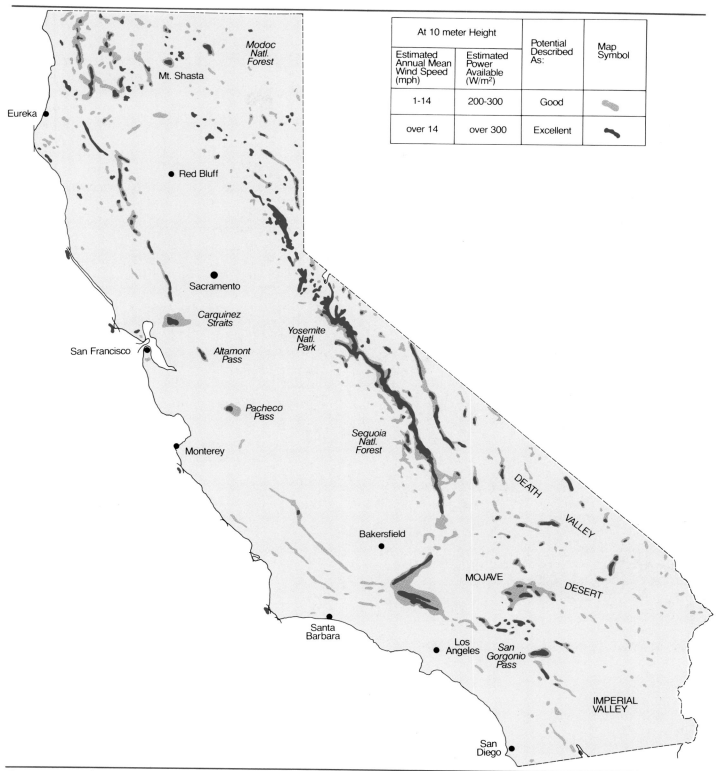

At 10 meter Height		Potential Described As:	Map Symbol
Estimated Annual Mean Wind Speed (mph)	Estimated Power Available (W/m²)		
1-14	200-300	Good	
over 14	over 300	Excellent	

Fig. 7–24 Wind resource areas of California.

Source: Miller and Simon, 1978, reported in California Energy Commission, March 1984.

potential at identified high-temperature geothermal sites in the state (see Chapter 5). Another parallel exists in the capacity developed to date. When the planned 1,700 Megawatts of wind electric capacity (see Table 7–11) is reached, it will almost match the geothermal generating capacity now installed in the state.

Information is not readily available on the amounts of *output* from the young wind farms or on the long-term

Various two- or three-bladed horizontal-axis machines in a wind farm in the Tehachapi Mountains of Southern California. Those in the foreground are rated at 65 Kilowatts each. (Courtesy of Southern California Edison.)

economics of wind farm projects. If, and when, wind farms are deprived of their special status with regard to tax credits and depreciation rates, then cold economics will determine the future of this means of providing wind power. It may be reasonable to expect that, by that time, the effects of competition and mass production in wind machine manufacturing will have lowered capital costs enough to make a major difference.

NOTES

1. Indispensible to subsequent wind power studies, the data on magnetic tape is now known as "the SANDIA tape."

2. Studies by SANDIA, Lockheed-California, and Batelle all include maps of seasonal as well as annual average Power Available.

3. The factor for height effects dramatically increases calculated Power Extractable for blades of large radius: the value of $(R/10)^{3/7}$ is only 0.428 for a blade radius of 10 meters, but is 1.811 for a radius of 40 meters.

4. More understanding of capacity factor can be gained from records of Rapid City, South Dakota, considered by G. E. to represent the Low regime, with design speed of 12 mph. Records for that station (Reed, 1975, p. 133) show winds exceeding 12 mph for 64 percent of the year at the surface. If conversion efficiency of 0.40 is assumed, the resulting "composite" capacity factor would be $(0.64) (0.40) = 0.256$, which is consistent with the factor used by G. E. for the Low regime (Table 7–7).

5. These are consistent with rotor diameters of 218, 278, and 331 feet, respectively, which are larger than those in Table 7–5 and thus lead to a more conservative estimate of number of machines per square mile.

6. Shape of curves is based on the Fischer-Prye method of predicting market penetration of new products (G. E. *Final Report*, 1977.

References

American Wind Energy Association. *AWEA Listing of Windfarm Projects.* Alexandria, Virginia, 1984.

Batham, Mike. California Energy Commission. Personal communication, Aug. 1984.

Betz, A.D. "Windmills in the Light of Modern Research." *Die Naturwissenschaften,* 15:46, Nov. 18, 1927.

California Energy Commission. *Wind Energy: Investing in Our Energy Future.* Sacramento, Cal. March 1984.

Crutcher, H. L. *Upper Wind Statistics Charts of the Northern Hemisphere.* NAVAER 50 IC-535, Vol. 1, Chief of Naval Operations, Washington, DC, Aug. 1959.

Eldridge, Frank R. *Wind Machines.* Report prepared for National Science Foundation and RANN, with cooperation of Energy Research and Development Administration, Oct., 1975.

Elliot, Dennis. *Synthesis of National Wind Energy Assessments.* Richland, Washington: Batelle, Pacific Northwest Laboratories, July, 1977.

Elliot, D. L. *et al.* *Wind Energy Resource Atlas,* Volumes 1 through 12 (for 12 regions). Richland, Washington: Pacific Northwest Laboratory. Available through National Technical Information Service, Springfield, Virginia.

General Electric Company. *Wind Energy Mission Analysis, Executive Summary.* Contract No. EY-76-C-02-2578. Philadelphia, Pa.: G.E. Space Division, Feb. 18, 1977.

————. *Wind Energy Mission Analysis, Final Report.* Contract No. EY-76-C-02-2578. Philadelphia, Pa.: G.E. Space Division, Feb. 18, 1977.

Heronemus, W. E. "Power from Offshore Winds." Proceedings of the 8th Annual Marine Technology Society Conference. Washington, DC, 1972.

————. "The United States Energy Crisis: Some Proposed Gentle Solutions." Paper at joint meeting of American Society of Mechanical Engineers, and Institute of Electrical and Electronics Engineers, W. Springfield, Mass., Jan. 12, 1972.

Lockheed–California Co. *Wind Energy Mission Analysis, Final Report.* Burbank, Calif.: Lockheed–California Co., 1976.

Marier, Donald. "Wind Farms, Boom or Bust," *Alternative Sources of Energy,* May–June, 1983, 9–17.

Malloy, Molly. "The Solar Age 1984 State Tax Credit Survey" *Solar Age,* June, 1984, 24–26.

METREK. *Solar Energy: A Comparative Analysis to the Year 2020.* Prepared for E.R.D.A. under Contract No. E-(4918)-2322, Technical Report MTR-7579. McLean, Va.: The Mitre Corporation, Metrek Division, Aug., 1978.

Miller, A., and R. Simon. "Wind Power Potential in California," San Jose State University, prepared for California State Energy Commission, May 1978.

Moore, Taylor. "Wind Power: a Question of Scale," *Electric Power Research Institute Journal,* May 1984, 6–16.

Reed, J. W. *Wind Power Climatology of the United States.* SAND 74-0348. Albuquerque, N. Mex.: SANDIA Laboratories, June, 1975.

————, **R. C. Maydew, and B. F. Blackwell.** *Wind Energy Potential in New Mexico.* SAND 74-0071. Albuquerque, N. Mex.: SANDIA Laboratories, July 1974.

Trewartha, Glenn T. *An Introduction to Climate,* 4th Ed. N.Y.: McGraw-Hill, 1968.

U.S. Department of Energy. *Annual Energy Outlook, 1983.* Washington, DC, May 1984.

U.S. Department of Energy. *The MOD-Two Wind Turbine Development Project.* Washington, DC, October 1980.

U.S. Department of Energy. *Wind Energy Systems Program.* Washington, DC, 1980.

U.S. Department of the Interior. Bureau of Mines. *United States Energy Through the Year 2000.* 1975.

U.S. Energy Research and Development Administration (E.R.D.A.). Wind Systems Branch Division of Solar Energy. *Federal Wind Energy Program.* Washington, DC, Jan. 1, 1977.

————. *Federal Wind Energy Program: Summary Report.* Washington, DC: Jan. 1, 1977. *Federal Wind Energy Program: Summary Report,* Jan. 1, 1978.

U.S. Geological Survey. *The National Atlas of the United States of America.* Washington, DC, 1970.

Hydroelectric Power

8

- World Hydroelectric Power Resources
- United States Hydroelectric Power Resources

Hydroelectric power is produced from the natural movement or flow of water, as in ocean tides or rivers. The power of water can be harnessed by taking advantage of its natural fall from one level to another or by creating this fall with dams. The falling water can do work in water wheels or modern turbines, and can then be converted into other energy forms, such as electrical.

The hydrologic cycle explains how solar energy is converted to mechanical energy. Water is continually evaporated from the oceans by the sun. The wind then transports it, sometimes thousands of miles, from tropical to mid-latitude regions. This moisture is then dropped as precipitation on both oceans and continents. Finally, the water runs off the land back to sea level. In this way, solar radiation is converted into the kinetic, or mechanical energy, of a flowing fluid.

The use of falling or swiftly running water to produce energy dates back to ancient Chinese and Egyptian civilizations, where the falling water was used to turn water wheels. During the early period of the Industrial Revolution, mechanical work was generated by this type of water power in both North America and Europe.

Prior to electricity, power from water wheels could be transported only for very short distances, and its use was severely restricted. The development of the electrical generator and transmission lines greatly increased the ability to transmit power over a wider area. In spite of these advances, however, hydroelectricity is still generally consumed in the region of its production. Its limited use is due partly to problems of transmission and partly to competition from cheaper forms of energy.

As with all major sources of energy, there are advantages and disadvantages associated with the development and operation of hydroelectric power plants. The major *advantages* are as follows:

1. Hydroelectricity is a renewable resource powered by the hydrologic circle. Therefore the water that drives the turbines is not affected by the supply and cost of fossil fuels.

2. Hydroelectric power plants can save fuel by replacing coal, oil, gas, and uranium fueled plants.

←Old flour mill at Pigeon Forge, Tennessee, demonstrating the mechanical use of water power. (Courtesy of the American Petroleum Institute.)

3. Since the burning of a fuel is not involved in the production of hydroelectricity, the plants do not produce thermal or atmospheric pollution.

4. Hydroelectric power plants are flexible in meeting changing power demands. They can adjust to the changing demand for power at a given time and can supply the maximum power needed during a stated period.

5. Hydroelectric power plants are very efficient producers of electricity. They have more than twice the efficiency of coal plants, operating at 85 to 95 percent efficiency.

6. The fact that heat is not involved in the process contributes to a long life and low level of maintenance of the hydro equipment.

7. Hydroelectric power has a long history and the engineering involved is well-known.

8. With no fuel cost, hydroelectric power plants are relatively inexpensive to operate.

The major *disadvantages* of hydroelectric power facilities are:

1. The high cost of dam construction coupled with high interest rates often makes it difficult to obtain financing for hydroelectric power facilities.

2. Unless there is a sufficient storage of water behind a dam the output of electricity can fluctuate seasonally depending on stream flow.

3. Hydroelectric power plants can affect the quantity and quality of water flowing in a river. Fish migrations are restricted by dams and fish health is affected by changes in water temperature and introduction of excess nitrogen at spill ways.

4. Reservoirs flood large areas of land that could be used for other purposes including irreplaceable wilderness land and farm land.

WORLD HYDROELECTRIC POWER RESOURCES

In 1981, the total world hydroelectric power potential was estimated at more than 2.5 million Megawatts. Only 19 percent of this total has been developed (United Nations, 1983). Asia houses almost 30 percent of the total world capacity, followed by Africa and North America with 17 and 16 percent respectively (Fig. 8–1). The most intensive development of hydroelectricity has occurred in Europe and North America. These two regions have developed 44 and 32 percent of their respective capacities and over 56 percent of the world's developed capacity. On the other hand, Asia, Africa, and South America have over 57 percent of the total world capacity, but less than 30 percent of the total developed capacity. The Soviet Union holds approximately 11 percent of both world total and world developed capacity. Presently, hydroelectric power production accounts for 7 percent of the world's primary energy production (Department of Energy, November 1983). This figure represents a 2 percent increase from 1973. In the future, hydropower seems destined to play a relatively minor role in the development of energy resources in most parts of the world. In some underdeveloped and developing countries, however, where fuel resources are scarce and hydroelectric potential great, hydropower is likely to be of increasing importance.

World's Ten Largest Hydroelectric Facilities

Name of Dam	Country	Present Capacity (Megawatts)	Ultimate Capacity (Megawatts)	Initial Operation
Itaipu	Brazil Paraguay	—	12,600	UC[1]
Grand Coulee	United States	5,463	9,780	1941
Paulo Afonso	Brazil	1,299	6,744	1955
Guri	Venezuela	524	6,500	1967
Tucurui	Brazil	—	6,480	UC
Sayanskaya	U.S.S.R.	—	6,400	UC
Krasnoyarsk	U.S.S.R.	6,096	6,096	1968
La Grande	Canada	—	5,416	UC
Churchill Falls	Canada	5,225	5,225	1971
Bratsk	U.S.S.R.	4,100	4,600	1964

[1] UC indicates under construction.

Sources: Grathwohl, 1982; and Federal Energy Regulatory Commission. January 1980.

UNITED STATES HYDROELECTRIC POWER RESOURCES

There are two basic types of hydroelectric power plants in the United States: *conventional* and *pumped storage*. Conventional hydroelectric plants produce power from natural stream flow, and are limited by the hydrostatic head created by the dam and reservoir (Fig. 8–2). Pumped storage

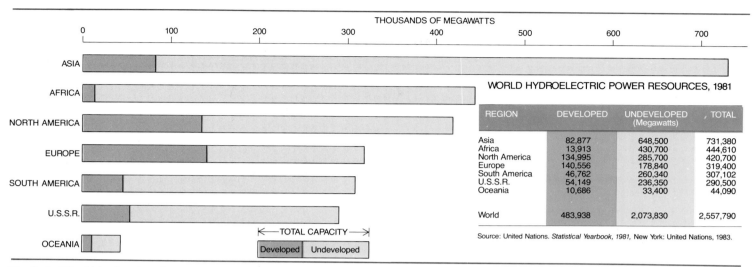

THOUSANDS OF MEGAWATTS

WORLD HYDROELECTRIC POWER RESOURCES, 1981

REGION	DEVELOPED	UNDEVELOPED (Megawatts)	TOTAL
Asia	82,877	648,500	731,380
Africa	13,913	430,700	444,610
North America	134,995	285,700	420,700
Europe	140,556	178,840	319,400
South America	46,762	260,340	307,102
U.S.S.R.	54,149	236,350	290,500
Oceania	10,686	33,400	44,090
World	483,938	2,073,830	2,557,790

Source: United Nations. *Statistical Yearbook, 1981,* New York: United Nations, 1983.

Fig. 8–1 World hydroelectric power resources, 1981.

installations use the same generation principles as conventional systems, but produce power during peak load periods.

Pumped storage projects are essentially a means of storing energy in that low-cost off-peak energy is used to pump water to an upper reservoir where it is stored as potential energy. The water is then released when needed to produce high-value peak load power. For every 3 Kilowatts of energy generated during peak periods, approximately 4 Kilowatts of off-peak energy is required for pumping. Although ordinarily designed for peak load operation, pumped storage

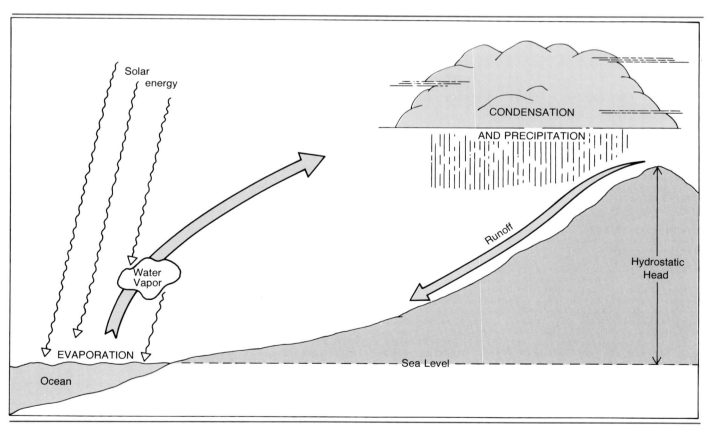

Simplified view of the hydrologic cycle in which water is raised by solar energy and can perform work as it falls back to sea level.

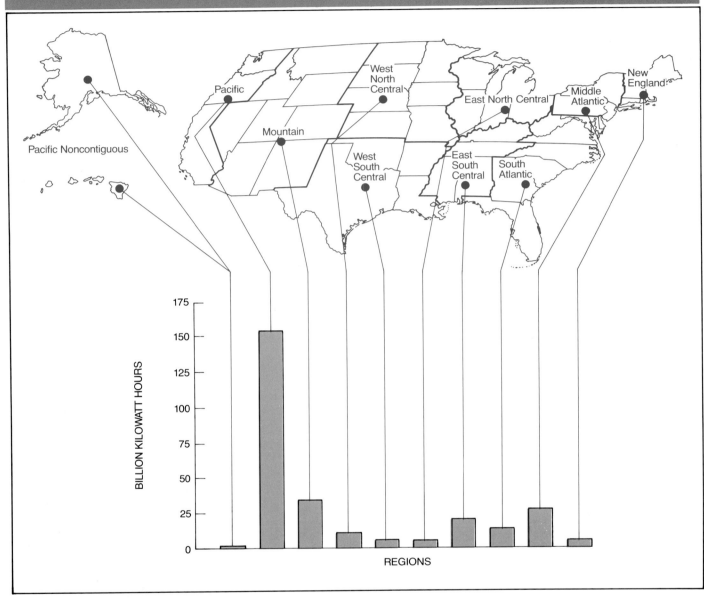

HYDROELECTRIC POWER—NET GENERATION, BY REGION AND STATE, 1980

Pacific Noncontiguous

Pacific

West North Central

Mountain

West South Central

East North Central

East South Central

South Atlantic

Middle Atlantic

New England

BILLION KILOWATT HOURS

REGIONS

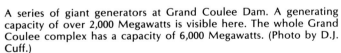

Part of the Grand Coulee Dam on the Columbia River in the state of Washington. Each of the curved penstocks serves a turbine in the generator house at the left. (Photo by D.J. Cuff.)

A series of giant generators at Grand Coulee Dam. A generating capacity of over 2,000 Megawatts is visible here. The whole Grand Coulee complex has a capacity of 6,000 Megawatts. (Photo by D.J. Cuff.)

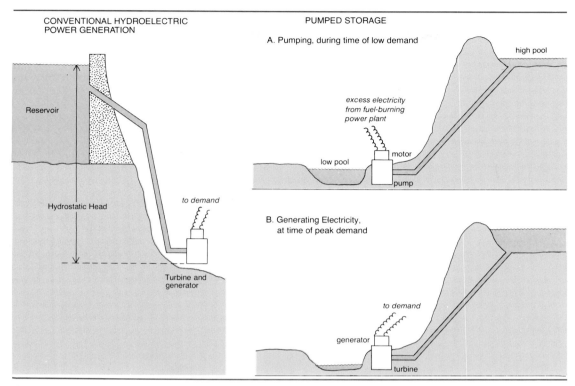

CONVENTIONAL HYDROELECTRIC
POWER GENERATION

Reservoir

Hydrostatic Head

to demand

Turbine and
generator

PUMPED STORAGE

A. Pumping, during time of low demand

high pool

excess electricity
from fuel-burning
power plant

low pool

motor

pump

B. Generating Electricity,
at time of peak demand

to demand

generator

turbine

Fig. 8-2 The pumped storage plant, showing two phases of operation.

plants may be used in emergency situations such as during unscheduled outages of thermal units. The advantage of pumped storage plants is the use of excess steam-generated power to pump water to the upper reservoirs during lower electricity demand periods. During these periods of lowered demand, it is not practical to shut down thermal (coal or nuclear) base load plants because of the high cost and lengthy time period required to start them up again. Instead, this excess base power is used to operate the pump in a pumped storage plant. The stored energy (water) can then be released during periods of peak power demands when base load plants are already under full load. Currently, pumped storage is the only method available for storing and releasing large amounts of electrical energy over short periods of time (Federal Energy Regulatory Commission, 1980). The pumped storage facility is sometimes located at an existing hydroelectric dam. When it is not associated with any natural hydrostatic head, it is considered a "pure" pumped storage facility. Figure 8-2 illustrates the pure type.

Trends in Development of Hydroelectric Power Resources

Hydroelectric power development in the United States began with the electrical age, around the year 1878. The first hydroelectric plant in the United States was a water-wheel generator at Niagara Falls, used to illuminate the Falls by arc lamps in 1879. The Appleton, Wisconsin hydroelectric plant provided electricity for incandescent lighting in 1882

and became the world's first hydroelectric central station. It was equivalent to an installed capacity of 12.5 Kilowatts.

By 1907, 15 percent of the U.S. electric generating capacity was provided by water power, and by 1930 hydroelectric power plants provided 30 percent of the nation's generating capacity. In 1984, hydroelectric installations provided approximately 13 percent of the nation's electric energy. The Grand Coulee project is the largest hydroelectric project in the United States. With the completed addition of the third power plant, its installed generating capacity will be 6,170 Megawatts. Figure 8-3 shows that the total conventional hydroelectric capacity in 1980 was approaching 65,000 Megawatts. The expected capacity to be developed by the year 2000 is 105,000 Megawatts. The projection for conventional hydroelectric capacity in the year 2000 is based primarily on data from the Second National Water Assessment (Federal Energy Regulatory Commission, January 1983) which reflects an average annual increase of 1.25 percent.

Rocky River, located on the Housatonic River in Connecticut, was the first pumped storage project constructed in the United States. It has an installed capacity of 31 Megawatts and was first operated in 1929. In the early 1970s the installed capacity of pumped storage hydroelectricity jumped tremendously due to the start-up of three large facilities: Northfield Mountain, Massachusetts (1972), Ludington, Michigan (1973), and Blenheim-Gilboa, New York (1973). These three facilities added almost 4,000 Megawatts to the national total. In 1981, the installed pumped storage capacity of the United States was approxi-

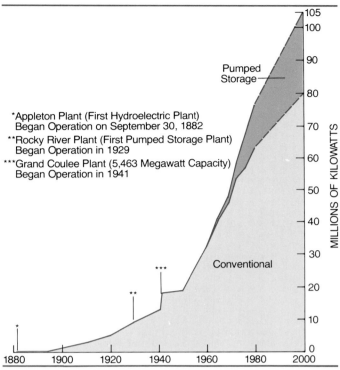

Fig. 8-3 Growth in U.S. developed hydroelectric capacity, 1880–2000, showing conventional and pumped storage types.
Source: Federal Energy Regulatory Commission. *Hydroelectric Power Resources of the United States,* January 1, 1980.

mately 16,000 Megawatts; this capacity is expected to more than double by the year 2000.

Conventional Hydroelectric Power Capacity in the United States

The U.S. Army Corps of Engineers' hydroelectric power resource study of 1981 estimated that the potential conventional hydroelectric power capacity of the United States was almost 129,000 Megawatts. At that time almost 65,000 Megawatts had been developed (U.S. Army Corps of Engineers, September 1981). Of the 65,000 Megawatts of developed capacity, over 5,400 Megawatts of capacity was located at the Grand Coulee plant on the Columbia River in Washington state. This is the largest facility in the country and has an ultimate potential capacity of over 9,700 Megawatts (Grathwohl, 1982). Other conventional hydroelectric facilities with installed generating capacity greater than 1,500 Megawatts are the Chief Joseph and Dalles plants on the Columbia River in Washington, the John Day Plant on the Columbia River in Oregon, and the R. Moses plant on the Niagara River in New York.

Although it appears that about one-half of the nation's estimated hydroelectric power capacity remains undeveloped, these statistics can be misleading. First, the estimated undeveloped potential has been reduced by 44,000 Megawatts since 1978 (Federal Energy Regulatory Commission, 1978). After taking into account consideration

of the economic, social, and environmental factors, the number of acceptable undeveloped sites on large rivers is probably small.

The developed hydroelectric power capacity represents roughly 13 percent of the nation's total electrical generating capacity. In the unlikely event of full development of the undeveloped potential, the 129,000 Megawatts would constitute 18 percent of the national generating capacity projected for 1995 Department of Energy, May 1984). The most recent forecasts for hydroelectric power usage in the United States indicate a decline to between 10 and 11 percent of the total electric generating needs in 1995. It is clear that unless the national need for electrical generating capacity is reduced significantly, conventional hydroelectric power alone cannot provide a large part of it.

Distribution of Hydroelectric Power Capacity by Economic Region and by State

The U.S. Army Corps of Engineers has assessed the conventional hydroelectric power potential of the United States according to 11 geographic regions as delineated by the U.S. Bureau of the Census (Table 8-1 and Fig. 8-4). This assessment allows comparison of the hydroelectric potential with demographic, economic, and other energy resource data.

The Pacific region dominates all others with 41 percent of the total hydroelectric potential, 48 percent of the developed potential, and 35 percent of the undeveloped potential. The Mountain and South Atlantic regions rank second and third in terms of total hydroelectric potential with 17 and 11 percent, respectively, of the United States' total. These two regions also house 9 and 11 percent respectively of the nation's total developed and 24 and 11 percent of its undeveloped potential.

Five states, Washington, Oregon, California, Idaho, and New York, have 54 percent of the total hydroelectric capacity of the country (Table 8-2 and Fig. 8-5). These same five states have 57 percent of the developed capacity and 54 percent of the undeveloped potential. Washington and Oregon are the most extensively developed states, containing almost 40 percent of the nation's hydroelectric capacity.

In the eastern United States, the most major hydropower development has occurred in Alabama and Tennessee where 7 percent of the nation's capacity is located. Much of the development in these two states has been along the Tennessee River and its tributaries. The principal agency involved in developing hydroelectricity in this region is the Tennessee Valley Authority (TVA).

Conventional Hydroelectric Power Under Construction, Planned

In 1980, approximately 86 new hydroelectric plants or additions were under construction or planned in the United States (Table 8-3 and Fig. 8-6). When completed, these plants and additions will have an installed capacity of 5,342

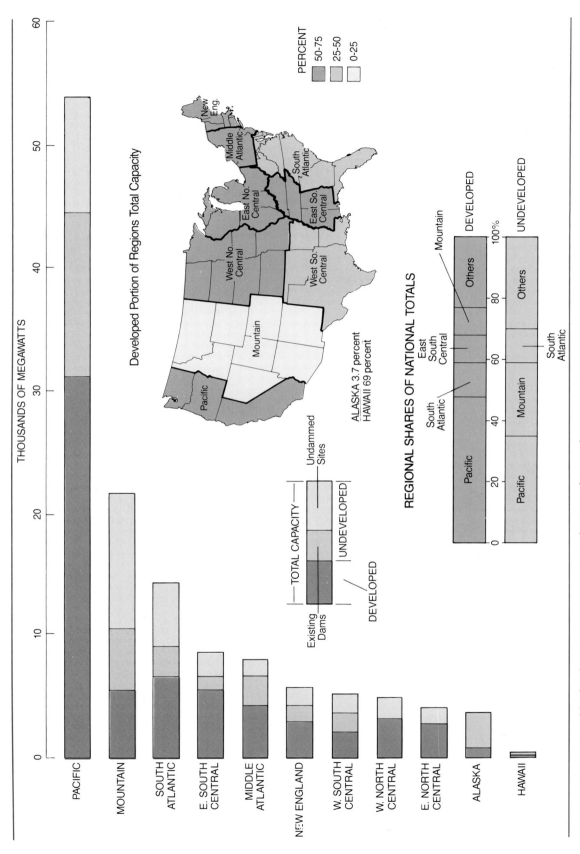

Fig. 8-4 Conventional hydroelectric power capacity in the United States, by economic region as of September 1981.

Source: U.S. Army Corps of Engineers. *National Hydroelectric Power Resources Study, Volume 12, Data Base Inventory,* 1981.

TABLE 8–1
Conventional Hydroelectricl Power Capacity in the United States, by Economic Region, 1981

| | | CAPACITY (Megawatts) | | | | |
| | | UNDEVELOPED | | | | |
REGION	DEVELOPED	At Exist-ing Sites	At Un-dammed Sites	TOTAL UNDEVELOPED	TOTAL POTENTIAL	PROPORTION DEVELOPED (PERCENT)
New England	2,978.8	1,064.6	1,761.9	2,826.5	5,859.3	50.8
Middle Atlantic	4,141.2	2,207.0	1,383.0	3,590.0	7,731.2	53.6
East North Central	2,724.9	1,247.5	15.0	1,262.6	3,987.6	68.3
West North Central	3,275.0	1,419.9	69.9	1,471.7	4,764.7	68.7
South Atlantic	7,096.1	1,991.0	5,289.2	7,280.3	14,376.6	49.4
East South Central	5,734.9	1,740.0	867.5	2,647.5	8,382.4	68.7
West South Central	2,099.2	1,385.0	1,827.4	3,212.4	5,311.5	39.5
Mountain	5,588.2	4,127.5	11,733.9	15,861.3	21,449.6	21.4
Pacific	31,037.7	13,770.4	8,723.6	22,493.9	53,481.7	58.0
Alaska	136.4	68.6	3,466.6	3,535.2	3,671.6	3.7
Hawaii	17.4	1.4	6.4	7.8	25.2	69.0
U.S. Total	64,883.8	29,022.9	35,144.3	64,994.2	128,878.4	50.3

Source: U. S. Army Corps of Engineers. *National Hydroelectric Power Resources Study*, Volume 12, *Data Base Inventory, Table 7–3*, September 1981.

TABLE 8–2
Summary of Conventional Hydroelectric Capacity by State and Economic Region showing Amounts Developed, Undeveloped (Undammed Sites), and Additional Capacity at Existing Dams as of September 1981 (in Megawatts)

REGION AND STATE	DEVELOPED	ADDITIONAL CAPACITY AT EXISTING DAMS	TOTAL PO-TENTIAL AT EXIST-ING DAMS	POTENTIAL AT UNDEVEL-OPED SITES	TOTAL NEW POTENTIAL (COLS. 2 and 4)	TOTAL POTENTIAL (COLS. 1 and 5)	PERCENT DEVELOPED	ADDITIONAL AS A PORPORTION OF TOTAL UNDEVELOPED (COLS. 2 and 5)
New England								
Connecticut	125.4	59.6	185.0	0.0	59.6	185.0	67.8	100.0
Maine	539.5	416.8	1,011.8	1,642.6	2,059.4	2,598.9	20.8	20.2
Massachusetts	1,753.6	111.7	1,865.3	0.0	111.7	1,865.3	94.0	100.0
New Hampshire	382.1	318.2	700.3	20.3	338.5	720.6	53.0	94.0
Rhode Island	1.5	11.8	13.3	0.0	11.8	13.3	11.3	100.0
Vermont	230.7	146.5	377.2	99.0	245.5	476.2	48.4	59.6
Total	2,978.8	1,064.6	4,098.9	1,761.9	2,826.5	5,859.3	50.8	37.7
Middle Atlantic								
New York	3,712.7	1,332.4	5,045.1	631.4	1,963.8	5,676.5	65.4	67.8
New Jersey	2.4	25.2	27.6	0.0	25.2	27.6	8.6	100.0
Pennsylvania	426.1	849.4	1,275.5	751.6	1,601.0	2,027.1	21.0	53.0
Total	4,141.2	2,207.0	6,348.2	1,383.0	3,590.0	7,731.2	53.6	61.5
East North Central								
Indiana	3.4	72.0	75.4	0.0	72.0	75.4	4.5	100.0
Illinois	27.2	249.6	276.8	0.0	249.6	276.8	9.8	100.0
Michigan	2,266.0	396.4	2,662.4	6.9	403.3	2,662.4	84.9	98.3
Ohio	0.0	151.2	151.2	4.1	155.4	155.4	0.0	97.3
Wisconsin	428.3	378.3	806.6	4.0	382.3	810.6	52.8	99.0
Total	2,724.9	1,247.5	3,972.4	15.0	1,262.6	3,987.6	68.3	98.8

TABLE 8-2 *(continued)*

REGION AND STATE	DEVELOPED	ADDITIONAL CAPACITY AT EXISTING DAMS	TOTAL PO- TENTIAL AT EXIST- ING DAMS	POTENTIAL AT UNDEVEL- OPED SITES	TOTAL NEW POTENTIAL (COLS. 2 and 4)	TOTAL POTENTIAL (COLS. 1 and 5)	PERCENT DEVELOPED	ADDITIONAL AS A PORPORTION OF TOTAL UNDEVELOPED (COLS. 2 and 5)
West North Central								
Iowa	135.0	461.3	596.3	0	461.3	596.3	22.6	100.0
Kansas	1.9	41.0	42.9	0	41.0	42.9	4.4	100.0
Minnesota	168.9	329.2	498.1	6.9	336.1	505.0	33.5	97.9
Missouri	986.2	287.0	1,183.2	62.9	349.9	1,246.1	71.9	82.0
Nebraska	182.0	25.7	207.7	0.0	25.7	207.7	87.6	100.0
North Dakota	400.0	275.7	675.7	0.0	257.7	675.7	59.2	100.0
South Dakota	1,491.0	0.0	1,491.0	0.0	0.0	1,491.0	100.0	—
Total	3,275.0	1,419.9	4,694.9	69.8	1,471.7	4,764.7	68.7	96.5
South Atlantic								
Florida	41.7	21.8	63.5	9.0	30.8	72.5	57.5	70.7
Georgia	2,261.7	253.4	2,515.1	389.4	642.8	2,904.5	77.9	39.4
Maryland	496.6	23.6	520.2	0.0	23.6	520.2	95.5	100.0
North Carolina	1,918.7	466.8	2,385.5	2,298.8	2,765.6	4,684.3	41.0	16.9
South Carolina	1,606.6	309.2	1,915.8	1,285.5	1,594.7	3,201.3	50.2	19.4
Virginia	616.9	188.5	805.4	589.6	778.1	1,395.0	44.2	24.2
West Virginia	153.9	727.7	881.6	716.9	1,444.7	1,598.5	9.6	50.4
Total	7,096.1	1,991.0	9,087.1	5,289.2	7,280.3	14,376.6	49.4	27.3
East South Central								
Alabama	2,841.6	341.5	3,183.1	24.0	365.5	3,207.0	88.6	93.4
Kentucky	797.3	973.4	1,770.7	325.0	1,298.4	2,095.7	38.0	75.0
Mississippi	0.0	69.5	69.5	12.0	81.4	81.4	0.0	85.3
Tennessee	2,096.0	355.6	2,451.6	506.5	862.1	2,958.1	70.1	41.2
Total	5,734.9	1,740.0	7,474.9	867.5	3,474.9	8,343.2	68.7	50.1
West South Central								
Arkansas	928.0	516.8	1,444.8	460.0	976.8	1,904.8	48.7	52.9
Louisiana	81.0	141.6	222.6	94.8	236.4	317.4	25.5	59.9
Oklahoma	769.0	313.4	1,082.4	52.6	366.0	1,135.0	67.8	85.6
Texas	321.2	413.2	734.4	1,220.0	1,633.2	1,954.3	16.4	25.3
Total	2,099.2	1,385.0	3,484.2	1,827.4	3,212.4	5,311.5	39.5	43.1
Mountain								
Arizona	217.3	87.3	304.6	1,518.2	1,605.5	1,822.8	11.9	5.4
Colorado	224.0	355.6	579.6	897.2	1,252.7	1,476.7	15.2	24.1
Idaho	2,662.4	1,532.2	4,194.6	6,767.9	8,300.2	10,962.5	24.3	18.5
Montana	2,059.0	904.8	2,963.8	1,057.1	1,961.9	4,021.0	51.2	46.1
Nevada	3.2	3.0	6.2	0.0	3.0	6.2	51.6	100.0
New Mexico	24.3	381.8	406.1	1.5	383.2	407.6	6.0	99.6
Utah	171.4	473.4	644.8	1,051.1	1,524.5	1,695.9	10.1	31.1
Wyoming	226.6	389.4	616.0	440.9	830.3	1,056.9	21.4	46.9
Total	5,588.2	4,127.5	9,715.7	11,733.9	15,861.3	21,449.6	21.4	35.2
Pacific								
California	5,108.7	3,854.3	8,963.4	2,429.4	6,283.7	11,342.4	45.0	61.3
Oregon	6,932.6	3,032.2	9,964.8	1,883.3	4,915.4	11,848.1	58.5	61.7
Washington	18,996.4	6,883.9	25,880.3	4,410.9	11,294.8	30,291.2	62.7	60.9
Total	31,037.7	13,770.4	44,808.5	8,723.6	22,493.9	53,481.7	58.0	61.2
Alaska	136.4	68.6	205.0	3,466.6	3,535.2	3,671.6	3.7	1.9
Hawaii	17.4	1.4	18.8	6.4	7.8	25.2	69.0	17.9
U.S. Total	64,883.8	29,022.9	93,908.6	35,144.3	64,167.2	129,051.0	50.3	45.2

Source: U. S. Army Corps of Engineers. *National Hydroelectric Power Resources Study,* Volume 12, *Data Base Inventory Table 7–3,* September 1981.

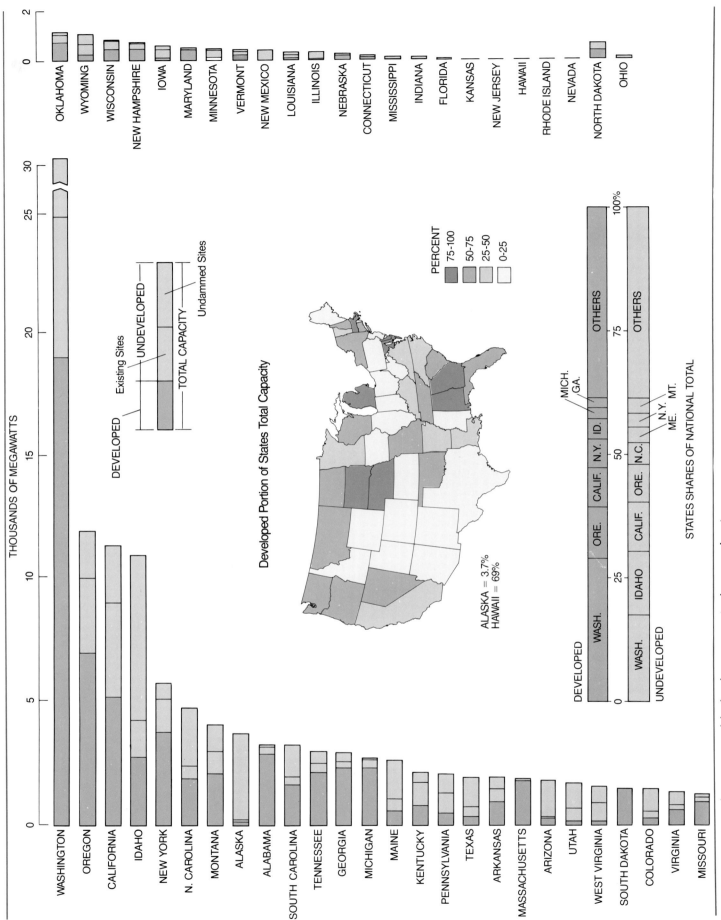

Fig. 8–5 Summary of conventional hydroelectric capacity by state showing amounts developed, undeveloped (undammed sites), and additional capacity at existing dams as of September 1981.

Source: U.S. Army Corps of Engineers. *National Hydroelectric Power Study, Volume 12, Data Base Inventory,* 1981

TABLE 8–3

Conventional Hydroelectric Plants or Additions under Construction or Planned as of January 1, 1980[1]
(in Kilowatts)

STATE AND PLANT	UNDER CONSTRUCTION	PLANNED	STATE AND PLANT	UNDER CONSTRUCTION	PLANNED
Alabama			**Michigan**		
R.L. Harris	135,000		Sturgis	1,620 (A)	
Walter Bouldin	225,000		Totals	1,620	
Mitchell	97,000 (A)		**Minnesota**		
Martin Dam		171,000 (A)	Cloquet	996 (A)	
Harris		117,000 (A)	Totals	996	
Totals	457,000	288,000	**Missouri**		
Alaska			Clarence Cannon	27,000	
Soloman Gulch	12,000		Totals	27,000	
Hidden Falls	200		**Montana**		
Green Lake	16,500		Libby	420,000	
Snettisham	27,000 (A)		Libby Rereg		88,000
Chester Lake	2,500		Totals	420,000	88,000
Totals	58,200		**New Hampshire**		
California			Sawmill	3,174	
Auburn	300,000		Totals	3,174	
Rollins	11,000		**New York**		
Inskip	1,650 (A)		Colliersville	1,450	
South	2,750 (A)		Lower Beaver	1,000	
Kerckhoff No. 2	140,000		Union		2,000 (A)
Big Creek No. 3	35,000 (A)		Kensico		3,000
Santa Barbara Hy	1,500		Ashokan		4,750
Pyramid	75,000		Dolgeville		12,600
Cottonwood	14,100		Trenton		30,200 (A)
Volta 2		1,000	Fort Edward		10,000
Pine Flat		165,000	Hudson Falls		60,000
San Luis Obispo		5,900	South Glen Falls		12,200 (A)
Sepulveda Canyon		9,000	Feeder Dam		2,000 (A)
Venice		10,000	Sherman Island		8,000 (A)
Foothill Feeder		9,100	Glen Park		20,000
Yorba Linda		5,100	Granby		10,000
Rio Hondo		2,000	Totals	2,450	164,750
San Dimas		9,900	**Ohio**		
Santiago Creek		3,000	Greenup	70,560	
Coyote Creek		3,000	Racine L & D	40,000	
Corona		3,000	Totals	110,560	
Temescal		3,000	**Oregon**		
Lake Matthews		4,900	Bull Run No. 1	23,750	
Totals	581,000	233,900	Pelton Rereg Dam	15,000	
Colorado			Totals	38,750	
Loveland	900		**South Carolina**		
Totals	900		St. Stephen	84,000	
Georgia			Totals	84,000	
Richard B. Russell		300,000	**Texas**		
Bartletts Ferry		100,000 (A)	Amistad	32,000	
Totals		400,000	Totals	32,000	
Idaho			**Vermont**		
Brownlee	225,000 (A)		White Current	432	
Lower New	7,200 (A)		Great Falls	600 (A)	
City	5,200 (A)		Totals	1,032	
Upper Plant	7,200		**Washington**		
Cascade		12,800	Mayfield	40,000 (A)	
Swan Falls		90,000	Bonneville 2nd P	532,000	26,000 (a)
Dike Project		50,000	Potholes E. Canal	5,200	
A.J. Wiley		75,000	Potholes E. Canal	2,600	
Totals	244,600	227,800	Grand Coulee	707,000 (A)	
Kentucky			High Ross		372,000
Cannelton	70,560		Totals	1,286,800	398,200
Totals	70,560		**Wisconsin**		
Maine			Black Brook	650	
Brunswick/Topsha	12,000		Totals	650	
Barkers Mill	1,500		**Wyoming**		
Cold Stream		83,000	Seminoe	8,400 (G)	
Totals	13,500	83,000	Totals	8,400	
Massachusetts					
Lawrence Hydro	14,800				
Totals	14,800		U.S. Totals	3,457,992	1,883,650

[1]Capacity at undeveloped sites, except "A" denotes an addition to an existing plant and "a" an addition to an existing plant for which the initial installation is presently underway. "G" denotes a rewind addition.

Source: Federal Energy Regulatory Commission, 1980.

Fig. 8-6 Conventional hydroelectric plants or additions under construction of planned.
Source: Federal Energy Commission, *Hydroelectric Power Resources of The United States,* January 1, 1980.

Megawatts. Of the 86 total hydroelectric projects, 64 were new plants and 22 were additions to existing installations.

Three-fourths of the new hydroelectric plants or additions under construction are located in the Mountain and Pacific regions of the country. Of that three-fourths, 67 percent approximately are located in Washington, Idaho, and California. In the eastern part of the country, the most intensive construction activity is taking place in Alabama.

Approximately 50 percent of the planned activity will be located in the Mountain and Pacific regions, with Washington, California, and Idaho accounting for over 45 percent. Outside the western United States, the major facilities planned will be located in Alabama, Georgia, and New York, accounting for approximately 45 percent of the planned hydroelectric projects.

Along with the new hydroelectric plants or additions under construction or planned, 384 installations have been projected. Projected hydroelectric plants or additions are defined as potential developments not under construction or included in reports at the Regional Electric Reliability Council. These projected plants have obtained licenses or permit status from the Federal Power Commission and have been authorized or recommended for federal construction, or they have structural provisions for plant additions. If completed, these projects will add 18,582 Megawatts to the nation's generating capacity. Approximately 200 of these projects would have capacities of less than 15 Megawatts.

Pumped Storage Hydroelectric Capacity

As of 1981, the total developed pumped storage capacity was approximately 16,000 Megawatts (Table 8-4). Between 1978 and 1981 the total pumped storage capacity grew by approximately 6,000 Megawatts. However, by the end of the century it is possible that the nation's hydroelectric pumped storage capacity will have doubled.

DISTRIBUTION OF PUMPED STORAGE HYDRO-ELECTRIC CAPACITY BY STATE In 1981, 23 states held all of the developed pumped storage capacity of the United States (Fig. 8-7). Two-thirds of the total developed capacity is located in states east of the Mississippi

TABLE 8-4
Pumped Storage Projects in the United States as of November 1, 1981

STATE	PLANT NAME	IN OPERATION[a]	PROJECTED[b]	CAPACITY (MW)	DATE OF INITIAL OPERATION
Arizona	Mormon Flat	X		58	1971
	Horse Mesa	X		130	1972
	Montezuma[1]		X	505	1988
Arkansas	Degray	X		68	1971
California	Senator Wash	X		7	1966
	O'Neill	X		25	1967
	Thermalito	X		115	1968
	Edward G. Hyatt	X		644	1968
	San Luis	X		424	1968
	Castaic	X		1,331	1973
	Helms	X		1,050	1983
Colorado	Flatiron	X		72	1954
	Cabin Creek	X		300	1966
	Mt. Elbert[1]		X	200	1985
	Azure[2]		X	240	UA[3]
	Oak Creek[2]		X	3,600	UA[3]
Connecticut	Rocky River	X		31	1929
Georgia	Carters	X		500	1975
	Wallace	X		324	1980
	Rocky Mountain[1]		X	675	1986
	Richard B. Russell[1]		X	600	UA[3]
Massachusetts	Northfield Mountain	X		1,000	1972
	Bear Swamp	X		600	1974
Michigan	Ludington	X		1,978	1973
Missouri	Taum Sauk	X		408	1963
	Harry S. Truman	X		160	1981
	Clarence Cannon	X		58	1983
New Hampshire	Mud Pond[2]		X	400	UA[3]
New Jersey	Yards Creek	X		387	1965
New Mexico	Seboyeta[1]		X	600	1991
New York	Lewiston	X		240	1962
	Blenheim-Gilboa	X		1,000	1973
	Cornwall[1]		X	2,000	1991
	Prattsville[2]		X	1,000	UA[3]
North Carolina	Hiawassee	X		117	1956
Oklahoma	Salina	X		260	1968
Pennsylvania	Muddy Run	X		800	1967
	Kinzua	X		396	1970
South Carolina	Jocassee	X		610	1974
	Fairfield	X		511	1979
	Bad Creek		X	1,000	1991
South Dakota	Gregory County[2]		X	1,180	UA[3]
Tennessee	Raccoon Mountain	X		1,530	1979
Texas	Buchanan	X		34	1950
	Village Bend[2]		X	730	UA[3]
Virginia	Smith Mountain	X		536	1956
	Bath County[1]		X	2,100	1985
	Brumley Gap[2]		X	3,000	UA[3]
West Virginia	Davis[1]		X	1,025	1988
Washington	Grand Coulee	X		314	1973
Totals		34	16	34,873	

Notes: [a] Plants in operation have a capacity of 16,018 Megawatts
    [b] Plants licensed, under construction, or under study have a capacity of 18,855 Megawatts.
    [1] Under construction or licensed.
    [2] Preliminary permit, application pending, or under study
    [3] Unannounced

Source: U.S. Army Corps of Engineers. *National Hydroelectric Power Study*, Volume 10, *An Assessment of Hydroelectric Pumped Storage*, November 1981.

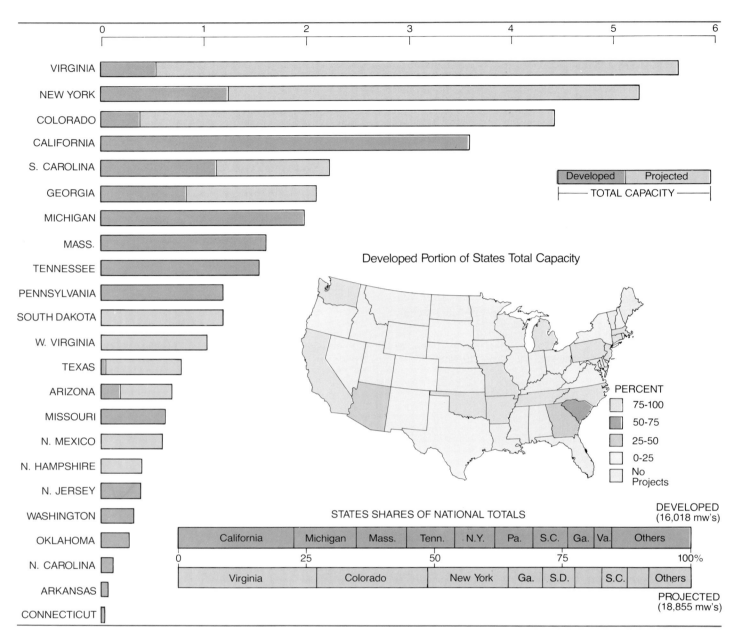

Fig. 8-7 Pumped storage hydroelectric generating capacity in 23 states showing amounts developed as of November 1, 1981 and amounts projected in 1991.
Source: U.S. Army Corps of Engineers, *National Hydroelectric Power Study*, Volume 10, *An Assessment of Hydroelectric Pumped Storage*, 1981.

where the population is dense and the demand for peak load power is large. The largest U.S. pumped storage hydroelectric facility is located at Ludington, Michigan (Fig. 8-8). This facility has a capacity of almost 2,000 Megawatts. Two other pumped storage plants in the United States have capacities greater than 1,500 Megawatts: the Northfield Mountain Plant on the Connecticut River in Massachusetts and the Raccoon Mountain Plant on the Tennessee River in Tennessee.

Five states, California, Michigan, Massachusetts, Tennessee, and New York, house more than two-thirds of the nation's developed pumped storage capacity. California, alone, has slightly more than 20 percent of the total. In the eastern part of the country, Michigan is the leader in developed pumped storage capacity with over 12 percent of the total. Three southern states, Tennessee, South Carolina, and Georgia house over 21 percent of the country's total.

PROJECTED PUMPED STORAGE CAPACITY

Presently, there is an estimated 18,855 Megawatts of pumped storage capacity under construction, licensed, or under study (Table 8-4). Should all of these installations be completed, pumped storage capacity would amount to almost 35,000 Megawatts. Almost 90 percent of the projected pumped storage capacity would reside in seven states: Virginia, Colorado, New York, Georgia, South Dakota, West Virginia, and South Carolina. As a region, the southeastern states hold nearly 40 percent of the projected pump storage capacity. Sixty percent of the projected capacity is located east of the Mississippi River.

River Segments Precluded from Hydroelectric Development

Federal law precludes development of hydroelectric power in many parts of the United States. For example, as a result of concern for the scenic qualities of the Falls and the navigability of the Niagara River, the development of Niagara Falls was subject to constraints when New York state acquired, in 1886, land along the river in an attempt to preserve the natural beauty of the site (Federal Energy Regulatory Commission, 1980).

Power development in national parks and national monuments, for example, is prohibited unless specifically authorized by Congress. Furthermore, the Wild and Scenic Rivers Act of 1968 and subsequent amendments designate segments of 28 rivers as parts of a National Wild and Scenic Rivers System thereby excluding hydroelectric development (Fig. 8–9). Also, segments of 75 additional rivers have been authorized for study. In 1980 there were 76 sites in Wild and Scenic and Special Act areas excluded from hydroelectric development. The undeveloped hydroelectric potential at these sites amounts to 11,696 Megawatts. Moreover, there is a moratorium on development at 107 additional sites that have a hydroelectric potential of 8,373 Megawatts.

Finally, the Wilderness Act of 1964 bars incompatible land and water uses in areas designated by Congress as "wilderness areas." The President, however, may authorize certain activities in these areas deemed to be in the national interest.

United States' Hydroelectric Potential at Existing Dams

For more than a decade, rising fuel costs, rapidly escalating construction costs for thermal generating facil-

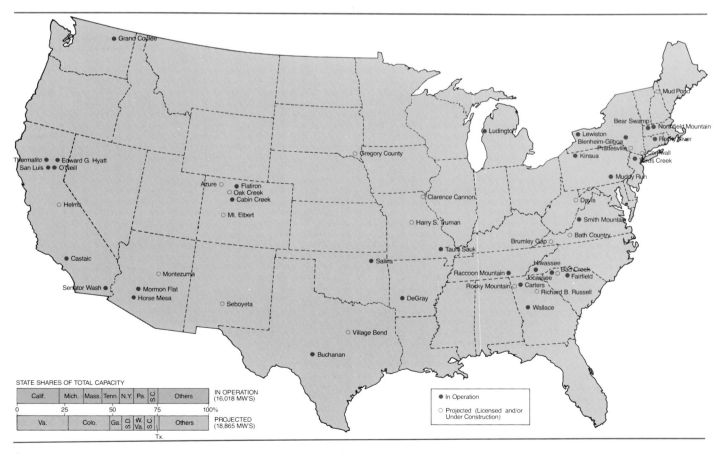

Fig. 8–8 Pumped storage projects in the United States as of November 1, 1980.
Source: U.S. Army Corps of Engineers. *National Hydroelectric Power Resources Study*, Volume 10, *An Assessment of Hydroelectric Pumped Storage*, 1981.

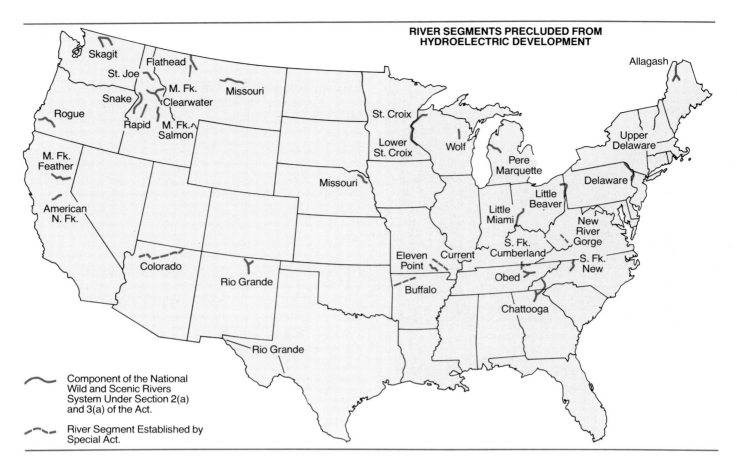

RIVER SEGMENTS PRECLUDED FROM HYDROELECTRIC DEVELOPMENT

Skagit, Flathead, St. Joe, M. Fk. Clearwater, Missouri, Snake, Rogue, Rapid, M. Fk. Salmon, St. Croix, Lower St. Croix, Wolf, Missouri, Pere Marquette, M. Fk. Feather, American N. Fk., Colorado, Rio Grande, Eleven Point, Current, S. Fk. Cumberland, Buffalo, Obed, Chattooga, Rio Grande, Little Miami, Little Beaver, New River Gorge, S. Fk. New, Delaware, Upper Delaware, Allagash

⁓ Component of the National Wild and Scenic Rivers System Under Section 2(a) and 3(a) of the Act.

⁓ River Segment Established by Special Act.

RIVER SEGMENTS DESIGNATED FOR STUDY UNDER SECTION 5(a) OF THE WILD AND SCENIC RIVERS ACT

Priest, Moyie, Snake, John Day, Illinois, Owyhee, Salmon, Bruneau, Clarks Fork, Snake, Sweetwater, Encampment, Green, Elk, Yampa, Cache La Poudra, Big Thompson, Colorado, Gunnison, Los Pinos, Dolores, Piedra, Conejos, Tuolumne, Verde, Kern, San Francisco, Kettle, Upper Mississippi, Manistee, Ausable, Wisconsin, Upper Iowa, Gasconade, Illinois, Buffalo, Sipsey Fork, Cahaba, Escatawpa, Black Creek, Nolichuckey, Bluestone, Greenbriar, Gauley, Allegheny, Youghiogheny, Cacapon, Pine Creek, Fish Creek, Penobscot, Housatonic, Shepaug, Ogeechee, Suwannee, Myakka, Loxahatchee

⁓ River Segment Designated for Study Under Section 5(a) of the Wild and Scenic River Act.

Fig. 8–9 River segments of the United States presently precluded from hydroelectric development.

Source: Federal Energy Regulatory Commission, *Hydroelectric Power Resources of the United States*, January 1, 1980.

The Raccoon Mountain pumped storage plant on the Tennessee River, six miles west of Chattanooga, Tennessee. At night and on weekends, excess power from the grid is used to run turbines in reverse—as pumps—to raise water from the river to the man-made lake on the mountain top. During periods of peak demand, water is released to fall 1,100 feet through tunnels in the mountain to the powerhouse hidden in its base. The four turbines in the power house have a combined capacity of 1.5 Megawatts. (Courtesy of Tennessee Valley Authority.)

ities, and increased public concern over the safety of nuclear plants have not only made it necessary to seek energy alternatives, but also to re-examine old energy options. In his message of April 20, 1977, President Carter indicated that new or additional hydroelectric capacity at existing dams could be installed at a cost less than the cost of building new coal or nuclear capacity. The U.S. Army Corps of Engineers has recently completed a study which assessed the potential for developed additional capacity at existing dam sites.

STATE AND REGIONAL ASSESSMENT

In their preliminary report of 1979, the U.S. Army Corps of Engineers estimated that over 1,400 existing dam sites have the potential of producing an additional 29,000 Megawatts of capacity, which represents only 30 percent of the additional capacity reported. This estimate of additional capacity at existing dam sites represents approximately 45 percent of the total undeveloped capacity of the United States.

Three states, Washington, California, and Oregon, have 47 percent of the total hydroelectric potential at existing dams (Table 8-5). The Pacific, Mountain, and Middle Atlantic regions have 69 percent of the additional capacity at existing dams. The Pacific region alone is estimated to have almost 14,000 Megawatts of additional capacity at existing dam sites, and the Mountain and Middle Atlantic regions

have approximately 4,100 and 2,200 Megawatts respectively.

In the South Atlantic Region, almost 83 percent of the total undeveloped capacity is the additional capacity at existing dams (Fig. 8-10). Other regions in which the additional capacity at existing dams accounts for more than one-half of the total undeveloped capacity are the Middle Atlantic with 69 percent, the Mountain with 64 percent, New England with 62 percent and the West South Central with 57 percent. The East North Central and the West North Central regions both have less than 4 percent of their undeveloped capacity at existing dams.

STATE AND REGIONAL ASSESSMENT BY SCALE OF SITES

The estimates by the Army Corps of Engineers of existing and additional hydropower at existing dams have been grouped into three categories based on potential Megawatts capacity. These include *small-scale* (0.05–15 Megawatts); *intermediate-scale* (15–25 Megawatts); and *large-scale* (greater than 25 Megawatts). For the United States, 4 percent of the existing hydropower capacity is at small-scale sites; 2 percent at intermediate sites; and 96 percent at large-scale sites (Table 8-5).

The estimates for additional capacity at existing dam sites is somewhat different with 12 percent of the total

Niagara Power Project at Niagara Falls, New York. Water diverted from a point above the Falls (out of the view, to the right) is allowed to drop to the level of the lower portion of the Niagara River. As it drops, it drives turbines in the Robert Moses Plant in the foreground. At times of reduced demand, excess power is used to pump diverted water up into the artificial lake in the background. When needed, the water falls through turbines in the distant plant—the Lewiston Pump Generating Plant—and then falls again through the main power plant. Building the artificial lake and accompanying power plant was a way of expanding the generating capacity at this site to deal with peak power loads. (Courtesy of New York Power Authority.)

TABLE 8–5
Installed Generating Capacity at Existing Dams as of September 1981, and Potential Additional Capacity

REGION AND STATE	SCALE OF SITE POTENTIAL						CAPACITY TOTALS		NUMBER OF DAMS WITH POTENTIAL FOR ADDITIONAL CAPACITY
	Small (0.05–15 MW)		Intermediate (15–25 MW)		Large (over 25 MW)				
	Existing	Additional	Existing	Additional	Existing	Additional	Existing	Additional	
New England									
Connecticut	33.7	39.5	24.0	19.9	67.6	0.0	125.0	59.5	43
Maine	261.0	225.0	92.2	18.1	186.0	174.0	540.0	417.0	109
Massachusetts	86.5	35.6	34.0	42.5	1,633.0	33.5	1,754.0	112.0	48
New Hampshire	70.2	137.0	31.0	0.0	281.0	182.0	382.0	318.0	99
Rhode Island	1.5	11.7	0.0	0.0	0.0	0.0	1.5	11.7	19
Vermont	99.4	96.9	24.3	49.4	107.0	0.0	231.0	146.0	71
Total	482.1	545.7	205.5	129.9	2,274.6	389.5	3,033.5	1,064.2	389
Middle Atlantic									
New Jersey	2.4	25.2	0.0	0.0	0.0	0.0	2.4	25.2	4
New York	414.0	431.0	196.0	265.0	3,103.0	636.0	3,713.0	1,332.0	154
Pennsylvania	0.0	190.0	19.6	210.0	407.0	449.0	426.0	849.0	72
Total	416.4	646.2	215.6	475.0	3,510.0	1,085.0	4,141.4	2,206.2	230
East North Central									
Indiana	3.4	71.9	0.0	0.0	0.0	0.0	3.4	71.9	13
Illinois	27.1	95.8	0.0	34.3	0.0	119.0	27.1	250.0	14
Michigan	177.0	71.6	50.4	51.7	2,029.0	273.0	2,266.0	396.0	18
Ohio	0.0	119.0	0.0	32.4	0.0	0.0	0.0	151.0	30
Wisconsin	216.0	142.0	112.0	158.0	99.9	78.2	428.0	378.0	28
Total	423.5	500.3	162.4	276.4	2,138.9	470.2	2,724.5	1,246.9	103
West North Central									
Iowa	6.9	50.3	0.0	89.6	128.0	321.0	135.0	461.0	12
Kansas	1.8	19.1	0.0	21.8	0.0	0.0	1.8	41.0	4
Minnesota	99.2	98.8	0.0	87.1	69.5	143.0	169.0	329.0	18
Missouri	3.0	16.2	16.0	141.0	877.0	130.0	896.0	287.0	11
Nebraska	11.9	3.6	18.0	21.9	152.0	0.0	182.0	25.6	3
North Dakota	0.0	3.6	0.0	0.0	400.0	272.0	400.0	276.0	2
South Dakota	8.0	0.0	0.0	0.0	1,483.0	0.0	1,491.0	0.0	0
Total	130.8	191.6	34.0	361.4	3,109.5	866.0	3,274.8	1,419.6	50
South Atlantic									
Florida	11.7	2.0	0.0	19.8	30.0	0.0	41.7	21.8	2
Georgia	43.2	18.2	160.0	40.5	2,113.0	195.0	2,262.0	253.0	11
Maryland	2.9	23.5	19.2	0.0	474.0	0.0	497.0	23.5	7
North Carolina	54.2	66.1	103.0	82.2	1,762.0	318.0	1,919.0	467.0	27
South Carolina	76.4	123.0	51.8	63.4	1,478.0	122.0	1,607.0	309.0	29
Virginia	65.1	118.0	0.0	70.7	552.0	0.0	617.0	188.0	28
West Virginia	51.8	40.2	0.0	86.4	102.0	601.0	154.0	728.0	22
Total	305.3	391.1	280.0	363.0	6,511.0	1,236.0	7,097.7	1,990.3	126
East South Central									
Alabama	7.5	35.2	0.0	64.5	2,834.0	242.0	2,842.0	341.0	15
Kentucky	0.0	130.0	0.0	90.7	797.0	753.0	797.0	973.0	38
Mississippi	0.0	10.8	0.0	58.6	0.0	0.0	0.0	69.4	4
Tennessee	10.6	9.4	39.0	21.7	2,046.0	324.0	2,096.0	356.0	4
Total	18.1	185.4	39.0	235.5	5,677.0	1,319.0	5,735.0	1,739.4	61
West South Central									
Arkansas	11.0	59.3	0.0	144.0	917.0	313.0	928.0	517.0	18
Louisiana	0.0	50.6	0.0	56.9	81.0	34.0	81.0	142.0	9
Oklahoma	0.0	44.1	0.0	73.3	769.0	196.0	769.0	313.0	13
Texas	51.6	47.7	45.0	32.1	225.0	333.0	321.0	413.0	26
Total	62.6	201.7	45.0	306.3	1,992.0	876.0	2,099.0	1,385.0	66

(continued)

Table 8–5 (*continued*)

| REGION AND STATE | SCALE OF SITE POTENTIAL | | | | | | CAPACITY TOTALS | | NUMBER OF DAMS WITH POTENTIAL FOR ADDITIONAL CAPACITY |
| | Small (0.05–15 MW) | | Intermediate (15–25 MW) | | Large (over 25 MW) | | | | |
	Existing	Additional	Existing	Additional	Existing	Additional	Existing	Additional	
Mountain									
Arizona	28.2	10.2	0.0	0.0	189.0	77.0	217.0	87.2	9
Colorado	54.7	110.0	0.0	60.7	169.0	185.0	224.0	356.0	38
Idaho	105.0	160.0	16.5	139.0	2,541.0	1,233.0	2,662.0	1,532.0	47
Montana	32.1	95.0	35.0	18.7	1,992.0	791.0	2,059.0	905.0	27
Nevada	3.2	3.0	0.0	0.0	0.0	0.0	3.2	3.0	2
New Mexico	0.0	33.4	24.2	22.9	0.0	325.0	24.2	382.0	11
Utah	33.3	42.9	0.0	39.4	138.0	391.0	171.0	473.0	30
Wyoming	18.1	37.9	56.0	37.3	152.0	314.0	227.0	389.0	14
Total	271.4	491.5	131.7	318.0	5,181.0	3,316.0	5,587.4	4,127.2	178
Pacific									
California	168.0	178.0	154.0	146.0	4,787.0	3,530.0	5,109.0	3,854.0	87
Oregon	103.0	148.0	157.0	202.0	6,673.0	2,682.0	6,933.0	3,032.0	43
Washington	117.0	145.0	0.0	67.1	18,880.0	6,671.0	18,996.0	6,884.0	53
Total	388.0	471.5	311.0	415.1	30,340.0	12,883.0	31,038.0	13,770.0	183
Alaska	44.2	45.3	15.0	23.2	77.1	0.0	136.0	68.6	19
Hawaii	17.3	1.3	0.0	0.0	0.0	0.0	17.3	1.3	4
U.S. Total	2,559.7	3,671.1	1,439.2	2,093.8	60,811.1	22,440.7	64,884.6	29,018.7	1409

Source: U. S. Army Corps of Engineers. *National Hydroelectric Power Study*, Volume 12, *Data Base Inventory*, September 1981.

The water falls less than 10 feet at this site near the old waterworks on the Schulkill River in Philadelphia, but it will be used by the City of Philadelphia for a low-head hydroelectric installation. (Photo by W.J. Young.)

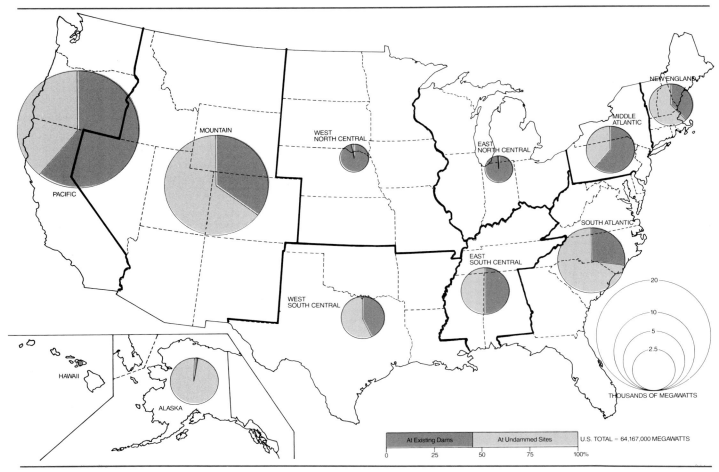

Fig. 8-10 Undeveloped conventional hydroelectric generating potential, showing proportion at existing dams by economic region.

Source: U.S. Army Corps of Engineers. *National Hydroelectric Power Resources Study,* Volume 12, *Data Base Inventory, Table 7-3,* 1981.

assigned to small-scale sites; 10 percent to intermediate-scale sites; and 78 percent to large-scale sites. Regionally, the amounts of additional capacity at small-scale dam sites varies from 51 percent in New England to 3 percent in the Pacific Region. New England's additional capacity at small dams accounts for less than 2 percent of the total additional capacity of the United States.

POTENTIAL FOR INCREASING THE OUTPUT OF EXISTING HYDROELECTRIC PLANTS

In order for there to be additional potential hydropower at an existing plant, an opportunity must exist for (1) passing more of the annual volume through the powerhouse; (2) increasing the effective operating head; or (3) technical opportunity to generate more efficiently from available head and flow (U.S. Army Corps of Engineers, July 1981). The primary measures for increasing energy output are: adding new generating units, rehabilitating or replacing existing units, modifying water handling facilities and altering existing operating policies (reallocation of existing storage and/or change of annual and seasonal operation rule curves).

By evaluating the above options at almost 1,300 sites nationwide, the Corps has estimated the potential for increasing output at these plants. The test for "achieveability" of the energy increase consisted of comparing the calculated benefit to cost (B/C) ratio. If the B/C ratio at a site was calculated to be equal to or greater than 1.0, the energy increase was considered "achieveable" (U.S. Army Corps of Engineers, July 1981). The Corps' evaluation revealed that the potential for increasing energy output at existing dam sites was relatively modest. It indicated that with an increased capacity of 15,453 Megawatts, the annual energy increase would be approximately 32 million Megawatt hours (Figure 8-11). Although this amounts only to an 11 to 12 percent increase in energy output, it would provide a considerable amount of fuel savings over a 20- or 30-year period. The Army Corps estimated that the 11 to 12 percent increase in hydroelectric output would replace 60 million barrels of fuel oil annually. Over a 30-year period this would amount to 1.8 billion barrels or 75.6 billion gallons of fuel oil.

The estimate of "achieveable" increase in output at existing hydroelectric plants is summarized in Figure 8–12 and Table 8–6 by North American Electric Reliability Council Regions (NERC). The electric industry created the Council in 1968 to augment the reliability and adequacy of bulk power supply in the electric utility systems of North America. NERC consists of nine regional reliability councils and encompasses essentially all the power systems of the contiguous United States and the Canadian systems in Ontario, British Columbia, Manitoba, and New Brunswick (Department of Energy, June 1983).

The regional results indicate that the WSCC region located in the West and Northwest has the most "achieveable" potential with 16 million Megawatt hours—50 percent of the total. The NPPC region located in the Northeast houses 10.3 million Megawatt hours of "achieveable" energy increase or 32 percent of the total. No other region has more than 4 percent of the total "achieveable" energy increase at existing dam sites.

Retired Hydropower Plants

During the 40-year period between 1930 and 1970 the unit cost of electricity decreased steadily. The increased efficiencies of steam-electric plants, aided by the use of larger units, combined with relatively stable fuel costs and resulted in the inability to justify the operation and maintenance of many small hydropower plants. Consequently, many of these small hydropower plants were abandoned, dismantled, or not replaced when they became inoperable because of wear, flooding, or other damage. More than three-fourths of the hydropower plants with a generating capacity of 15 Megawatts or less have been retired.

Since the early 1970s, however, prices of oil and coal have tripled, and price of gas quadrupled. Average efficiencies of steam-electric plants have leveled off at about 33 percent. Renewed attention has been given to the hydropower potential at existing dams, existing standby power plants, and small hydro plants in general.

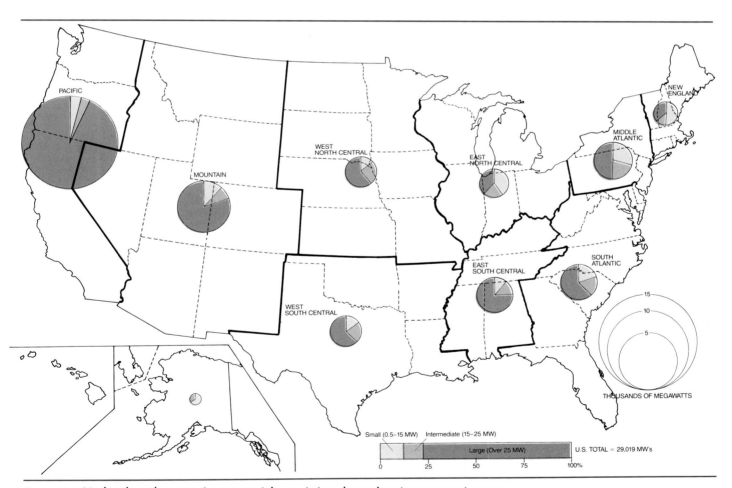

Fig. 8-11 Undeveloped generating potential at existing dams showing proportions at sites of small, intermediate, and large potential by economic region.
Source: U.S. Army Corps of Engineers. *National Hydroelectric Power Resource Study*, Volume 12, *Data Base Inventory*, 1981.

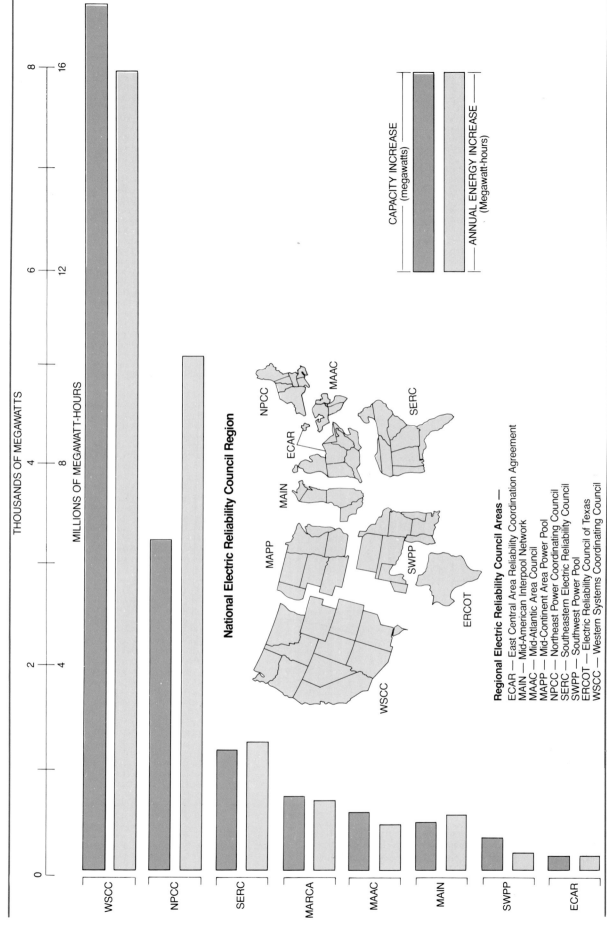

Fig. 8-12 Achievable potential for increasing hydroelectric power output at existing dam sites.

Source: U.S. Army Corps of Engineers. *National Hydroelectric Power Resources Study, Volume 9, Potential for Increasing the Output of Existing Hydroelectric Plants,* 1981.

THOUSANDS OF MEGAWATTS

MILLIONS OF MEGAWATT-HOURS

National Electric Reliability Council Region

NPCC

MAAC

ECAR

SERC

MAIN

MAPP

WSCC

SWPP

ERCOT

Regional Electric Reliability Council Areas —

ECAR — East Central Area Reliability Coordination Agreement
MAIN — Mid-American Interpool Network
MAAC — Mid-Atlantic Area Council
MAPP — Mid-Continent Area Power Pool
NPCC — Northeast Power Coordinating Council
SERC — Southeastern Electric Reliability Council
SWPP — Southwest Power Pool
ERCOT — Electric Reliability Council of Texas
WSCC — Western Systems Coordinating Council

CAPACITY INCREASE
(megawatts)

ANNUAL ENERGY INCREASE
(Megawatt-hours)

WSCC

NPCC

SERC

MARCA

MAAC

MAIN

SWPP

ECAR

TABLE 8-6
Achievable Potential for Increasing Hydroelectric Power Output at Existing Dam Sites by North American Electric Reliability Region

REGION[1]	EXISTING		POTENTIAL	
	Installed Capacity (MWs)	Average Annual Energy (10^6 MWh)	Capacity Increase (MWs)	Annual Energy Increase (10^6 MWh)
ECAR	533.6	2.031	118.2	0.248
ERCOT	314.7	0.716	3.2	0.002
MAAC	906.4	3.533	556.0	0.883
MAIN	1,025.8	3.341	463.0	1.064
MARCA	3,074.9	13.021	701.9	1.319
NPCC	3,198.3	16.286	3,291.1	10.288
SERC	10,818.9	39.286	1,579.3	2.500
SWPP	2,688.1	6.431	307.1	0.308
WSCC	40,540.0	187.150	8,675.5	16.086
Other[2]	294.4	.757	—	—
U.S. Total	63,375.4	272.552	15,452.9	32.391

[1] Regions are spelled out in Figure 8-12.
[2] Alaska, Hawaii, and Puerto Rico.
Source: U.S. Army Corps of Engineers. *National Hydroelectric Power Resources Study*, Volume 9, *Potential for Increasing the Output of Existing Hydroelectric Plants*, July 1981.

TABLE 8-7
Retired Hydropower Plants with Some Potential for Redevelopment

STATE AND GEOGRAPHIC DIVISION	NUMBER OF SITES	PREVIOUSLY INSTALLED CAPACITY (KW)
New England	1,397	348,549
Maine	176	75,766
New Hampshire	591	70,505
Vermont	144	31,991
Massachusetts	208	109,097
Rhode Island	178	23,271
Connecticut	100	37,919
Middle Atlantic	314	170,536
New York	265	147,206
New Jersey	19	3,484
Pennsylvania	30	19,846
E. North Central	445	212,319
Ohio	22	19,755
Indiana	33	15,603
Illinois	46	28,825
Michigan	172	82,004
Wisconsin	172	66,132
W. South Central	255	140,023
Minnesota	72	82,134
Iowa	115	25,397
Missouri	6	10,272
North Dakota	1	67
South Dakota	7	5,421
Nebraska	38	10,596
Kansas	16	6,136
South Atlantic	185	138,933
Delaware	3	1,104
Maryland	6	1,445
Dist. of Columbia	2	3,785
Virginia	38	45,307
West Virginia	2	1,120
North Carolina	78	35,499
South Carolina	35	24,598
Georgia	18	13,582
Florida	3	12,493
E. South Central	31	21,261
Kentucky	5	712
Tennessee	25	19,069
Alabama	1	1,480
W. South Central	22	11,893
Arkansas	2	1,040
Louisiana	1	2,000
Oklahoma	4	1,690
Texas	15	7,163
Mountain	138	132,993
Montana	14	5,133
Idaho	39	84,048
Wyoming	14	8,756
Colorado	38	15,528
N. Mexico	6	1,181
Arizona	5	2,795
Utah	16	12,192
Nevada	6	3,360
Pacific	94	69,105
Washington	34	34,708
Oregon	33	16,771
California	27	17,626
Alaska	16	4,550
Hawaii	11	1,857
U.S. Total	2,908	1,252,019

Source: Federal Energy Regulatory Commission, 1980.

The Department of Energy and the U.S. Army Corps of Engineers have identified over 2,900 sites where retired hydropower facilities are located. Most of these sites appear to have some potential for redevelopment (Federal Energy Regulatory Commission, 1980).

The previously installed generating capacity at these retired sites amounts to 1,252 Megawatts. Every state except Mississippi has one or more retired hydropower plant sites. New York state has the largest number, 259, and the largest amount of previously installed capacity, 147 Megawatts (Fig. 8-13 and Table 8-7). Massachusetts follows closely with 109 Megawatts at 208 sites. Idaho, Minnesota, and Michigan have 39, 72, and 172 sites respectively, with 84, 82, and 82 Megawatts.

The New England Region has the largest number of sites with 1,397 and the largest previously installed capacity at these sites, 349 Megawatts. The East North Central Region ranks second with 445 sites and 212 Megawatts. The Merrimack, Connecticut, Providence, and Hudson river basins in the Northeastern states have a combined total of 997 sites and 254 Megawatts.

Hydropower, which was previously not economically feasible, has become cost competitive at many sites including those where plants have been retired. Although hydroelectric power production is not expected to grow significantly through the remainder of this century, it will continue to be an important regional source of energy. If small existing sites are developed in New England, hydropower could contribute significantly to the energy requirements of that region.

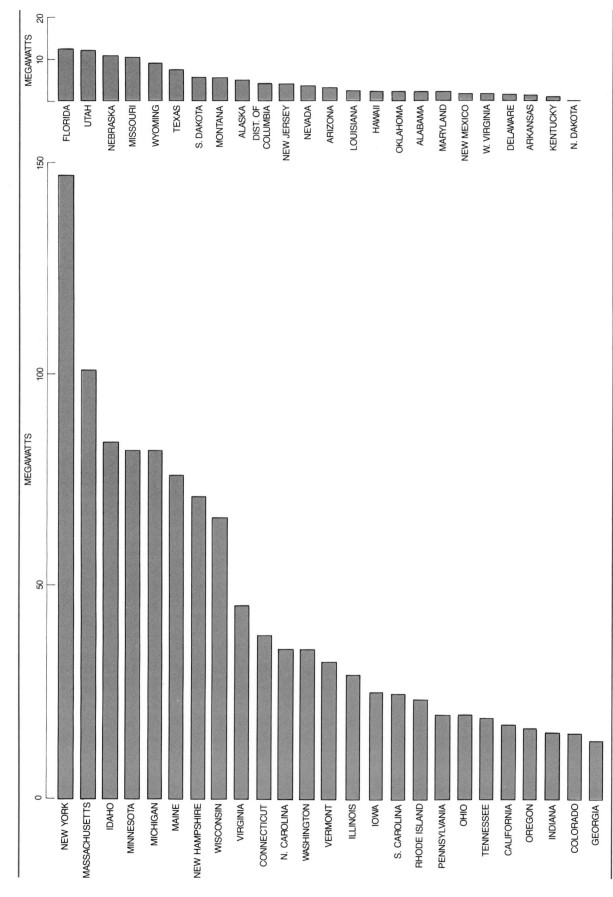

Fig. 8–13 Previously installed capacity at retired hydropower plants with some potential for redevelopment by state.

Source: Federal Energy Regulatory Commission, *Hydroelectric Power Resources of the United States,* January 1, 1980.

References

Grathwohl, Manfred. *World Energy Supply—Resources, Technologies, Perspectives.* New York: Walter de Gruyter, 1982.

Federal Energy Regulatory Commission. Data on conventional and reversible hydroelectric potential (Tables III and IV). Supplied by Neal J. Jennings, Director, Division of River Basins, 1978.

Federal Regulatory Commission. *Hydroelectric Power Resources of the United States, 1980.* Washington: Government Printing Office, January 1, 1980.

United Nations. Statistical Yearbook, 1981. New York, NY: United Nations, 1981.

U.S. Army Corps of Engineers. *Estimate of National Hydroelectric Power Potential at Existing Dams.* Washington: Government Printing Office, July 20, 1977.

U.S. Army Corps of Engineers. *National Hydroelectric Power Resources Study,* Volume 10, *An Assessment of Hydroelectric Pumped Storage.* Washington, DC: Government Printing Office, November 1981.

U.S. Army Corps of Engineers. *National Hydroelectric Power Resources Study,* Volume 12, *Data Base Inventory.* Washington, DC: Government Printing Office, September 1981.

U.S. Army Corps of Engineers. *National Hydroelectric Power Resources Study,* Volume 2, *National Report.* Washington, DC: Government Printing Office, May 1983.

U.S. Army Corps of Engineers. *National Hydroelectric Power Resources Study,* Volume 9, *Potential for Increasing Output of Existing Hydroelectric Plants.* Washington, DC: Government Printing Office, July 1981.

U.S. Army Corps of Engineers. *National Hydroelectric Power Resources Study,* 6 Volumes, *Preliminary Inventory of Hydropower Resources.* Washington, DC: Government Printing Office, July 1979.

U.S. Department of Energy. *Annual Energy Outlook, 1983.* Washington, DC: Government Printing Office, May 1984.

U.S. Department of Energy. *1982 International Energy Annual.* Washington, DC: Government Printing Office, November 1983.

U.S. Department of Energy. *Inventory of Power Plants in the United States.* Washington, DC: Government Printing Office, June 1983.

U.S. Department of the Interior. *Energy Resources on Federally Administered Lands.* Washington, DC: Government Printing Office, November 1981.

Energy from the Ocean

9

- Thermal Gradients
- Tidal Power
- Currents and Waves
- Salinity Gradients

This chapter brings together *five* quite different forms of energy that exist in or at the edge of the oceans: ocean thermal gradients, tidal power, currents, waves, and salinity gradients. What these five have in common is the potential to generate electrical power—but the methods for doing that are greatly varied, as are the magnitudes of the potential in each and their technological readiness.

Only one of the energy forms, thermal gradients, entails heat stored in the ocean. The others, in one way or another, offer mechanical means for driving a generator by flowing or falling water. An important trait that distinguishes among the five forms is a geographic dimension: Waves and salinity are more or less universal in distribution, but the other three can be utilized only in very specific locations.

On the basis of the magnitudes of energy potential that *may* be harnessed, ocean thermal gradients and salinity rank highest; waves rank second; currents are the next; and tidal power has the smallest potential. However, when they are ranked by *technological readiness*, the order changes: tidal and ocean thermal are first; waves and currents second; while the method for using salinity gradients is by far the most remote. This order is reflected in the following discussion of character and locations.

THERMAL GRADIENTS OR OCEAN THERMAL ENERGY CONVERSION (OTEC)

Each year about 1.5 quadrillion Megawatt hours of solar energy arrive at the earth's outer atmosphere (Hayes, 1977). Roughly 47 percent of this reaches the earth's surface—an energy flow that matches the output from 80 million very large electrical generating plants (1,000 Megawatts each) running at full capacity. Because 70 percent of the earth's surface is oceans, and because disproportionate amounts of the solar radiation are received in low latitudes, a great deal of this energy is absorbed by tropical oceans.

Waves off the coast of Carmel, California. (Photo by Nick Barber.)

One of the results of this energy absorption is that surface waters attain temperatures up to 30 degrees Celsius (85 degrees Fahrenheit). This is definitely low-grade energy: The temperatures are lower than those in a solar collector and are not adequate for heating a building. At the same time, temperatures in the deep ocean water are about 5 degrees Celsius (40 degrees Fahrenheit) because melting polar icecaps provide a constant supply of cold water that runs under the warm surface water. It is the *contrast in temperatures* (sometimes called ocean thermal gradient) that can be utilized to drive a heat engine in Ocean Thermal Energy Conversion (OTEC).

Efforts to make use of this resource were first made by the French. In 1881, J. D'Arsonval suggested that the temperature difference could be used to run an engine. In the 1920s, Georges Claude persuaded the French government to build several OTEC plants in hopes they would provide inexpensive electrical power to France's tropical colonies. As the overseas empire faded, so did the French interest in the project (Hayes, 1977). In the United States, OTEC plants were studied seriously in the 1960s. Since 1972, funds for OTEC research and development have been provided by the Solar Energy Program of the federal government.

Two different types of heat engine are being considered for driving a generator: *open cycle* and *closed cycle*. In the first type, sea water itself is evaporated at reduced pressure (hence low temperatures) and its vapor drives very large turbine blades. In the closed cycle type, which appears more economically attractive, a fluid with low boiling point, such as ammonia or propane, is evaporated by heat from surface waters, allowed to expand into a turbine, condensed by cool bottom waters, and then returned to the evaporator (Fig. 9–1). The cooling and condensation stage is essential in a closed system to provide lower pressures downstream from the turbine: Without lowered pressures there, the vapor would not be motivated to rush through the turbine. In this kind of system very large volumes of both warm and cool ocean waters must be circulated through heat exchangers. Therefore, corrosion of the exchangers and fouling by marine organisms are serious problems to be overcome.

In 1979, a small generating plant of the closed cycle type was established on a barge off the west coast of the island of Hawaii. It used ammonia as the working fluid, and generated 50 Kilowatts of electricity. Previous experiments (including the French) have been shore-based and have used the open cycle. This Hawaiian test, therefore, is important as a first trial of a floating closed cycle OTEC installation. It was funded by the state of Hawaii and by Lockheed Missiles and Space Company.

In the ocean the volumes of warm and cool waters are very large, and the temperature difference is renewable, since it depends on solar radiation. Nevertheless, it would be possible, by installing too many OTEC plants in one area, to degrade the thermal resource and then have to wait until the original temperature contrasts were re-established.

There are some ecological effects, even in a well-managed program. Disruption of the local environment

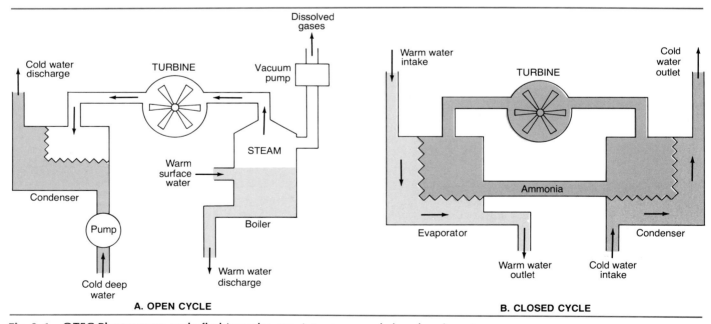

Fig. 9–1 OTEC Plants: open cycle flashing salt water into steam and closed cycle, using ammonia as working fluid.
Source: Cohen, 1978.

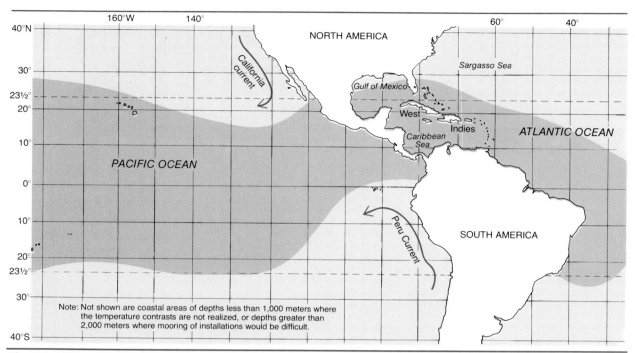

Fig. 9-2 Ocean areas where temperature contrast between surface and bottom waters (1,000 meters depth) is more than 20 degrees Celsius.
Source: redrawn from map by Ocean Data Systems Incorporation, 1978.

for marine life is unavoidable. In addition, the cold bottom water is rich in carbon dioxide, and when brought to the surface in large volumes, and warmed, could add substantial amounts of CO_2 to an atmosphere already enriched in that gas as a result of burning fossil fuels. One compensating factor is the increased photosynthesis that would occur when the nutrients in the bottom waters are deposited at the surface. The surge in phytoplankton growth could be used, in fact, as the basis for commercial mariculture systems.

A very attractive feature of the ocean thermal resource is its constancy. The temperature contrast is unaffected by the nighttime interruption of solar energy; and it is only slightly altered by seasonal changes. OTEC electrical generators, if feasible, could produce power round-the-clock and make very effective use of the installation. In the language of the utility industry, they would have a very large *capacity factor*. Electricity produced at sea could be transmitted ashore by cable, or it could be used, instead, to produce various energy-intensive products. Aluminum, for instance, if refined at sea, could free some generating capacity on the continent. As an alternative, energy could be brought to shore in the form of ammonia or hydrogen (both derived from sea water) for use in fuel cells that produce electricity. Another option would depend on the wide use of new lithium-air batteries. When the batteries

are run down they contain lithium hydroxide. This could be shipped to the OTEC platforms where it would be processed for pure lithium which would then be used to replenish the batteries.

Favorable Areas and the Resource Potential

Temperature contrasts must be 20 degrees Celsius (36°F) or greater in order to power heat engines, and such contrasts occur only in tropical and subtropical areas. Figure 9-2 shows how the general pattern of favorable areas is affected by cold currents. Both the Peru and the California currents, which carry cold water from higher latitudes toward the equator, interrupt the tropical band where surface waters are sufficiently warm for a contrast of 20 degrees or more. The map is deliberately simple and does not show that near all shorelines the waters are too shallow for the contrasts to be realized. Areas with bottom depths less than 1,000 meters (3,200 feet) usually are unsuitable, and such areas of shallow water overlying continental shelves are quite extensive in the Caribbean.

A number of United States coastal areas have access to OTEC resources. The Gulf of Mexico is within the favorable area, and both Puerto Rico and the Virgin Islands have access to temperature contrasts of the Gulf-Caribbean

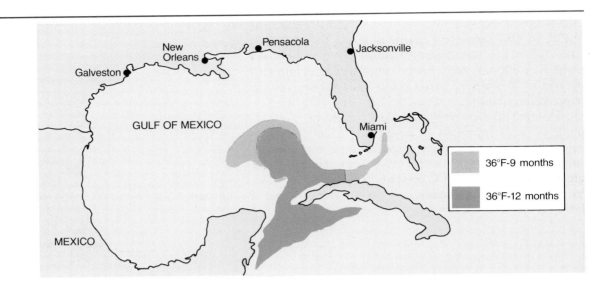

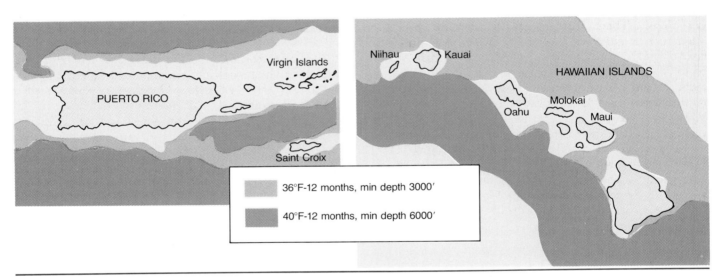

Fig. 9–3 Resource areas for OTEC development in Gulf of Mexico, Puerto Rico, and Hawaii areas.
Source: Department of Energy, *Program Summary*, 1979.

area. The islands of Hawaii fall within the favored area in the Pacific, and the island of Guam (Lat. 14° North, Long. 143° East) is in a favorable area not mapped on Figure 9–2 because of the complexity of continental shelf patterns around Indonesia.

Figure 9–3 examines the three major resource areas more closely. In the Gulf of Mexico, a surface to bottom contrast of 36 degrees Fahrenheit is attainable in areas quite close to the Florida coast. In parts of that area the contrast is weakened in winter months, but for very large sections the 36 degree contrast persists throughout the year. In the vicinity of Puerto Rico and the Hawaiian

Islands, the maps show where a more desirable 40 degree contrast can be found. Off the Puerto Rico coast, greater contrast can be obtained by siting a plant farther offshore and using bottom water from the greater depth (minimum 6,000 feet). In Hawaii, steep bottom slopes toward the southwest make the 40 degree contrast quite accessible. The challenge, in either Puerto Rico or Hawaii, is to bring that very cold water to the surface without expending too much energy.

POTENTIAL IN GULF OF MEXICO   One estimate of the generating potential in the Gulf suggests a capacity of 200 to 600 thousand Megawatts, that is, the equivalent of 200 to 600 generating plants of the 1,000 Megawatt size—typical of a very large coal-burning or nuclear power plant. The estimate takes into account that in continuous operation the projected power plants must not degrade the temperature resources (Cohen, 1978).

This potential is very large and attractive. Furthermore, it varies seasonally in a way that matches varying demand in the southeastern states. During the summer, surface waters are warmer, temperature contrasts are greater, and greater OTEC power output is possible. Also, in the same summer season, air conditioning causes a heavy demand for electricity.

Prospects for Development

Despite the very low conversion efficiencies that are inevitable when small temperature difference exists on opposite sides of a turbine, the possibilities for development look promising—especially in seas surrounding the Islands, that is, Hawaii, Guam, Puerto Rico, and the Virgin Islands. There, the resource is accessible, and the alternative is imported fuel oil for power generation. In favor of OTEC is its large capacity factor of 85 percent versus 65 to 70 percent for fuel-fired plants. And, freed from fuel costs, OTEC plants are expected to have operating and maintenance costs that are only 1.5 percent of capital investment. As with any alternative to fuel burning, the viability of OTEC will depend heavily upon fuel prices.

One projection by the Department of Energy (Fig. 9-4) shows that after 1990 OTEC generating costs will be lower than those for oil-fired plants, and they will gradually fall as the technology matures. On this basis the projection foresees a rapid growth in OTEC generating

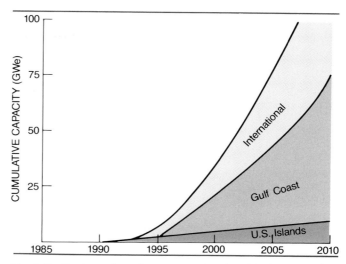

Fig. 9-5 Projected market penetration for OTEC power plants in world and United States.
Source: Department of Energy, *Program Summary*, 1979.

capacity worldwide as well as in the United States and its possessions. The larger part of the U.S. growth will be in Gulf Coast areas where demand is very high (Fig. 9-5).

Research sponsored by the Department of Energy is proceeding on technologies and materials for both floating and land-based OTEC plants. One floating platform is the converted tanker ship, now called OTEC-One. A Land-based plant is being built on the island of Oahu (see photos).

A heat exchanger unit being lowered into *OTEC-1*, a former tanker ship, now converted to an experimental floating platform for OTEC power generation. The cold-water pipe from this kind of plant will extend straight down to reach the cold bottom water. (Courtesy of TRW Inc.)

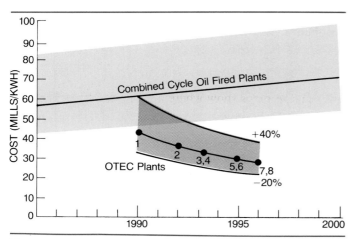

Fig. 9-4 Projected OTEC electricity costs versus those for oil-fired power plants.
Source: Department of Energy, *Program Summary*, 1979.

Artist's rendering of a 40 Megawatt OTEC plant planned for Oahu, Hawaii, showing three water-carrying pipes, roughly 30 feet in diameter, that are crucial to the plant's operation. One pipe is relatively short, bringing warm water from the ocean surface nearby. A second pipe, laid on the ocean floor, may be two miles long, in order to bring cold bottom water from depths of 3,000 feet. The third pipe is slightly longer than the first, and is used to discharge cool water from the plant to the ocean bottom just offshore. (Courtesy of TRW Inc.)

TIDAL POWER

The energy for generating electricity from tides originates in the motions of the earth, moon, and sun. The gravitational force of the moon and sun cause a bulging of the ocean surface, and the earth's rotation leads to a rising and falling of the water as the bulge encounters coastlines.

Although the use of tides for mechanical power (water wheels) dates to eleventh-century England, it was not revived as a serious alternative to other forms of electrical power generation until recently. In principle, it is akin to hydroelectric power generation, because it makes use of a hydrostatic head created by the rising and falling tides (see box).

A minimum tidal range (difference between high and low tide levels) is needed if tidal power is to be practical. A range of 15 feet, or 5 meters, often is cited as the minimum. On this basis, around 40 sites exist in the world according to one assessment (Riva *et al.*, 1978, reported in Pryde, 1983). Only two of these are developed: one at La Rance, France, and the other, only a pilot plant, at Kislaya Guba, 40 miles north of Murmansk in the Soviet Union. The La Rance installation is a working commercial power plant with generating capacity of 240 Megawatts.

In the United States only three coastal areas are promising (see Fig. 9-6). One is the Bay of Fundy area where the highest tides in the world occur. The other two are in Cook Inlet and Bristol Bay, Alaska.

This is one section of a cold-water pipe being laid onto the ocean floor for testing in research sponsored by the U.S. Department of Energy. The power plant that used the cold water would be fixed at the shore of an island, not floating at sea. (Courtesy of TRW Inc.)

THE PRINCIPLE OF TIDAL POWER GENERATION

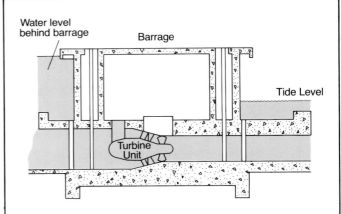

The barrage, or dam, holds back the water in the estuary as the tide falls. Then, gates are opened and the water rushes seaward through the turbine. Later, the rising tide will be held back by the barrage, then released to flow through another turbine into the river estuary.

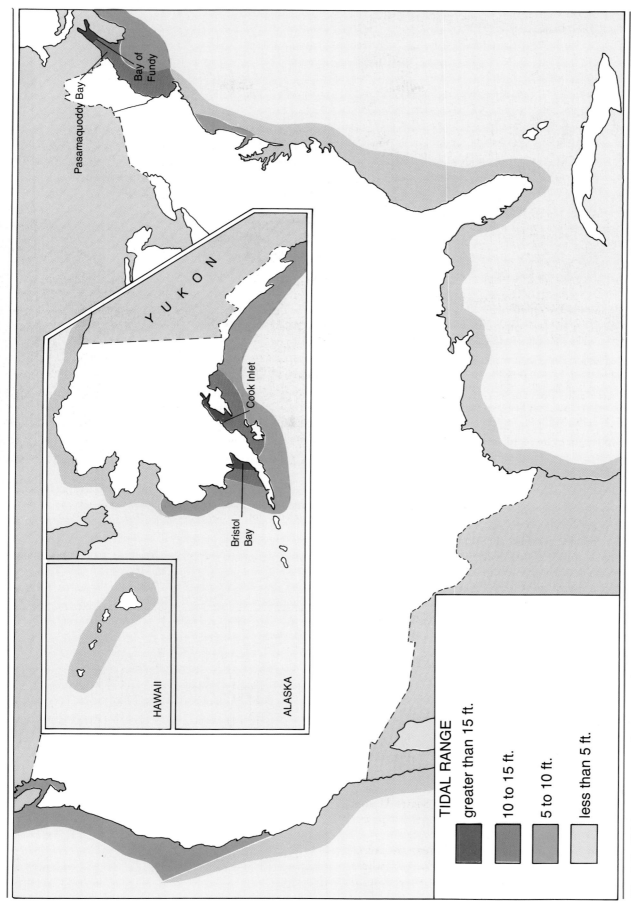

Fig. 9-6 Tidal ranges on U.S. coasts. Where range exceeds 15 feet, the coast may be suitable for power plant installations.

Source: U.S.G.S., *National Atlas of the United States of America*, 1970.

TIDAL RANGE

greater than 15 ft.

10 to 15 ft.

5 to 10 ft.

less than 5 ft.

YUKON

Pasamaquoddy Bay

Bay of Fundy

Cook Inlet

Bristol Bay

HAWAII

ALASKA

Looking seaward across the tidal power dam spanning the mouth of La Rance River on the coast of France.

The two prime areas, Cook Inlet and Pasamaquoddy, Maine near the Bay of Fundy, are enlarged in Figures 9–7 and 9–8 to show locations of possible power plants, but not the dams that would accompany them. Table 9–1 provides vital information for evaluating the areas. An important point to note is that the sites are alternatives. Since not all plants would be feasible at one time, it is *not appropriate to add* the generating capacities listed in the table.

Cook Inlet shows much greater tidal range and, correspondingly, greater generating capacities and annual output. One method of comparing Cook Inlet with Pasamaquoddy, is to assume only the largest feasible power plant is built in each area. In Cook Inlet, this would be a

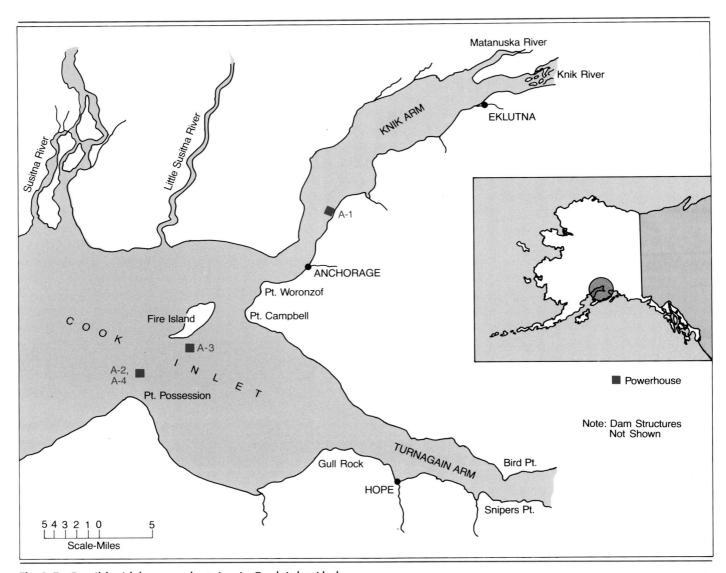

Fig. 9–7 Possible tidal power plant sites in Cook Inlet Alaska.
Source: ERDA, *Tidal Power Study*, 1977.

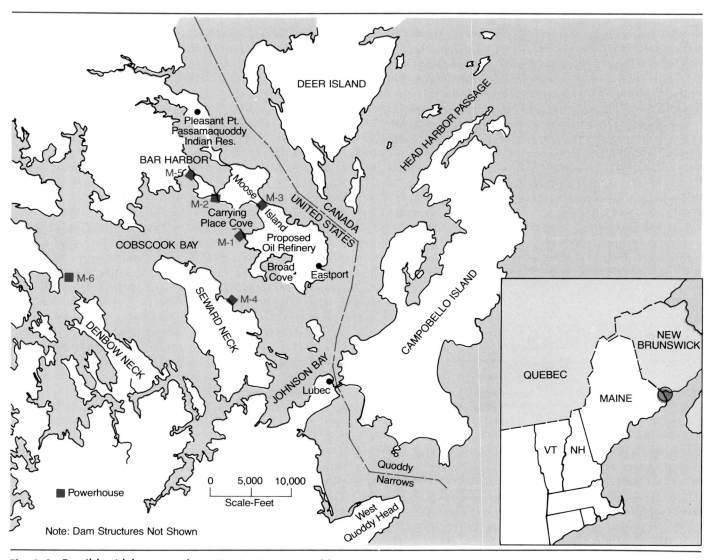

Fig. 9–8 Possible tidal power plant sites in Pasamaquoddy Bay area, Maine and New Brunswick.

Source: ERDA, *Tidal Power Study,* 1977.

plant with 3,550 Megawatts—a giant ranking with Grand Coulee Dam. In Maine, the plant would have generating capacity of 1,000 Megawatts—considerably smaller, but nevertheless comparable to the largest of coal-fired or nuclear power plants.

In the Pasamaquoddy area, construction was actually begun for a project on one site in the 1930s. Since then, the possibility of a joint U.S.-Canadian project has been studied repeatedly, but no firm plans have emerged. For perspective, it should be realized that, given the limited number of sites, tidal power—even if fully developed in the United States—will never make a very large contribution to the nation's electrical needs.

CURRENTS AND WAVES

Although a great deal of energy resides in ocean currents in various parts of the world, most of those currents flow too slowly for power generation. It happens that the most attractive ocean current for power generation is just off the U.S. coast: the *Florida Current* flowing between the mainland and the Bahamas (Fig. 9–9). This current contributes to what is usually called the Gulf Stream or the North Atlantic Drift.

In the stretch between Florida and the Bahamas the current is around 75 miles wide and flows at speeds up to 2.9 knots (3.3 miles per hour). In that area, water is flowing at 30 million cubic meters per second, which leads to a total power of 25,000 Megawatts. The world's second most powerful current, the *Kuroshio* southeast of Tokyo, is rated at 8,000 Megawatts.

Only a small portion of that power can ever be harnessed—but the possibilities are very appealing. The flow is continuous, with none of the periodic interruptions that make tidal power questionable. In this sense, it is like OTEC power generation. Furthermore, the Florida current is immediately adjacent to a densely populated coastal area with great demand for electricity.

Although no electrical power has yet been generated from ocean currents, the technology seems simple and attainable. One method is to suspend a turbine in the current flow, using special turbines invented for this purpose (Department of Energy, 1979). Another, less expensive, method is the drogue chute system (Fig. 9–10).

The Department of Energy estimates that the total current electrical energy potential for the United States is only one one-hundredth that of the OTEC potential and only one-tenth that in ocean waves. Because of this low return, the technology is not given great emphasis in Department-funded research.

Energy In Waves

Since much of the United States lies in the zone of westerly winds, the greatest wave energy is found on the coasts of Washington, Oregon, and California. But recent evaluations have suggested that much of the coastal waters within 25 miles of U.S. shores could be exploited (Department of Energy, 1979). A great variety of jointed rafts, bobbing buoys, and nodding floats have been invented to harness some of the energy in waves—with the British leading the way. However, an economically viable technology has not yet been identified.

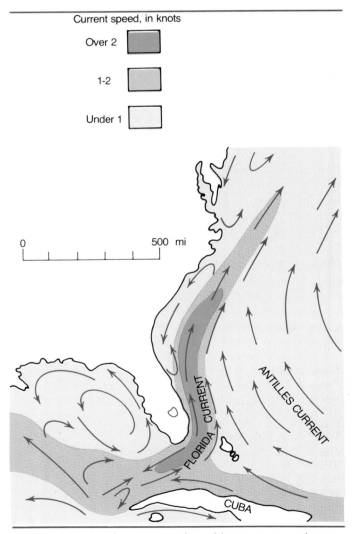

Fig. 9–9 The Florida current and neighboring waters, showing approximate current speeds.
Source: U.S.G.S., *National Atlas of the United States of America*, 1970.

SALINITY GRADIENTS

This rather mysterious ocean energy possibility has potential which is very difficult to assess. In one document the possible resource is described as very large and comparable to that of OTEC, then later as relatively small compared to OTEC (Department of Energy, 1979, p. 3 and p. 189). There are two broad methods of utilization under the umbrella of salinity gradients, and both make use of having salt water at a site where fresh water is also available. Contrast in salinity is essential.

One method is chemical, essentially a battery that creates an electrical potential when certain membranes permit one type of salt to pass through in one direction and another salt to pass through in the other. The result is said to be a *dialytic battery*.

The other method also makes use of membranes, but uses the contrast between fresh river water and salty ocean water to create a hydrostatic head. In theory, a dam is built at the mouth of a river on the ocean. River water is allowed to fall through a turbine into a special lake that lies *below sea level* and thereby creates a hydrostatic head without need for a very high dam. The level in this special lake is *held* below sea level through the osmotic pressure that exists between two solutions of widely differing salinity—when these two solutions are allowed to meet at a special membrane.

The requirements for such a scheme are rather imposing. There must be a large enough piping and membrane system to draw off fresh water at a rate that matches the flow through penstock and turbine. And there must be a river mouth with canyon or bluffs that will permit the building of two dams—one on the river side of the pond, and one on the ocean side. These requirements, plus the fact that no large-scale working model has been built, suggest that salinity gradients will not be used for generating electrical power in the near future.

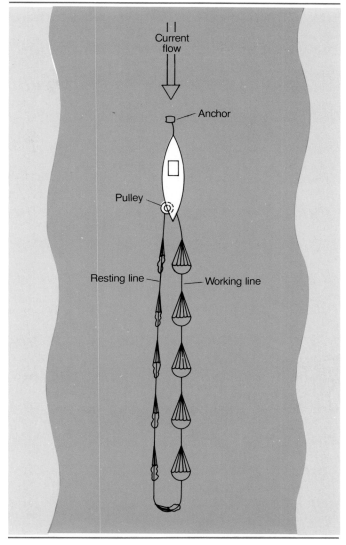

Fig. 9-10 Drogue chute device for using ocean current energy.
Source: DOE/SERI, Annual Report, 1981.

References

Cohen, Robert. "An Overview of the U.S. OTEC Development Program." Paper delivered to American Society of Mechanical Engineers, Engineers Energy Technology Conference, Houston, Texas, Nov. 6-9, 1978.

Hayes, Dennis. *Rays of Hope: The Transition to a Post-Petroleum World.* New York; Norton, 1977.

Ocean Data Systems Inc. "Large-Scale Distribution of OTEC Thermal Resource." Maps prepared for Division of Solar Technology, U.S. Department of Energy, 1978.

Pryde, Philip, R. *Nonconventional Energy Resources.* New York: Wiley-Interscience, 1983.

Riva, Joseph P., *et al.* *Energy from the Ocean.* Prepared for House Committee on Science and Technology by Science Policy Research Division, Library of Congress. Washington, DC: Government Printing Office, 1978.

U.S. Department of Energy. *Ocean Energy Systems, Program Summary Fiscal Year, 1979,* DOE/ET-0118. Washington, DC: Government Printing Office, 1979.

U.S. Department of Energy/Solar Energy Research Institute (SERI). *Ocean Energy Conversion Systems, Annual Research Report.* March 1981.

U.S. Energy Research and Development Administration (ERDA). *Tidal Power Study for the U.S. Energy Research and Development Administration.* Washington: Government Printing Office, March 1977. (Study Conducted by Stone and Webster Engineering Corp., Boston.)

U.S. Geological Survey. *National Atlas of the United States of America.* Washington, DC: Government Printing Office, 1970.

10 Biomass

- Wood and Its Use as Fuel
- Synthetic Fuels and Their Feedstocks
- Energy versus Food
- Future Biomass Production

The term *biomass* is applied to a great variety of plants or plant-derived materials that may be used directly or indirectly for energy supplies. Table 10-1 is an overview of the two major sources of biomass and two kinds of its applications.

The sources of energy are wastes or residues from activities not directly related to energy production and materials from crops grown specifically for their energy content. In the first category are crop residues such as corn stalks, manures, municipal trash, sewage, and tree tops from forestry operations. These materials are physically available now (if not economically available) for use. Crops grown specifically for energy include trees of various species, sugar cane, sorghum, ocean kelp, and water hyacinth. The economic viability of such crops will depend upon prices for other energy supplies and could be encouraged through government incentives. Most energy crops will compete for land with food crops and with trees used for lumber and paper.

◄Wood contains solar energy stored through many decades by the process of photosynthesis. The wood may be burned directly for heat or may be converted into charcoal plus liquid and gaseous fuels. (Courtesy of the American Petroleum Institute and Olympic Resource Management.)

TABLE 10-1
Biomass Sources and Applications

SOURCES		APPLICATIONS
Wastes from NonEnergy Activities	**Energy Crops**	
Urban trash	Trees used as wood or as charcol	Used directly as fuels
Tree tops and limbs from forestry operations		
Animal manures	Trees for wood	Used as raw materials for synthetic fuels and products, such as ammonia
Urban garbage	Rubber from trees and guayule bush	
Sewage	Water hyacinth	
Crop residues	Sunflower	
Tree tops and limbs	Sugar cane	
	Sea kelp	
	Corn	
	Sorghum	

All plant materials store solar energy in the form of chemical energy created by the process of photosynthesis. In that process, light energy is used to convert water and carbon dioxide into energy-rich carbohydrates plus oxygen. This energy may be released through combustion of materials such as firewood, wood chips from forest wastes, or even urban trash, which typically contains a great deal of paper (Table 10–1). When burned, dry wood will supply roughly the same amount of energy per ton as lignite coal, the lowest of the four ranks of coal discussed in Chapter 1.

An alternative to combustion is to use plants or wastes as raw materials for the production of synthetic fuels with high energy content such as alcohols, methane, gas, or fuel oil. As raw material for synthetic fuels, biomass will compete with coal, oil shales, and tar sands. In the production of ammonia for fertilizer use and as feed for fuel cells (producing electricity) biomass will compete with ocean thermal gradient installations (OTEC).

Information in this field of greatly varied possibilities is scattered and difficult to obtain. The U.S. Department of Energy manages a research program on various aspects of biomass potential and processes, but necessarily the work itself is conducted at many different institutions including government-supported laboratories such as Oak Ridge and Argonne National Laboratories. The U.S. Forest Service is another center for information. A great deal of the research underway deals with either a specific process, a certain feedstock, an exotic plant, or the potential for a certain county or region of the country. National studies that permit a grasp of the geographic patterns of the relevant resources are rare indeed.

Recognizing that a comprehensive picture of biomass resources and activity is impossible, this chapter nevertheless offers information on the following topics: wood, and its use as a fuel; synthetic fuels from biomass, especially alcohol; and two views of the future of biomass in the United States.

WOOD AND ITS USE AS A FUEL

Trees may be used in a number of ways. They may be grown expressly for fuel purposes, then chipped or turned into charcoal. Or, the portions of trees now rejected by the timber industry can be recovered as a by-product of that activity. Whatever the method of gathering wood, the nation's forest resource is a very important part of the total biomass potential. Also, the national patterns of forest biomass indicate: (1) what parts of the nation are, by virtue of temperature and rainfall, the more productive areas, and (2) what parts have productive land areas not devoted to agricultural production.

The Forest Service of the U.S. Department of Agriculture recently completed an inventory of tree biomass on commercial forest land throughout the country (Department of Agriculture, 1981). This study considers the *whole tree,* not only the marketable portion, and it offers a great deal of information on the resource, state by state, divided according to ownership, class of timber, species group (hardwood versus softwood), and actual species.

Here, we present only part of that information. Figure 10–1, shows the total tree biomass in millions of green tons for every state, with states ranked according to that total. Each bar is divided according to softwoods versus hardwoods. To consider the fact that some states are larger than others, we also present *green weight per acre* in the accompanying map.

On the basis of total green weight, Oregon and Washington, with dominantly softwoods (needleleafed trees) are the leading states. But southeastern states— Alabama, Georgia, and North Carolina—are not far behind. What is striking about the bar graph is the gradual dwindling of total biomass from top of the ranking to the bottom, where the states with smallest resource are either small, arid, or both. The map, showing *intensity* of forest resource by *tonnage per acre,* makes clear the moist areas of the West Coast, the Southeast, and the Northeast, in contrast to arid Southwest and the semi-arid Great Plains.

The illustrations in Figure 10–1 reveal only the *static* picture of forest resource, that is, the tonnage that exists at the time of the inventory, without any explicit information about an area's capacity to *regenerate* biomass. Regeneration is dealt with in Figure 10–2, which shows an estimate of how much woody biomass was grown *in the year 1980* within each of ten U.S. Department of Agriculture regions. The Southeast was most productive, with 116 million dry tons. Appalachian and Pacific regions both rank as number two, followed closely by Delta states and the Northeast.

Wood Burning

In 1860, wood contributed around 80 percent of total U.S. energy needs. Its contribution shrank as coal fired the nation's industries—and shrank farther as cheap oil and natural gas made inroads and the national energy budget ballooned. The low point for wood was in 1973, when it provided only 2 percent of the total. However, the international oil crisis that year sparked an interest in stoves and firewood, so that now 3 percent of the total is provided by wood, and the porportion is still rising (see Fig. 10–3). Wood burned by industries (mostly forest products industries) has, in fact, been rising steadily over the past 30 years: it is commercial and residential use that has risen dramatically in the last few years, with wood use in residences doubling between 1973 and 1983 (Flavin, 1984).

This increase in residential wood burning has created a phenomenon peculiar to the 1980s—woodsmoke air pollution. In Missoula, Montana, in January of 1984 there

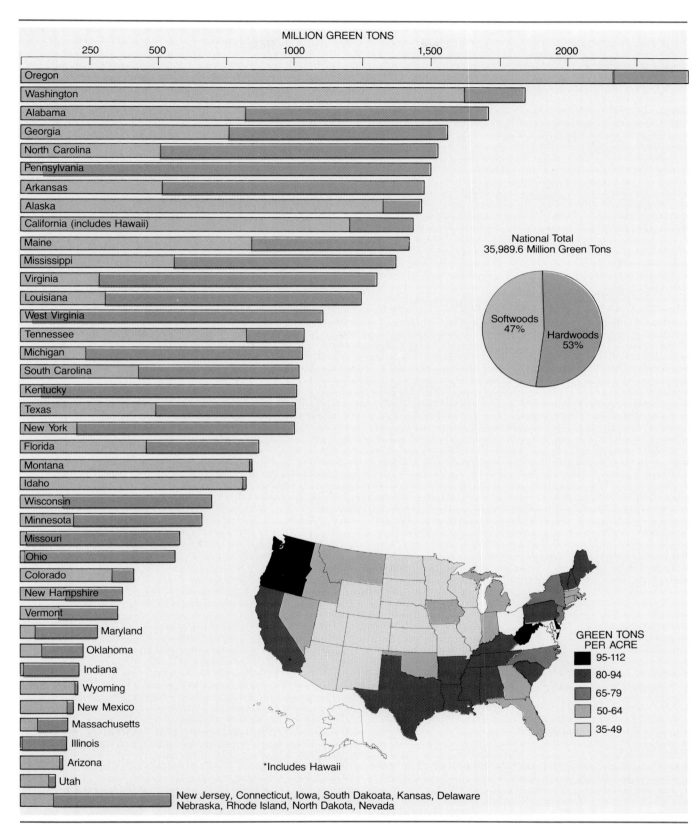

MILLION GREEN TONS

Oregon
Washington
Alabama
Georgia
North Carolina
Pennsylvania
Arkansas
Alaska
California (includes Hawaii)
Maine
Mississippi
Virginia
Louisiana
West Virginia
Tennessee
Michigan
South Carolina
Kentucky
Texas
New York
Florida
Montana
Idaho
Wisconsin
Minnesota
Missouri
Ohio
Colorado
New Hampshire
Vermont
Maryland
Oklahoma
Indiana
Wyoming
New Mexico
Massachusetts
Illinois
Arizona
Utah
New Jersey, Connecticut, Iowa, South Dakoata, Kansas, Delaware Nebraska, Rhode Island, North Dakota, Nevada

National Total
35,989.6 Million Green Tons

Softwoods 47%
Hardwoods 53%

GREEN TONS PER ACRE
95-112
80-94
65-79
50-64
35-49

*Includes Hawaii

Fig. 10–1　Tree biomass on commercial forest lands, by state.
Source: U.S. Department of Agriculture, Nov. 1981.

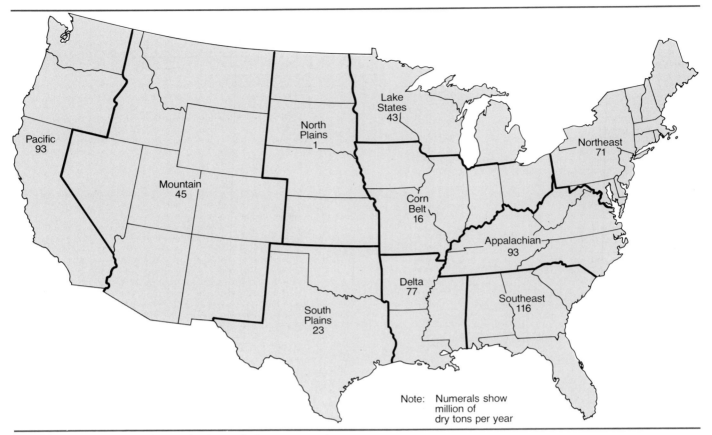

Fig. 10–2 Annual rate of growth of woody biomass, 1980.
Source: Map from Shen, Argonne National Laboratory, 1984.

was a week-long pollution episode blamed, largely, on wood burning compounded by a temperature inversion that restricted air circulation (*Time,* Jan. 16, 1984).

The extent of wood burning in 1981 for *all purposes*— residential, industrial, and commercial—in various parts of

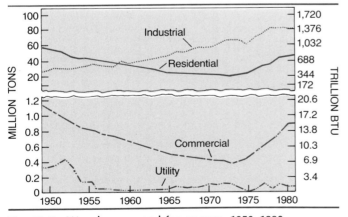

Fig. 10–3 Wood consumed for energy, 1950–1980.
Source: Department of Energy, Aug. 1982.

the country is shown in Figure 10–4. Measured just by tonnage consumed, the leaders are California and North Carolina. In both of these states, the industrial component accounts for 65 to 70 percent of the total burned, so they are consistent with the national average. Low on the list is a group of states that includes Vermont and New Hampshire, well-known for wood-burning! The reason is population size, which influences both residential and industrial use of wood for fuel. To consider this factor, Figure 10–3 includes a map showing *per capita tonnage* burned. On this measure, California falls in the least intensive category, though North Carolina is in the second-highest. Most intensive wood burners are Maine (outstanding, with 2.4 tons per capita), New Hampshire, Vermont, and groups of southeastern and northwestern states. Least intensive are those states in arid and semi-arid areas where forest is scarce (see Fig. 10–1).

In the industrial sector, the great majority of wood burned is waste materials, such as bark and scraps, produced by the forest products industry, and then burned at the site to provide heat or steam for the operation of the mill. A small, but growing, part of industrial wood burning is generation of electrical power by utility companies.

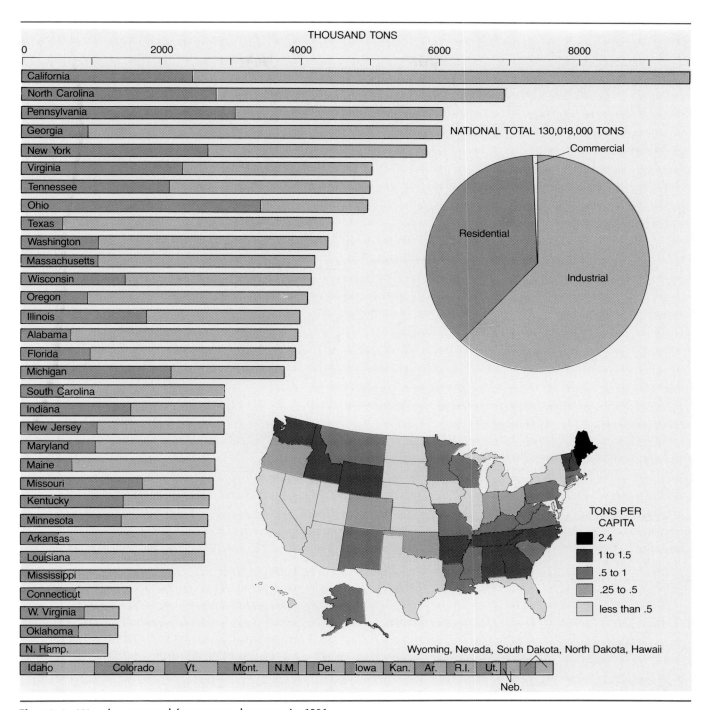

Fig. 10–4 Wood consumed for energy, by state, in 1981.
Source: Department of Energy, Aug. 1982. *Note:* this is considerable by many biomass experts to be an underestimate of wood burned.

Table 10-2 lists a selection of such projects, with Burlington, Vermont, one of the first large projects in the nation, at the top of the list. The firm responsible for the power plants at Burney and Westwood in California is planning to develop 20 more such power plants across the country by the year 2000 (Flavin, 1984).

SYNTHETIC FUELS AND THEIR FEEDSTOCKS

A great variety of plant materials can be used as raw materials for production of liquid and gaseous synthetic fuels. Four major types have been identified: herbaceous

TABLE 10–2
Selected Plants, Other than Forest Products Industry, generating Electricity from Wood Fuel

PROJECT	CAPACITY	WOOD SOURCE	START-UP
	(Megawatts)		
Burlington Electric Dept. Burlington, Vermont	50.0	private forestland harvests	Spring 1984
W.P. Energy Kettle Falls, Washington	46.0	mill residues	Fall 1983
Dow Corning Midland, Michigan	22.4	private forestland harvests	December 1982
Proctor and Gamble Long Beach, California	13.5	industrial waste wood	June 1983
Ultrapower, Inc. Burney, California	11.0	forest residues	September 1984
Ultrapower, Inc. Westwood, California	11.0	forest residues	September 1985
Proctor and Gamble Staten Island, New York	10.5	industrial waste wood; wood chips	June 1983

Not included are utilities and industries blending wood with coal.
Source: Based on personal communications with utility and industry officials, as reported in Flavin, 1984.

plants, woody plants, aquatic plants, and manure (see Table 10–3).

Herbaceous plants, that is, plants that do not produce persistent woody material, are divided into those with low moisture content and those with high moisture content. Some of those materials, herbaceous high moisture and woody materials, can be obtained as residues and also from energy crops. Manure and herbaceous low moisture

plants are residues only, while aquatic plants are produced only as energy crops.

Conversion of biomass materials may be considered under two broad headings: thermochemical processes and biochemical processes. Thermochemical processes can be used for all five types of biomass, but the low-moisture herbaceous and the woody materials are most suitable and are the only raw materials shown on the simplified view of

TABLE 10–3
Selected Biomass materials used as Feedstocks for Synthetic Fuel Processes

BIOMASS TYPE	OBTAINED AS RESIDUES FROM AGRICULTURAL AND FORESTRY OPERATIONS	GROWN AS ENERGY CROPS
Herbaceous, Low Moisture	Small grain field residues	
Herbaceous, High Moisture	Wastes from vegetable fields and packing sheds. Residues from sugar cane, sugar beet, corn, sorghum, cotton	Sugar cane, corn, sorghum
Woody Plants	Mill bark, mill wood, logging residues, orchard prunings, standing vegetation not suitable for lumber	Various species
Manure	Only beef manure from feedlots is considered	
Aquatic Plants		Microalgae in fresh water, and giant kelp in the ocean

Source: Ernest, et al. 1979.

Wood chips in the foreground are being fed into this prototype unit for converting farm and forest residues into methane gas. Developed by Professor John Goss at the University of California, Davis, the unit supplies gas to a boiler that heats one of the campus buildings. (Courtesy of U.S. Department of Energy.)

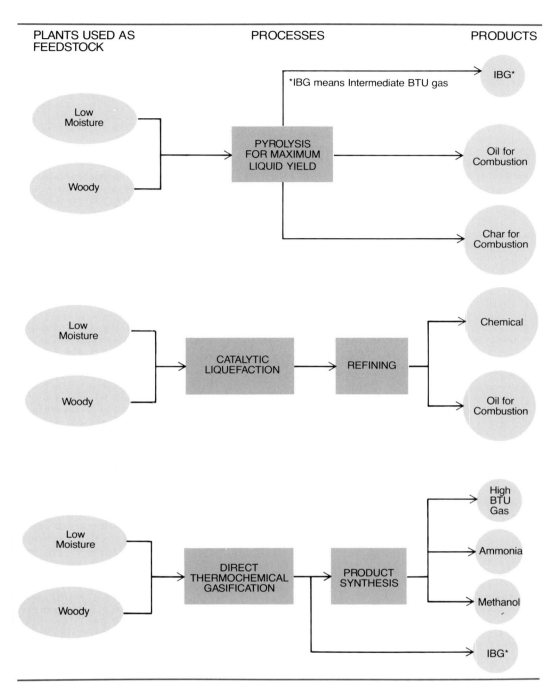

Fig. 10-5 Three thermochemical processes (simplified) for converting biomass into oil, gas, and chemicals.
Source: Stanford Research Institute, Vol. 5, 1978.

the three processes (Fig. 10-5). *Pyrolysis* is a process that converts organic materials into gases or liquids at temperatures of 500-900 degrees Celsius by heating in a closed vessel in the absence of oxygen. The process may be adjusted to favor the production of either liquid fuel or gases. In the version shown, liquid fuels are the major product, while intermediate-BTU gas and char (charcoal, if the raw material is wood) are lesser products. The same two raw materials, low moisture herbaceous plants and woody plants, are favored for *catalytic liquefaction* which

yields oil for combustion and chemicals, and for *direct gasification* which yields gases of high and intermediate energy content, along with methanol and ammonia.

Biochemical processes are essentially *anaerobic digestion* and *fermentation* (Fig. 10-6). Both processes are best served by the same raw materials, high moisture herbaceous plants, manure, and marine crops such as giant kelp. Anaerobic digestion yields high-BTU gas, which is methane, and intermediate-BTU gas, which is methane mixed with CO or CO_2. Though not included in Table

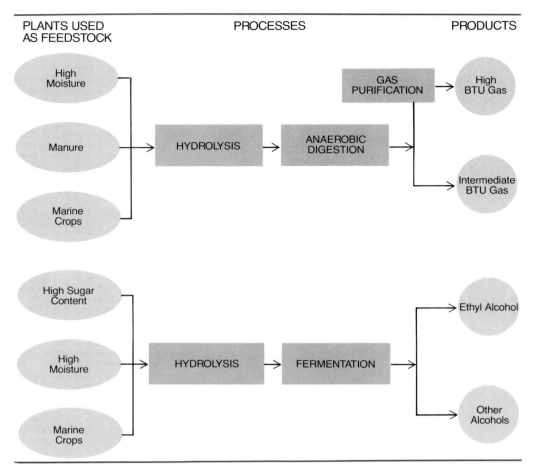

PLANTS USED AS FEEDSTOCK   PROCESSES   PRODUCTS

Fig. 10-6 Two biochemical processes (simplified) for converting biomass materials into gas and alcohols.
Source: Stanford Research Institute, Vol. 5, 1978.

10-3, sewage and garbage are suitable raw materials for this digestion process: In fact, anaerobic digestion once was widely used at sewage plants and the resulting methane burned to provide heat at the plants. Fermentation is the well-known process that yields alcohols from raw materials such as sugar cane, corn, or algae.

Alcohol

Alcohol (ethanol and methanol) is the only alternative liquid fuel that is commercially available now, and is the only one likely to be available in quantity before 1990. It can be produced from coal as well as from biomass, using existing technology (see Chapter 1). When added to gasoline (to make *gasohol*) alcohol can serve both to extend the supply of gasoline and to improve the octane rating.

Reductions in petroleum needs will be realized *only if large amounts of petroleum are not used in the production of alcohol.* The question of net energy balance is part of this issue, that is, whether the energy value of the alcohol produced will exceed the energy consumed in the growing, transportation, and conversion of the raw mate-

The alternative. Gasohol being sold along with gasoline at Wheaton, Maryland. (DOE photo by Jack Schneider.)

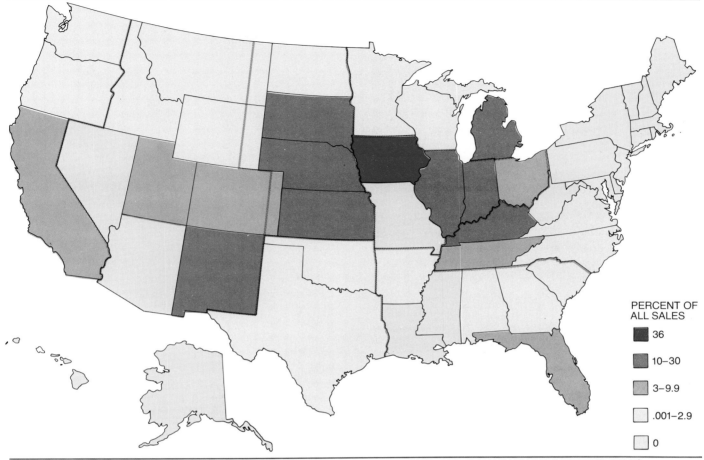

Fig. 10-7   Ethanol–gasoline blends (gasohol) as proportion of all gasoline sales, February, 1984.
Source: Information Resources, Inc., *Alcohol Outlook*, June 1984.

PERCENT OF
ALL SALES

- 36
- 10–30
- 3–9.9
- .001–2.9
- 0

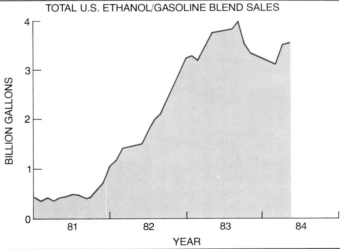

TOTAL U.S. ETHANOL/GASOLINE BLEND SALES

rials. Considering energy spent in the growing of corn, its transport, and its conversion, and taking "energy credit" for the dried grain by-product, the ethanol energy produced from corn is 1.05 times the energy expended. If that energy credit for the by-product is eliminated, then the energy produced is only 0.98 of the energy expended (Department of Energy, June 1979). Evidently the balance is close. The *savings of petroleum itself* can be greatly enhanced by not burning petroleum fuels in the distilling process. Coal, wood, or agricultural residues can be burned instead; or in some areas solar energy can be used for heat. Furthermore, waste heat from utilities or from industries could be employed to minimize fuel use in distillation.

THE CURRENT ALCOHOL PICTURE   The production of alcohol is being promoted now by federal policy that exempts gasohol from the excise tax on motor fuels, and by similar tax exemptions in some states.

Sales of gasohol grew rapidly in 1982 and early 1983, but leveled off and fell in late 1983 (see Fig. 10-7). The distribution of sales is represented here by data for one month—February 1984—showing proportion of all gaso-

line sales that were gasohol. Highest porportions are in the Corn Belt and neighboring Plains states.

The location of present alcohol plants has been deduced from two sources: a September 1981 Department of Energy survey of plants, published in 1982 (Department of Energy, Feb. 1982), and current data from the major alcohol industry publication (Information Resources Inc., 1984). Figure 10-8 shows the plants scattered across the

Fig. 10-8 Locations of operating alcohol fuels plants, July, 1983.
Sources: Department of Energy, Feb. 1982, and Information Resource Inc. brochure, 1984.

country, with concentrations in the Corn Belt, Plains states, and the Northwest. Very large plants that dominate the industry are in Iowa, Illinois, Indiana, Kentucky, and Tennessee.

Feedstocks for all these alcohol plants have been generalized for each state in an attempt to show the national pattern. The major feedstock in a state is identified on Figure 10-9, though in some states the dominance is not clear and could change with one or two new plants using a different feedstock. The great variety of feedstocks is evident, with corn prevailing in the expected states, but also in Alabama and Florida. It is the feedstock for all the very large alcohol plants shown on Figure 10-8.

ALCOHOL IN THE FUTURE A very complete projection of alcohol production and feedstock supplies in the year 2000 was made by the Department of Energy in 1979. Raw materials production is summarized in Table 10-4, which shows the types of feedstocks expected in each region: There are no absolute values in that table, only percentages. The Corn Belt and the Southeast are seen to be the largest producing regions. The Corn Belt will supply grains, sorghum, and agricultural residues. The Southeast region will supply wood and cane sugar. The three regions that each are expected to supply roughly 10

percent of the raw materials are varied in their character. The Pacific supply is mostly municipal solid wastes and wood. The North Plains and the Lake states are expected to provide a mixture of materials, with grains being most prominent. In the Delta States and South Plains regions, sugars and wood play a large role. The Northeast region supplies only 7 percent of the national total of raw materials in this projection, but is distinguished by its large contribution to the national total of municipal solid waste.

Projected volumes of alcohol produced in the year 2000 are listed, by region, in Table 10-5, and mapped in Figure 10-10. As expected on the the basis of the feedstock projection, greatest production will be in the Corn Belt and the Southeast. In every region, methanol constitutes well over half the production, and, in the national total, it accounts for 74 percent.

ENERGY VERSUS FOOD

Any discussion of crops produced expressly for energy must confront the fact that energy crops may compete with food crops for land and water resources. Two questions arise. Is it wise to devote agricultural resources to

Fig. 10–9 Primary feedstocks for alcohol fuels, by state, as of 1981.
Source: Department of Energy, Feb. 1982.

TABLE 10–4
Biomass Materials for Alcohol, in the year 2000, showing Proportions in Each Region
(in percentages)

U.S. DEPARTMENT OF AGRICULTURE FARM PRODUCTION REGIONS	WOOD	AGRIC. RESIDUES	GRAINS	SUGARS	MUNIC. SOLID WASTE	FOOD WASTES	TOTAL
Northeast	10	1			29	6	7
Southeast	17	2		27	14	30	13
Appalachian	12						5
Lake States	9	10	18	16	6	15	10
Corn Belt	8	42	43	10	17	22	21
Delta States	12	8		12	5	5	9
North Plains	1	18	26	16			10
South Plains	9	8	8	15	8		9
Mountain	9	5	5		4	2	6
Pacific	13	6		4	17	20	10
	100%	100%	100%	100%	100%	100%	100%

Source: U.S. Department of Energy, June 1979, p. 52.

TABLE 10–5
Projected Maximum Alcohol Production (Ethanol and Methanol) from U.S. Biomass Resources for the Year 2000, by Ten USDA Regions (in billions of gallons per year)

REGION	ETHANOL	METHANOL	TOTAL
Northeast	3.7	13.3	17.0
Southeast	7.8	18.7	26.5
Appalachian	3.1	11.4	14.5
Lake States	5.7	14.1	15.8
Corn Belt	10.1	29.8	39.9
Delta States	5.3	15.8	21.1
Northern Plains	4.7	9.6	14.3
Southern Plains	5.1	13.3	18.4
Mountain	3.2	11.5	14.7
Pacific	5.3	17.2	22.5
TOTAL	54.0	154.7	208.7

Source: Department of Energy, June 1979.

energy production? And, in a free market, how will certain energy crops compete with food crops—and on what kinds of land will they tend to be produced? The following materials provide only partial answers to these questions.

Dryland Crops

One solution, which is appropriate only for certain synthetic fuels, is to grow the energy crops on land not suitable for food production. There are some drought-loving plants that could be produced in arid areas of the Southwest (see Table 10–6). Two of those four plants, buffalo gourd and jojoba, yield edible oils. Jojoba oil can be used, in addition, for production of alcohols, lubricants, and waxes. Guayule and the gopher plant can be used for production of rubber, and can, therefore, reduce dependence on crude oil as a raw material for synthetic rubber. The gopher plant (Euphorbia lathyris) is especially interesting because it makes *hydrocarbons* as well as carbohy-

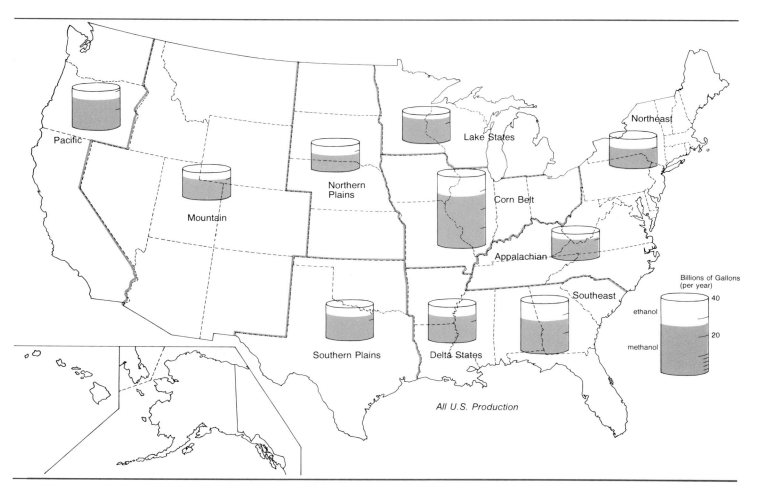

Fig. 10–10 Maximum ethanol and methanol production from biomass resources in year 2000, according to Department of Energy Projection.
Source: Department of Energy, June 1979.

TABLE 10–6
Potentially Commercial Dryland Crops for Oils, Rubber, and Crude Oil Substitute

PLANT	SPECIES	YIELD AND APPLICATIONS
Buffalo Gourd	Cucurbita	Edible oils, proteins, starch
Guayule	Parthenium argentatum	Rubber to reduce dependence on foreign natural rubber and on synthetic rubber made from petrolum
Jojoba	Simmondsia chinensis	Edible oils for use in foods. Source of alcohols, lubricants, waxes
Gopher Plant	Euphorbia lathyris	Hydrocarbons for rubber and as crude oil substitute

Source: Johnson and Hinman, 1980.

drates by photosynthesis. Among 200 species of desert plants analyzed, only Euphorbia lathyris appears to have the hydrocarbon content and biomass potential that justify development for the sake of hydrocarbons (Johnson and Hinman, 1980).

The areas appropriate for cultivation of the gopher plant are limited by sunlight and moisture as shown on Figure 10–11. Areas of intense sunlight needed for best growth of the crop are arbitrarily defined as those areas where average daily solar radiation (on an annual basis) is greater than 450 Langleys, that is, greater than 450 calories of energy received per square centimeter. The moisture limitations are as follows. Any area receiving less than 50 centimeters (roughly 20 inches) of precipitation per year is considered too dry for food crops. Areas receiving less than 15 centimeters are too dry for the gopher plant. Areas receiving 15 to 50 centimeters, therefore, are most suitable for growing the plant.

Figure 10–11 shows that suitable moisture coincides with required solar radiation in parts of California, Nevada, southern Utah, western Colorado, Arizona, New Mexico, and western Texas. Studies at the Office of Arid Land Studies at the University of Arizona indicate that roughly 8 to 12 million hectares of this land could be used with little or no irrigation needed. On the basis of present yields of 25 barrels of oil per hectare, 11 million hectares would supply about 4 percent of the nation's need for crude oil each year. If genetic and agronomic improvements were to increase the yield to 65 barrels per hectare, the production would be 10 percent of the present crude oil need (Johnson and Hinman, 1980).

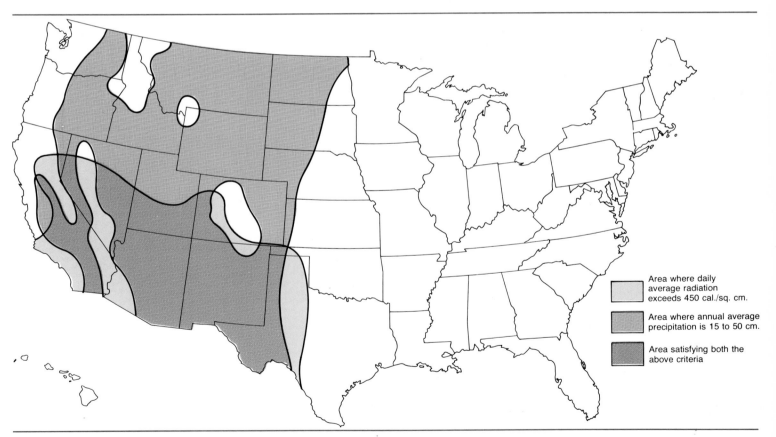

Area where daily average radiation exceeds 450 cal./sq. cm.

Area where annual average precipitation is 15 to 50 cm.

Area satisfying both the above criteria

Fig. 10–11 Areas suitable for growing gopher plant (Euphorbia lathyris) for hydrocarbons.
Source: "Oils and Rubber from Arid Land Plants" by J.D. Johnson and C.W. Hinman, *Science*, Vol. 208, May 2, 1980, pp. 460–464.

Competition of Energy Crops with Food Crops

For every part of the nation, it is possible to assess what crops would be grown on certain classes of land, if assumptions are made about crop prices, energy prices, and costs of production for each of the many food and energy crops considered. Such an ambitious study was conducted at Argonne National Laboratory, under contract to the U.S. Department of Energy, by a team led by Dr. Sinyan Shen (see Shen, 1983).

This study made use of a linear programming model for agricultural crops developed at Iowa State's Center for Agricultural and Rural Development (the CARD Model). The United States is divided into 105 crop-producing regions, each of which contains varying amounts of land in five different productivity classes ranging from marginal to most productive.

The model, when fed all the relevant data on costs and prices of crops, as well as demand levels for specified crops, will identify what parcels of land can be used for certain crops. In this application of the model, three energy crops were added to the possibilities: herbaceous grasses, Kenaf (a hemp-like plant), and a form of tree orchard.

One of the most interesting results of the study is the identification of 125 million acres of land in the United States that *is now considered marginal* but could be brought into production for energy crops. In fact, it could be used for some food crops as well, but is not used for food crops now. This marginal land is now forest, rangeland, and unmanaged pasture lands.

The distribution of this marginal land that could be converted to energy crops without displacing any food production is shown in Figure 10–12. The largest acreage totals per state appear in Texas, a number of Great Plains states, and other scattered states. Initially, 375 million

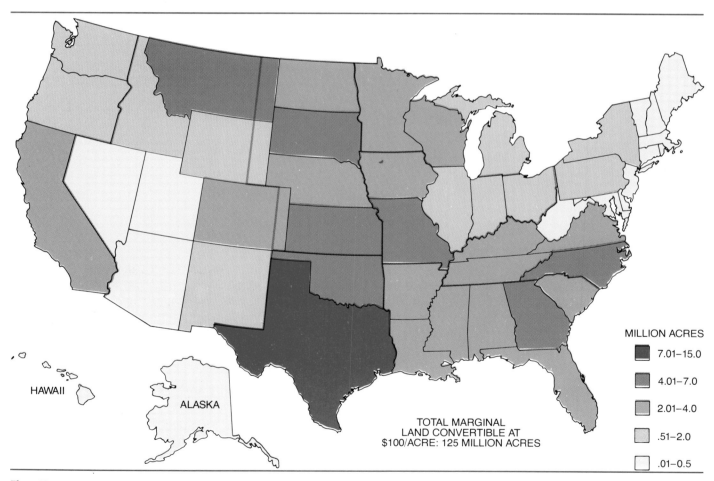

MILLION ACRES

7.01–15.0

4.01–7.0

2.01–4.0

.51–2.0

.01–0.5

TOTAL MARGINAL
LAND CONVERTIBLE AT
$100/ACRE: 125 MILLION ACRES

HAWAII

ALASKA

Fig. 10–12 Marginal agricultural land that may be converted to energy crops if conversion costs are less than $100 per acre.
Source: Shen, Argonne National Laboratory, 1984.

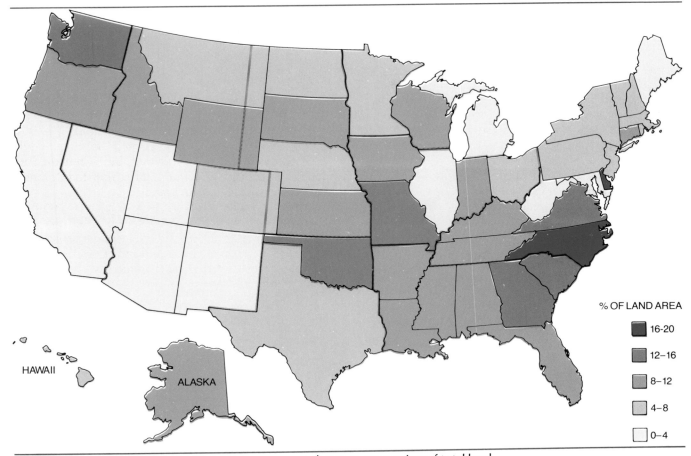

Fig. 10-13 Marginal land convertible to energy crops shown as proportion of total land area of each state.
Source: Shen, Argonne National Laboratory, 1984.

% OF LAND AREA

16-20

12–16

8–12

4–8

0–4

acres of marginal land were identified, but costs of converting much of the land (by clearing forest, for instance) were prohibitive. When conversion costs were kept to $100 per acre and less, the qualifying marginal lands total 125 million acres, and are distributed as shown on Figure 10-12.

To consider the fact that large acreages are likely to occur in large states, we map, in Figure 10-13, the convertible marginal land as a proportion of the total land area of each state. On this basis, Texas is no longer ranked high. Instead, states such as North Carolina and Delaware stand out; Washington, South Carolina, and Georgia are ranked next; and Texas is in the second-lowest category.

If energy crops were grown on these converted marginal lands, the areas occupied by specific crops would depend on their moisture and temperature requirements. For the herbaceous grasses, there is no restriction, since they can be grown anywhere in the 48 states. Possible areas for irrigated Kenaf, nonirrigated Kenaf, and tree orchards are shown in Figure 10-14. Kenaf is limited to southern—subtropical—areas, while tree orchards are limited to areas receiving over 20 inches of rain per year.

FUTURE BIOMASS PRODUCTION

The potential contribution of biomass to future energy needs is problematic—given the complexities of economic factors, competition with food crops, unknown future energy prices, and the possibility of government policies that might provide a favorable economic climate.

The foregoing analysis showed that substantial amounts of now-marginal land could be devoted to energy crops. But what about the total of energy crops on all land?

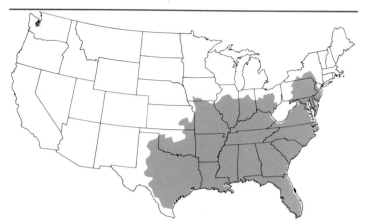

A. NON-IRRIGATED KENAF

B. IRRIGATED KENAF

C. TREE ORCHARDS

Fig. 10-14 Areas suitable for three of the energy crops considered in determining convertible marginal lands.
Source: Shen, Argonne National Laboratory, 1984.

TABLE 10-7
Two Federal Estimates of Biomass Production
in the year 2000

| FEEDSTOCK | GROSS ENERGY (QUADS) | |
	ERAB	OTA
Wood, plants, plant residues animal wastes	9.05 - 9.42	6.0 - 16.6
(Wood only)	(4.77)	(5.0 - 10.0)
Other wastes and sewage	1.67 - 1.89	[a]
Total	10.72 - 11.31	6.0 - 16.6

[a] Not considered in the OTA report.
Sources: U.S. Department of Energy, Nov. 1981, and U.S. Office of Technology Assessment, 1980.

TABLE 10-8
Current and Projected Biomass Energy Production
in the United States (Quads)[1]

| | YEAR | | | | |
	1970	1980	1990	2000	2010
Wood[2]	1.5	2.3	2.5	3.1	3.6
Herbaceous Biomass and Wastes[3]	0.1	0.4	0.7	0.9	1.1
Total	1.6	2.7	3.2	4.0	4.7

[1] Projections are based on the NEPP model assuming no unusual decline in world oil demand, a world oil price in 1982 dollars per barrel of $26 in 1985 to $84 in 2010, and 2.8% economic growth in the U.S. from 1982 to 2000.
[2] Primarily direct combustion and electricity production.
[3] Includes production of alcohol, landfill gas and anaerobic digestion of herbaceous biomass and wastes.
Source: Shen, Argonne National Laboratory, 1984.

At present, biomass resources of all kinds supply 2.8 Quads of energy, that is, 3.7 percent of all energy supplies. In the year 2000, two government projections suggest total biomass energy of 6 to 16 Quads (see Table 10-7). A more modest projection, giving past years for perspective, is presented in Table 10-8 according to Argonne National Laboratory, 1984. This view suggests only 4 Quads of biomass energy output in the year 2000.

References

Ernest, Kent, *et al.* *Mission Analysis for the Federal Fuels From Biomass Program, Volume III: Feedstock Availability, Final Report,* Menlo Park, California: Stanford Research Institute, January, 1979.

Flavin, Christopher. "Developing Renewable Energy." In Lester R. Brown, *State of the World.* New York: Norton, 1984.

Fowler, John M. and Kathryn Fowler. *Fuels from Plants: Bioconversion,* Factsheet No. 1. Oak Ridge: National Science Teachers' Association, 19XX.

Information Resources Inc. *Alcohol Outlook,* published monthly.

———. Brochure advertising a publication, *The U.S. Alcohol Fuels Industry Data Base.*

Johnson, Jack D. and C. Wiley Hinman. "Oils and Rubber from Arid Land Plants." *Science,* 208:4443, 1980, 460–464.

Shen, Sinyan. Argonne National Laboratory. "Regional Impacts of Herbaceous and Woody Biomass Production on U.S. Agriculture." Paper in *Energy from Biomass and Wastes VII,* Institute of Gas Technology, Chicago, Illinois, 1983.

———. Unpublished map. "Amount of Marginal Land Convertible to Biomass Production at $100 per acre," and unpublished tabulations. Personal communication, Sept. 1984.

Time magazine. Jan. 16, 1984.

U.S. Department of Agriculture, Forestry Service. *Tree Biomass—A State of the Art Compilation.* Washington, DC: Government Printing Office, Nov. 1981.

U.S. Department of Energy. *Biomass Energy.* Report of the Energy Research Advisory Board Panel on Biomass (ERAB) Washington, DC, Nov. 1981.

U.S. Department of Energy. *Biomass Energy Technology Research Program Summary FY 1983.* DOE/CE–0032/1 Washington, DC: Government Printing Office, 1983.

U.S. Department of Energy. *Estimates of U.S. Wood Energy Consumption from 1949 to 1981.* DOE/EIA–0341. Washington, DC: Government Printing Office, August 1982.

U.S. Department of Energy. *Large-scale Alcohol Fuels Plants Directory.* DOE/SERI SP–290–1467. Golden, Colo: Solar Energy Research Institute, Feb. 1982.

U.S. Department of Energy. *The Report of the Alcohol Fuels Policy Review.* DOE/PE–0012. Washington, DC: Government Printing Office, June 1979.

U.S. Office of Technology Assessment. *Energy from Biological Processes,* Washington, D.C.: Government Printing Office, 1980.

Part Three

OVERVIEW

The energy sources reviewed in the foregoing chapters offer supplies of energy that, taken together, have very great potential and wide variety in character. It is reasonable to hope that some mineral fuels can be used in the near term as a bridge to the more remote future in which a combination of renewable energy forms will provide clean and sustainable energy. The fuel options will be narrowed by decisions about environmental safety and about the net energy return of one fuel versus another. The renewable energy forms of the future will depend upon which are proven workable by current development work.

Matters of environmental policy or net energy analysis cannot be dealt with here. Neither can this book speculate on the ultimate workability of the various forms of renewable energy. Part Three, however, does offer something that may be helpful in planning: a summary of the energy potentials in each of the energy sources studied earlier. Both national totals and regional characters are compared with a view toward a better understanding of the options that exist within the United States.

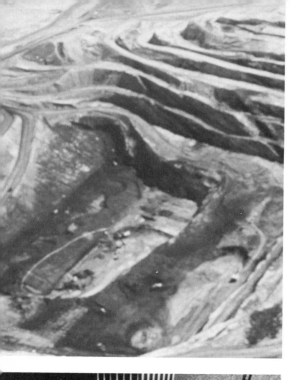
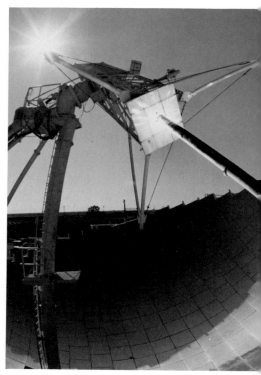

11 Conclusion

NONRENEWABLE RESOURCES

The seven nonrenewable energy sources discussed in this atlas—coal, crude oil, natural gas, shale oil, oil from tar sands, nuclear fuels, and geothermal heat—can all be represented by energy amounts thought to be recoverable from identified and from hypothetical or undiscovered deposits. However, the amounts of energy economically recoverable from identified deposits, that is, Reserves, are not firmly definable for shale oil, tar sands, and geothermal heat.

Summary By State

The resource character of states is reviewed in two parts:

1. The *reserves* of fossil fuels and uranium oxide are added together for each state.
2. The energy *possibly recoverable* from identified oil shales and geothermal occurrences is treated in a similar way.

RESOURCES RECOVERABLE ECONOMICALLY AT PRESENT: TRADITIONAL FOSSIL FUELS AND URANIUM OXIDE Reserves of coal, conventional crude oil, conventional natural gas, and uranium oxide are roughly equivalent in their economic feasibility. They may be added, therefore, to show how each one contributes to state totals of recoverable energy.

Coal Reserves are those energy amounts thought to be recoverable from the Demonstrated Reserve Base on the assumption of 90 percent recovery in surface mining and 50 percent recovery in underground mining. Since mineable tonnages are known for the four major ranks separately, the estimates of recoverable coal are translated into energy amounts by using the four appropriate BTU per ton values.

Conventional crude oil and natural gas must be represented by rather conservative numbers. For crude oil, only data for Measured (Proved) plus Indicated reserves are available by state. Missing from this summary are:

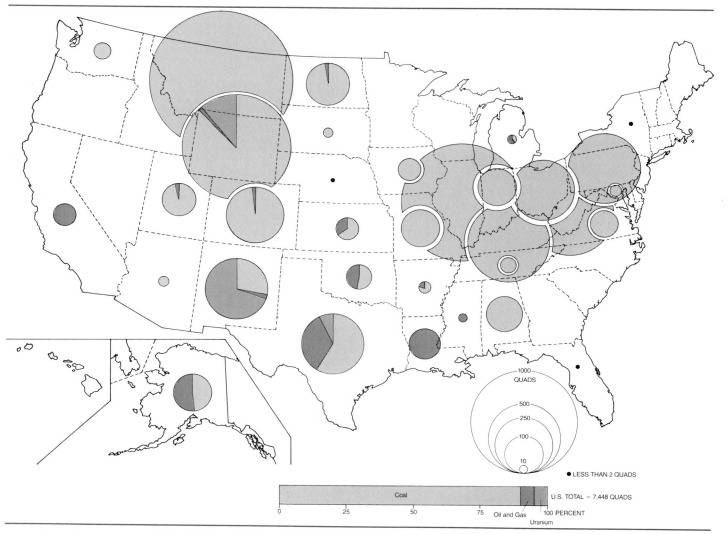

Fig. 11-1 State totals of Recoverable energy in *reserves* of coal, conventional crude oil, and natural gas, and uranium oxide, showing proportions in each resource.

1. Inferred amounts in known reservoirs that will likely be added to Reserves as young fields are developed by further drilling.

2. Future Reserves in undiscovered reservoir rocks.

3. Oil that will be recovered from known and undiscovered reservoirs by advanced recovery techniques.

The latter two are rightly excluded as they do not qualify as Reserves. Exclusion of Inferred amounts is less justifiable, but cannot be avoided since data for these amounts are not available by state. For natural gas, Measured (Proved) reserves are used, and again the Inferred and Undiscovered amounts are missing.

Uranium oxide is represented by Reserves that are recoverable at costs of 100 dollars per pound, and therefore embrace the smaller amounts recoverable from richer ores at costs of 30 and 50 dollars per pound. The Department of Energy reports uranium Reserves by state only for New Mexico, Wyoming, and Texas. They lump together the Reserves in Arizona, Colorado, Utah and lesser Reserves in other states, to avoid disclosing the Reserves of mining

companies. For this summary by state, the energy in uranium oxide Reserves is assumed to be only that in the fissile U_{235} isotope: the greater energy in the more abundant U_{238} isotope is reflected in tabulations and maps by economic region.

Excluded from this state summary are unconventional or synthetic crude oil from oil shales or tar sands, and unconventional natural gas from tight gas sands of the West or the Devonian shales of Appalachian states. The Reserve numbers selected here are conservative for oil and gas and also for coal. For oil and gas, further drilling will probably locate more new economically recoverable amounts than further exploration for coal. Rising prices and changing technology, however, will open up more potential for coal than for oil and gas. Improved mining techniques could add substantially to the coal that is mineable; and the advent of in-place gasification could add a similar amount of energy recoverable from deep beds.

Table 11-1 lists energy reserves for coal, petroleum (conventional oil and gas), and uranium oxide for each state. Figure 11-1 shows how the state totals are divided

342

among those three components. On a national basis coal Reserves constitute 90 percent of the total. Furthermore, coal overshadows petroleum in most states where the two coincide. Oklahoma and Texas both regarded traditionally as oil states, are dominated by coal, though inclusion of Inferred amounts of crude oil would alter the picture substantially in states where fields are young. Louisiana is clearly an oil and gas state, but the Texas total includes substantial contributions from coal and uranium oxide. In New Mexico and Wyoming the large amounts of uranium oxide contribute heavily to total energy Reserves.

RESOURCES LARGELY NONECONOMIC: SHALE OIL, TAR SANDS, AND GEOTHERMAL HEAT

Energy amounts in oil shales, tar sands, and geothermal occurrences are not necessarily comparable in economic feasibility. Nevertheless, they all require either a rise in fuel prices, a development effort, or both, before the amounts thought recoverable from identified occurrences will be realized. The rather generous amounts defined below are not Reserves, and may not be future Reserves: They are better called *maximum potentially recoverable energy from identified occurrences.*

TABLE 11–1
State Shares of Energy Recoverable from Coal, Natural Gas, Conventional Crude Oil, and Uranium Oxide Reserves (Quads)

STATE	COAL[1]	CONVENTIONAL CRUDE OIL[2]	CONVENTIONAL NATURAL GAS[3]	URANIUM OXIDE[4]	TOTAL
Alabama	92.1	0.3	0.9		93.3
Alaska	69.6	40.6	35.0		145.2
Arizona	7.9	0.0	0.0		7.9
Arkansas	6.7	0.7	2.1		9.5
California	0.0	29.7	6.1		35.8
Colorado	217.9	1.0	3.4		222.3
Florida	0.0	0.4	0.1		0.5
Illinois	1,186.9	0.8	0.0		1,187.7
Indiana	153.1	0.2	0.0		153.3
Iowa	33.3	0.0	0.0		33.3
Kansas	23.2	1.9	10.4		35.5
Kentucky	519.8	0.2	0.6		520.6
Louisiana	0.0	15.0	44.6		59.6
Maryland	11.6	0.0	0.0		11.6
Michigan	1.7	1.2	1.4		4.3
Mississippi	0.0	1.1	1.6		2.7
Missouri	126.3	0.0	0.0		126.3
Montana	1,517.6	1.3	0.9		1,519.8
Nebraska	0.0	0.2	0.0		0.2
New Mexico	75.7	3.2	12.6	183.2	274.7
New York	0.0	0.0	0.3		0.3
North Dakota	124.8	1.4	0.7		126.9
Ohio	307.9	0.7	2.1		310.7
Oklahoma	25.1	5.2	17.6		47.9
Pennsylvania	403.7	0.2	1.9		405.8
South Dakota	4.6	0.0	0.0		4.6
Tennessee	15.7	0.0	0.0		15.7
Texas	169.8	41.9	54.9	19.1	285.7
Utah	86.3	1.0	2.5		89.8
Virginia	51.8	0.0	0.2		52.0
Washington	16.6	0.0	0.0		16.6
West Virginia	565.0	0.3	2.4		567.7
Wyoming	910.3	5.3	10.9	111.0	1,037.5
Others	0.4[a]	0.2[b]	0.1[b]	42.2[c]	42.9
Total	6,725.1	154.0	213.3	355.6	7,448.0

[1] Recoverable from Demonstrated Reserve Base as of January 1, 1982
[2] Proved Reserves as of January 1, 1984
[3] Proved Reserves as of January 1, 1984
[4] Reserves up to $100 per pound as of January 1, 1983
[a] Georgia, Idaho, Oregon, and North Carolina
[b] Arizona, Missouri, Nevada, New York, South Dakota, Tennessee, and Virginia
[c] Most is found in Colorado, Utah, and Arizona

Shale oil amounts recoverable are only those from the Green River formation of Colorado, Utah, and Wyoming. Lower-grade deposits from the Appalachian states and Alaska are omitted completely. Assuming that future fuel prices will justify extracting oil from the leaner as well as the richer beds, this estimate includes oil recoverable from all identified beds averaging 15 gallons of oil per ton of rock. Potential oil in-place in such beds is about 1,800 billion barrels, of which roughly 610 billion (one-third of oil in-place) may be recoverable.

Deposits of tar sands and other bitumen bearing rocks occur in more than a dozen states. However, more than 50 percent of the measured resources containing an estimated 66 billion barrels of oil are found in Utah. Roughly a third of the oil in the tar sands is considered recoverable.

Geothermal heat recoverable may be estimated for only two of the three major exploitable occurrences, because extraction of heat from the volcanic type has not yet been demonstrated. For *hydrothermal* occurrences heat recoverable at well-head is roughly 25 percent of identified heat in-place to a depth of 3 kilometers. For geopressured occurrences in deep sand bodies along the Gulf Coast, both thermal and methane energy are included to a depth of 6.86 kilometers (22,500 feet). For this summary the most ambitious Gulf Coast development plan is assumed, which implies recovery of roughly 3.3 percent of the heat in-place. Geothermal heat recoverable at well-head without regard to application is loosely analogous to energy in crude oil or natural gas at well-head. The difference is that oil and gas provide very high temperatures and much higher conversion efficiencies than geothermal heat. The comparison with energy in crude oil from shales, therefore, is not entirely appropriate.

Table 11-2 and Figure 11-2 note the energy amounts recoverable. For the country as a whole, energy in Green River oil shales and in geopressured reservoirs is far greater than that in hydrothermal occurrences and tar sands. Western and southwestern states monopolize the uncon-

TABLE 11-2
All Energy Recoverable from Identified Green River Oil Shales, Tar Sands, and Exploitable Geothermal Occurrences (Quads)

| STATE | RECOVERABLE WELL-HEAD GEOTHERMAL ENERGY | | | | | RECOVERABLE GREEN RIVER OIL SHALE | | |
	Hydrothermal 150°C	Hydrothermal 90–150°C	Hydrothermal Less than 150°C	Geopressured	Total	Over 15 Gal/Ton	Over 30 Gal/Ton	TAR SANDS
Alabama	0.00	0.00	0.00					8.30
Alaska	1.94	5.52	0.83		8.29			
Arizona	0.28	0.87	3.42		4.57			
California	134.14	8.02	5.29		147.45			11.10
Colorado	0.28	2.30	3.42		6.00	2,222.20	655.60	
Georgia	0.00	0.00	0.01		0.01			
Hawaii	1.95	0.00	0.05		2.00			
Idaho	4.21	121.36	5.79		131.36			
Kansas	27.03	0.00	3.13		30.16			
Kentucky	0.00	0.00	0.00					7.20
Louisiana	0.00	0.00	0.00	722.54	722.54			
Montana	0.00	2.58	16.40		18.90			
Nebraska	0.00	0.00	0.94		0.94			
Nevada	0.00	8.46	5.22		13.68			
New Mexico	28.88	1.03	1.25		32.06			
North Carolina	0.00	0.00	0.14		0.14			
North Dakota	0.00	0.00	247.50		247.50			
Oklahoma	0.00	0.00	5.41		5.41			
Oregon	21.44	12.94	2.56		36.94			
South Dakota	0.00	0.00	5.41		5.41			
Texas	0.00	0.00	0.81	1,909.57	1,910.38			19.40
Utah	11.39	1.43	1.99		14.81	583.30	94.40	66.60
Virginia	0.00	0.00	0.23		0.23			
Washington	0.04	0.00	1.06		1.10			
West Virginia	0.00	0.00	0.01		0.01			
Wyoming	0.00	0.60	9.02		9.62	583.30	22.20	
Other	0.00	0.00	0.00	1,576.09[2]	1,576.09			8.90[1]
Total	231.58	165.11	319.90	4,208.20	4,923.84	3,388.80	772.20	121.50

[1] Kansas, Missouri, Oklahoma, Wyoming, and New Mexico.
[2] Federal Offshore areas in Gulf of Mexico.

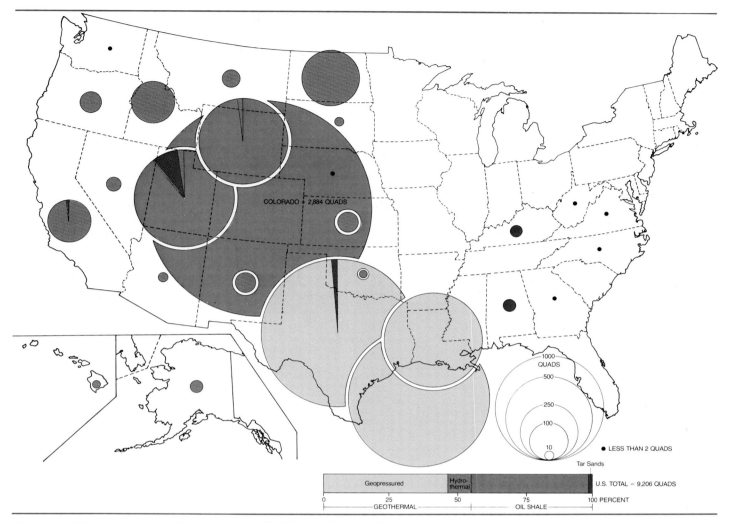

Fig. 11–2 All energy recoverable from *Identified* Green River oil shales, tar sands, and exploitable geothermal occurrences. (Quads)

ventional crude oil and geothermal resources. It is apparent that the two broad types of resource rarely combine in any one state. There are the "shale oil states," Colorado, Utah, and Wyoming, and there are "geothermal states," among which California and Idaho are outstanding for hydrothermal occurrences, and Texas and Louisiana for geopressured reservoirs. Wyoming would be a dual "shale oil and geothermal state" on the basis of *energy in-place,* but the geothermal energy occurs in Yellowstone National Park and cannot be exploited.

This summary would be more complete if unconventional natural gas from the tight sands of western states and from Devonian shales of the Appalachian states were included. These energy amounts are only vague estimates now, and cannot be clearly assigned to states or to economic regions. An estimate does appear later in the national overview of energy recoverable from identified and hypothetical sources.

Summary By Region

In this summary recoverable energy amounts thought to be *economic or near-economic* are brought together in 11 economic regions. As in the foregoing summary by state, the energy amounts from various sources are added for each region using Quads as the energy unit.

FOSSIL FUELS AND GEOTHERMAL HEAT Reserves of coal, conventional crude oil, and conventional natural gas may be compared with Reserves of nuclear fuels (defined below) because they are all roughly equivalent in economic feasibility. Oil from tar sands is ignored, in part because amounts are small, but mainly because extraction is not taking place currently, and seems unlikely in the face of competition from oil shales. For oil shales and geothermal heat, no Reserve amounts are recognized by industry or government, however, but the current activity in both fields suggests there are some economically obtainable energy amounts. These important resources must be estimated here for inclusion in the summary by region.

For oil shales, only the Green River formation in western states is considered, and only the beds whose crude oil yield is expected to meet or exceed 30 gallons per ton of rock are included. Table 11-2 shows that all three of the western shale oil states have some beds of this richness, but Colorado has the largest share.

For geothermal heat, volcanic occurrences are excluded because no estimate of recoverable heat can be made. Also excluded is heat recoverable from geopressured reservoirs because the proposed application, electric power generation, has yet to be demonstrated. Only hydrothermal systems are considered to hold economic or near-economic energy that is comparable to that in the rich beds of Green River oil shales. For purposes of comparison some portion of the well-head heat estimated recoverable by the U.S. Geological Survey must be selected, but the portion is uncertain. While an earlier Survey report suggested that *half* the hydrothermal energy recoverable may be called Reserves for electrical applications, the 1978 Survey report declined to make an estimate of Reserves. The experimental power plants being built in California and space-heating applications in Idaho and Oregon justify the inclusion of some of the hydrothermal well-head heat. For the purposes of this summary and for want of a better number, 50 percent of this heat is called Reserves. The 50 percent ratio gives only a very rough estimate of possible Reserves in each region because of the following reasons. (1) The systems with highest temperatures and shallow depths would be most attractive for electrical power generation. (2) Only the systems with nearby settlements would be used for space heating. (3) These characteristics do not occur equally in all areas that have hydrothermal resources. To say, for instance, that in each state half the recoverable heat is economically attractive could be most misleading. The summary is most meaningful when state amounts are agglomerated into regions. Regions containing only one state, however, such as Alaska or Hawaii, suffer the most distortion in this comparison.

NUCLEAR FUELS To represent economically recoverable uranium energy the Reserves of uranium oxide producible at forward costs of $100 per pound are assigned to economic regions from physiographic regions in which data are collected. Thorium energy is represented by Reserves of thorium oxide.

Because transmutation of uranium to plutonium in breeder reactors makes use of the abundant isotope, U_{238}, while nonbreeders use the naturally fissile U_{235}, the energy content of uranium Reserves used for breeders is roughly 140 times greater than if nonbreeders are assumed. National totals of 355.6 Quads versus 49,756 Quads demonstrate the difference (Table 11-3). Therefore, for the purpose of this overview two regional summaries are pre-

sented: one without breeder fuels and one with—the latter including thorium transmuted to fissile U_{233}. Obviously, the comparison of national resources with and without breeder fuels is important only if breeder reactors were to become a competitive energy producing technology. Presently they are not and will not be in the foreseeable future.

REGIONAL PATTERNS Table 11-3 summarizes the energy in all the amounts defined above, for 11 regions and shows regional totals both with and without breeder fuels. Figure 11-3 expresses those regional totals in a pair of unconventional maps in which the sizes of regions reveal their total recoverable energy. Since one scaling system applies to both maps, the impact of including or excluding breeder fuels is readily noted.

Figure 11-4 maps the regional totals and their various resource components assuming that nonbreeders are used for conversion. On this basis uranium oxide does not make a conspicuous contribution. In the Mountain Region, clearly the richest in energy, uranium contributes only 8 percent of the 4,148 Quads, while coal reserves account for 68 percent. The East North Central Region, with a total of 1,656 Quads, is next largest and owes 99.9 percent of its energy to coal Reserves, mostly those in Illinois. Most other regions are, likewise, dominated by coal energy, however, oil and gas are important in the West South Central Region (Texas, Lousiana, and Oklahoma). In this region oil and gas account for about 28 percent of the total resources. In the Pacific Region geothermal energy represents almost 60 percent of the total.

Figure 11-5 shows that with breeders assumed, uranium and thorium oxide account for 88 percent of the nation's energy Reserves. In those regions where uranium or thorium occur they play the leading role, constituting 94 percent of the Mountain region's immense total (largely because of uranium) as well as 87 percent and 64 percent in South Atlantic and East North Central regions. Uranium in the Colorado Plateau and Wyoming Basis and thorium in Idaho and Montana are responsible for the Mountain region's character. The South Atlantic and East North Central regions emerge as leading energy suppliers because of thorium Reserves in placer deposits along the Atlantic coast in Georgia and Florida, and in ancient conglomerate rocks in Michigan. Coal is still the dominant resource in the Middle Atlantic, West North Central, and East South Central regions, but those regional totals are eclipsed by breeder fuel energy in the three regions affected.

RENEWABLE SOURCES OF ENERGY

In this section, the energy from solar radiation itself, and its manifestations in wind, ocean thermal gradients, and biomass, are summarized. Electrical applications are

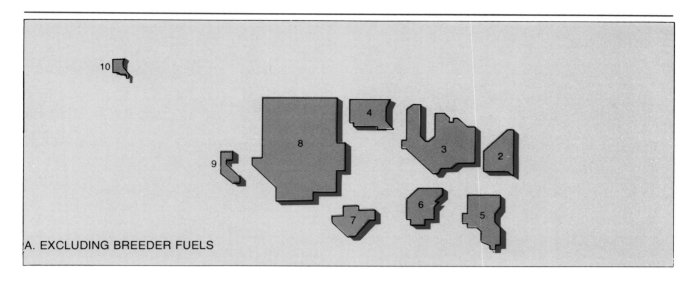

REGIONS

1 New England   6 E. So. Central
2 Middle Atlantic 7 W. So. Central
3 E. No. Central   8 Mountain
4 W. No. Central   9 Pacific
5 South Atlantic   10 Alaska

1000 Quads

New England Region has no recoverable energy; Hawaii has
0.95 quadrillion BTUs from geothermal sources

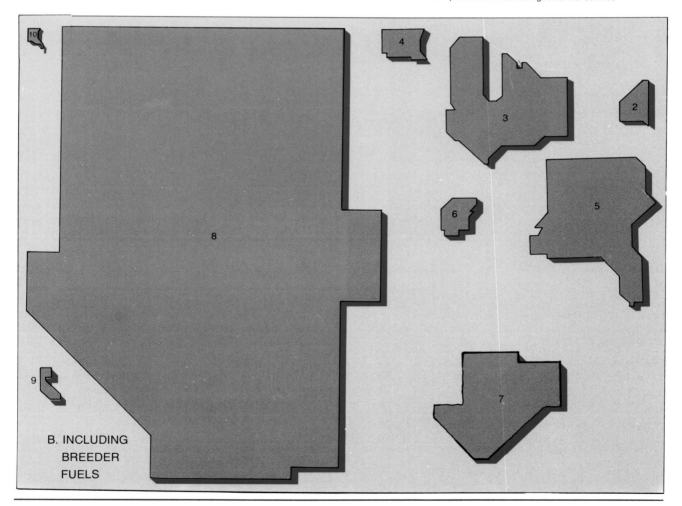

B. INCLUDING
BREEDER
FUELS

Fig. 11–3 Total energy Recoverable from estimated reserves of nonrenewable resources
(A excludes breeder reactors and B does not). The area of each economic region is made
proportional to its total recoverable energy in Quads.

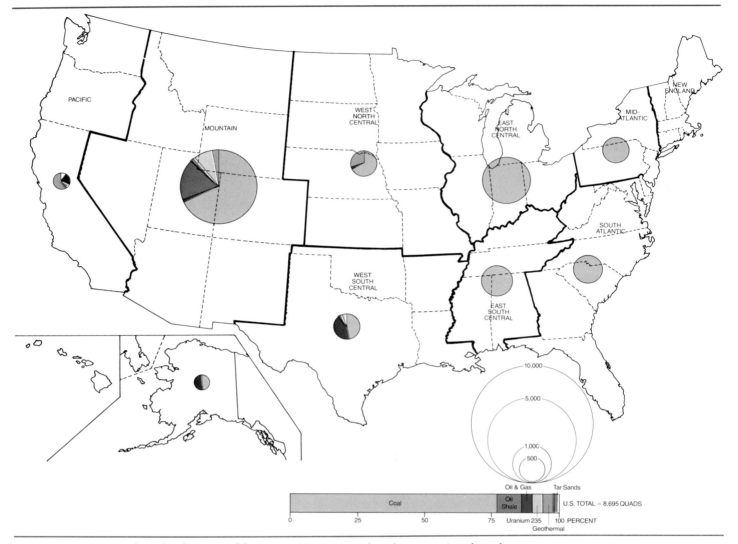

Fig. 11–4 Regional totals of recoverable energy in reserves of coal, conventional crude oil, and natural gas, and uranium oxide (U-235), combined with energy in possible reserves of Green River oil shales, tar sands, and hydrothermal-type geothermal occurrences. Breeder fuels are excluded.

TABLE 11–3
Regional Shares of Energy Recoverable from Estimated Reserves of Nonrenewable Resources, including Geothermal (Quads)

ECONOMIC REGION	COAL	GAS	OIL	OIL SHALE[1]	TAR SANDS	URANIUM 235	URANIUM 238	THORIUM	GEOTHERMAL[2]	REGION TOTALS	REGION TOTALS WITHOUT U₂₃₈ OR THORIUM
New England	0.00	0.00	0.00	0.00	0.00	0.00	0.00	0.00	0.00	0.00	0.00
Middle Atlantic	403.70	2.20	0.20	0.00	0.00	0.00	0.00	0.00	0.00	403.70	403.70
East North Central	1,649.60	3.50	2.90	0.00	0.00	0.00	0.00	2,899.00	0.00	4,555.01	1,656.00
West North Central	312.20	11.10	3.50	0.00	0.00	0.00	0.00	0.00	144.44	463.24	463.24
South Atlantic	628.40	3.00	0.70	0.00	0.00	0.00	0.00	4,437.00	0.20	5,069.29	632.29
East South Central	627.60	3.10	1.60	0.00	15.50	0.00	0.00	0.00	0.00	647.80	647.80
West South Central	201.60	119.20	62.80	0.00	19.40	19.10	2,674.00	0.00	3.11	3,035.11	425.21
Mountain	2,815.70	30.30	11.80	772.20	66.60	336.60	47,083.00	11,122.00	115.50	62,016.10	4,148.40
Pacific	16.60	6.10	29.70	0.00	11.10	0.00	0.00	0.00	92.77	156.27	156.27
Alaska	69.60	35.00	40.60	0.00	0.00	0.00	0.00	0.00	4.15	149.34	149.34
Hawaii	0.00	0.00	0.00	0.00	0.00	0.00	0.00	0.00	1.00	1.00	1.00
Other	0.40[a]	0.10[b]	0.20	0.00	8.90[c]	0.00	0.00	0.00	0.00	9.60	9.60
U.S. Total	6,725.10	213.30	154.00	772.20	121.50	355.40	49,756.00	18,458.00	361.17	76,508.86	8,695.25

[1] Using only recoverable oil from beds averaging 30 gallons per ton or more.

[2] Near future geothermal reserves *excluding* all geopressured but *including* one-half of recoverable well-head energy.

[a] Georgia, Idaho, Oregon, North Carolina

[b] Arizona, Missouri, Nevada, New York, South Dakota, Tennessee, and Virginia

[c] Kansas, Missouri, Oklahoma, and New Mexico

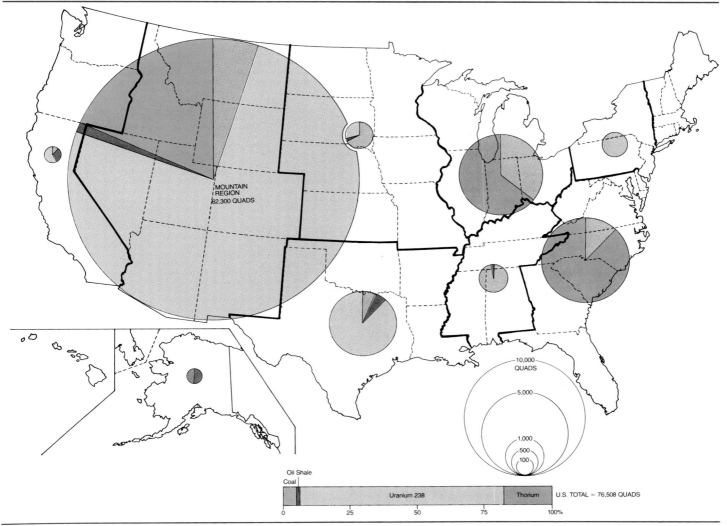

Fig. 11-5 Regional totals of recoverable energy in reserves of coal, conventional crude oil and natural gas, uranium oxide (U-238) and thorium oxide, combined with energy in possible reserves of Green River oil shales, tar sands, hydrothermal-type geothermal occurrences. The inclusion of breeder fuels drastically changes the amount and the distribution of recoverable energy.

dealt with first, then the electrical and nonelectrical together.

Electrical Generation

Wind, hydroelectric, and ocean thermal gradients (OTEC) are the only renewable energy flows that lend themselves to estimates of *maximum* generating potential, that is, the installed generating capacity consistent with saturation of the areas suitable. Tables 11-4 and 11-5 summarize those maximum installed capacities for comparison. It is not realistic, of course, to assume any one of the sources will be used to its fullest extent.

Maximum installed capacity for wind machines is

greater than for hydroelectric and OTEC, despite the exclusion of very substantial wind potential in Atlantic and Pacific offshore areas. Maximum annual *output,* however, is slightly greater from hydroelectric, and is greatest from OTEC. This change in ranking is because of widely differing *capacity factors:* Wind machines are expected to run at their rated capacity only 0.22 to 0.38 of the time, whereas OTEC installations are expected to run steadily at or near their capacity, as expressed by the factor 0.85. If the OTEC assessment is realistic and the technology proves workable, this source of electrical power could be extremely important. Figure 11-6 compares the maximum generating capacities, and provides some estimates of what proportion of those potentials may be installed by the year 2000.

TABLE 11–4
Total Potential Hydroelectric Capacity and Output for Economic Regions 1981

REGION	TOTAL POTENTIAL CAPACITY[1] (MEGAWATTS)	CAPACITY FACTOR (PERCENT)	OUTPUT (QUADS)
New England	5,851.3	46	0.081
Middle Atlantic	7,731.2	69	0.159
East North Central	3,987.6	54	0.064
West North Central	4,764.7	48	0.068
South Atlantic	14,376.6	31	0.133
East South Central	8,343.2	45	0.112
West South Central	5,311.5	31	0.049
Mountain	21,449.6	49	0.314
Pacific	53,481.7	58	0.927
Alaska	3,671.6	46	0.051
Hawaii	25.2	66	0.001
U.S. Total	129,051.0	52	2.00

Note: Output in Quads equals $\dfrac{\text{(Total potential in KW)(8760 hours)(Capacity Factor)}}{2.93 \times 10^{11}}$

[1]Total potential is the total potential installed plus the additional capacity at existing dams as reported by the Army Corps of Engineers.

A more complete review of projected electrical contributions, expressed as output, not installed capacity, is in Table 11–6. For the sake of uniformity, the projection by METREK, assuming incentives of the National Energy Plan, is the source of all this information. According to this view of the future, wind makes the largest contribution, followed by solar and OTEC. Wind's contribution is suprisingly large in light of the maximum potential noted above: The projected installed capacity is roughly 10 percent of the maximum, yet the estimated output is roughly 62 percent of the maximum output which assumes capacity factors of 0.22 and 0.38.

Fuel Savings from Electrical and Non-electrical Applications

The energy value of fuels saved is considerably greater than the energy value of electrical output or heat delivered from a renewable source. If, for instance, the renewable energy displaces coal-fired boilers for electrical generation, the coal energy saved is roughly 2.9 times the output energy because only 38 percent of coal energy in the boiler is converted to electrical energy.

Figure 11–7 show estimates of fuel saved in the year 2020 according to assumptions of the National Energy Plan (METREK, 1978), and according to three scenarios based on three different sets of assumptions (Stanford Research Institute, 1977). For electrical applications the fuel savings projected by METREK appear as a portion of the maximum or saturation savings tabulated earlier. Savings due to wind appear very large in relation to maximum potential, and may assume offshore installations not considered in the maximum potential or may assume much higher capacity factors than those assumed in estimating potential savings. For nonelectric applications there is considerable variety among the projections. A striking example is that the projection of fuel saved by solar process heat according to METREK exceeds the most optimistic projection by the Stanford Research Institute, but for space heating and biomass fuels the METREK projections are more modest. Figure 11–7 shows that the estimate of the total fuel energy saved according to METREK lies somewhere between the Low Demand and the Solar Emphasis scenarios of the Stanford Research Institute. METREK's ambitious 6.6 Quads saved by wind dynamos has no counterpart in the Stanford scenarios because wind power was excluded in the latter as noncompetitive.

Energy saved through using renewable sources can be a very substantial part of the total need. In the Reference case, the 15 Quads saved through all renewable sources is

TABLE 11–5
National Summary of Maximum Annual Electrical Output (and Fuel Saved) from Hydroelectric, Wind, and Ocean Thermal Gradient Installations, assuming Saturation

SOURCE	INSTALLED CAPACITY (THOUSANDS OF MEGAWATTS)	ANNUAL OUTPUT (BILLIONS OF KILOWATT-HOURS)	QUADS[4]	ANNUAL FUEL SAVED (QUADS[5])
Wind	470.3	1,069.5[1]	3.65	9.59
Hydroelectric	129.1	701.0[2]	2.00	5.26
OTEC	200–600	1,489–4,468[3]	5.08–15.24	13.36–40.08
Mean OTEC	400	2,998	10.16	26.70

[1]Assuming capacity factors of 0.22, 0.33, and 0.38 in Low, Medium and High regimes respectively (General Electric, 1977).

[2]Assuming capacity factor of 0.62 (Army Corps of Engineers, 1979).

[3]Assuming capacity factor of 0.85 (Cohen, 1978).

[4]Using equivalence, 1 Quad = 2.93×10^{11} Kilowatt hours output.

[5]Displacing coal-fired generators with efficiency of 0.38. Fuel saved is 2.63 × (output energy).

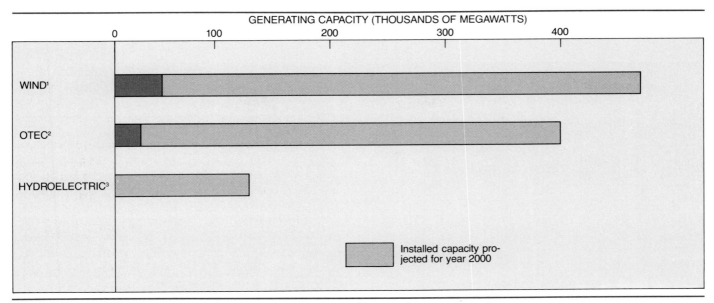

Fig. 11–6 Maximum or saturation generating capacity estimated for wind, ocean thermal gradients (OTEC) and hydroelectricity, showing projected installed capacities by the year 2000.

[1]Potential at saturation of land areas only, 470 thousand MW according to General Electric, 1977. Installed capacity by year 2000, 40.5 thousand MW according to METREK, 1978.

[2]Potential at saturation, 200 to 400 thousand MW, and installed capacity by year 2000 of 6 to 35 thousand MW, according to Cohen, 1978.

[3]Total potential of 129 thousand MW is developed and undeveloped capacity according to U.S. Army Corps of Engineers, September, 1981.

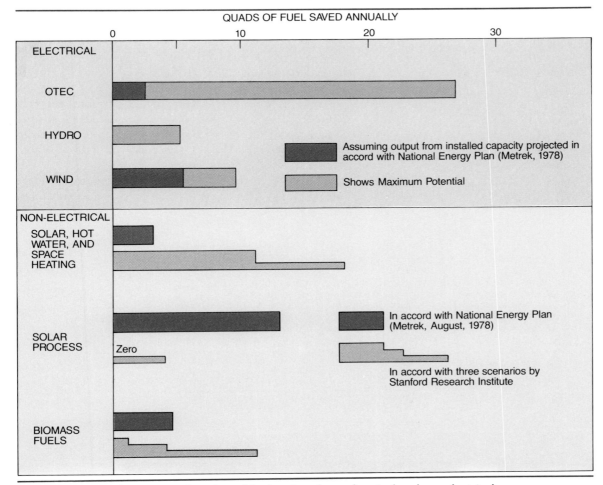

Fig. 11–7 National summary of projected fuel savings from electrical and nonelectrical applications of renewable energy flows. Amounts are quads of fuel saved annually as of the year 2000.

TABLE 11–6
Projected Electrical Output in the Year 2020
from Five Expressions of Solar Energy

ENERGY SOURCE	ANNUAL OUTPUT		
	Billions of Kilowatt hours	Quads[1]	FUELS[2] SAVED
Wind	662	2.26	6.6
Solar radiation, thermal	293	1.00	2.9
Solar radiation, photovoltaic	21	0.07	0.2
Ocean thermal gradients	242	0.82	2.4
Biomass	47	0.16	0.6
Total	1,265	4.31	12.7

Note: Figures assume incentives of the National Energy Plan.

[1] Using equivalence: 1 Quad = 10^{11} Kilowatt hours output.

[2] Except for the biomass entry, Metrek's fuel saved is roughly 2.9 times the output energy. The 0.6 Quads of fuel saved by biomass is 3.75 times the output energy, and may be erroneous.

Source: Output in Kilowatt hrs. and Fuels Saved in Quads are taken from Metrek, Aug., 1978.

only 7.5 percent of the national need; in the Low Demand case, though, the 18 Quads saved is 18 percent of national need; and in the Solar Emphasis case, the 48 Quads saved is about 23 percent of the national need.

SUMMARY OF NONRENEWABLE ENERGY AMOUNTS AND PROJECTED ENERGY DEMAND

Figure 11–8 summarizes energy recoverable from Identified and Undiscovered (hypothetical) occurrences of nonrenewable resources. In addition, it allows comparison of the amounts of various resources with the cumulative demand for those resources to the year 2000. The cumulative demand to the year 2000 is that projected by the National Energy Policy Plan (NEPP), Scenario B. The key assumptions of NEPP, Scenario B projection are outlined in Table 11–7. Figure 11–9 shows the projected amounts for each resource need to the year 2000.

It is obvious that nuclear fuels predominate overwhelmingly in the resource supply if breeders are assumed. With no breeders, nuclear energy shrinks to an amount slightly greater than the energy recoverable from conventional oil or gas. With the no breeder assumption, coal energy is the largest amount, followed by geothermal heat (assuming ambitious development in the Gulf Coast area). Shale oil energy, recoverable from all beds averaging 15 gallons per ton or better, is more than twice the energy recoverable from conventional crude oil and natural gas together. If the energy amount were the only criterion, the cumulative energy needed to the year 2000, as shown in Figure 11–8, apparently could be met easily by relying on coal, geothermal, or shale oil energy.

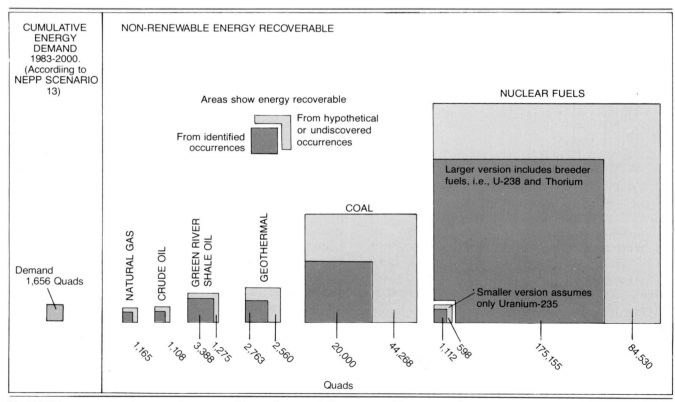

Fig. 11–8 National summary of recoverable and potentially recoverable energy from nonrenewable sources, compared with total cumulative demand for the nation to the year 2000.

TABLE 11-7
Key data and assumptions for the Department of Energy's National Energy Policy Plan (NEPP)
projection of *Energy Demand* to the year 2000 (Scenario B), 1983

| YEAR | SCENARIO B WORLD OIL PRICE[1] | | | DOMESTIC POLICIES |
	Nominal Dollars Per BBL	1982 Dollars Per BBL	GNP 1982 BILLION DOLLARS	
HIST.				• Implementation of the Administration's Natural Gas Consumer Regulatory Reform Legislation.
1960	N/A	N/A	1527	
1965	N/A	N/A	1925	• No major changes in current environmental laws.
1970	2.96	6.70	2249	
1975	13.93	22.94	2551	• No major changes to tax incentives provided under current law.
1980	33.89	39.30	3053	
1981	37.05	39.26	3113	• Continuation of Federal land leasing programs at current levels.
ESTI.				
1982	33.59	33.59	3056	• Continuation of Federal support for long-term Research and Development efforts.
PROJ.				
1983	28.60	27.40	3126	• Continuation of Synthetic Fuels Corporation efforts at current levels.
1984	28.60	25.90	3271	
1985	30.10	25.90	3439	
1986	32.20	25.90	3553	
1987	35.90	27.10	3665	
1988	41.10	29.20	3771	

ECONOMICALLY RECOVERABLE ENERGY RESERVES AS OF 1980

1989	46.00	30.90	3874	Resource	Discovered[2]	Undiscovered	Total
1990	50.00	31.90	3978				
1995	N/A	46.50	4526	Crude oil (Billion Barrels)	29.8	28–73	58–103
2000	N/A	57.40	5065	Natural gas (Trillion Cu. Ft.)	199	393–689	592–888
2005	N/A	72.20	5671	Coal (Billion Tons)	246	200–400	446–646
2010	N/A	83.60	6275				

[1]Refiner acquisition cost of crude oil imports.
[2]Excludes resources already recovered. Also excludes natural gas liquids (NGL) estimated at about 4.9 billion barrels of oil equivalent.
Source: Department of Energy. *Energy Projections to the Year 2010*, October 1983.

Demands Upon Nonrenewable Energy Sources According to the National Energy Policy Plan—Scenario B

Figure 11-10 arranges recoverable energy amounts from Table 11-8 in bar graph form. Because of the size of the coal resource base, linear scaling is not the ideal medium for a comparison of energy supplies by resource type: The energy thought to be eventually recoverable from coal would be more than twice that indicated by the length of the bar if hypothetical resources were included (see Table 11-8). The arrangement does allow, however, a comparison of energy amounts with possible demands on certain fuels between 1983 and 2000.

Energy demands are based on the NEPP Scenario B projection provided by the Department of Energy. This projection calls for a cumulative demand of approximately 1,656 Quads to the year 2000. The demand for each resource is shown in the table inset in Figure 11-10. Apparently the projected demand will consume less than

6 percent of the Demonstrated Reserve Base of coal and about one-third of the Reserves of uranium oxide. Very little of the resource base for geothermal will be consumed and only small amounts of shale oil are expected to be used. However, the projected demand for crude oil and natural gas will consume very large portions of the recoverable and potentially recoverable resources. In the case of crude oil, all of the Identified resources would be consumed and approximately 200 Quads of Undiscovered resources. However, approximately 200 Quads of the cumulative demand for crude oil is expected to be met by imports. Nevertheless, the demand for domestic supplies of crude will deplete all Identified resources. There seems to be little doubt that crude oil resources in the United States will become extremely scarce by the end of this century and that the nation will have no choice but to rely more heavily on synthetic fuels from coal and shale oil, and on renewable energy resources.

The demand on the natural gas supply will not be quite as severe as that for crude oil. The cumulative demand for natural gas according to the NEPP projection

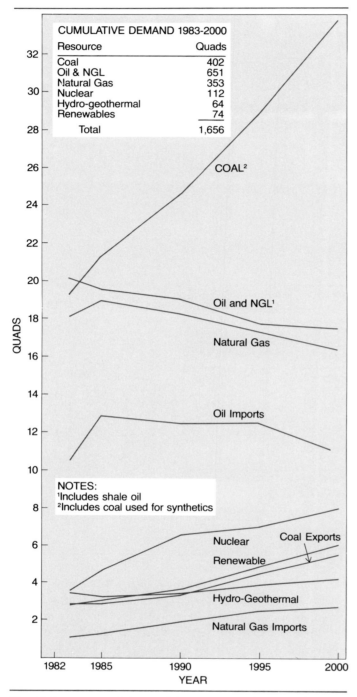

CUMULATIVE DEMAND 1983-2000

Resource	Quads
Coal	402
Oil & NGL	651
Natural Gas	353
Nuclear	112
Hydro-geothermal	64
Renewables	74
Total	1,656

NOTES:
[1]Includes shale oil
[2]Includes coal used for synthetics

Fig. 11-9 Projected demand for energy in the United States, 1983-2000, according to the Department of Energy's NEPP Scenario B Assumptions.

Source: Department of Energy. *Energy Projections to the Year 2010*, October 1983.

Identified Resources and Cumulative Demand to Year 2000 According to NEPP Scenario B Projection

Resource	Identified Resources Supply (Quads)	NEPP Scenario B Projection Demand (Quads)
Coal	20,000	402
Hydro-geothermal	2,756	64
Shale Oil	3,388	small
Uranium-235	1,112	112
Crude Oil	459	651
Natural Gas	498	353
Totals	28,213	1,656

is 353 Quads, about 70 percent of the presently identified resource base. No doubt, as domestic supplies of natural gas become increasingly more depleted, the United States will have to lean more heavily on imports from Canada and Mexico as well as on synthetic gas made from coal.

AREAS OF RESOURCE OCCURRENCE AND COINCIDENCE

Nonrenewable Sources

Figure 11-11 maps the literal areas of occurrence of coal of all ranks, conventional oil and gas fields, Green River oil shales, uranium oxide, and identified geothermal occurrences. Excluded for the sake of simplicity are the oil sands of Utah, tight gas sands of the western states, and Devonian (gas) shales of the Appalachians.

Most coincidences of resource location occur in the West, where current activity in subbituminous coal mining, Green River shale processing, and geothermal exploitation foreshadow a future of rapid development. Generally the region is short of both surface and ground water, while at the same time is susceptible to air pollution because of upper-air inversions that accompany the semi-arid and arid conditions. Some of the anticipated development will result in fuels that are transportable and hence will be burned beyond the region; still, surface mining damage and water pollution in the region must be dealt with. In the case of geothermal energy, especially for space heating, the energy consumption must be near the source, so some migration to geothermal sites may occur.

Another interesting area is the Gulf Coast, where conventional oil and gas fields and future prospects coincide with uranium oxide deposits and with geothermal energy in geopressured reservoirs.

Renewable Sources

Figure 11-12 combines areas of potential for ocean thermal gradients, wind, and solar radiation. Ocean gradients are suitable for electrical generation only in the Gulf of Mexico (and in Hawaii, Virgin Islands, Puerto Rico, and Guam). Favorable areas of wind (Fig. 11-12) are a combination of High, Moderate, and Low wind regimes. Areas favorable for use of solar radiation are represented by two measures. First, the average total daily radiation, direct and diffuse, is used to define areas where various applications, including photovoltaic, would enjoy an abundant resource. Second, the more restricted areas of abundant direct (focusable) radiation, suitable for high-temperature applications, are mapped. Areas where the two coincide are recognized by a darker tone on the map.

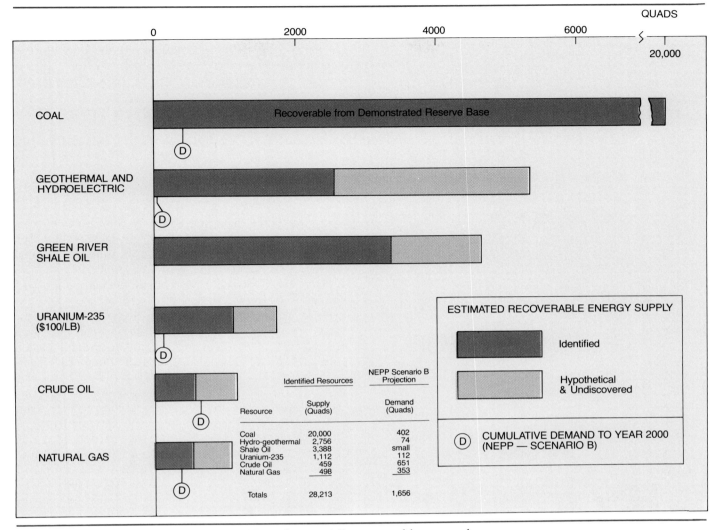

Fig. 11-10 National summary of recoverable and potentially recoverable energy from nonrenewable sources combined with cumulative demand on each fuel to the year 2000.

Not represented here are the areas in Wyoming, Colorado, and Nevada where exceptional net savings are possible through the use of solar heating systems for buildings. The definition of such areas depends upon specific assumptions about the particular application rather than upon the raw resource.

Especially interesting in Figure 11-12 is the area embracing southern California, Nevada, Utah, southern Wyoming, Colorado, New Mexico, and west Texas, where favorable wind regimes coincide not only with exceptional total solar radiation but also with greatest direct radiation. As shown in Figure 11-11, it is apparent that many southwestern areas favored by the sun and wind are short of surface water. Not shown is the fact that southern plains areas through Kansas, Oklahoma, and west Texas, which are rich in wind and sun, are also areas of overdrawn ground water.

Biomass resource areas and areas of hydroelectric potential are excluded from the summary because of difficulty in defining literal areas of occurrence. In the case of hydroelectric, the areas of greatest undeveloped potential are in the states of Alaska, Washington, California, Idaho, and Oregon.

The 11 Economic Regions and Their Energy Resources

Figure 11-13 summarizes the resources occurring in the 11 economic regions of the country according to two viewpoints. First, a *region's role or contribution with regard to particular resources* is noted. The letter U indicates that the region is unique in holding the resource,

TABLE 11–8
National Summary of Energy thought to be Recoverable from Identified and Undiscovered (or Hypothetical) Nonrenewable Sources (Quads)

RESOURCE	IDENTIFIED	HYPOTHETICAL OR UNDISCOVERED	TOTAL	PERCENT OF TOTAL	PERCENT TOTAL WITHOUT BREEDER FUELS[1]
Geothermal	2,763	2,560	13,662	4.0	17.0
Geopress.	2,307	623	2,930		
Hydro. 150° + C	206	917	1,123		
Hydro. 90–150°C	167	957	1,124		
Hydro. Under 50°C	83	63	146		
Nuclear Fuels[2]			241,160	72.0	2.0
Uranium-235	1,112	598	1,710		
Reserves	357		357		
Probable	755		755		
Possible		337	237		
Speculative		261	261		
Uranium-238	155,455	83,932	239,450		
Reserves	49,444		49,444		
Probable	106,011		106,011		
Possible		47,303	47,303		
Speculative		36,629	36,629		
Thorium	18,588	0	18,588		
Crude Oil	459	649	1,108	0.003	1.0
Reserves	154	459	613		
Indicated	8		8		
Inferred	130		130		
Sub-economic	167	190	357		
Natural Gas	498	605	1,165	0.003	1.4
Reserves	213	605	818		
Inferred	181		181		
Subeconomic	104	62	166		
Shale Oil[3]	3,388	1,275	4,663	1.3	6.0
Coal	20,000	44,268	64,268	19.0	82.0
From DRBase	6,725		6,725		
Other	13,926	44,268	58,194		
Total	202,263	133,303	335,606		
TOTAL WITHOUT BREEDER FUELS	28,822	49,371	78,193		

[1] Breeder fuels are U-238 and thorium. Total energy recoverable for these is 258,196 Quads.

[2] Includes uranium oxide resources up to cost of 50 dollars per pound. Reserves here include 120,000 tons of uranium oxide recovered as by-product in other refining operations.

[3] Using all beds that together average over 15 gallons per ton of rock. In beds yielding over 30 gallons per ton, the recoverable energy is estimated at 772 Quads.

while the letter O indicates the region is outstanding. Outstanding status is assigned for mineral fuels on the basis of a region's holding more than 10 percent of the nation's energy Reserves as defined earlier (Table 11–3). For renewable energy, the assignment of outstanding status is more subjective but still considers the region's share of the resource potential. The second viewpoint recognizes the *importance of certain resources to the region.* A block of color in a space shows that, in light of other alternatives, the resource indicated may play a large role in the future of the region, and may, on the basis of economic projections, deliver more energy in that region than in regions favored with a greater potential but having smaller demand.

A useful distinction may be made between *national* resources which are transportable fuels that can be shipped throughout the nation, and *regional* resources, such as solar radiation, wind, hydroelectric, OTEC power, and geothermal heat, which are likely to be used locally or within short distances of their extraction. It is recognized, however, that electrical power produced from such renewable or semirenewable (geothermal) energy flows can be converted into products such as hydrogen or ammonia and shipped from remote sites to the marketplace for consumption in fuel cells or engines.

MOUNTAIN REGION Because of its size and geological variety this region embraces a remarkable collection of energy sources. It is unique in holding rich oil shale deposits; it is outstanding in its share of the nation's coal energy, nuclear fuels, and geothermal Reserves. It is also outstanding with regard to solar radiation, wind regimes, and hydroelectric potential. Some of the resources, such as oil shales and intense direct radiation, are concentrated in small parts of the region, but some, such as coal Reserves and wind potential, are accessible in many parts of the region. Because of its mineral energy which can be processed and shipped, the Mountain region appears as a storehouse of national energy, but at the same time is well-endowed with geothermal, solar, wind and hydroelectric energy which are better regarded as resources for local or regional use.

WEST SOUTH CENTRAL REGION
Because the region includes Texas, Louisiana, and Oklahoma, it holds 51 percent of the nation's oil and gas Reserves, and it also holds a large portion of the oil and gas that is not yet proven. Substantial, though not outstanding, amounts of uranium oxide occur in the Texas coastal plain along with

large amounts of energy in geopressured reservoirs. Although geopressured reservoirs probably exist elsewhere, the large and well-studied geopressured resource is unique to the Gulf Coast and, therefore, to the West South Central region. Energy in these reservoirs was not included in the summary of energy Reserves (Table 11–3) because of some uncertainty about the economics of extraction. However, it probably will prove large and economically feasible. As a southern region, this one has abundant total radiation as a resource, and as a Gulf region it will have access to electrical energy from OTEC installations if they prove workable. Like the Mountain region, therefore, this one is important for national fuel resources and has outstanding regional resources as well.

EAST NORTH CENTRAL REGION
Because of deposits in Illinois and Ohio, this region holds 24 percent of the nation's coal Reserve energy (Table 11–3). In addition, 16 percent of thorium oxide Reserves are in the Palmer district of upper Michigan. The region is outstanding in no other national resource, and is not well-endowed

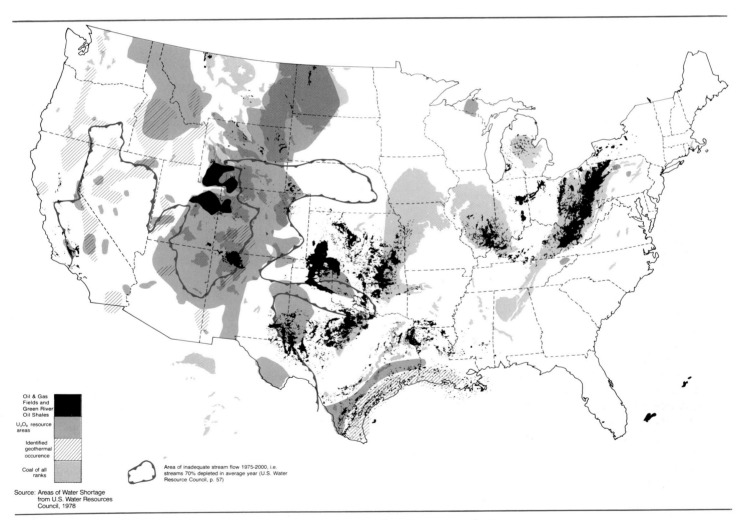

Oil & Gas Fields and Green River Oil Shales

U₃O₈ resource areas

Identified geothermal occurence

Coal of all ranks

Area of inadequate stream flow 1975-2000, i.e. streams 70% depleted in average year (U.S. Water Resource Council, p. 57)

Source: Areas of Water Shortage from U.S. Water Resources Council, 1978

Fig. 11–11 Nonrenewable energy sources: locations superimposed to show areas that are energy-rich and areas of coincidence with shortage of surface water.
Source: Areas of water shortage from U.S. Water Resources Council, 1978.

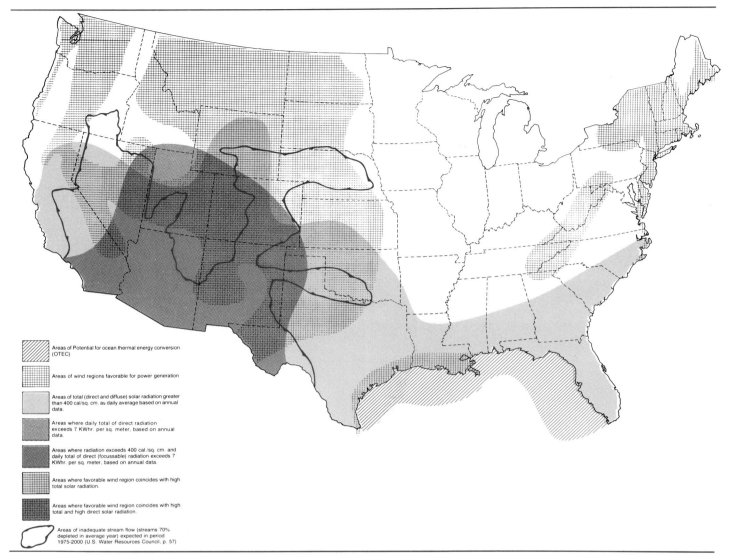

Areas of Potential for ocean thermal energy conversion (OTEC)

Areas of wind regions favorable for power generation

Areas of total (direct and diffuse) solar radiation greater than 400 cal/sq. cm. as daily average based on annual data.

Areas where daily total of direct radiation exceeds 7 KWhr. per sq. meter, based on annual data.

Areas where radiation exceeds 400 cal./sq. cm. and daily total of direct (focussable) radiation exceeds 7 KWhr. per sq. meter, based on annual data.

Areas where favorable wind region coincides with high total solar radiation.

Areas where favorable wind region coincides with high total and high direct solar radiation.

Areas of inadequate stream flow (streams 70% depleted in average year) expected in period 1975-2000 (U.S. Water Resources Council, p. 57)

Fig. 11-12 Three renewable energy sources: areas favorable to solar radiation, wind, and ocean thermal gradients superimposed to show areas that are energy-rich and areas of coincidence with shortage of surface water.

with solar radiation, wind, or hydroelectric energy. Nevertheless economic projections (METREK, 1978) suggest that electrical output from wind generators may be very important to the region—presumably at sites on the shores of the Great Lakes where local winds are more attractive than those for the region as a whole.

ALASKA Because of its coal and petroleum, this state (and region) holds a large part of the nation's transportable energy. For oil and gas, 22 percent of the nation's Reserves are in Alaska, as well as large parts of anticipated Reserves. While Reserves of coal are not outstanding, Hypothetical amounts of coal are extremely large and may be considered part of the resource for coal gasification. This technology, and/or coal liquefaction, could facilitate long-distance shipment of Alaska's coal energy. In geothermal energy, Alaska holds moderate amounts of hydrothermal heat in exploitable occurrences, and very large amounts of heat in Aleutian volcanic occurrences, many of which have not been measured. Ostensibly, this is a regional resource, but may in the future be shipped southward in the form of products such as hydrogen. The same may be said for the very attractive coastal wind power not included in the wind chapter assessment, and for the hydroelectric potential which is 13 percent of the national total and is now only 0.4 percent developed.

SOUTH ATLANTIC REGION In the scheme of Figure 11-13 this is the last ranked region with any outstanding portion of the nation's mineral energy reserves. Thorium oxide deposits in beach placers and offshore deposits give this region a resource that could be very important if future nuclear power makes use of thorium breeders. The region includes the state of West Virginia,

REGIONS	ENERGY SOURCES													
	NON-RENEWABLE (1)									RENEWABLE				
	Fossil fuels					Nuclear fuels		Geothermal		Hydro-electric (3)	Solar Radiation		Wind (6)	OTEC
	Coal on the basis of			Conv. oil and gas	Green River Oil Shales	U₃O₈	ThO₂	Hydro-Thermal & Volcanic	Geo-pres-sured		High Temp	Low Temp (5)		
	Low Sulfur	Gasifi-cation	Total energy											
Mountain	75%	51%	39%		▓ (unique)	75%	60%	55%		13%				
West So. Central				51%					▓ (unique)					
East So. Central			24%				16%						▨	
Alaska		36% (2)		22%				▨		13%			▨	
South Atlantic			▨				24%						▨	▨
Mid-Atlantic			▨							▨			▨	
West No. Central			▨									▨		
East So. Central			▨											
Pacific								43%		36% ▨ (4)		▨		
New England										▨		▨	▨	
Hawaii										▨	▨	▨	▨	▨

THE REGION'S ROLE IN THE NATIONAL PICTURE

▓ Region is unique in holding the resource indicated.

☐ Region is outstanding in its share of national total. For non-renewable and hydroelectric the region holds more than 10% of national total.

IMPORTANCE OF THE RESOURCE WITHIN A REGION.

▨ Resource important to region's self-sufficiency, though region may not hold outstanding share of national total.

1. In most cases, a region's character is based on its share of energy *reserves* as shown by percentages entered in resource columns.

2. Considering hypothetical coal as well as reserves.

3. Percentages show region's proportion of total hydroelectric potential, developed and undeveloped.

4. Mostly in southern California.

5. Regions noted as outstanding are those spanning the southern part of the country where total radiation is abundant. For solar *space heating* the West North Central region and northern part of the mountain region are outstandingly attractive on the basis of net savings estimated by simulation studies.

6. Very attractive wind power available exists off the Pacific, mid-Atlantic, and South Atlantic coasts, though it was not included in the wind chapter assessment of total potential generating capacity.

Fig. 11–13 Eleven economic regions and their energy resource highlights.

whose coal Reserves are very large, though not enough to make the region outstanding: In a ranking by state shares of coal Reserve energy, West Virginia would be number four (Table 11-1). Offshore wind regimes could be important to the region in the future, as could the potential for the OTEC power shared by two other regions bordering the Gulf of Mexico.

MID-ATLANTIC, WEST NORTH CENTRAL, AND EAST SOUTH CENTRAL REGIONS
These three regions hold similar totals of mineral energy Reserves in which coal is the dominant resource (Table 11-3). In the Mid-Atlantic, offshore wind potential and New York and Pennsylvania's hydroelectric potential are distinguishing features. In West North Central the oil and gas energy of North Dakota is significant; in East South Central, hydroelectric potential is slightly greater than that of Mid-Atlantic and earns it a marginally outstanding rating in Figure 11-13.

PACIFIC REGION
Because of small coal deposits, the region does not have a large portion of the nation's mineral energy Reserves. Yet, on a state-by-state basis, California holds 16 percent of the nation's crude oil Reserves, while Alaska and Texas each hold 28 percent, and Louisiana only 9 percent. The distinctive character of the Pacific region, though, is due to semirenewable and renewable resources. In hydrothermal energy Reserves, the region is clearly outstanding with 43 percent, mostly in California and Oregon: These occurrences are largely of the high-temperature type suitable for electrical power generation. Southern California has areas of high total and direct solar

radiation suited, therefore, to both low- and high-temperature application. The attractive wind potential is mostly off the Washington coast, while outstanding hydroelectric potential is concentrated in Washington state, but is also very large in Oregon and California.

NEW ENGLAND
New England stands out as the only region completely without mineral fuels or geothermal occurrences. This is largely because it is small in area and lacks the sedimentary rocks necessary for fossil fuels and also the recent volcanic activity that confers western-style geothermal potential. Some low-grade uranium ores exist here, but they exist in other regions as well. Because of high fuel prices, the region can profitably use solar radiation for space heating and other low-temperature applications. It has outstanding wind potential offshore, and very substantial hydroelectric potential in Maine and New Hampshire.

HAWAII
Because of its simple volcanic nature, there are no mineral fuels in Hawaii, but fortunately it has a variety of semirenewable and renewable energy sources upon which to draw. While its geothermal resources appear small when compared to the national total they are very substantial in relation to the region's needs. It receives large amounts of total and direct solar radiation, as well as predictably steady trade winds. As with many tropical volcanic islands, the steeply sloping ocean floors mean that floating OTEC platforms can find suitable temperature contrasts quite near to the shore. If the technology is workable, OTEC-generated power may be very important to Hawaii.

References

METREK. *Solar Energy: A Comparative Analysis to the Year 2020.* Mitre Technical Report MTR-75-79. Prepared for ERDA under Contract No. E-(4918)=2322 by the Mitre Corp., METREK division, McLean, Virginia, 1978.

Stanford Research Institute. *Solar Energy in America's Future: A Preliminary Assessment, Executive Summary.* Washington, DC: U.S. Energy Research and Development Administration, March 1977.

U.S. Department of Energy. *Energy Projections to the Year 2010,* Washington, D.C., Government Printing Office, Oct., 1983.

U.S. Water Resources Council. *The Nation's Water Resources, 1975-2000, Volume I: Summary.* Washington, DC: Government Printing Office, December 1978.

Suggested Readings

Introduction

Allen, Edward Lawrence. *Energy and Economic Growth in the United States.* Cambridge, Mass.: M.I.T. Press, 1979.

American Physical Society. *Efficient Use of Energy.* New York: American Institute of Physics, 1975.

An Analysis of Federal Incentives Used to Stimulate Energy Production. Richland, Washington, Pacific Northwest Laboratory (PNL-2410), March 1978.

Banks, Ferdinand E. *Scarcity, Energy, and Economic Progress.* Lexington, Mass.: Lexington Books, 1977.

Bergman, Elihu, Hans A. Bethe, and Robert E. Marshak. *American Energy Choices Before the Year 2000.* Lexington, Mass.: Lexington Books, 1978.

Boffey, P. M. "How the Swedes Live Well While Consuming Less Energy." *Science,* 196, May 20, 1977, 856.

Boslough, John. "Rationing a River." *Science,* 81, June 1981, 26.

Barnet, Richard J. *The Lean Years: Politics in an Age of Scarcity.* New York: Simon & Schuster, 1980.

Brown, Lester R., and Pamel Shaw. *Six Steps to a Sustainable Society.* Washington, DC: Worldwatch Society, 1982.

Brown, Lester R. *et al. State of the World 1984.* New York: W. W. Norton, 1984.

California Energy Commission. *Comparative Evaluation of Nontraditional Energy Resources.* Sacramento, Cal.: February 1980.

Carr, Donald E. *Energy and the Earth Machine.* New York: W. W. Norton, 1976.

Christensen, John W. *Energy, Resources, and Environment.* Dubuque, Iowa: Kendall/Hunt, 1981.

Colorado Energy Research Institute. *Net Energy Analysis: An Energy Balance Study of Fossil Fuel Resources.* Golden, Colorado: Colorado Energy Research Institute, 1976.

Commoner, Barry. *The Politics of Energy.* New York: Knopf, 1979.

Commoner, Barry. *The Poverty of Power: Energy and the Economic Crisis.* New York: Knopf, 1976.

Council On Environmental Quality. *Solar Energy: Progress and Promise.* Washington, DC: April 1978.

Council on Environmental Quality. *The Good News about Energy.* Washington, DC: 1979.

Crabbe, David and Richard McBride. *The World Energy Book: An A–Z Atlas, and Statistical Source Book.* Cambridge, Mass.: MIT Press, 1979.

Criner, Douglas, E. "Fuel Cells-An Industrial Cogeneration Perspective." *Energy Technology VII: Expanding Supplies and Conservation.* Washington, DC: Government Institutes, 1980, pp. 983–991.

Culp, Archie W. *Principles of Energy Conversion.* New York: McGraw-Hill, 1979.

Dallaire, G. "Transportation Innovations That Would Banish America's Energy Crisis." *Civil Engineering,* November 1981, 47–50.

Darmstadter, Joel. *How Industrial Societies Use Energy: A Comparative Analysis.* Baltimore, Mar.: Published for Resources for the Future by Johns Hopkins University Press, 1977.

Demand and Conservation Panel of the Committee on Nuclear and Alternative Energy Systems, National Academy of Sciences. "U.S. Energy Demand: Some Low Energy Futures," *Science,* 200: 1978, 142–152.

"Energy Conservation: Spawning a Billion Dollar Business." *Business Week,* April 6, 1981, 58–69.

"Slants and Trends." *Energy Resources and Technology,* 351, September 5, 1980.

Energy in Transition, 1985–2010. Committee on Nuclear and Alternative Energy Systems, National Academy of Sciences, 1980.

Energy: The Next Twenty Years. Report by a study group sponsored by the Ford Foundation and administered by Resources for the Future, Cambridge, Mass.: Ballinger Publishing, 1979.

Executive Office of the President. *The National Energy Plan,* Office Energy Policy and Planning. Washington, DC: Government Printing Office (040-000-00380-1), 1977.

Fickett, A. P. "Fuel-Cell Power Plants." *Scientific American,* 219:6, December 1978, 70–76.

"Fly wheels: Energy-Saving Way to Go," *Environmental Science and Technology,* 10:7, July 1976, 636–639.

Georgescu-Rogen N. "The Crisis of Resources: Its Nature and Its Unfolding." In G. Daneke (ed.) *Energy, Economics and the Environment: Toward a Comprehensive Perspective.* Lexington, Mass.: D.C. Heath, 1982.

Government Institutes, Inc. *Energy Technology Series,* Volumes I–XI. Rockville, Mar.: Government Institutes, Inc., 1984.

Gray, C., and F. von Hippel. "The Fuel Economy of Light Vehicles." *Scientific American,* 244:5, May 1981, 48–60.

Griffin, James M. *Energy, Economics and Policy.* New York: Academic Press, 1980.

Hafele, W. "A Global and Long-Range Picture of Energy Developments." *Science,* 209:4452, July 4, 1980, 174–182.

Hayes, E. T. "Energy Resources Available to the United States." *Science,* 203 January 19, 1979, 233.

Hayes, Denis. *Energy: The Case for Conservation.* Washington, DC: Worldwatch Institute, 1976.

Hayes, Denis. *Rays of Hope: The Transition to a Post-Petroleum World.* New York: W. W. Norton, 1977.

Hirst, E., and J. C. Moyers. "Efficiency of Energy Use in the United States." *Science,* 179, March 30, 1973, 1299–1304.

Hoffman, Peter. *The Forever Fuel: The Story of Hydrogen.* Boulder, Col: Westview Press, 1981.

Hoyle, Fred Sir. *Energy or Extinction?: The Case for Nuclear Enerlgy.* London: Heinemann, 1977.

Jorgensen, J. G., et al. (Eds.) *Native Americans and Energy Development.* Cambridge, Mass.: Anthropology Resource Center, 1978.

Kash, D., et al. Energy Alternatives: A Comparative Analysis, for the President's Council on Environmental Quality. Published as *Our Energy Future,* Norman: University of Oklahoma Press, 1976.

Kohrl, William. *Water and Power.* University of California Press, 1982.

Krutilla, John V. Economic and Fiscal Impacts Coal Development: Northern Great Plains, Baltimore: Published for *Resources for the Future* by Johns Hopkins University Press, 1979.

Landsberg, H. H. "Low-Cost Abundant Energy: Paradise Lost?" *Science,* 184, April 19, 1974, 247–253.

Lapedes, D. N. (Ed.) *Encyclopedia of Energy.* New York: McGraw-Hill, 1976.

Lave, L. "Conflicting Objectives in Regulating the Automobil." *Science,* 212, May 22, 1981, 893–899.

Loftness, Robert L. *Energy Handbook* (2nd ed.). New York: Van Nostrand Reinhold, 1984.

Lovins, Amory. *Non-Nuclear Energy: The Case for Ethical Strategy.* Cambridge, Mass.: Ballinger, 1975.

Lovins, Amory. *Soft Energy Paths: Toward a Durable Peace.* San Francisco: Friends of the Earth International, 1977.

Lovins, Amory. *World Energy Strategies: Fact, Issues, and Options.* San Francisco: Friends of the Earth International, 1975.

Lovins, Amory and L. Hunter Lovins. *Brittle Power: Energy Strategy for National Security.* Andover, Mass., Brick House, 1982.

Lovins, Amory and L. Hunter Lovins. *Energy Unbound: Your Invitation to Energy Abundance.* San Francisco: Sierra Club Books, 1984.

Mazur, A., and E. Rosa. "Energy and Life-Style." *Science,* 186, November 15, 1974, 607–610.

McGown, Linda B, and John Bockris. *How to Obtain Abundant Clean Energy.* New York: Plenum Press, 1980.

McKinley, A., and Swisher, J. (Eds.) *Alternative Energy Futures: An Assessment of U. S. Options to 2025.* Prepared for Department of Energy by the Institute for Energy Studies, Stanford, Cal.: Stanford University, 1979.

McMullen, John T. *Energy Resources.* New York: Wiley, 1977.

Meador, Roy. *Future Energy Alternatives: Long Range Prospects for America and the World.* Ann Arbor, Mich.: Ann Arbor Science Publishers, 1978.

Nash, Hugh (Ed.) *The Energy Controversy: Soft Path Questions and Answers by Amory Lovins and His Critics.* San Francisco: Friends of the Earth, 1979.

National Academy of Science. *Criteria for Energy Storage Research and Development.* Washington, DC: National Academy Press, 1976.

National Energy Strategies Project. *Energy in America's Future: The Choices Before Us.* Baltimore: Published for *Resources for the Future* by Johns Hopkins University Press, 1979.

National Geographic Society. *Energy* (special report). Washington, DC: National Geographic Society, February 1981.

National Research Council. *Alternative Energy Demand Futures to 2010.* Committee on Nuclear and Alternative Energy Systems, Demand and Conservation Panel. Washington, DC: National Academy of Sciences, 1979.

Nordhaus, William D. *The Efficient Use of Energy Resources.* New Haven, Conn.: Yale University Press, 1979.

Nuclear Energy Policy Study Group. *Nuclear Power Issues and Choices.* Cambridge, Mass.: Ballinger, 1977.

Perman, Jonathan. *State Energy Factbook.* Washington, DC: Northeast–Midwest Institute, 1984.

Petroleum Publishing Company. *International Petroleum Encyclopedia.* Tulsa, Oklahoma, (Published annually), 1980.

Phillips, Owen M. *The Last Chance Energy Book.* Baltimore, Mar.: Johns Hopkins University Press, 1979.

Pimental, David. *Food, Energy and Society.* New York: Wiley, 1979.

Price, Kent A. (Ed). *Regional Conflict and National Policy.* Baltimore, Mar.: For *Resources for the Future,* Johns Hopkins University Press, 1982.

Pryde, Philip R. *Nonconventional Energy Resources.* New York: Wiley-Interscience, 1983.

Ross, M., and R. Williams. *Our Energy: Regaining Control.* Highstown, N.J.: McGraw-Hill, 1981.

Ross, M., and R. Williams. "The Potential for Fuel Conservation." *Technology Review,* February, 1977, 49–57.

Schipper, L., and A. J. Lichtenberg. "Efficient Energy Use and Well Being: The Swedish Example." *Science,* 194, December, 1976, 1001–1013.

Schurr, S. (Ed.) *Energy In America's Future: The Choices Before Us* (A Resources for the Future Study). Baltimore, Mar.: Johns Hopkins University Press, 1979.

Siever, R. (Ed.) *Energy and Environment: Readings from Scientific American.* San Francisco: W. H. Freeman, 1980.

Sonenblum, Sidney. *The Energy Connections: Between Energy and the Economy.* Cambridge, Mass.: Ballinger 1978.

Starr, C. (Ed.) *Current Issues in Energy: A Selection of Papers.* New York: Pergamon Press, 1979.

Starr, C. "Is Sweden More Energy-Efficient?" *Science,* 196, April 1977, 121–124.

Steinhart, John S. *et al. Pathway to Energy Sufficiency: The 2050 Study.* San Francisco: Friends of the Earth, 1979.

Stobaugh, Robert and Daniel Yergin. *Energy Future: Report of the Energy Project of the Harvard Business School.* New York: Random House, 1979.

"The Big Shortfall in Auto Fuel Economy," News and Comments, *Science,* 205, September 21, 1979, 1232–1235.

"The Nation's Water Resources, 1975–2000." *Second National Water Assessment by the U.S. Water Resources Council,* Volume 4: Lower Colorado Region, Upper Colorado Region, California Region, 1978. Washington, DC: Government Printing Office, 1978.

Van Tassel, Alfred J. (Ed.) *The Environmental Price of Energy.* Lexington, Mass.: Lexington Books, 1975.

Veziroglu, T. Nejat and Walter Seifritz (Eds.) Hydrogen Energy System. *Proceedings of the 2nd World Hydrogen Energy Conference.* Zurich Switzerland, August 21–24, 1978. Elmsford, N.Y.: Pergamon Press, 1979.

Warkov, Seymour (Ed.) *Energy Policy in the United States: Social and Behavioral Dimensions.* New York: Praeger, 1978.

Wholey, Jane. "Hawaii: New Dynasty in Renewable Energy." *Solar Age,* 6:5, May 1981, 22–28.

Workshop On Alternative Energy Strategies. *Energy Global Prospects, 1985–2000.* New York: McGraw-Hill, 1977.

World Energy Conference. *World Energy Conference of Energy Resources, 1983.* London: World Energy Conference, 1983.

Yokell, M. D. *Environmental Benefits and Costs of Solar Energy.* Lexington, Mass.: Lexington, Books, 1980.

Coal

"Acid Rain." United States Environmental Protection Agency, Office of Research and Development, EPA-600/9-79-036, July 1980.

Ackerman, Bruce A., and William T. Hassler. *Clean Coal/Dirty Air.* New Haven, Conn.: Yale University Press, 1981.

Alexander, Tom. "A Promising Try at Environmental Detente for Coal." *Fortune,* February 13, 1978, 94.

Allar, Bruce. "No More Coal-Smoked Skies?" *Environment,* 26:2, 1984, 25–30.

Averitt, Paul. "Coal Resources of the United States, January 1, 1974." U. S. Geological Survey Bulletin 1412. Washington, DC: Government Printing Office, 1975.

Berry, L. G., *et al. National Coal Utilization Assessment.* Oak Ridge, Tenn.: Oak Ridge National Laboratory, 1978.

Burton, P. "Acid Rain: The Water That Kills." *National Parks,* July/August, 1982, 9.

Coal—Bridge to the Future, Report of the World Coal Study. Cambridge, Mass.: Ballinger Publishing, 1980.

"Coal Slurry Pipelines." *Coal Traffic Annual, 1979.* National Coal Association, 1980.

Cowling, E. "Acid Precipitation in Historical Perspective." *Environmental Science and Technology,* 16:2, 1982, 110A.

Denver Post. "Town Revived in Coal Boom." July 13, 1976.

Derbyshire, Frank. "Coal: Phoenix of the Twentieth Century." *Earth and Mineral Sciences,* 52:3 Spring 1983. (Pennsylvania State University, University Park, Pennsylvania, 1983.)

Electric Power Research Institute. *Coal Gasification Systems: A Guide to Status, Applications, and Economics.* Palo Alto, Cal.: Electric Power Research Institute, June 1983.

Flavin, Christopher. *The Future of Synthetic Materials: The Petroleum Connection.* Washington, DC: Worldwatch Institute, 1980.

Fluor Engineering and Constructors. *Effects of Sulfur Emission Controls on the Cost of Gasification Combined Cycle Power Systems.* Palo Alto, Cal.: Electric Power Research Institute (AF–196), 1978.

Gorham, E. "What to Do About Acid Rain." *Technology Review,* October, 1982, p. 59.

Gray, W., and P. Mason. "Slurry Pipelines: What the Coal Man Should Know in the Planning Stage." *Coal Age,* August, 1975.

Green, B. B. *Biological Aspects of Surface Coal Mine Reclamation, Black Mesa and San Juan Basin.* Regional Studies Program, Argonne National Laboratory. Argonne, Il: Argonne National Laboratory, 1977.

Grim, E. C. "Environmental Assessment of Western Coal Surface Mining." In *Proceedings of: National Conference on*

Health, Environmental Effects, and Control Technology of Energy Use. Washington, DC: Environmental Protection Agency, pp. 177–178, 1976.

"Gulf Successful in Gasifying Steeply Dipping Coal Beds." *Oil and Gas Journal,* December 28, 1981, 71.

Hall, T. A., I. L. White, and S. C. Ballard. "Western States and National Energy Policy: The New State's Rights." *American Behavioral Scientist, 22,* November/December 1978, 191–212.

Hohememser, Christopher, John Flower, and Robert Goble. "Power Plant Performance." *Environment, 20,* 1978, 25–32.

"How Many More Lakes Have to Die." *Canada Today,* 12:2, 1981. Available from Canadian Embassy, 1771 N Street, N.W., Room 300, Washington, D.C., 20036.

Kienlen, K. W. "Extensive Coal Exploration Seen Provided Ports Can Offer Facilities." *Rocky Mountain News,* March 1, 1981, 80.

Kirschten, J. D. "Converting to Coal—Can It Be Done Cleanly?" *National Journal,* 9, May 21, 1977, 781–784.

"Landfill Made by Scrubbers." *Business Week,* January 16, 1978.

Likens, G. "Acid Precipitation." *Chemical and Engineering News,* 54, November 22, 1976, 29.

Likens, G. "The Not So Gentle Rain." In *Yearbook of Science and the Future,* Encyclopedia Brittanica. 1981, pp. 212–227.

Myhra, D. "Colstrip, Montana—The Modern Company Town." *Coal Age,* 80, 1975, 54–57.

National Research Council. *Assessment of Technology for the Liquefaction of Coal.* Washington, DC: National Academy of Sciences, 1977.

"New Carter Plan Places Emphasis on Synfuels." *Oil and Gas Journal,* July 23, 1979.

"New Fears Surround Shift to Coal." *Fortune,* November 28, 1978.

Oak Ridge National Laboratory. *Environmental and Health Aspects of Disposal of Solid Wastes from Coal Conversion: An Information Assessment.* Oak Ridge, Tenn.: Oak Ridge National Laboratory, 1978.

Office of Technology Assessment. *The Direct Use of Coal: Prospects and Problems of Production and Combustion.* Washington, DC: Government Printing Office, 052-003-00664-2, 1979.

Ralph M. Parsons Co. *Preliminary Design Study for an Integrated Coal Gasification Combined Cycle Power Plant,* prepared for the Southern California Edison Company, Palo Alto, Calif.: Electric Power Research Institute, 1978.

Perry, Harry. "Clean Fuels from Coal." In *Advances in Energy Systems and Technology,* (Vol. 1). New York: Academic Press, 1978.

Richards, B. "Powder River Basin: New Energy Frontier." In "Energy: Facing Up to the Problem, Getting Down to Solutions," a special report published by *National Geographic,* pp. 96–113, February 1981.

Rudolph, P. "Synfuels from Coal—How and at What Cost." *Energy Technology VII: Expanding Supplies and Conservation.* Washington, DC: Government Institutes, Inc., 1980, pp. 642–650.

"Sulfur Emission: Control Technology and Waste Management." *U. S. Environmental Protection Agency Decision Series,* EPA 600/9-9-79-019, May 1979.

Surface Mining: Soil, Coal and Society. National Research Council, 1981.

"Synfuels: Uncertain and Costly Fuel Option." *Chemical and Engineering News,* August 29, 1979.

"Synthetic Fuels Report." *Oil and Gas Journal,* June 29, 1981.

"The American Coal Miner." *The President's Commission on Coal.* Washington, DC: Government Printing Office, 1980.

"The Great Black Hope: Coal." *Colorado Business,* September/October 1976.

"Underground Coal Gasification Expands U. S. Energy Options." *UA Journal,* 41:3, March 1979.

Wasp, Edward J. "Slurry Pipelines." *Scientific American,* November 1983.

Young, Gordon. "Will Coal Be Tomorrow's Black Gold." *National Geographic Magazine,* August 1975.

Crude Oil and Natural Gas

Cook, James. "The Great Oil Swindle," *Forbes,* March 15, 1982.

Ezzati, Ali. *World Energy Markets and OPEC Stability.* Lexington, Mass.: Lexington Books, 1978.

"The Geopolitics of Oil." Prepared for Senate Committee on Energy and Natural Resources (November 1980). (This book is summarized in an article in *Science,* 210, December 19, 1980, p. 1324.)

Hubbert, M. K. "Energy Resources." In Preston Cloud (Chairman), *Resources and Man.* National Academy of

Sciences and National Research Council: San Francisco: W. H. Freeman, 1969, pp. 157–242.

Metz, W. "Mexico: The Premier Oil Discover in the Western Hemisphere." *Science,* 202, pp. 1262, December 22, 1978.

Nehring, Richard. *Giant Oil Fields and World Oil Resources.* Santa Monica, Cal.: Rand Corporation (R-2284-CIA), June 1978.

Odell, Peter R. *Oil and World Power: Background to the Oil Crisis.* (4th Ed.) Baltimore, Mar.: Penguin Books, 1975.

Petroleum Publishing Company. *International Petroleum Encyclopedia.* Tulsa, Ok. Published annually.

Plattner, A. "Congress Approves Waivers to Help Secure Financing for Alaska Gas Pipeline." *Congressional Quarterly Weekly Report,* December 12, 1981, 2431–2432.

Satriana, M. *Unconventional Natural Gas: Resources, Potential, and Technology.* Park Ridge, N.J.: Noyes Data Corp., 1980.

Stobaugh, Robert. "After the Peak: The Threat of Imported Oil." In R. Stobaugh and D. Yergin. *Energy Future.* New York: Random House, 1979.

Tissot, B. *Petroleum Formation and Occurrence: A New Approach to Oil and Gas Exploration.* New York: Springer-Verlag, 1978.

U. S. Department of Energy. *The Natural Gas Market Through 1990,* DOE/EIA-0366. Washington, DC: Government Printing Office, May 1983.

U. S. Department of Energy. *The Petroleum Resources of the Middle East,* DOE/EIA-0395. Washington, DC: Government Printing Office, May 1983.

U. S. Department of Energy. *The Petroleum Resources of the North Sea,* DOE/EIA-0381. Washington, DC: Government Printing Office, February 1983.

Wildawsky, Aaron, and Ellen Tennenbaum. *The Politics of Mistrust: Estimating America Oil and Gas Resources.* Beverly Hills: Sage Publications, 1981.

Oil Shales and Tar Sands

"An Assessment of Oil Shale Technology." Office of Technology Assessment, Congress of the United States, Washington, DC: Government Printing Office, 1980.

Andersen, J. C., and J. E. Keith. "Energy and the Colorado River." *Natural Resources Journal,* April 17, 1977.

Arrnadale, T. "Rocky Mountain West: Unfinished Country," In H. Gimlim (ed), *American Regionalism.* Washington, DC: Congressional Quarterly, Inc., 1980.

"Controversy, Sluggish Markets Slow U.S. Synthetic Fuels Corporation." *Oil and Gas Journal,* April 23, 1984.

Culbertson, William C., and Janet K. Pitman. "Oil Shale." U.S. Geological *Professional Paper,* 820. Washington, DC: Government Printing Office, 1973.

Donnell, John. "Oil Shale Resources: How Much?" *Shale Country,* February 1976.

"Exxon Halts Colony Project." *Oil and Gas Journal,* May 10, 1982.

Federal Energy Administration. *Project Independence: Potential Future Role of Oil Shale,* U.S. Department of Interior. Washington, DC: Government Printing Office, 1974.

Fletcher, K., and M. Baldwin (Eds.). *A Scientific and Policy Review of the Final Environmental Impart Statement for the Prototype Oil Shale Leasing Program of the Department of the Interior.* Washington, DC: The Institute of Ecology, 1973.

Fradkin, Philip. *A River No More: The Colorado River and the West.* New York: 1981.

Grathwohl, Manfred. *World Energy Supply—Resources, Technology, Perspectives.* New York: Walter de Gruyter, 1982.

Maugh, T. "Tar Sands: A New Fuels Industry Takes Shape." *Science,* 199, February 17, 1978.

Morse, Jerome G. *Energy Resources in Colorado.* Colorado Energy Research Institute, Colorado School of Mines, Boulder, Co: Westview Press, 1979.

Murdock, S. H. and F. L. Leistritz. *Energy Development in the Western United States.* New York: Praeger, 1979.

"Oil from Shale Is Still a Distant Hope." *Business Week,* 23, April 1979.

"Oil Shale and The Environment." Office of Research and Development, U.S. Environmental Protection Agency. Springfield: NTIS, October 1977.

"Oil Shale: Prospects on the Upswing." *Science,* 190:4321, December 9, 1977.

"One Oil Shale Plan Advancing, Another Stalls." *Oil and Gas Journal,* July 2, 1984.

Petzric, P.A. "Oil Shale—An Ace in the Hole for National Security." *Shale County,* October 1975.

"Rio Blanco Seeks Commercial Operation Delay." *Oil and Gas Journal,* January 17, 1977.

"Shale Closest among Liquid Fuels." *Oil and Gas Journal,* January 17, 1977.

"Shale Oil Finally Rocking off Dead Center." *Oil and Gas Journal,* June 18, 1979.

"Sohio Withdraws From Utah Shale Project." *Oil and Gas Journal,* March 5, 1984.

Shohinpoor, Moshen. "Making Oil from Sand." *Technology Review,* February/March 1982.

"Synfuels Offer Challenging Future." *Oil and Gas Journal,* June 29, 1981.

Tippee, R. "Tar Sands, Heavy Oil Push Building Rapidly in Canada." *Oil and Gas Journal,* January 30, 1978.

"TOSCO: Oil Shale Projects Still Viable." *Oil and Gas Journal,* March 22, 1982.

"Uncertainty, Cost/Price Squeeze Hit Fledgling Synfuels Industry." *Oil and Gas Journal* May 24, 1982.

U.S. Department of Interior. *Energy Resources on Federally Administered Land.* Washington, DC: Government Printing Office, November 1981.

U.S. Office of Technology Assessment. *An Assessment of Oil Shale Technologies.* Washington, DC: Government Printing Office, June 1980.

"Utah Shale Project Begins Start-up Tests." *Oil and Gas Journal,* January 30, 1984.

Van West, Frank P. "Green River Oil Shale." *Geologic Atlas of the Rocky Mountain Region.* Denver: Rocky Mountain Association of Geologists, 1972.

Welles, Christopher. *The Elusive Bonanza—The Story of Oil Shale—America's Richest and Most Neglected National Resource.* New York: E.P. Dutton, 1970.

"White River Advances Oil Shale Project." *Oil and Gas Journal,* October 25, 1982.

Nuclear Fuels

Ball, John M. *An Atlas of Nuclear Energy: A Non-technical World Portrait of Commercial Nuclear Energy.* Atlanta: George State University, Department of Geography, 1984.

Bebbington, W. P. "The Reprocessing of Nuclear Fuels." *Scientific American,* December 1976, 30–41.

Beckmann, Peter. *The Health Hazards of Not Going Nuclear.* Boulder, Co.: Golem Press, 1976.

Berger, John J. *Nuclear Power: The Unviable Option.* New York: Dell, 1977.

Browne, Corinne, and Robert Munroe. *Time Bomb: Understanding the Threat of Nuclear Power.* New York: William Morrow, 1981.

Bupp, I., and J. Derian. *Light Water: How the Nuclear Dream Dissolved.* New York: Basic Books, 1978.

Caldicott, Helen. *Nuclear Madness.* New York: Bantam Books, 1981.

Cohen, Bernard L. *Before It's Too Late: A Scientist's Case for Nuclear Power.* New York: Plenum Press, 1983.

Ford, Daniel F. *Three Mile Island: Thirty Minutes to Meltdown.* New York: Penguin Books, 1983.

Flowers, Brian J. "Nuclear Power: A Perspective on the Risks, Benefits, and Options." *Bulletin of the Atomic Scientists,* 34:3, March 1977.

Gofman, John W., and Arthur R. Tamplin. *Poisoned Power: The Case Against Nuclear Power* (2nd ed). Emmaus, Penn.: Rodale Press, 1979.

Gofman, John W. *Radiation and Human Health.* San Francisco: Sierra Club Books, 1981.

Harding, Jim. "Lights Dim for Nuclear Power." *Not Man Apart,* April 1984, pp. 21–22.

Hewlett, Richard G. *Federal Policy for Disposal of Radioactive Wastes from Commercial Nuclear Power Plants.* Washington, DC: Department of Energy, March 1979.

Hohenemser, D., R. Kasperson, and R. Kates. "The Distrust of Nuclear Power." *Science,* 196, April 1, 1977, 25–34.

Holdren, J. P. "Fusion Energy in Context: Its Fitness for the Long Term." *Science,* 200, April 14, 1978, 168.

Jakimo, Alan, and Irvin C. Bupp. "Nuclear Waste Disposal: Not in My Backyard." *Technology Review,* March/April 1978, 64–72.

Kaku, Michio, and Jennifer Trainer. *Nuclear Power: Both Sides.* New York: W. W. Norton, 1982.

Kemeny, John G. "Saving American Democracy: The Lessons of Three Mile Island." *Technology Review,* June/July 1980, 65–75.

Kulcinski, G. L., et al. "Energy for the Long Run: Fission or Fusion." *American Scientist,* 67, 1979. 78–89.

Lester, R. K., and D. J. Rose. "The Nuclear Wastes at West Valley, New York." *Technology Review,* May 1977, 20–29.

Lindholm, Ulf, and Paul Gnirk. *Nuclear Waste Disposal: Can We Rely on Bedrock?* New York: Pergamon Press, 1982.

Lidsky, Lawrence M. "The Trouble with Fusion." *Technology Review,* October 1983, 32–34.

Lipschultz, Ronnie. *Radioactive Waste: Politics, Technology, and Risk.* Cambridge, Mass.: Ballinger, 1980.

Lovins, Amory. "The Case against the Fast Breeder Reactor." *Bulletin of Atomic Scientists,* March 1973.

Lovins, Amory B., and L. Hunter Lovins. *Energy/War: Breaking the Nuclear Link,* San Francisco: Friends of the Earth, 1980.

Martin, Daniel W. *Three Mile Island: Prologue or Epilogue?* Cambridge, Mass.: Ballinger, 1980.

McCracken, Samuel. *The War Against the Atom.* New York: Basic Books, 1982.

Metz, W. D. "Laser Fusion: One Milepost Passed—Millions More to Go." *Science,* 186, December 27, 1974.

Metz, W. D. "Magnetic Containment Fusion: What Are the Prospects?" *Science,* 178 October 20, 1972, 291.

National Academy of Sciences. *Energy in Transition 1985–2010: Final Report of the Committee on Nuclear and Alternative Energy Systems.* Washington, DC: National Academy Press, 1980.

National Electric Reliability Council. *Fossil and Nuclear Fuel for Electric Utility Generation: Requirements and Constraints.* Princeton, N.J.: National Electric Reliability Council, August 1977.

National Research Council's Committee on Nuclear and Alternative Energy Systems. *Energy in Transition, 1985–2010.* San Francisco: Freeman, 1980.

National Research Council. *Risks and Impacts of Alternative Energy Systems,* Committee on Nuclear and Alternative Energy Systems, Risk and Impact Panel Washington, DC: National Academy of Sciences, in preparation.

"Nuclear Energy: How Bright a Future." *Environmental Science and Technology,* 11:2 February 1977, 128–130.

"Nuclear Dilemma: The Atom's Fizzle in an Energy-Short World." *Business Week,* December 25, 1978, p. 54.

Nuclear Plant Reactor Safety. "NRC Panel Renders Mixed Verdict on Rasmussen Reactor Safety Study." *Science,* 201, September 29, 1978, 1196.

The Nuclear Energy Policy Study Group, S. Keeny, Jr., Chairman. *Nuclear Power, Issues and Choices.* Cambridge, Mass.: Ballinger, 1977.

"Nuclear Power: Can We Live With It?" A panel discussion including D. Kleitman, N. Rasmussen, R. Stewart, and J. Yellin, *Technology Review,* 81:7 June/July 1979, 32.

Nuclear Safety After Three Mile Island. *Electric Power Research Institute Journal* 5, 1980, 24–34.

"Reactor Safety Study: An Assessment of Accident Risks in U. S. Commercial Nuclear Power Plants." *Nuclear Regulatory Commission,* Wash-1400 (the Rasmussen Report), October 1975.

"Risk Assessment Review Group Report to the U. S. Nuclear Regulatory Commission." Prepared for the U. S. Nuclear Regulatory Commission, NUREG/CR-0400, (review of the Rasmussen Report), September 1978.

Rose, D., and R. Lester. "Nuclear Power Weapons, and International Stability." *Scientific American,* 238:4 April 1978, 45–57.

Rose, D. J. and M. Feirtag. "The Prospect for Fusion." *Technology Review,* December 1976, 21–43.

Shapely, D. "Reactor Safety: Independence of Rasmussen Study Doubted." *Science,* 197, July 1977, 29–31.

Stobaugh, Robert and Daniel Yergin (eds.) *Energy Future.* See "The Nuclear Stalemate," I. C. Bupp. New York: Random House, 1979.

"The Accident at Three Mile Island (The Need for Change: The Legacy of Three Mile Island)." *Report of the President's Commission,* J. Kemeny, Chairman, October 1979.

U. S. Department of Energy. *Fusion Energy.* Washington, DC: Government Printing Office, 1980.

Waldrop, M. "Compact Fusion: Small Is Beautiful." *Science,* 219, January 1983, 154.

Weinberg, Alvin M. "Is Nuclear Energy Necessary?" *Bulletin of the Atomic Scientists,* March 1980, 31–35.

Weinberg, A. "Salvaging the Atomic Age." *The Wilson Quarterly,* Summer 1979, 88–112.

Weinberg, A. "Social Institutions and Nuclear Energy." *Science,* 177, July 7, 1972, 27–34.

Willrich, M., and T. Taylor. *Nuclear Theft: Risks and Safeguards.* Cambridge, Mass.: Ballinger, 1974.

Yergin, Daniel. "The Terrifying Prospect: Atomic Bombs Everywhere." *Atlantic Monthly,* April 1977.

Zaleski, C. P. "Breeder Reactors in France." *Science,* 208:4440, 1980, 137–144.

Geothermal Heat

American Association of Petroleum Geologists. *Geothermal Gradient Map of North America.* Tulsa, Ok., 1976.

Armstead, H., Ed. *Geothermal Energy: Its Past, Present, and Future Contributions to the Energy Needs of Man.* New York: Wiley, 1978.

Axtman, R. C. "Environmental Impact of a Geothermal Power Plant." *Science,* 187:4179, 1975, 785–803.

Bowen, Robert. *Geothermal Resources.* London: Applied Science Publishers, 1979.

Bowen, Robert. "Net Energy Delivery from Geothermal Resources." *Geothermal Energy,* 5:2, February 1977, 15–19.

Cummings, Ronald G., et al. "Mining Earth's Heat: Hot Dry Rock Geothermal Energy." *Technology Review,* February, 1979. 58–78.

Department of Energy. *Geothermal Energy and Our Environment.* Washington, DC: Department of Energy, 1980.

DiPippo, R. *Geothermal Energy as a Source of Electricity* (prepared for the U. S. Department of Energy). Washington, DC: Government Printing Office, 1980.

Ellis, Albert J. *Chemistry and Geothermal Systems.* New York: Academic Press, 1977.

Howard, J. H. (ed.) *Present Status and Future Prospects for Nonelectrical Uses of Geothermal Resource.* (Berkeley, Cal.: Lawrence Livermore Laboratory (UCRL-51926), 1975.

Kerr, R. A. "Geopressured Energy Fighting Uphill Battle." *Science,* 207:4438, 1980, 1455–1456.

Milora, S. L., and J. W. Tester. *Geothermal Energy as a Source of Electric Power.* Cambridge, Mass.: M.I.T. Press, 1976.

Muffler, L. J. P. "Geothermal Resources." In *United States Mineral Resources,* U. S. Geological Survey *Professional Paper No. 820.* Washington, DC: Government Printing Office, 1973, pp. 251–261.

Pollack, Henry N., and D. S. Chapman. "The Flow of Heat from the Earth's Interior." *Scientific American,* 237:2, August 1977, 60–76.

U. S. Department of Energy. *An Assessment of Geothermal Development in the Imperial Valley of California.* Washington, DC: Government Printing Office July 1980.

U. S. Library of Congress. *Energy from Geothermal Resources.* Prepared for Committee on Science and Technology of House of Representatives by Science and Policy Research Division of the Library of Congress. Wahington, DC: Government Printing Office, 1978.

Urban, T. C., W. H. Diment, J. H. Sass, and I. M. Jamieson. "Heat Flow at The Geysers, California, U.S.A." In *Proceedings, Second United Nations Symposium on the Development and Use of Geothermal Resources.* Available from Superintendent of Documents, Washington, DC: Government Printing Office, 1975.

Wahl, Edward F. *Geothermal Energy Utilization.* New York: Wiley, 1977.

White, D. W. and D. L. Williams (eds.) "Assessment of Geothermal Resources of the United States—1975." U. S. Geological Survey Circular 726. Washington, DC: Government Printing Office, 1975.

Solar Radiation

Anderson, B. and Riordan, M. *The Solar Home Book.* Harrisville, N.H.: Chesire Books, 1976.

Anderson, Bruce. "Low Impact Solution." *Solar Age,* September 1976.

Beckman, W. A., *et al. Solar Heating Design by the F-Chart Method.* New York: Wiley, 1977.

Bockris, J. O. *Energy Options: Real Economics and the Solar-Hydrogen System.* London: Taylor & Francis. 1980.

Bockris, J. O. *Energy—The Solar Hydrogen Alternative.* New York: Wiley, 1976.

Butti, Ken, and John Perlin. *A Golden Thread: 2,500 Years of Solar Architecture and Technology.* New York: Chesire Books, 1980.

Crowley, John and L. Zurie Zimmerman. *Practical Passive Solar Design: A Guide to Homebuilding and Land Development.* New York: McGraw-Hill, 1984.

Davidson, M., D. Grether, and K. Wilcox. *Ecological Considerations of the Solar Alternative.* Berkeley, Cal.: Lawrence Berkeley Laboratory (LBL-5927), February 1977.

Davis, Norah D., and L. Lindsey. *At Home in the Sun.* Charlotte, Vt.: Garden Way, 1979.

Doane, J. W., *et al., A Government Role in Solar Thermal Repowering* Golden, Col.: Solar Energy Research Institute (SERI/TP-51-340), 1979.

Duffie, John, and W. A. Beckman. "Solar Heating and Cooling." *Science,* 191, 1976, 143–149.

The First Passive Solar Home Awards. Franklin Research Service, GPO Stock No. 023–000–00571–4, 1978.

Flavin, Christopher. *Electricity from Sunlight: The Future of Photovoltaics.* Washington, DC: Worldwatch Institute, 1982.

Hotton, Peter. *So You Want to Build an Energy-Efficient Addition.* New York: Little, Brown, and Co., 1983.

Howell, D. *Your Solar Energy Home.* Elmsford, N.Y.: Pergamon Press, 1979.

Howell, Y., and J. A. Bereny. *Engineer's Guide to Solar Energy,* 1979. Available from Solar Information Services, P. O. Box 204, San Mateo, California 94401.

Jayadev, T. and Edesess, M. *Solar Ponds* (SERI/TR 731-587). Golden, Col: Solar Energy Research Institute, April 1980.

Kendall, H. and Nadis, S. *Energy Strategies Toward a Solar Future* (prepared for the Union of Concerned Scientists). Cambridge, Mass. Ballinger Publishing Co. 1980.

Kreider, J. and F. Kreith (eds.) *Solar Energy Handbook.* New York: McGraw-Hill, 1981.

Lebens, R. M. *Passive Solar Heating Design.* New York, N.Y.: Halsted Press, 1980.

Levy, M. Emmanuel. *The Passive Solar Construction Handbook.* Emmaus, Pa., Rodale Press, 1983.

Lovins, Amory. "Soft Energy Technologies." In *Annual Review of Energy,* Jack M. Hollander (ed.) (Vol. 3) Palo Alto, Cal.: Annual Reviews, Inc., pp. 477–517, 1978.

Maidique, Modesto A. "Solar America." In R. Stobaugh and D. Yergin, *Energy Future.* New York: Random House, 1979.

Mazria, E. *The Passive Solar Energy Book.* Emmaus, Pa.: Rodale Press, 1979.

McDaniels, David K. *The Sun: Our Future Energy Source.* New York: Wiley, 1979.

McPhillips, M. Ed. *The Solar Age Resource Book.* New York: Everest House, 1979.

Metz, W. D., and A. L. Hammond (ed.) *Solar Energy in America.* Washington, DC: American Association for the Advancement of Sciences, 1978.

Moore, Jerry, and Richard Heinemeyer. *Solar Index* (a bibliography). From Solar Index Inc., Box 6933, Denver, Colorado 80206.

1983 Passive Solar Design Awards. Available from American Gas Association, Arlington, Virginia.

Passive Solar Design Handbook, Volume Three: Passive Solar Design Analysis. Los Alamos National Laboratory. Stock number 061-00-00598—6. Order from Superintendent of Documents, Washington, DC, 20402

Paul, J. K. (ed.) Passive Solar Energy: Design and Materials. In *Energy Technology Review no. 41.* Park Ridge, N.J.: Noyes Data Corporation, 1979.

Popular Science; Solar Energy Handbook 1979. New York Times Mirror Magazines, 1979.

Smith, J. L. "Photovoltaics." *Science,* 212:4502, June 1981, 1472–1478.

Solar Energy in America's Future: A Preliminary Assessment. (ERDA, GPO Stock No. 060-000-00051-4, March 1977.)

"Solar in New Construction." *Solar Engineering and Contracting,* January/February, 1984.

Solar Heating and Cooling Demonstration Program: A Descriptive Summary of HUD Cycle 4 and 4A Solar Residential Projects. AIA Research Corp., GPO No. 023--000-531-0, 1979.

SERI, *The Solar Radiation Data Directory,* Golden, Col.: Solar Energy Research Institute. From Superintendent of Documents, Washington, D.C. 20402, Stock No. 061-000-00619, 1984.

U. S. Department of Energy. *Technology Assessment of Solar Energy.* Washington, DC: Department of Energy, 1979.

Ward, D. S. *Solar Heating and Cooling Systems Operational Results Conference, Summary.* Golden, Col.: Solar Energy Research Institute (SERI/TP-49-209), 1979.

Wells, M., and I. Spetgang. *How to Buy Solar Heating without Getting Burnt.* Emmaus, Pa.: Rodale Press, 1978.

Whipple, C. "The Energy Impacts of Solar Heating." *Science,* 208:4441, April 1980, 262–266.

Yokell, M., *et al. Environmental Benefits and Costs of Solar Energy.* Golden, Col.: Solar Energy Research Institute (SERI/TR-52-074), 1979.

Wind Power

Changery, M. J. *Initial Wind Energy Assessment Study.* National Oceanic and Atmospheric Administration. Springfield, Va.: NTIS, May 1975.

Cheremisnoff, N. *Fundamentals of Wind Energy.* Ann Arbor, Mich.: Ann Arbor Science Publishers, 1978.

Coty, U. A., *et al.* *United States Wind Speed and Wind Power Duration Tables, By Months.* Lockheed-California Company, Los Angeles, California, October 1975.

Curtis, E. H. "A National Wind Energy Construction Program: Its Energy and Economic Impact." *Energy Technology VII: Expanding Supplies and Conservation.* Washington, DC: Government Institutes, Inc., June 1980, pp. 1504–1523.

DeRenza, D. J. (ed.) "Wind Power: Recent Developments." *Energy Technology Review no. 46.* Park Ridge, N.J.: Noyes Data Corporation, 1979.

Eldridge, Frank R. *Wind Machines.* Report prepared for the National Science Foundation and Research Applied to National Needs (RANN). Washington, DC: Government Printing Office, 1975.

Elliot, D. L. *Sythesis of National Wind Energy Assessments.* Battelle Pacific Northwest Laboratories, Richland, Washington, July 1977.

Flavin, Christopher. *Wind Power: A Turning Point.* Washington, DC: Worldwatch Institute, July 1981.

Gustavson, M. R. "Limits to Wind Power Utilization." *Science,* 204:4388, April 1979, 13–17.

Heronemus, W. E. "Power from Offshore Winds.: *Proceedings of the 8th Annual Marine Technology Conference,* Washington, DC, 1972.

Hunt, V. D. *Windpower: A Handbook on Wind Energy Conversion Systems.* New York: Van Nostrand Reinhold, 1981.

Inglis, David R. *Wind Power and Other Energy Options.* Ann Arbor, Mich.: University of Michigan Press, 1978.

Justus, Carl J. *Winds and Wind System Performance,* Philadelphia, Pa.: Franklin Institute Press, 1978.

Kahn, E. "The Compatibility of Wind and Solar Technology with Conventional Energy Systems." In *Annual Reviews of Energy,* Jack M. Hollander (ed.) Palo Alto, Calif.:

Marier, Donald. *Windpower for the Homeowner.* Emmaus, Pa.: Rodale Press, 1981.

Marsh, W. D. *Requirements Assessment of Wind Power Plants in Electric Utility Systems* (vol. II). Palo Alto, Cal.: Electric Power Research Institute (ER-978 V.2), 1979.

McGuigan, Dermot. *Harnessing the Wind for Home Energy.* Charlotte, Vt.: Garden Way, 1978.

Merriam, M. "Wind Energy for Human Needs." *Technology Review,* 79:3, January 1977, 28.

Metz, W. D. "Wind Energy: Large and Small Systems Competing." *Science,* 197, 1977, 971–973.

Naar, John. *The New Wind Power.* New York: Penguin Books, 1982.

Part, Jack. *The Wind Power Book.* Palo Alto, Cal.: Cheshire Books, 1981.

Putnam, Palmer C. *Power from the Wind.* New York: Van Nostrand, 1948. (Reprinted, Van Nostrand Reinhold, 1974.)

Reed, J. W. *Wind Power Climatology of the United States.* Albuquerque, N.M.: Sandia Laboratories, June 1975 (original), April 1979 (supplement).

Sandia Laboratories. *Wind Energy: A Revitalized Pursuit.* Albuquerque, N.M.: Sandia Laboratories, March 1975.

Smith, R. J. "Wind Power Excites Utility Interest." *Science,* 207:4432, February 1980, 739–742.

Sorensen, Brent. "Energy and Resources." *Science,* 189:4199, July 1975, 255–260.

Torrey, Volta. *Wind-Catchers, American Windmills of Yesterday.* Brattleboro, Vt.: The Stephen Green Press, 1976.

Wind Energy Weekly. Newsletter of the American Wind Energy Association, 1516 King Street, Alexandria, Virginia 22314.

Zelby, Leon W. "Don't Get Swept away by Wind Power Hopes." *Bulletin of the Atomic Scientists,* March 1976, 59.

Hydroelectric Power

Deudney, D. "Hydropower: An Old Technology for a New Era." *Environment,* 23:7, September 1981.

Federal Energy Regulatory Commission. *Hydroelectric Power Resources of the United States.* Washington, DC: Government Printing Office, January 1, 1980.

Federal Power Commission. *Hydroelectric Power Resources of the United States, 1976.* Washington, DC: Government Printing Office, 1976.

Hayes, Dennis. "The Solar Prospect." *Worldwatch Paper 11.* Washington, DC: Worldwatch Institute, March 1977.

Johnson, Mike. *Information package on Low-head Small Hydro.,* from Little Spokane Hydroelectric, Box 82, Chattaroy, Washington 99003.

Lilienthal, David F. "Lost Megawatts Flow over Nation's Myriad Spillways." *Smithsonian Magazine,* 8:6, September 1977.

McGuigan, Dermot. *Harnessing Water Power for Home Energy.* Charlotte, Vt.: Garden Way, 1978.

Metz, William D. "Solar Thermal Energy: Bringing the Pieces Together." *Science,* 197, August 12, 1977.

Noyes, R. *Small and Micro Hydroelectric Power Plants.* Park Ridge, N.J.: Noyes Data Corp., 1980.

Smith, B. P. "Power from Yesterday's Dams." *Environment,* 20:9, November 1978, pp. 16–20.

U. S. Army Corps of Engineers. *National Hydroelectric Power Resources Study,* Vol. X, *An Assessment of Hydroelectric Pumped Storage.* Fort Belvoir, Virginia, November 1981.

U. S. Army Corps of Engineers. *National Hydroelectric Power Resources Study,* Vol. II, *National Report.* Fort Belvoir, Virginia, May 1983.

U. S. Army Corps of Engineers. *National Hydroelectric Power Resources Study,* Vol. IX, *Potential For Increasing the Output of Existing Hydroelectric Plants.* Fort Belvoir, Virginia, July 1981.

U. S. Army Corps of Engineers. *National Hydroelectric Power Resources Study,* Vol. XII *Data Base Inventory.* Fort Belvoir, Virginia, September 1981.

U. S. Comptroller General. *Hydropower—An Energy Source Whose Time Has Come Again.* Washington: DC: General Accounting Office, January 1980.

U. S. Department of Energy. *Inventory of Power Plants in the United States.* Washington, DC: Government Printing Office, June 1983.

Energy from the Ocean

Charlier, R. H. "Energy from the Ocean: A Look at Tidal Power." *Alternative Sources of Energy,* 50, July/August 1981, 23–27.

Charlier, R. H. *Tidal Energy.* New York: Van Nostrand Reinhold, 1982.

"Energy from the Sea." *Popular Science,* June, 1975.

Haber, George. "Solar Power from the Oceans." *New Scientist,* March 10, 1977.

Hagen, Arthur W. *Thermal Energy from the Sea; Energy Technology Review No. 8.* Park Ridge, N.J.: Noyes Data Corp., 1975.

Hartline, Beverly K. "Trapping Sun-Warmed Ocean Water For Power." *Science* 209, 794–96.

Issacs, J., and W. Schmitt. "Ocean Energy: Forms and Prospects." *Science,* 207:4428, January 1980, 265–273.

Metz, W. D. "Ocean Temperature Gradients: Solar Power from the Sea." *Science,* 180 1973, 1266–1267.

Metz, W. D. "Ocean Thermal Energy the Biggest Gamble in Solar Power." *Science,* 198, 1977, 178–180.

Perlman, Eric. "Kilowatts in Paradise." *Science 82,* 3:1, January/February 1982, 78.

Riva, Joseph P., *et al.* *Energy from the Ocean.* Washington, DC: Government Printing Office, 1978.

Zener, Clarence. "Solar Sea Power." *Bulletin of the Atomic Scientists,* January 1976.

Biomass

Anderson, Russel, E. *Biological Paths to Self-Reliance: A Guide to Biological Solar Energy Conversion.* New York: Van Nostrand Reinhold, 1979.

Benemann, J. R., *et al.* "Energy Production by Microbial Photosynthesis." *Nature,* 268, 1977, 19–23.

Brown, L. "Food versus Fuel: Competing Uses for Cropland." *Environment,* 22:4, May 1980, 32–41.

Calvin, Melvin. "Petroleum Plantations for Fuels and Materials." *BioScience,* 29:9, 1979, 533–538.

Calvin, Melvin. "Photosynthesis as a Resource for Energy and Materials." *American Scientist,* 64:3, 1976, 270–278.

Chambers, R., *et al.* "Gasohol: Does It or Doesn't It produce Positive Net Energy?" *Science,* 206, November 16, 1979, 789–795.

Commoner, Barry. "A Reporter at Large: Ethanol." *New Yorker,* October 10, 1983.

Cote, Wilfred A. (ed.) *Biomass Utilization.* New York: Plenum Press, 1983.

Eckholm, Erik P. "The Other Energy Crisis: Firewood." *Worldwatch Paper No. 1,* Washington, DC: Worldwatch Institute, 1976.

Grantham, J. B. "Potentials of Wood for Producing Energy." *Journal of Forestry,* 72:9, 1974.

Hammond, A. L. "Alcohol: A Brazillian Answer to the Energy Crisis." *Science,* 195, 1977, 564–66.

High, C. "New England Returns to Wood." *Natural History,* 89:2, February 1980, 41–32.

Mears, Leon G. "Energy From Alcohol." *Environment* 20, 17–20.

Mitchell, J. G. "Whither the Yankee Forest?" *Audubon,* 83:2, March 1981, 76–99.

National Academy of Science. *Alcohol Fuels: Options for Developing Countries.* Washington, DC: National Academy Press, 1983.

Pimental, David, *et al.* "Biological Solar Energy Conversion and U. S. Energy Policy." *BioScience,* 28:6, 1978, 376–381.

Pinental, David, *et al.* "Environmental and Social Costs of Biomass Energy." *BioScience,* February 1984, 89–93.

Plotkin, S. E. "Energy from Biomass." *Environment,* 22:9, November 1980.

Poole, Alan and Robert H. Williams. "Flower Power: Prospects for Photosynthetic Energy." *Bulletin of the Atomic Scientist,* May 1976.

Sanderson, F. H. "Benefits and Costs of the U. S. Gasohol Program." *Resources* (journal of Resources for the Future), 67, July 1981, pp. 2–13.

Slesser, Malcolm. *Biological Energy Resources.* New York: Wiley, 1979.

Smil, Vaclav. *Biomass Energies—Resources, Links, Constraints,* New York: Plenum Press, 1983.

Smith, Nigel. *Wood: An Ancient Fuel with a New Future.* Washington, DC: Worldwatch Institute, 1981.

Veziroglu, T. Nejat *et al.* (eds.) *Hydrogen Energy Progress IV.* New York: Pergamon Press, 1984.

Veziroglu, T. Nejat *et al.* (eds.) *Hydrogen Energy Progress V.* New York: Pergamon Press, 1984.

Weisz, Paul B., and J. Marshall. *Fuel from Biomass: A Critical Analysis of Technology and Economics.*

Weisz, P. and J. Marshall. "High-Grade Fuels from Biomass Farming: Potentials and Constraints," *Science,* 206, October 5, 1979, 24–29.

Zeimetz, K. Growing Energy: Land for Biomass Farms *(Agricultural Economic Report No. 425).* Washington, DC: Department of Agriculture, June 1979.

Appendix

GEOLOGIC TIME SCALE

ERA	PERIOD	EPOCH	MILLIONS OF YEARS BEFORE PRESENT
Cenozoic	Quarternary	Recent Pleistocene	2.5–Present
	Tertiary	Pliocene Milocene Oligocene Eocene Paleocene	63–2.5
Mesozoic	Cretaceous Jurassic Triassic		135–63 181–135 230–181
Paleozoic	Permian		280–230
	Carboniferous	Pennsylvanian Mississippian	345–280
	Devonian		405–345
	Silurian		425–405
	Ordovician		400–425
	Cambrian		600–500
Precambrian			3,500–600

ENERGY EQUIVALENTS

One Quad is Equivalent to the Following:

Fundamentally
1×10^{15} BTUs
252×10^{15} calories or 252×10^{12} K calories

In Fossil Fuels
180 million barrels of crude oil
0.98 trillion cu. ft. of natural gas
37.88 million tons of anthracite coal
38.46 million tons of bituminous coal
50.00 million tons of subbituminous coal
71.43 million tons of lignite coal

In Nuclear Fuels
2500 tons of U_3O_8 if only U_{235} is used
17.8 tons of U_3O_8 if all U_{238} is transmuted to PU_{239} and U_{235} is used as well.
15.87 tons of ThO_2 if all thorium is transmuted to U_{233}

In Electrical Output
2.93×10^{11} Kilowatt-hours electric

In Biomass
58.43×10^6 dry tons of wood

Energy Values of Single Fuel Units

Uranium Oxide	
U_{235} content only	1 ton equiv. to 4.0×10^{11} BTU
U_{235} and U_{238} content	1 ton equiv. to 5.618×10^{13} BTU
Thorium Oxide (as U_{233})	1 ton equiv. to 6.301×10^{13} BTU
Crude Oil	1 barrel equiv. to 5.56×10^6 BTU
Natural Gas	1 cu. ft. equiv. to 1.035×10^3 BTU
Coal	
Anthracite	1 ton equiv. to 26.4×10^6 BTU
Bituminous	1 ton equiv. to 26.0×10^6 BTU
Subbituminous	1 ton equiv. to 20.0×10^6 BTU
Lignite	1 ton equiv. to 14.0×10^6 BTU
Wood	1 dry ton equiv. to 17.1×10^6 BTU

CONVERSION FACTORS

TO CONVERT FROM:	INTO THESE UNITS:	MULTIPLY BY THIS FACTOR	TO CONVERT FROM:	INTO THESE UNITS:	MULTIPLY BY THIS FACTOR
Length			**Weight**		
Feet	Meters	0.305	Metric tons	Short tons	1.1025
Miles	Meters	1,609	Kilograms	Pounds	2.20
Microns	Meters	1×10^{-6}	Kilograms	Tons	0.0011
Area			**Radiation**		
Acres	Square feet	43,560	BTU/square feet	Langleys	0.271
Acres	Square meters	4,047	Langleys	BTU/square feet	3.69
Square centimeters	Square feet	0.00108	Langleys/minute	Watts/square centimeters	0.0698
Square centimeters	Square inches	0.155			
Square feet	Square inches	144	**Energy and Power**		
Square feet	Square meters	0.0929	BTUs	calories	252
Square inches	Square centimeters	6.45	Kilogram calorie (food calorie)	calories	1,000
Square meters	Square feet	10.8			
Square meters	Square miles	3.68×10^{-7}	BTU	Joules	1,055
Square miles	Acres	640	BTU	Kilowatt-hour	2.93×10^{-4}
Square miles	Square feet	2.79×10^{7}	BTU	Megawatt-year	3.34×10^{-11}
Square miles	Square meters	2.59×10^{6}	calories	BTU	3.97×10^{-3}
Volume			calories	Foot-pounds	3.09
Barrels	U.S. gallons	42	calories	Joules	4.18
U.S. gallons	Imperial gallons	0.8326	calories/minute	Watts	0.0698
Imperial gallons	U.S. gallons	1.201	Gigajoules	BTU	0.95×10^{15}
U.S. gallons	Cubic inches	231	Kilowatts	Watts	1×10^{3}
Imperial gallons	Cubic inches	277.42	Megawatts	Watts	1×10^{6}
U.S. gallons	Liters	3.79	Kilowatt-hour	BTU	3,413
Weight			Kilowatts	Horsepower	1.34
Short tons	Pounds	2,000	Kilowatt-hour	Foot-pounds	2.66×10^{6}
Short tons	Kilograms	907	Watt-hours	Joules	3,600
Short tons	Metric tons	0.907	Therm	BTU	1×10^{5}
			Joule	Therms	9.4782×10^{-7}
			Megajoules/ sq. mtr.	Langleys/ sq. cm.	24
			Megajoules/ sq. mtr.	BTU/sq. ft.	8.811×10^{1}
			Megajoules/ sq. mtr.	Kilowatt-hours	2.778×10^{-1}

Glossary

abundant metals   Those metals that constitute a relatively large portion of the earth's crust. Aluminum, for instance, is present as roughly 8 percent of ordinary rock. Distinguished from scarce metals whose crustal abundance is less than 0.1 percent.

active solar heating   Using a solar collector, a storage device, and pumps or fans to transfer the heat obtained from the sun. Such systems can be added (retrofitted) to existing buildings whether or not the buildings have been designed with using solar energy in mind.

ammonia (NH_3)   A colorless pungent gas composed of nitrogen and hydrogen and appearing in a number of energy contexts. It is manufactured for fertilizers because of its nitrogen content. OTEC installations and some biomass operations may produce it and thus save the natural gas (CH^4) now consumed in ammonia manufacture. Ammonia can also be used as a feedstock for fuel cells which produce electricity chemically. It may be used as the working fluid in OTEC generating plants.

aquifer   A rock, usually of the sedimentary type, that holds and transmits water.

associated-dissolved gas   Natural gas (dominantly methane) which occurs either as a "gas cap" overlying crude oil in the pores of a reservoir rock, or as gas dissolved in crude oil and released when the oil reaches atmospheric pressure.

basin, or geologic basin   A present-day area of thick sedimentary rocks which, in the past, accumulated in low areas. Surrounding areas at that time were high and subject to erosion. Examples are the Powder River Basin or the Western Gulf Basin.

binary power plant   Plant that uses heat to vaporize a fluid, such as freon, which has a low boiling point and serves as the working fluid to drive the turbine. A binary plant makes it possible to generate electricity with-low temperature energy found in some geothermal waters.

breeding ratio   In a breeder reactor, the ratio of Plutonium-239 produced to the Uranium-238 consumed. If the value is not greater than 1.0, the reactor is not truly a breeder.

British Thermal Unit (BTU)   The energy required to raise the temperature of one pound of water 1 degree Fahrenheit. Equivalent to 252 calories, or 0.252 Calories.

calorie   The energy required to raise the temperature of 1 gram of water 1 degree Celsius. Often called the gram calorie to distinguish it from the nutritionists' Kilocalorie or Calorie which is 1,000 calories.

capacity factor   As applied to electrical generators, the proportion of generating capacity (in Watts) that is produced on average. Or, the proportion of time during which the generator produces at capacity. Expressed as a fraction or percentage, the factor is multiplied by the generating capacity and by 8,760 hours in a year in order to express the annual output of energy in Kilowatt hours.

clastic sedimentary rocks   All sedimentary rocks that have been formed by accumulation of fragments and particles derived from pre-existing rock.

conglomerate   A sedimentary rock, of the clastic type, made up of clastic elements ranging in size from silt to pebbles.

conventional crude oil   Crude oil that exists as liquid oil in the pore spaces of a reservoir rock and will flow, with varying degrees of recovery effort, through the reservoir rock into a borehole. Distinguished from unconventional (synthetic) crude oils.

conventional hydroelectric power   Electrical power produced from water that by natural stream flow accumulates behind a dam of considerable height and thus holds potential energy.

conventional natural gas   Gas (dominantly methane) existing *as gas* in a reservoir rock, and able to flow into a borehole with little or no need for special techniques such as fracturing the reservoir rock. Distinguished from unconventional synthetic natural gas which is producible only with special efforts at recovery (from western tight gas sands and eastern shales) and from synthetic gases produced from various raw materials. Also distinguished from the methane which originates in coals and which is now trapped in fractures in coal beds.

conversion of energy, or energy conversion   The change of one energy form to another. 1. Chemical energy of coal may be changed into heat energy. 2. That heat energy may be converted into electrical, or 3. The kinetic energy of falling water may be converted into electrical. 4. Electrical energy may be converted into heat. In some conversions, especially conversion of heat into electrical, there is a very substantial loss of energy in the conversion process. See **efficiency.**

critical fuel cost   In the analysis of active solar space-heating systems, the cost of fuel (to the consumer) which is just high enough to make a solar installation economically attractive by virtue of the fuel saved.

crystalline rocks   A generalization applied to igneous and/or metamorphic rocks to distinguish them from sedimentary.

Cumulative Production   For some resources, the accumulation or total of all amounts that have been produced to date.

degree-days   The sum of degree differences between mean daily temperature and a desired indoor temperature. The total for each day may be added for a year or for any part of it. Heating degree-days use a base of 65 degrees Fahrenheit, while cooling degree-days use 80 degrees Fahrenheit.

Demonstrated Reserves (of crude oil or natural gas)   The total of Measured (Proved) reserves and Indicated additional amounts thought to be recoverable through secondary recovery techniques.

Demonstrated Reserve Base   All the coals, of various ranks, that are economically mineable according to criteria of the U.S. Bureau of Mines.

deuterium (Hydrogen-2)   An isotope of the element hydrogen that has a nucleus containing one proton and one neutron, and thus has a mass number of 2.

diffuse radiation   Solar radiation received under cloudy or hazy skies. Because it is diffuse it cannot be focused by mirrors or lenses for high temperature applications. It is effective, though, in warming solar collectors for space heating, and can also be utilized by photovoltaic cells. Distinguished from direct radiation.

direct-normal radiation   The direct (or that which can be focused) radiation falling onto a surface which is tilted to intercept radiation at a 90-degree angle. Thus the radiation is normal to the surface.

direct radiation   Solar radiation received under clear skies. Since it is not diffused, it can be focused by mirrors or lenses to cause high temperatures needed for power generation and some industrial tasks. Distinguished from **diffuse radiation.**

doubling time   In a breeder reactor, the number of years required to double the initial fuel load.

drainage basin   The land area encompassing all the tributary streams of the river or rivers that define the basin. The boundary of one basin comprises all the drainage divides that separate the included rivers from those that are excluded.

ductile   Capable of being hammered into thin layers or of being drawn out into wire. Describes certain metals.

economic   A short-hand expression describing a resource or a process that is economically attractive or feasible. Distinguished from noneconomic or subeconomic.

efficiency   As applied to a specific energy conversion device, such as a power plant, this is the ratio of the total output of useful energy from a process to the total energy input. Electrical energy delivered from a power plant is roughly 40 percent of coal energy input.

electrical resistance heating   An example is residential baseboard heating, in which heating elements convert electrical energy into heat. Since the goal is to make heat, the heating elements are 100 percent efficient. The overall efficiency from fuel through generator and transmission lines, however, is low. Distinguished from electrical **heat pumps.**

end-use energy   The amount of energy delivered by some process. Three examples are: heat produced by an electrical heating element; heat supplied to a building by a furnace; or mechanical energy that drives a vehicle.

energy   The ability to do work or to raise the temperature of a substance. Also equivalent in physics to the work accomplished. If the energy is expended more quickly (or work done more quickly then a device of greater power has been used. Energy is distinguished from power.

enhanced recovery techniques   Often called *tertiary* recovery techniques. Ambitious methods of recovering crude oil that remains in reservoir rock after other methods have ceased to be productive. Methods include use of steam, underground combustion, or gas injection in order to lower the oil's viscosity, or detergents to flush the oil out of the reservoir rock.

enrichment (of uranium oxide)   The process in which the abundance of the fissile isotope, U_{235}, is increased from the original 0.7 percent to 2–3 percent of the uranium in the U_3O_8 compound. The process depends on the slight differences in density between U_{235} and U_{238} which is the isotope that constitutes 99.3 percent of natural uranium.

equinox days   September 21 and March 21, when zenithal noon sun occurs at the Equator (Lat. 0). On these two days, length of day equals length of night at every latitude.

feathering of blades   In aircraft propellers or wind machine blades, the adjustment of blade pitch so that blade will bite the air at more or less effective angles. In high winds a wind machine may feather blades in order to slip-by the additional energy and thus maintain a fixed rpm. In very high winds the blades may be feathered completely and become idle for safety.

fission products   Unstable isotopes produced when an atomic nucleus is split. Typically, Uranium-235 plus a neutron splits into a heavier and a lighter nucleus; Strontium-90 and Xenon-138 are examples.

fixed carbon in coal   That part of the carbon that remains when coal is heated in a closed vessel until the volatile matter is driven off. It is the non-volatile matter in coal minus the ash.

flaring of natural gas   Burning of gas at the well-head when it is produced in association with crude oil, and no market or storage exists for the gas.

formation   A term applied to a rock unit that persists and is recognizable over many miles and is therefore a useful reference for rock units above and below it. Named for the location of the outcrop (exposure) where the rock was first thoroughly studied. An example is the Green River Formation.

fossil fuels   Coal, crude oil of various types, and natural gas: fuels which are derived from the accumulation of plant or animal remains in ancient sedimentary rocks. *Fossil* refers to the preservation of the products of ancient life (organic materials) which themselves depended on the sun's energy utilized through photosynthesis.

fuel saved   The amounts of fuel whose burning is averted by substituting some renewable energy flow such as solar radiation or wind. Amounts of fuel saved depend on what task the fuel would

have been used for. For instance, to accomplish the energy output of a solar-electric generator, roughly 2½ times the output energy would have been input as fuel because the conversion of heat to electrical energy is only 40 percent efficient. In space heating, however, as much as 85 percent of fuel's energy is delivered as heat to the building by means of a furnace.

generating capacity   The maximum number of Watts of electrical power that a generator can sustain. A rate, therefore, of converting some energy form into electrical energy. Expressed Kilowatts (1,000 Watts) or Megawatts (1,000,000 Watts) electric. Distinguished from **output**, and also distinguished from Kilowatts or Megawatts thermal (KWe as opposed to KWt) which express the rate of producing heat from the fuel.

geologic province   An area that is more or less uniform in its geologic history and characteristics and can be distinguished from other large areas with regard to its mineral prospects.

geophysical methods   Methods of mineral exploration that make use of seismic shocks, or that study the magnetic and gravity characteristics of an area to find areas with features favorable to mineral occurrence. For oil and gas the features usually sought are structures which might be a trap.

geopressured reservoirs   Deeply buried isolated sand bodies which have high pressure due to their inability to compact under the weight of overlying rock, and which hold saline waters whose high temperatures are due to insulating layers of overlying rock.

Gigawatts   Power, or electrical generating capacity equivalent to 1,000,000 Kilowatts, or 1,000,000,000 Watts.

half-life   The time period required for one-half the atoms in a given amount of a radioactive substance to decay into another isotope.

heat pump   A device like an air conditioner, which can move heat 'uphill' from an environment of lower temperatures (such as 50 degrees Fahrenheit air outdoors) to an environment of higher temperatures (such as 65 degrees air indoors). Efficiency of heat pumps can be surprisingly high because the machine is not converting electrical energy to heat (with an efficiency of 100 percent) but is using the electrical energy to *move* heat from one place to another. The energy delivered is more than 100 percent of the electrical energy expended.

heavy crude oil   Crude oil that is viscous, but actually low in specific gravity. Its viscosity leads to lower recovery rates than for 'light' (less viscous) crude oils. Extremely heavy crude oils are much like the bitumen in oil sands and must be extracted by special techniques which entail heating the oil-bearing rock in-place.

high-sulfur coal   Coal with sulfur content greater than 3 percent by weight. See **low-sulfur coal.**

hot water and space heating   The heating (by solar energy for example) of water for domestic uses and to provide heat for buildings.

hydrologic cycle   The processes by which water is evaporated from oceans and lakes, and eventually falls as precipitation, some of which infiltrates the ground, some of which evaporates, and some of which runs off in streams to oceans and lakes.

hydrostatic   Describes pressure resulting from a standing column of fluid, or a body of water held at some elevation, as by a dam.

hydrothermal   The hot spring type of geothermal occurrence in which waters heated at depth flow to the surface as hot water or steam. The water can be used for space heating, or for electrical power generation if temperatures exceed 150 degrees Celsius.

Hypothetical resources   Those mineral resources or resource amounts that are thought to exist on the basis of studied geologic predictions or explorations.

Identified resources   Those mineral resources or resource amounts known to exist on the basis of information supplied by closely spaced drill holes and/or surface exposures.

igneous rocks   Those rocks whose immediate origin is molten rock, either at the earth's surface (volcanic type) or within the earth (plutonic type).

igneous or volcanic geothermal occurrences   Relatively shallow molten or very hot rock which may be used as applied energy in the near future by introducing water through drill holes and then retrieving hot water or steam.

immiscible gases   Gases that do not mix with the oil but simply displace it from the reservoir rock in enhanced recovery of crude oil.

Inferred reserves   Those amounts of mineral resources that are less certain than *Measured* (Proved) resources but more certain than Hypothetical or Undiscovered.

in-situ recovery   Recovery of minerals from deeply buried rock without mining. The mineral-rich rock is left in place; the minerals are extracted at depth, and then brought to the surface in liquid or gas. For oil shales, this process entails distillation underground.

isobars   On weather maps, the lines joining points of equal atmospheric pressure. They define areas of high and low pressure and reveal, by their spacing, whether there is a rapid or gradual change in pressure across a given area.

isotopes   Different forms of an element that have the same number of protons in the nucleus of each atom but differ in the number of neutrons.

Joule   The amount of energy that would be used by a 1-Watt light bulb in 1 second.

kerogen   A waxy, organic material containing the remains of aquatic organisms. The constituent of Green River oil shales that is converted into crude oil by distillation.

Kilowatt   Power, or electrical generating capacity, of 1,000 Watts.

kinetic energy   The energy of some body that results from the body's motion. It is equivalent to $\frac{1}{2}mV^2$, where m is the mass and V is its velocity. Distinguished from *potential energy* whereby a body raised to some height has the potential to expend energy and do work as it falls.

lacustrine   Produced by or related to lakes, such as lacustrine sedimentary deposits, which are sediments deposited in lake basins. Distinguished from *fluviatile* (in rivers) and *marine* (in salt water).

Langley   Amount of solar radiation energy equivalent to 1 calorie received per square centimeter of surface.

leaching   The process of washing or draining by percolation of water. During the process material is picked up and carried or dissolved and carried.

linear scale   On graphs, the commonplace scale in which a 0–40 range of values will be double the length of 0–20.

liquefied natural gas (LNG)   Natural gas liquefied at very low temperatures and high pressure for long-distance transport by tanker to special terminals where it is allowed to warm and expand into pipelines.

lithium   A very light metal being used experimentally in the development of a rechargeable lithium-anode battery for automotive and peak-energy storage use. Also the source of tritium, a possible fuel for fusion-type nuclear reactors.

Lithium-6   An isotope of the metal lithium, which, under neutron bombardment, is converted into tritium, a radioactive isotope of hydrogen which is a possible fuel for fusion-type nuclear reactors.

logarithmic scale   On graphs, a scale in which the logarithms of values are evenly spaced, so the values themselves are more widely spaced at the low end and more crowded at the high end of the scale. Used for values which have a very large range from lowest to highest.

long ton   2,240 pounds or 1.016 metric tons or 1.120 short tons.

low-sulfur coal   Coal with sulfur content 1 percent or less by weight. See **high-sulfur coal**.

market penetration   The extent to which a new technology is expected to displace present ones. For solar energy, the proportion of some energy need that will be supplied by solar energy. For heating and cooling of buildings, for instance, the contribution from solar energy may be 10 percent in one region, but 30 percent in another region where economic or physical factors favor the use of solar energy rather than fuels.

median value   In any array of numbers the median is the number midway between the highest and the lowest. Distinguished from the *mean* value which is half the sum of the highest and lowest values, and the *mode,* which is the value that occurs most frequently.

Megawatt   Power, or electrical generating capacity equivalent to 1,000 Kilowatts, or 1,000,000 Watts.

metamorphic rocks   Those rocks derived through the alteration of some pre-existing rock. Marble, for instance, is altered limestone.

metric ton   1,000 kilograms or 1.10 short tons.

mineable   Terms applied (as with coal) to tonnages that are physically and economically accessible. Distinguished from amounts expected to be *recovered* (recoverable) through the mining.

miscible gases   Gases that mix with the oil and free it from the reservoir rock in enhanced recovery of crude oil.

moderators   A substance used in nuclear reactors to slow down neutrons, allowing them to be more readily captured by Uranium-235. Water deuterium (heavy water) and carbon in the form of graphite are commonly used moderators.

monazite   A mineral that is a phosphate of the cerium metals, but contains small quantities of thorium oxide. The mineral occurs as large crystals in crystalline rocks, and is also found in placer deposits.

natural gas liquids   Also called *condensate* and *liquid petroleum gases* (LPGs). Propane and similar heavy hydrocarbons that separate from some natural gases, either naturally through condensation in the reservoir rock, or in separators at the well-head. Distinguished from **liquefied natural gas (LNG)**.

neutron   An elemental particle present in all atomic nuclei. It has no electric charge, and a relative mass of 1.

non-associated gas   Natural gas (dominantly methane) which occurs as free gas trapped over salt water in the pores of a reservoir rock. Distinguished from the **associated-dissolved** type.

nonrenewable resources   Mineral resources that exist in finite quantities, especially mineral fuels which cannot be used again, but also geothermal occurrences which regenerate slowly. Distinguished from **renewable**.

off-peak period   Hours of the day when demand for electricity is low. This occurs principally between midnight and 7 A.M. During this period electrical energy is stored for use during peak hours of electrical demand.

oil shale   A sedimentary rock containing solid organic material which can be converted to a crude oil by distillation. In the Green River oil shales of Colorado the organic material is **kerogen** and the host rock is technically a marl or shaley limestone.

ore(s)   Rock(s) in which one or more valuable minerals exist in concentrations far greater than is average for crustal rocks. Sometimes applied only to rocks whose richness and accessibility make their mining economically attractive.

organic   Derived from plant or animal life. The term applies to a live plant, to the animals that eat the plant, and to the manures from and the carcasses of those animals. See **fossil fuels**.

outcrop   A surface exposure of all or part of some rock unit, which in other areas is buried and can be studied only by drill holes.

output of electrical energy   The estimated or actual electrical energy produced by a generating plant. It is the power produced (in Watts) multiplied by the time (in hours) during which that power is produced—and is expressed therefore in Kilowatt or Megawatt-hours, usually in a year's period. See **capacity factor**.

overburden   Soil and rock of little or no value, which overlie a buried deposit of economic value, such as coal, oil shale, or the ores of metals. In surface mining, the overburden is removed; in underground mining it is penetrated and supported during mining.

passive solar heating   Using structural elements of a building to admit, collect, and store solar energy for heating. In its pristine form, heating of this kind relies on convection, and uses no supplemental energy from fans or pumps to distribute heat through the building.

photosynthesis   In living plants, the production of carbohydrates from water and carbon dioxide (CO_2) by the action of solar radiation upon chlorophyll.

photovoltaic effect   The flow of electrons (hence potential difference, or voltage) that results when certain semi-conducting substances in solar cells are struck by radiation.

placer deposit   A water-borne deposit of heavy minerals that have been eroded from their original bedrock and eventually concentrated in gravels of streams or beaches.

polder   In the Netherlands, low-lying coastal areas turned into agricultural land by building dikes and pumping out ocean water.

power   A rate of doing work or expending energy. One Watt of power is the expenditure of one Joule of energy in one second.

power available in the wind   The power that exists instantaneously in winds of a given velocity. Expressed as Watts in each square unit of cross-section perpendicular to wind flow. Distinguished from **power extractable** and from **output**.

power density plot   For wind data at a particular site, a graphic expression of annual Kilowatt-hours of wind energy versus wind speeds. It reveals which winds at the site hold the greatest energy on an annual basis by virtue of their velocity and/or their frequency. May be more accurately called an *energy density plot*.

power duration plot   For wind data at a particular site, a graphic expression of the number of hours (during a year) that various wind velocities, and hence various values of *power available*, may be expected on the basis of wind records.

power extractable from wind   Theoretical Watts of power that can be gathered from wind of a given velocity, assuming: 1. a certain efficiency of extraction, and 2. extraction from a specified area perpendicular to wind flow. It is very different from actual power extracted because specific wind machines do not employ all the power available in areas of low, medium, high, and very high winds.

primary energy   The initial energy used for some process or chain of processes. Generally it is fuel energy, as used to generate power or drive engines. Distinguished from *end use energy* which, as the energy delivered, is usually less than the fuel energy input. For a country, the primary energy is the total of all primary energy entering the country's energy systems, and the end use energy is the total of all energy delivered to the consumers, whether as heat, electrical, or mechanical energy.

pumped storage   Electrical generating capacity that is due to the fall of water that has been pumped uphill by using excess electrical power—at a power plant that may not be primarily a hydroelectric facility.

Quad   One quadrillion (1×10^{15}) British Thermal Units.

rank of coal   A designation that groups coals according to their differing maturity and energy content. Those of higher rank are more mature and richer in energy. The ranks are, in decreasing order: anthracite, bituminous, subbituminous, lignite.

rejected heat   Amounts of heat thrown off in an energy conversion process such as the conversion of fuel energy to electrical energy in a power plant, or to mechanical energy in a motor. Although the rejection of heat is unavoidable, the waste is not; some rejected heat can be recovered and used for other tasks.

renewable resources   A resource which either flows continuously, (e.g., solar radiation and wind), or is renewed within a short period of time, (e.g., wood).

reprocessing (of nuclear fuel)   The recovery of uranium and plutonium from spent fuel rods by dissolving them in an acid solution.

Reserve Base for U.S. coal   Equivalent to the Demonstrated Reserve Base, which is comprised of the coals that are economically mineable.

reserves   For any mineral resource, the amounts that are *recoverable* (extractable) at current prices and using current technology. Differing degrees of certainty may be expressed by the terms Measured and Inferred reserves; but the amounts are those recoverable from *identified* occurrences.

reservoir (or reservoir rock)   A porous and permeable rock, such as sandstone, in which crude oil or natural gas accumulations are found.

resource   Anything of value to the society. Usually a raw material, such as forests or minerals, or an energy flow, such as wind. A great variety of certainty and economic worth can be embraced by the term "resource."

retort   A large pot or chamber in which oil shale is cooked (heated) in order to separate the oil from the shale rock.

retrofit   To install new equipment, such as solar heating apparatus, in an existing building. Distinguished from "new installations," or "installations in new buildings."

secondary recovery techniques   Methods of recovering additional crude oil after natural pressures in reservoir rock have been depleted. Entails injection of either water or gas into the reservoir rock to maintain pressures.

sedimentary rocks   Those rocks formed by the accumulation of grains or fragments of pre-existing rocks, or the accumulation of chemical precipitates such as $CaCO_3$, salt, or gypsum.

shale   A sedimentary rock, of the clastic type, made by accumulation and solidification of clay materials. Often important in oil and gas fields as the organic-rich source rock that provides the hydro-carbons and as the caprock that seals the reser-voir rock to make an accumulation possible.

shale oil   The crude oil resulting from distillation of kerogen in oil shales.

shield areas   Parts of continents in which crystalline rocks of the continent's basement are exposed in a non-mountainous region. Distinguished from mountainous areas and from areas where base-ment rocks are covered by sedimentary rocks.

short ton   Equivalent to 2,000 pounds. Distinguished from the **metric ton**, which is equivalent to 1,000 kilograms or 2,205 pounds and from the **long ton**, which is 1,016 kilograms or 2,240 pounds.

slurry   A mixture of solid and liquid. An example is coal slurry in which finely ground coal is mixed with water and moved by pipeline.

solar fraction   The proportion of a building's heating (or cooling) energy need that is provided by solar energy. Distinguished from **market penetration**.

solar-thermal power generation   The focusing of solar rays to achieve the high temperatures used to boil fluids for use in vapor-driven dynamos.

solstice days   June 21 and December 21, when ze-nithal noon sun occurs at Latitude 23½ North and Latitude 23½ South, respectively.

space heating   Heating of buildings, whether resi-dential, commercial, or industrial.

strata   The plural of *stratum* (rarely used). Layers or beds of rock. Usually applied to sedimentary rock lying relatively flat without severe folding.

strip mining   Loosely equivalent to surface mining. **Overburden** material is removed; then the valu-able rock is taken away in an operation that is everywhere open to the surface.

stratigraphic trap   In oil and gas fields, a reservoir rock which is discontinuous, such as a sand bar or a reef, thus permitting accumulation of oil or gas. Distinguished from **structural trap**.

structural trap   In oil and gas fields, a reservoir rock that is folded into structures such as domes, or faulted (broken), thus permitting an accumula-tion of oil or gas.

subduction   The process by which a part of the earth's crust is drawn down into the subcrustal zone. This occurs when two crustal plates collide and one of the two is forced under.

subsidence   A sinking, settling or otherwise lowering of parts of the crust of the earth.

sulfur oxides   Sulfur dioxide (SO_2) and sulfur trioxide. (SO_3). Gases produced when sulfur-rich coal and oil are burned. The gases can react with oxygen

and water to produce sulphuric acid (H_2SO_4) and thus are dangerous air pollutants.

syncrude   A shorthand expression for synthetic crude oils produced from mineral sources such as oil shales or from organic materials, which may be either wastes or crops especially grown for the purpose.

synthetic crude oils   Hydrocarbon fluids that can be produced from various raw materials. Mineral sources are the kerogen in oil shales and bitumen in oil sands. Non-mineral sources include urban trash and certain crops grown for the purpose.

synthetic gases   Gases produced from various sources including coal and a number of organic mate-rials such as garbage, manures, and certain energy crops.

temperature gradient   The rate at which tempera-ture increases with increasing depth below the earth's surface at a specific location. Steeper gra-dient refers to more rapid change with depth and denotes an area more favorable for developing geothermal resources.

Thorium-232   A radioactive isotope that can be transmuted under neutron bombardment into Uranium-233, which is fissionable in a chain reaction.

total gas reserves   Reserve amounts of both **non-associated** and **associated-dissolved** types of nat-ural gas.

total-horizontal radiation   The solar radiation (inso-lation) that is usually reported and mapped. It is the total of **direct** and **diffuse** radiation falling upon a horizontal surface of some particular loca-tion. Equivalent to insolation.

trillion   Used here with the United States' and French meaning of 1,000 billion, that is, 1×10^{12}. In Great Britain a trillion means 1,000,000 billion, or 1×10^{15}.

tritium (Hydrogen-3)   An isotope of the element hydrogen that has a nucleus containing one proton and two neutrons and thus has a mass number of 3.

Ultimate Recovery   For crude oil and natural gas the term is applied to the total of Cumulative Production and Proved reserves. It is "ultimate," therefore, only with regard to identified reservoirs.

underground mining   Mining in which the valuable rock to be removed is reached by shafts through the **overburden**.

Undiscovered Recoverable amounts (of crude oil or natural gas)   Estimates of amounts that may be recovered from reservoir rocks not yet dis-covered. Usually the historic recovery rates (pro-portions of oil in-place) are assumed.

uranium oxide (U_3O_8)   The compound that is seprated from uranium ores. Uranium resources are stated in tons of uranium oxide thought to be recoverable from ores of varying degrees of certainty and economic feasibility.

viscosity   The property of fluids that causes them not to flow easily because of the friction of their molecules. It is decreased by raising the fluid's temperature.

volatile matter in coal   The portion of the coal that is driven off as gas or vapor when heated. Much of it is combustible (compounds of hydrogen and carbon). Coals with a high proportion of volatile matter may be preferred for applications where a long flame is desirable.

Watt   A unit of power, that is, the rate at which energy is converted or consumed. Equivalent to 1 Joule of energy expended in 1 second.

well-head heat   Geothermal heat thought to be recoverable at the surface from exploitable occurrences at depth. Neither the application of the heat nor the efficiency of its use are taken into consideration.

zenithal noon sun   The condition when noon sun is at 90 degrees elevation angle as measured from any horizon. It occurs only in the latitude belt bounded by 23½ North and 23½ South.

Index